Complete Solutions Manual
for
PreCalculus
Third Edition

J. Douglas Faires
Youngstown State University

James DeFranza
St. Lawrence University

THOMSON
BROOKS/COLE
Australia • Canada • Mexico • Singapore • Spain • United Kingdom • United States

Printed in Canada
1 2 3 4 5 6 7 07 06 05 04 03

Printer: Webcom

ISBN: 0-534-46281-2

Brooks/Cole—Thomson Learning
10 Davis Drive
Belmont, CA 94002-3098
USA

Asia
Thomson Learning
5 Shenton Way #01-01
UIC Building
Singapore 068808

Australia/New Zealand
Thomson Learning
102 Dodds Street
Southbank, Victoria 3006
Australia

Canada
Nelson
1120 Birchmount Road
Toronto, Ontario M1K 5G4
Canada

Europe/Middle East/South Africa
Thomson Learning
High Holborn House
50/51 Bedford Row
London WC1R 4LR
United Kingdom

Latin America
Thomson Learning
Seneca, 53
Colonia Polanco
11560 Mexico D.F.
Mexico

Spain/Portugal
Paraninfo
Calle/Magallanes, 25
28015 Madrid, Spain

PREFACE

This Complete Solutions Manual for PreCalculus, Third Edition, by Faires and DeFranza, contains solutions to all the exercises presented in the book. Our approach to PreCalculus places a heavy emphasis on graphing techniques to help students visualize and master important topics. In this manual you will find more solutions accompanied by illustrations than you would find in other typical books at this level.

Also available for your students is a Student Study Guide. This Guide provides students with access to more extensive algebra, geometry and trigonometry review material. The algebra and trigonometry review is woven through the text making the presentation in the Guide suited for those students who need extra review and practice.

The Guide also contains supplemental examples for each section of the book, detailed solutions to all the odd-numbered exercises, and solutions for all of the exercises in the Chapter Tests.

Two short examinations that students can use to test their readiness for Precalculus and for Calculus can be downloaded in Adobe PDF form at the book web site

http://www.as.ysu.edu/~faires/PreCalculus3

We suggest that students work one of the examinations at the start of the PreCalculus course and the second at completion of the course. Students scoring approximately 16 or higher on the 40 question examinations are likely prepared to take a PreCalculus course based on our book. A score of approximately 28 or higher indicates that the student is ready for a University Calculus sequence.

If you have suggestions for improvements in future editions of the PreCalculus book or either of these supplements, we would be most grateful for your comments. Additional information about the book and any updates will be placed at the book web site.

Acknowledgements

We greatly appreciate the work of Estela S. Llinas, University of Pittsburgh at Greensburg, who reviewed and checked all the exercise solutions in this Manual. She made many helpful suggestions that will improve both the Manual and the textbook on which it is based. We would also like to sincerely thank Nicole Cunningham who, as our editorial assistant, did much of the LaTeX work on this and the other parts of the project, and Erika Faires, who redrew the art for the Manual.

Doug Faires — Youngstown State University
faires@math.ysu.edu

Jim DeFranza — St. Lawrence University
jdefranza@stlawu.edu

Table of Contents

Solutions to All Exercises

Exercise Set 1.2 (Page 12)

1. $-2 \le x \le 4$

2. $-3 \le x < -1$

3. $-\sqrt{3} < x \le \sqrt{2}$

4. $-\sqrt{5} \le x < -\sqrt{2}$

5. $x < 3$

6. $x \le -1$

7. $x \ge \sqrt{2}$

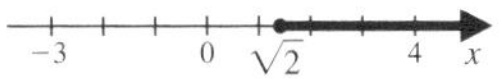

8. $x > -2$

9. $[-1, 2]$

10. $(2, 6]$

11. $[2, 5)$

12. $[-2, 4)$

13. $(-\infty, 2)$

14. $(-\infty, 2]$

15. $[3, \infty)$

16. $(-2, \infty)$

17. **a.** $|3 - 7| = |-4| = 4$ **b.** $\frac{1}{2}(3 + 7) = 5$

18. **a.** $|1 - 8| = |-7| = 7$ **b.** $\frac{1+8}{2} = \frac{9}{2}$

19. **a.** $|-3 - 5| = |-8| = 8$ **b.** $\frac{1}{2}(-3 + 5) = 1$

20. **a.** $|-4 - (-1)| = |-3| = 3$ **b.** $\frac{-4+(-1)}{2} = -\frac{5}{2}$

21. The inequality $x + 4 < 7$ implies that $x < 3$, so in interval notation $(-\infty, 3)$.

22. The inequality $x - 4 < 9$ implies that $x < 13$, so in interval notation $(-\infty, 13)$.

23. The inequality $2x - 2 \geq 8$ implies that $2x \geq 10$, so $x \geq 5$, and in interval notation $[5, \infty)$.

24. The inequality $3x + 2 \geq 8$ implies that $3x \geq 6$, so $x \geq 2$, and in interval notation $[2, \infty)$.

25. The inequality $-3x + 4 < 5$ implies that $-3x < 1$, so $x > -\frac{1}{3}$, and in interval notation $\left(-\frac{1}{3}, \infty\right)$.

26. The inequality $-2x - 3 \geq 7$ implies that $-2x \geq 10$, so $x \leq -5$, and in interval notation $(-\infty, -5]$.

27. The inequality $2x + 9 \leq 5 + x$ implies that $x \leq -4$, so in interval notation $(-\infty, -4]$.

28. The inequality $-3x - 2 < 3 - x$ implies that $-2x < 5$, so $x > -\frac{5}{2}$, and in interval notation $\left(-\frac{5}{2}, \infty\right)$.

29. The inequality $-2 < 3x - 2 < 5$ implies that $0 < 3x < 7$, so $0 < x < \frac{7}{3}$, and in interval notation $\left(0, \frac{7}{3}\right)$.

30. The inequality $-3 < 2x + 1 \leq 2$ implies that $-4 < 2x \leq 1$, so $-2 < x \leq \frac{1}{2}$, and in interval notation $\left(-2, \frac{1}{2}\right]$.

31. The inequality

$$(x + 1)(x - 2) \geq 0$$

is satisfied for x in $(-\infty, -1] \cup [2, \infty)$.

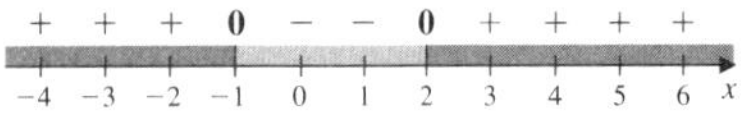

32. The inequality

$$(x - 1)(x + 3) < 0$$

is satisfied for x in $(-3, 1)$.

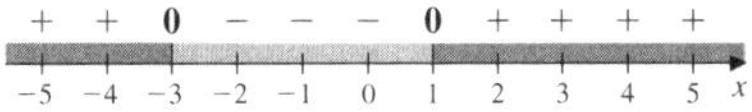

33. The inequality

$$x^2 - 3x + 2 = (x - 2)(x - 1) \leq 0$$

is satisfied for x in $[1, 2]$.

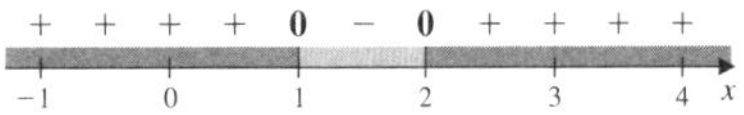

34. The inequality

$$x^2 - 2x - 3 = (x + 1)(x - 3) > 0$$

is satisfied for x in $(-\infty, -1) \cup (3, \infty)$.

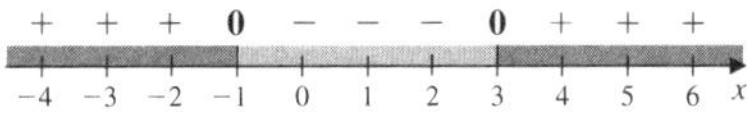

35. The inequality

$$(x-1)(x-2)(x+1) \le 0$$

is satisfied for x in $(-\infty, -1] \cup [1, 2]$.

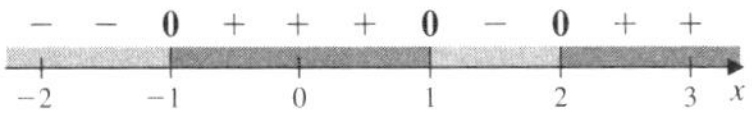

36. The inequality

$$(x-1)(x+2)(x-3) \ge 0$$

is satisfied for x in $[-2, 1] \cup [3, \infty)$.

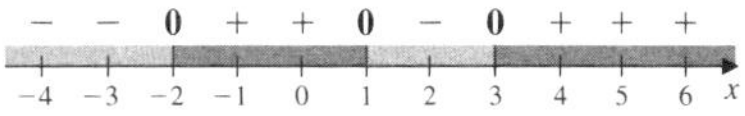

37. The inequality

$$x^3 - 3x^2 + 2x = x(x-1)(x-2) \ge 0$$

is satisfied for x in $[0, 1] \cup [2, \infty)$.

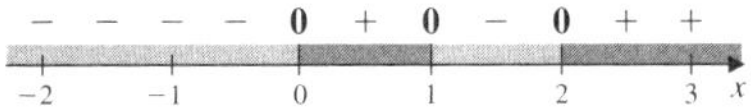

38. The inequality

$$x^3 - 2x^2 - 3x = x(x-3)(x+1) < 0$$

is satisfied for x in $(-\infty, -1) \cup (0, 3)$.

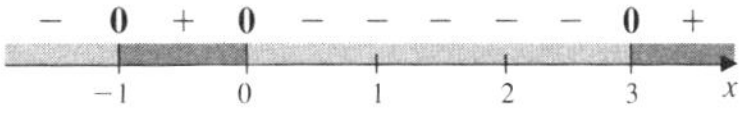

39. The inequality

$$x^3 - 2x^2 = x^2(x-2) < 0$$

implies that

$$(x-2) < 0 \text{ and } x \ne 0,$$

which is satisfied for x in $(-\infty, 0) \cup (0, 2)$.

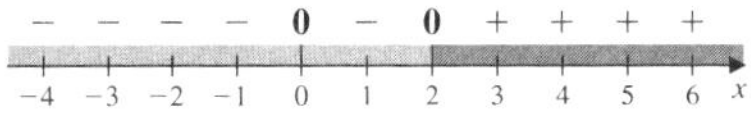

40. The inequality

$$x^3 - 2x^2 + x = x(x^2 - 2x + 1) > 0$$

implies that

$$x(x-1)^2 > 0 \text{ and } x \ne 0,$$

which is satisfied for x in $(0, 1) \cup (1, \infty)$.

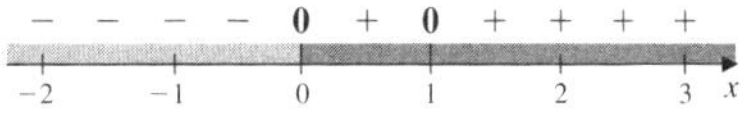

41. The inequality

$$\frac{x+3}{x-1} \ge 0$$

is satisfied for x in $(-\infty, -3] \cup (1, \infty)$.

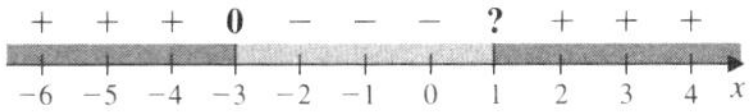

42. The inequality

$$\frac{x-2}{x+1} \le 0$$

is satisfied for x in $(-1, 2]$.

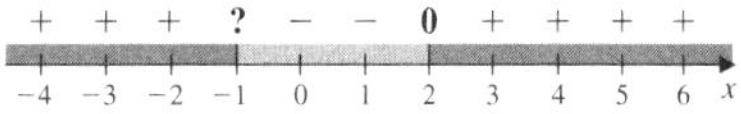

43. The inequality

$$\frac{x(x+2)}{x-2} \le 0$$

is satisfied for x in $(-\infty, -2] \cup [0, 2)$.

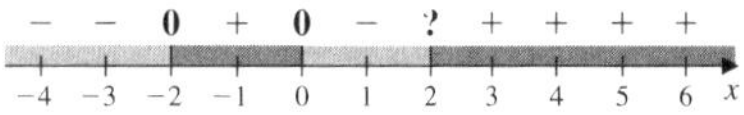

44. The inequality

$$\frac{x+2}{x(x-2)} > 0$$

is satisfied for x in $(-2, 0) \cup (2, \infty)$.

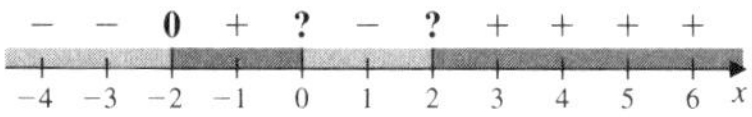

45. The inequality

$$\frac{(1-x)(x+2)}{x(x+1)} > 0$$

is satisfied for x in $(-2, -1) \cup (0, 1)$.

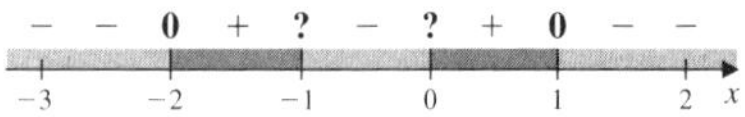

46. The inequality

$$\frac{(1-x)(x+3)}{(x+1)(2-x)} \le 0$$

is satisfied for x in $[-3, -1) \cup [1, 2)$.

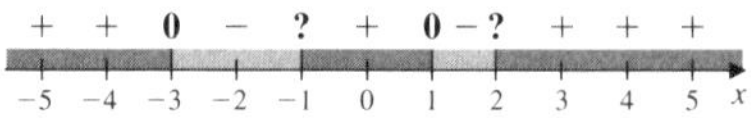

47. The inequality

$$\frac{1}{x} \le 5 \quad \text{implies that} \quad \frac{1}{x} - 5 \le 0,$$

which can be written as

$$\frac{1-5x}{x} \le 0,$$

so in interval notation $(-\infty, 0) \cup \left[\frac{1}{5}, \infty\right)$.

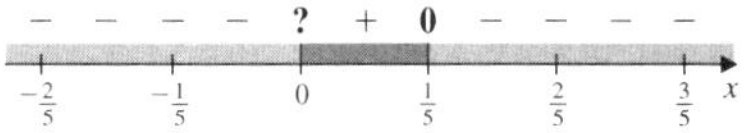

48. The inequality

$$-2 \le \frac{1}{x} \quad \text{implies that} \quad \frac{1}{x} + 2 \ge 0,$$

which can be written as

$$\frac{1+2x}{x} \ge 0,$$

so in interval notation $\left(-\infty, -\frac{1}{2}\right] \cup (0, \infty)$.

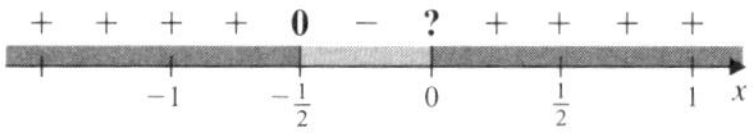

49. The inequality

$$\frac{2}{x-1} \geq \frac{3}{x+2}$$

implies that

$$\frac{2}{x-1} - \frac{3}{x+2} \geq 0,$$

and

$$\frac{2(x+2)-3(x-1)}{(x-1)(x+2)} \geq 0,$$

which can be written as

$$\frac{7-x}{(x-1)(x+2)} \geq 0,$$

so in interval notation $(-\infty, -2) \cup (1, 7]$.

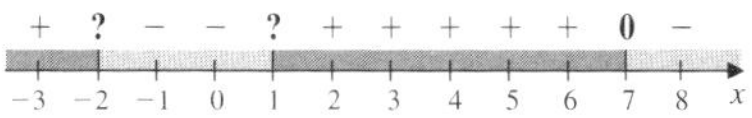

50. The inequality

$$\frac{2}{x-1} - \frac{x}{x+1} \leq -1$$

implies that

$$0 \geq \frac{2(x+1)-x(x-1)}{(x-1)(x+1)} + 1,$$

which can be written as

$$0 \geq \frac{3x+1}{(x-1)(x+1)},$$

so in interval notation $(-\infty, -1) \cup \left[-\frac{1}{3}, 1\right)$.

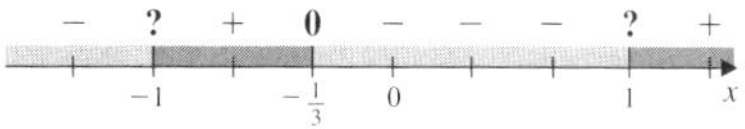

51. We have $|5x-3| = 2$ when $5x-3 = 2$ or $5x-3 = -2$, so $x = 1$ or $x = \frac{1}{5}$.

52. We have $|2x+3| = 1$ when $2x+3 = 1$ or $2x+3 = -1$, so $x = -1$ or $x = -2$.

53. We have $\left|\frac{x-1}{2x+3}\right| = 2$ when $\frac{x-1}{2x+3} = 2$ or $\frac{x-1}{2x+3} = -2$, which implies that $x-1 = 4x+6$ or $x-1 = -4x-6$ which implies $-7 = 3x$ or $5x = -5$, so $x = -\frac{7}{3}$ or $x = -1$.

54. We have $\left|\frac{2x+1}{x-3}\right| = 4$ when $\frac{2x+1}{x-3} = 4$ or $\frac{2x+1}{x-3} = -4$, which implies that $2x+1 = 4x-12$ or $2x+1 = -4x+12$ which implies $13 = 2x$ or $6x = 11$, so $x = \frac{13}{2}$ or $x = \frac{11}{6}$.

55. We have $|x-4| \leq 1$, which implies $-1 \leq x-4 \leq 1$, so $3 \leq x \leq 5$, and in interval notation we have $[3, 5]$.

56. We have $|4x-1| < 0.01$, which implies $-0.01 < 4x-1 < 0.01$, so $0.99 < 4x < 1.01$ which implies $0.2475 < x < 0.2525$, and in interval notation we have $(0.2475, 0.2525)$.

57. We have $|3-x| \geq 2$, which implies $3-x \geq 2$ or $3-x \leq -2$, so $-x \geq -1$ or $-x \leq -5$, which implies $x \leq 1$ or $x \geq 5$, and in interval notation we have $(-\infty, 1] \cup [5, \infty)$.

58. We have $|2x-1|>5$, which implies $2x-1>5$ or $2x-1<-5$, so $x>3$ or $x<-2$, and in interval notation we have $(-\infty,-2)\cup(3,\infty)$.

59. We have $\frac{1}{|x+5|}>2$, which implies $|x+5|<\frac{1}{2}$, so $-\frac{1}{2}<x+5<\frac{1}{2}$, which implies $-\frac{11}{2}<x<-\frac{9}{2}$, and in interval notation we have $\left(-\frac{11}{2},-5\right)\cup\left(-5,-\frac{9}{2}\right)$. We omit $x=-5$ since this is where $|x+5|=0$.

60. We have $\left|\frac{3}{2x+1}\right|<1$, which implies $\frac{3}{|2x+1|}<1$, so $|2x+1|>3$ which implies $2x+1>3$ or $2x+1<-3$, therefore $x>1$ or $x<-2$, and in interval notation we have $(-\infty,-2)\cup(1,\infty)$.

61. We have $\left|x^2-4\right|>0$, which implies $x^2-4\neq 0$, so $x\neq\pm 2$, and in interval notation we have $(-\infty,-2)\cup(-2,2)\cup(2,\infty)$.

62. We have $\left|x^2-4\right|\leq 1$, which implies $-1\leq x^2-4\leq 1$, so $x^2-3\geq 0$ which implies $x\leq-\sqrt{3}$ or $x\geq\sqrt{3}$ and $x^2-5\leq 0$ which implies $-\sqrt{5}\leq x\leq\sqrt{5}$, so $-\sqrt{5}\leq x\leq-\sqrt{3}$ or $\sqrt{3}\leq x\leq\sqrt{5}$. In interval notation we have $\left[-\sqrt{5},-\sqrt{3}\right]\cup\left[\sqrt{3},\sqrt{5}\right]$.

63. Since $0<a<b$ we have both $0<a^2<ab$ and $0<ab<b^2$. Hence, $0<a^2<ab<b^2$.

64. To solve $|x-1|=|2x+2|$, there are four possibilities:
(i) $x-1=2x+2$ which implies $x=-3$;
(ii) $x-1=-(2x+2)$ so $x-1=-2x-2$ which implies $x=-\frac{1}{3}$;
(iii) $-(x-1)=2x+2$ which implies $x=-\frac{1}{3}$;
(iv) $-(x-1)=-(2x+2)$ which implies $x=-3$;
So the solutions are $x=-3$ and $x=-\frac{1}{3}$. Note that $|-3-1|=4=|-6+2|$ and $|-\frac{1}{3}-1|=\frac{4}{3}=|2(-\frac{1}{3})+2|$.

65. **a.** Since $20\leq F\leq 50$ implies that $-12\leq F-32\leq 18$ we have $-12\left(\frac{5}{9}\right)\leq\frac{5}{9}(F-32)\leq 18\left(\frac{5}{9}\right)$, so $-\frac{20}{3}\leq C\leq 10$.

b. Since $20\leq C\leq 50$ we have $20\leq\frac{5}{9}(F-32)\leq 50$, which implies $\frac{9}{5}(20)\leq F-32\leq\frac{9}{5}(50)$ so $68\leq F\leq 122$.

66. We have $-16t^2+48t+128\geq 64$ which implies $-16t^2+48t+64=-16(t^2-3t-4)\geq 0$, so $(t-4)(t+1)\leq 0$ and $-1\leq t\leq 4$. Since t is time, it is positive and the solution is $0\leq t\leq 4$.

Exercise Set 1.3 (Page 19)

1. The points are on the graph.

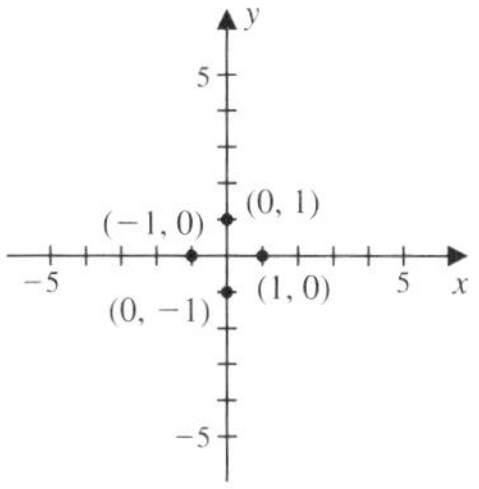

2. The points are on the graph.

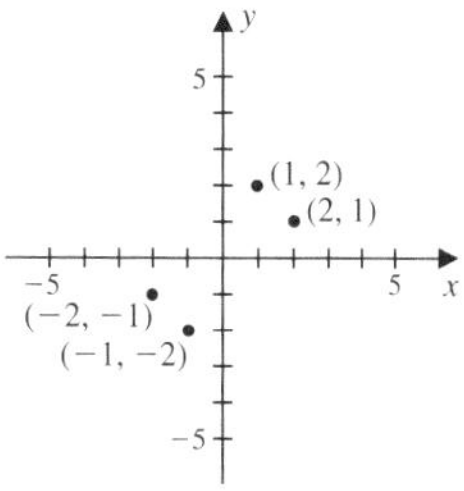

3. The points are on the graph.

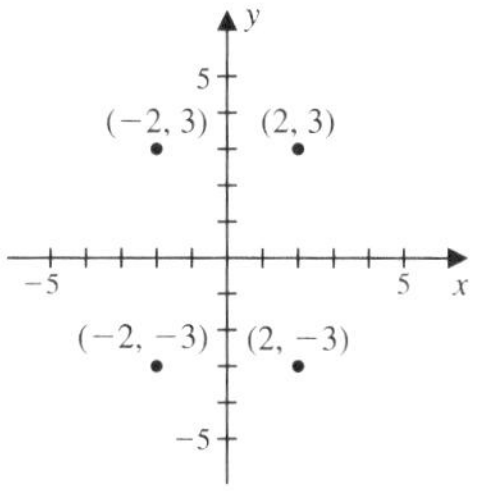

4. The points are on the graph.

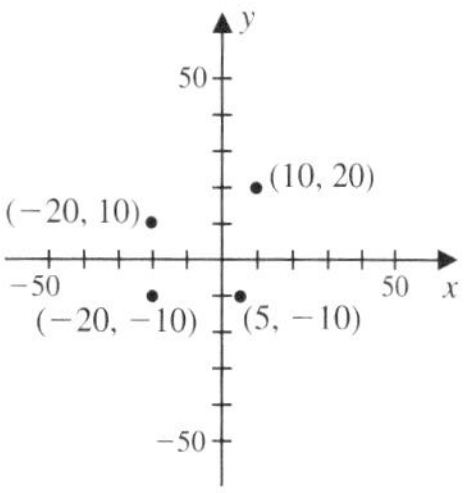

5. **a.** The distance is $\sqrt{(2-(-1))^2+(4-3)^2}=\sqrt{10}$.

b. The midpoint is $\left(\frac{2-1}{2}, \frac{4+3}{2}\right)=\left(\frac{1}{2}, \frac{7}{2}\right)$.

6. **a.** The distance is $\sqrt{(7-(-1))^2+(9-5)^2}=\sqrt{80}=4\sqrt{5}$.

b. The midpoint is $\left(\frac{-1+7}{2}, \frac{5+9}{2}\right)=(3,7)$.

7. **a.** The distance is $\sqrt{(\pi-(-1))^2+(0-2)^2}=\sqrt{(\pi+1)^2+4}=\sqrt{\pi^2+2\pi+5}$.

b. The midpoint is $\left(\frac{\pi-1}{2}, \frac{0+2}{2}\right)=\left(\frac{\pi-1}{2}, 1\right)$.

8. **a.** The distance is

$$\sqrt{\left(\sqrt{3}-\sqrt{2}\right)^2+\left(\sqrt{2}-\sqrt{3}\right)^2}=\sqrt{2\left(\sqrt{3}-\sqrt{2}\right)^2}=\sqrt{2}\left(\sqrt{3}-\sqrt{2}\right)=\sqrt{6}-2.$$

b. The midpoint is $\left(\frac{\sqrt{3}+\sqrt{2}}{2}, \frac{\sqrt{3}+\sqrt{2}}{2}\right)$.

9. The values when $x = 3$ are on the graph.

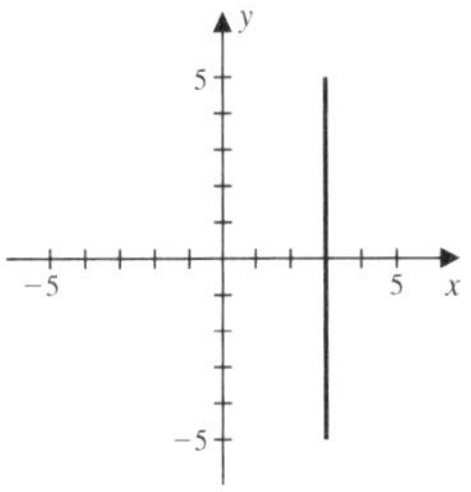

10. The values when $y = -2$ are on the graph.

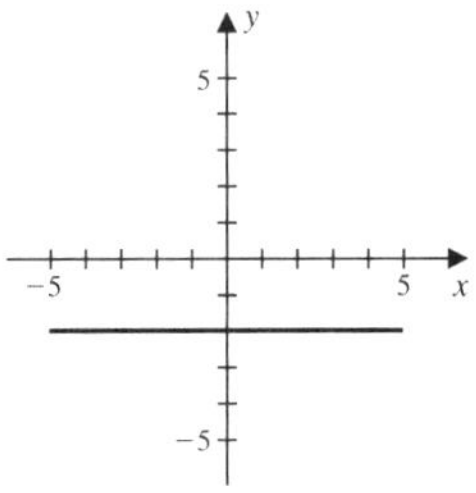

11. The values when $x > 1$ are on the graph.

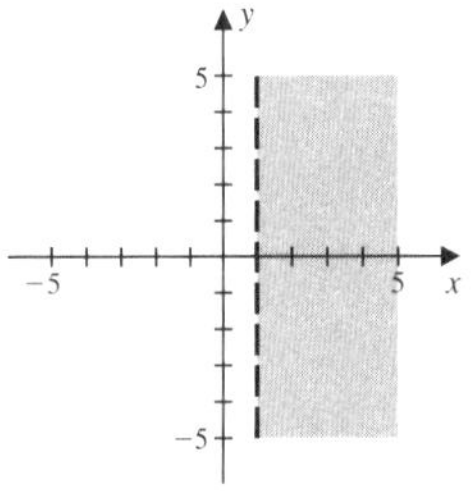

12. The values when $x < -2$ are on the graph.

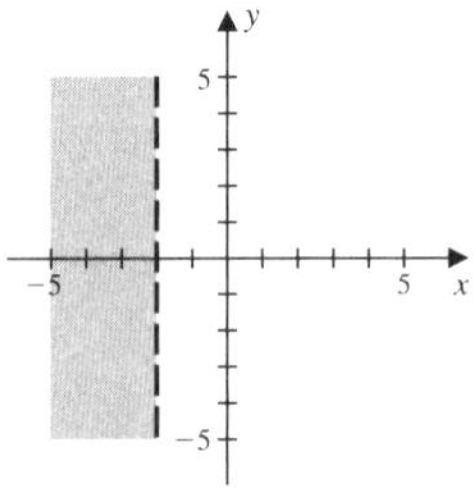

13. The values when $x \geq 1$ and $y \geq 2$ are on the graph.

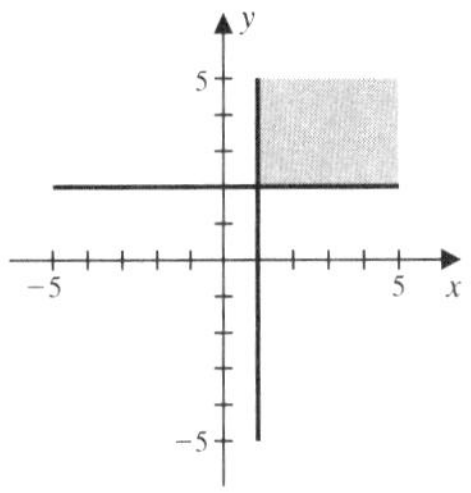

14. The values when $x < -2$ and $y < -3$ are on the graph.

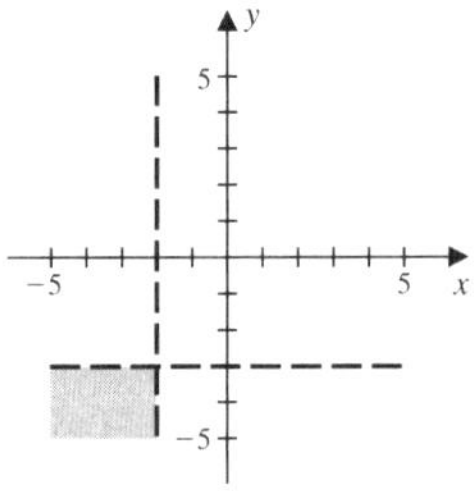

15. The values when $-3 < y \leq 1$ are on the graph.

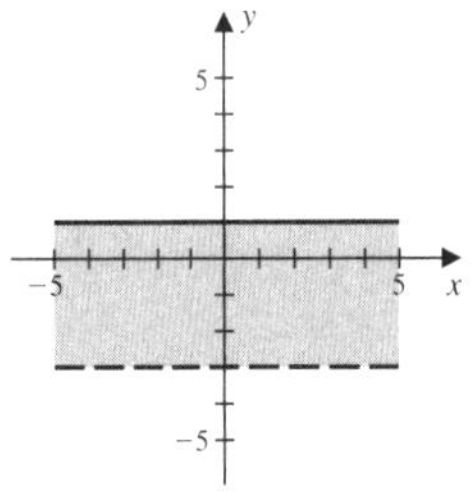

16. The values when $-1 \leq x \leq 2$ are on the graph.

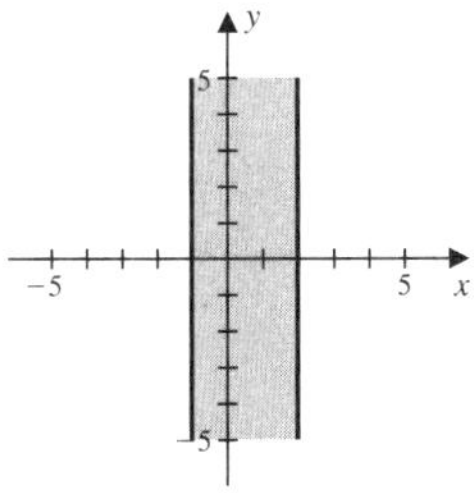

17. The values when $3 \le |x|$, that is, when $x \ge 3$ or $x \le -3$ are on the graph.

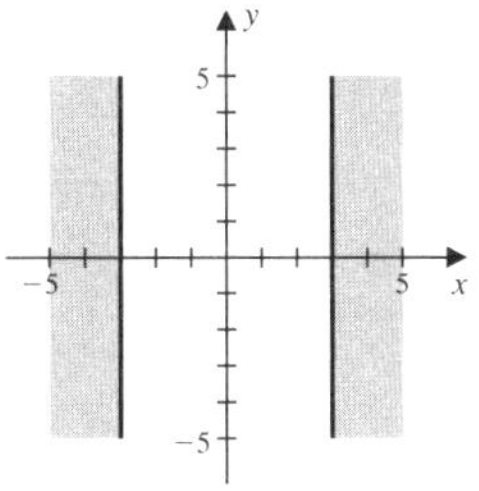

18. The values when $|y + 1| > 2$, that is, when $y > 1$ or $y < -3$ are on the graph.

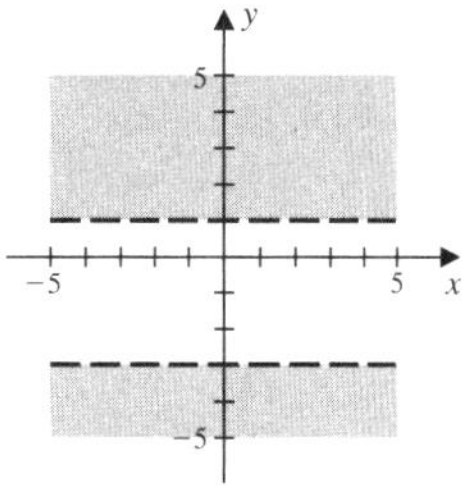

19. The values when $-1 \le x \le 2$ and $2 < y < 3$ are on the graph.

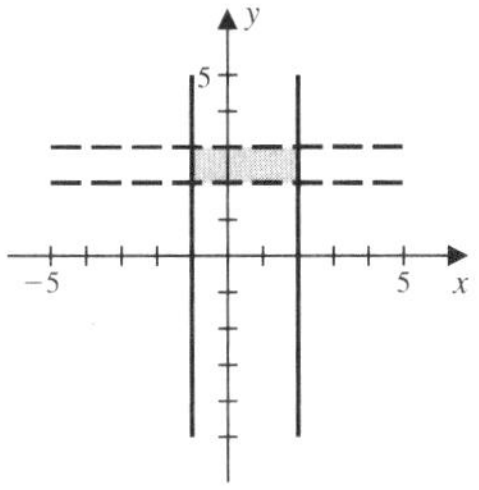

20. The values when $-3 < x \le 1$ and $-1 \le y \le 2$ are on the graph.

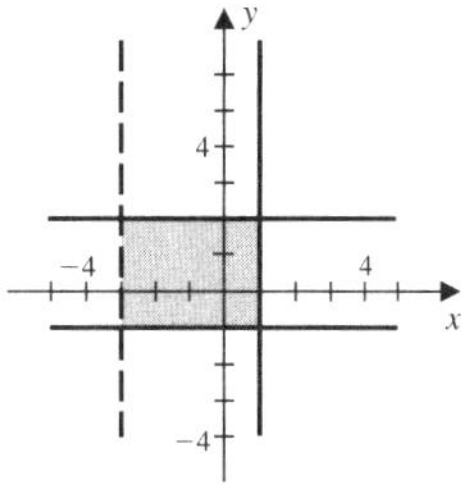

21. The values when $|x - 1| < 3$ and $|y + 1| < 2$, that is, when $-2 < x < 4$ and $-3 < y < 1$ are on the graph.

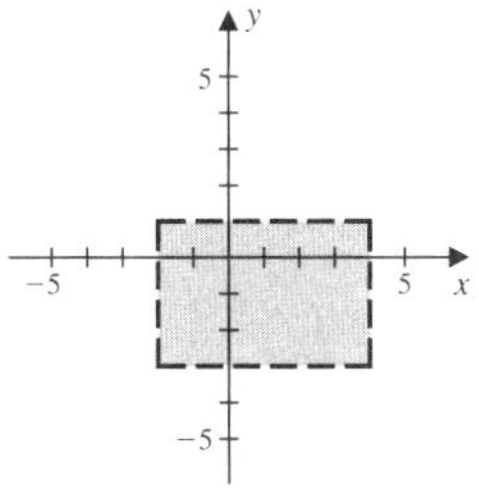

22. The values when $|x + 2| \le 1$ and $|y - 2| < 3$, that is, when $-3 \le x \le -1$ and $-1 < y < 5$ are on the graph.

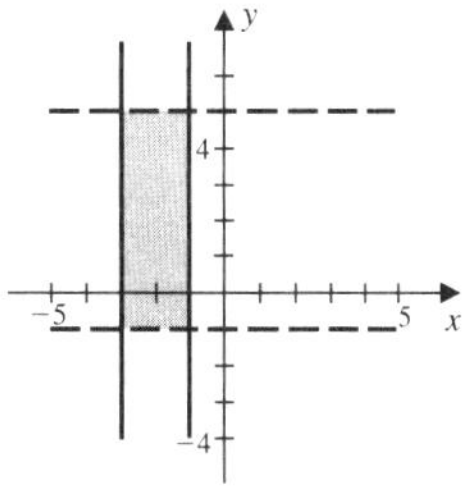

23. The circle with center (2, 0) and radius 3 has equation

$$9 = (x - 2)^2 + y^2.$$

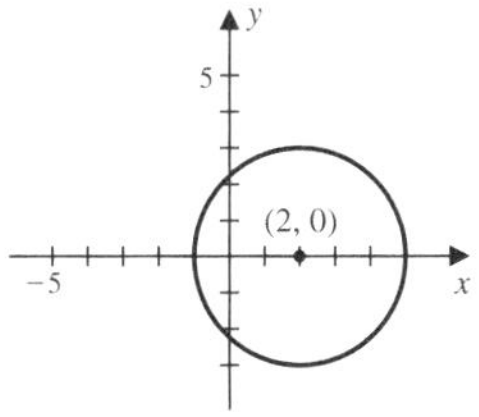

24. The circle with center (0, 2) and radius 3 has equation

$$9 = x^2 + (y - 2)^2.$$

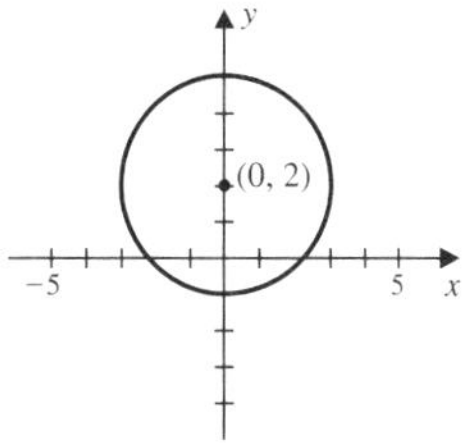

25. The circle with center $(-2, 3)$ and radius 2 has equation

$$4 = (x+2)^2 + (y-3)^2.$$

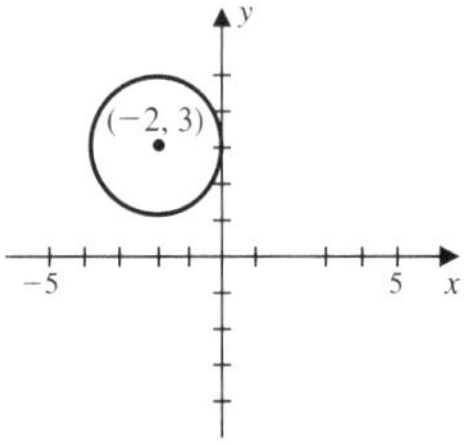

26. The circle with center $(1, -4)$ and radius 4 has equation

$$16 = (x-1)^2 + (y+4)^2.$$

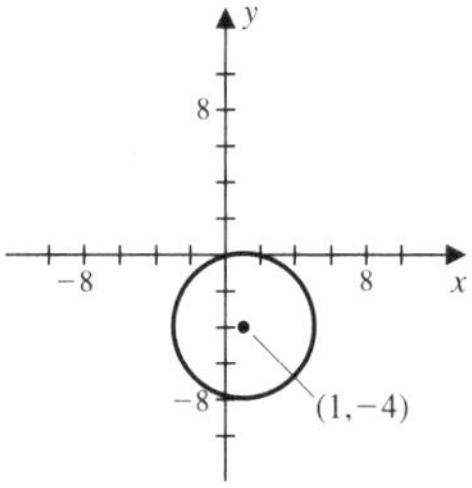

27. The circle with center $(-1, -2)$ and radius 2 has equation

$$4 = (x+1)^2 + (y+2)^2.$$

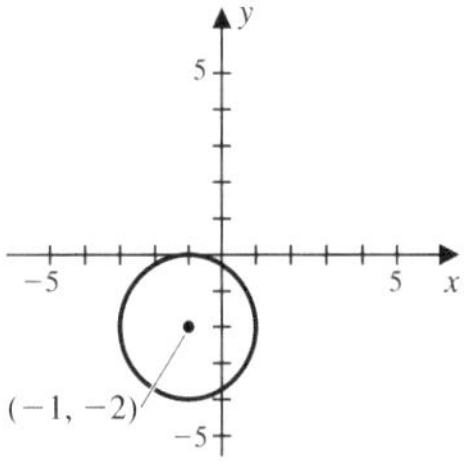

28. The circle with center $(-2, -1)$ and radius 3 has equation

$$9 = (x+2)^2 + (y+1)^2.$$

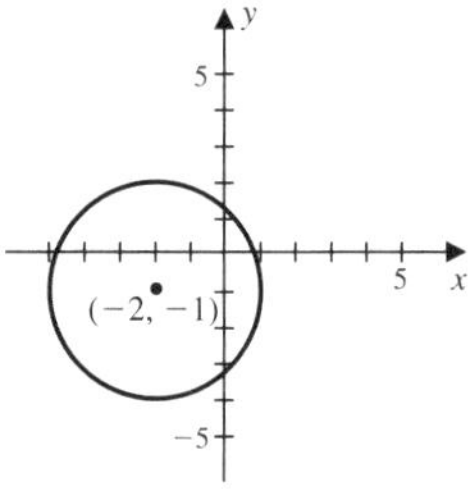

29. The circle has equation

$$4 = (x-2)^2 + (y-2)^2.$$

30. The circle has equation

$$16 = (x+2)^2 + (y-3)^2.$$

31. The circle has center $(0, 0)$ and radius $\sqrt{9} = 3$.

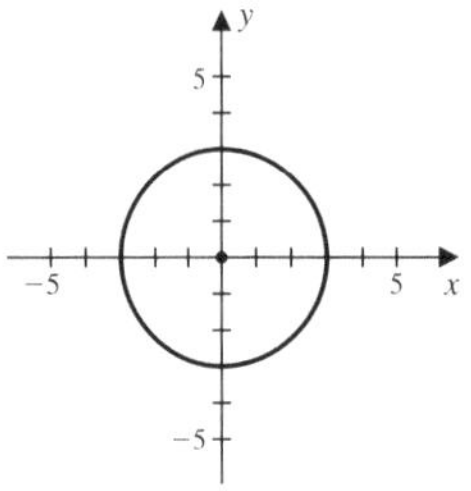

32. The circle has center $(0, 0)$ and radius $\sqrt{2}$.

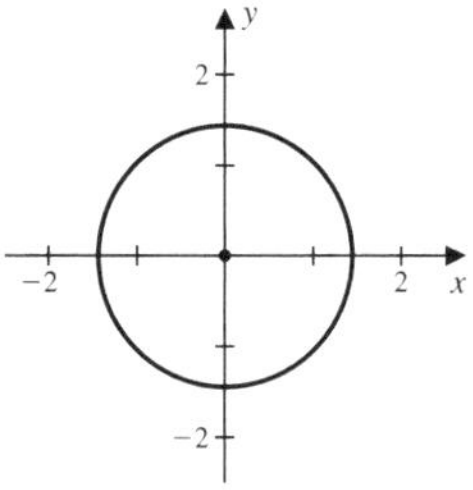

33. The circle has center (0, 1) and radius $\sqrt{1} = 1$.

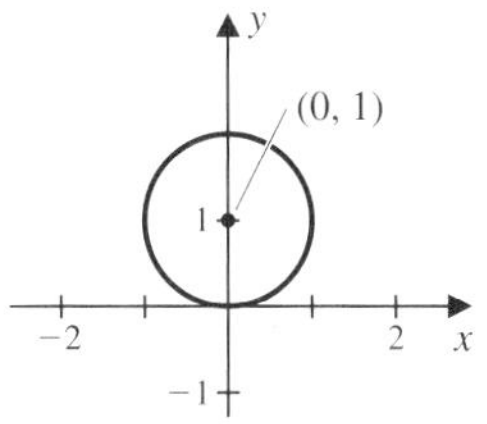

34. The circle with center (−1, 0) and radius $\sqrt{9} = 3$.

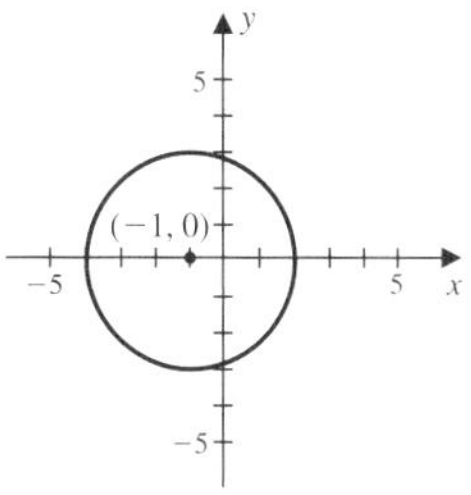

35. The circle has center (2, −1) and radius $\sqrt{9} = 3$.

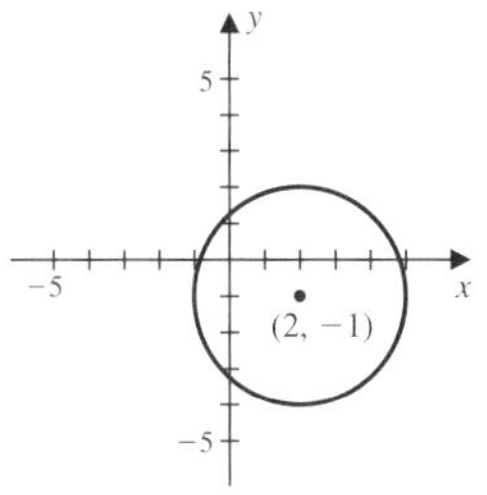

36. The circle has center (1, −2) and radius $\sqrt{16} = 4$.

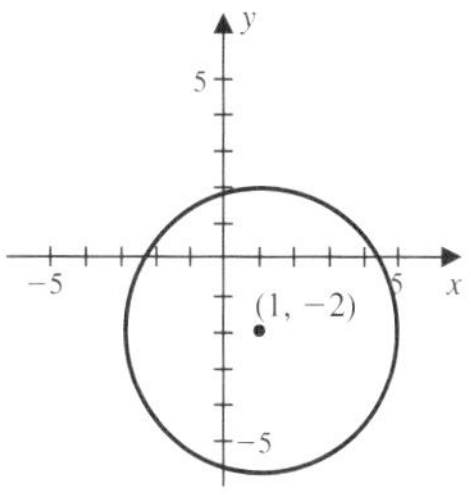

37. The circle has center (1, 0) and radius 2 since

$$3 = x^2 - 2x + y^2 = (x^2 - 2x + 1) - 1 + y^2,$$

so $(x-1)^2 + y^2 = 4$.

38. The circle has center (0, −2) and radius 1 since

$$-3 = x^2 + y^2 + 4y = x^2 + (y^2 + 4y + 4) - 4,$$

so $x^2 + (y+2)^2 = 1$.

39. The circle has center (−1, 2) and radius 1 since

$$\begin{aligned}-4 =& x^2 + 2x + y^2 - 4y \\ =& (x^2 + 2x + 1) - 1 + (y^2 - 4y + 4) - 4,\end{aligned}$$

so $(x+1)^2 + (y-2)^2 = 1$.

40. The circle has center (1, −2) and radius 3 since

$$\begin{aligned}4 =& x^2 - 2x + y^2 + 4y \\ =& (x^2 - 2x + 1) - 1 + (y^2 + 4y + 4) - 4,\end{aligned}$$

so $(x-1)^2 + (y+2)^2 = 9$.

41. The circle has center (2, 1) and radius 3 since

$$\begin{aligned}0 =& x^2 - 4x + y^2 - 2y - 4 \\ =& (x^2 - 4x + 4) - 4 \\ & + (y^2 - 2y + 1) - 1 - 4,\end{aligned}$$

so $(x-2)^2 + (y-1)^2 = 9$.

42. The circle has center (−2, −3) and radius 2 since

$$\begin{aligned}0 =& x^2 + 4x + y^2 + 6y + 9 \\ =& (x^2 + 4x + 4) - 4 + (y^2 + 6y + 9),\end{aligned}$$

so $(x+2)^2 + (y+3)^2 = 4$.

43. The values when $x^2 + y^2 \le 1$ are on the graph.

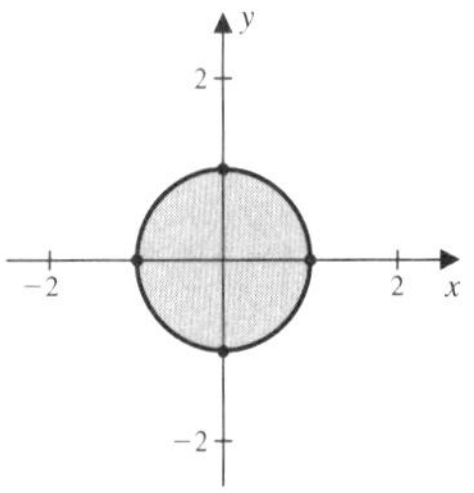

44. The values when $(x - 1)^2 + y^2 > 2$ are on the graph.

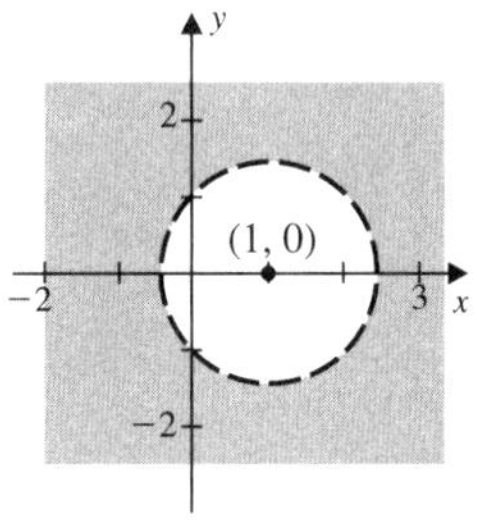

45. The values when $1 < x^2 + y^2 < 4$ are on the graph.

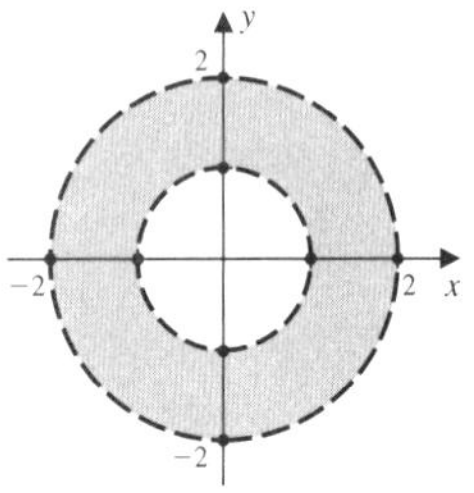

46. The values when $4 \le (x-1)^2 + (y-1)^2 \le 9$ are on the graph.

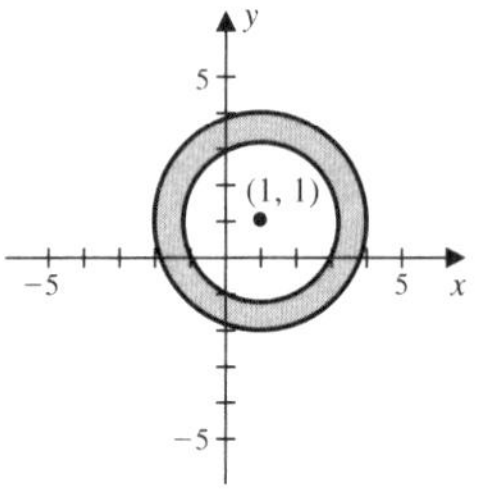

47. The values when $x^2 + y^2 \le 4$ and $y \ge x$ are on the graph.

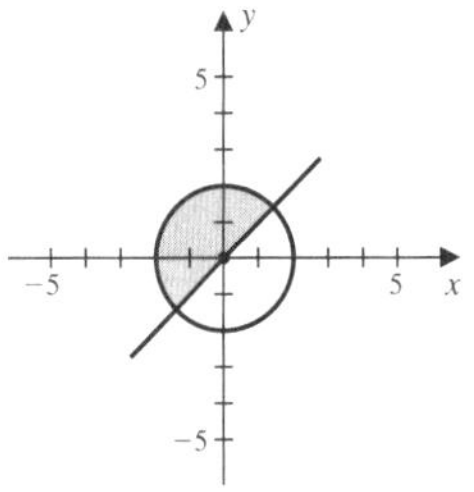

48. The values when $x^2 + y^2 \ge 4$ and $y \le x$ are on the graph.

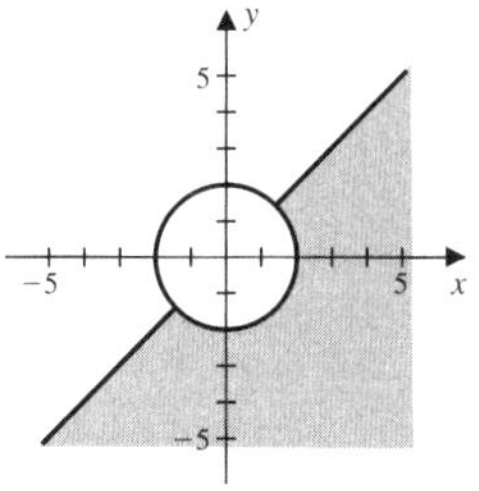

49. The point $(6, 3)$ is closer to the origin, since $d((6, 3), (0, 0)) = \sqrt{45}$ and $d((-7, 2), (0, 0)) = \sqrt{53}$.

50. The points are the same distance from $(3, 2)$, since

$$d((1, 6), (3, 2)) = \sqrt{(1-3)^2 + (6-2)^2} = \sqrt{4 + 16} = \sqrt{20}$$

and $d((7, 4), (3, 2)) = \sqrt{(7-3)^2 + (4-2)^2} = \sqrt{16 + 4} = \sqrt{20}$.

51. **a.** Let $a = d((-3, -4), (2, -1)) = \sqrt{34}$, $\quad b = d((-1, 4), (2, -1)) = \sqrt{34}$, and $c = d((-1, 4), (-3, -4)) = \sqrt{68}$.

Since $a^2 + b^2 = c^2$, the points are the vertices of a right triangle.

b. The right angle is located at vertex $(2, -1)$.

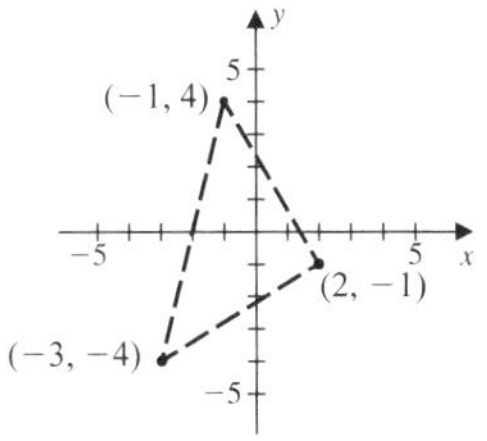

52. Since $d((2, 6), (-1, 2)) = d((2, 6), (2, 1)) = 5$, the triangle is isosceles.

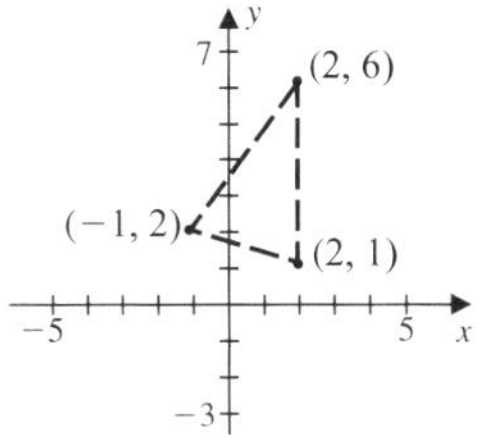

53. The point $(-3, -4)$ is 5 units left and 3 units down from the point $(2, -1)$. The unique point producing a square is the same distance from $(-1, 4)$, that is, $(-1 - 5, 4 - 3) = (-6, 1)$.

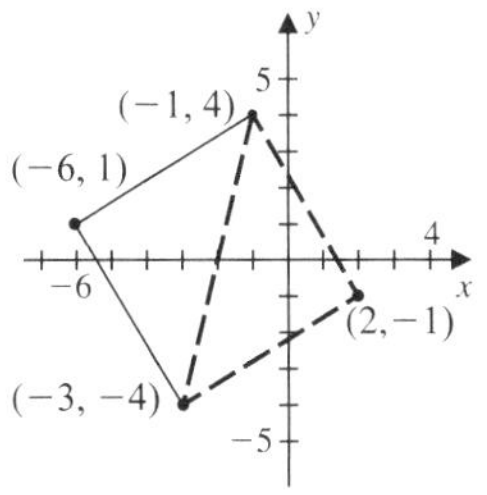

54. The points $(5, 5)$, $(-1, 7)$ and $(-1, -3)$ all work.

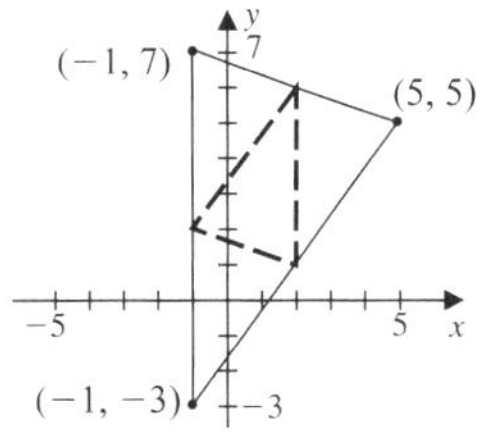

55. The radius $r = d((0,0),(2,3)) = \sqrt{13}$, so the equation of the circle is $x^2 + y^2 = 13$.

56. The radius $r = d((1,3),(-2,4)) = \sqrt{10}$, so the equation of the circle is $(x-1)^2 + (y-3)^2 = 10$.

57. The radius $r = d((3,7),(0,7)) = \sqrt{9} = 3$, so the equation of the circle is $(x-3)^2 + (y-7)^2 = 9$.

58. A point on the y-axis has the form $(0, y)$ and is equidistant from $(2, 1)$ and $(4, -3)$ provided

$$\sqrt{(0-2)^2 + (y-1)^2} = \sqrt{(0-4)^2 + (y+3)^2} \quad \text{or} \quad 4 + y^2 - 2y + 1 = 16 + y^2 + 6y + 9,$$

which simplifies to $-20 = 8y$ and $y = -\frac{5}{2}$. The point is $\left(0, -\frac{5}{2}\right)$.

59. A circle with radius 3, tangent to both the x- and y-axes, and center in the second quadrant, has center $(-3, 3)$. The equation is $(x+3)^2 + (y-3)^2 = 9$.

60. The point $(3, 4)$ is in the first quadrant and is a horizontal distance of 2 units from the line $x = 1$ and a vertical distance of 2 units from the line $y = 2$. So $(3, 4)$ is the center of the circle, and since the radius is 2, the equation is $(x-3)^2 + (y-4)^2 = 4$.

61. The area of the shaded region in the figure is $\pi(3)^2 - \pi(1)^2 = 8\pi$.

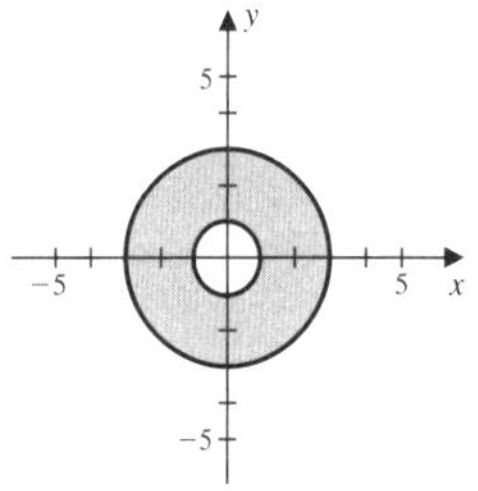

62. The area of the shaded region in the figure is $\pi(2)^2 - \pi(1)^2 = 3\pi$.

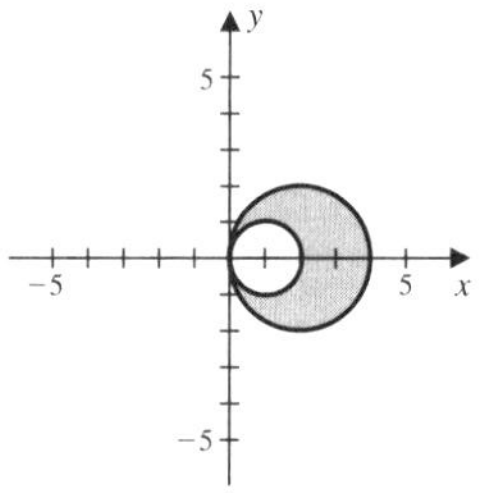

63. The values when $|x| + |y| \le 4$ are on the graph.

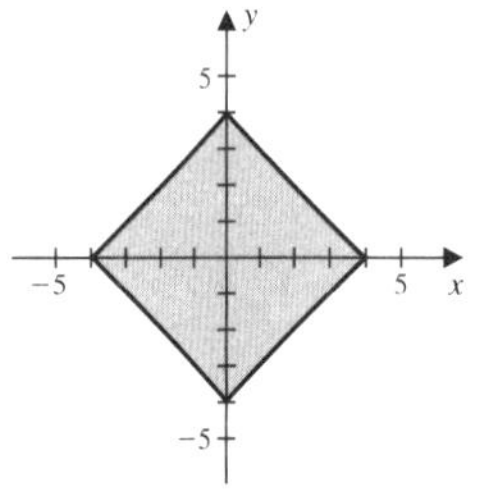

64. The values when $|x-1| + |y+2| \le 2$ are on the graph.

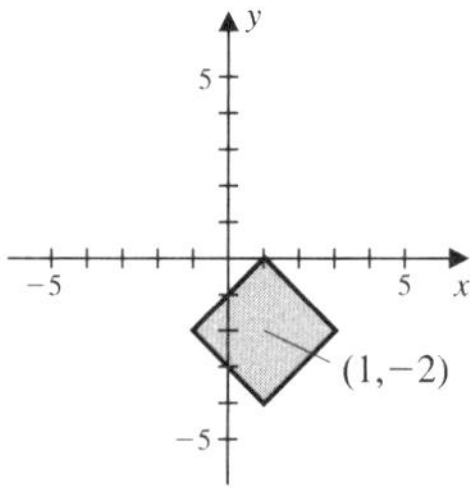

65. The area of the shaded region is the area of the circle minus the area of the square, that is, $\pi(1)^2 - \sqrt{2}\sqrt{2} = \pi - 2$.

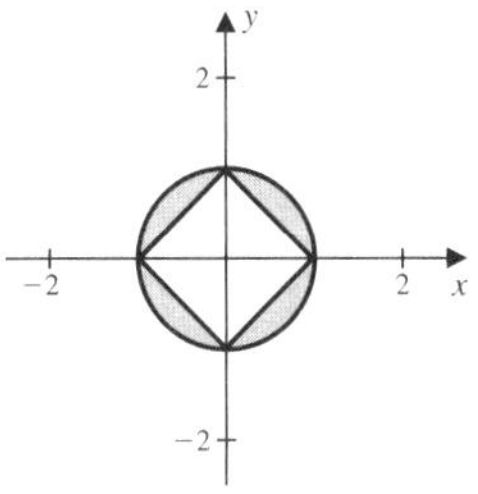

66. The area of the shaded region is the area of the circle minus the area of the square, that is, $\pi(2)^2 - \left(2\sqrt{2}\right)\left(2\sqrt{2}\right) = 4\pi - 8$.

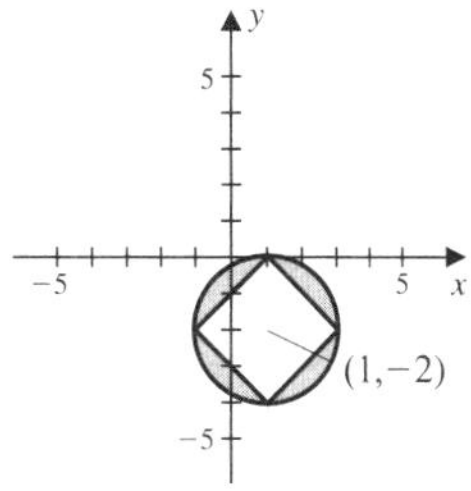

Exercise Set 1.4 (Page 25)

1. Symmetry with respect to the y-axis.

2. Symmetry with respect to the origin.

3. Symmetry with respect to the x-axis, y-axis, and origin.

4. No axis or origin symmetry.

5. Symmetry with respect to the origin.

6. Symmetry with respect to the y-axis.

7. Symmetry with respect to the x-axis.

8. Symmetry with respect to the origin.

9. For $y = x + 1$:
Intercepts: $(-1, 0)$, $(0, 1)$;
Symmetry: No axis or origin symmetry.

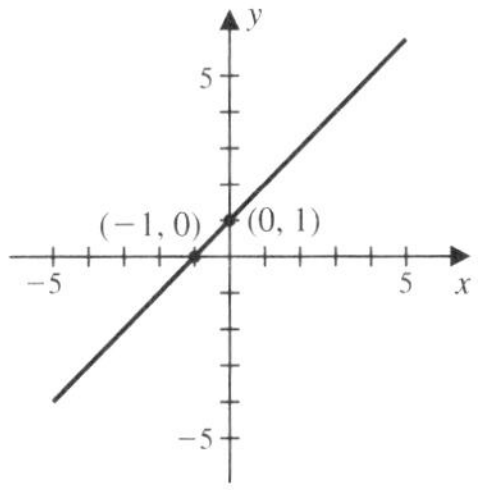

10. For $y = 2x - 3$:
Intercepts: $\left(\frac{3}{2}, 0\right)$, $(0, -3)$;
Symmetry: No axis or origin symmetry.

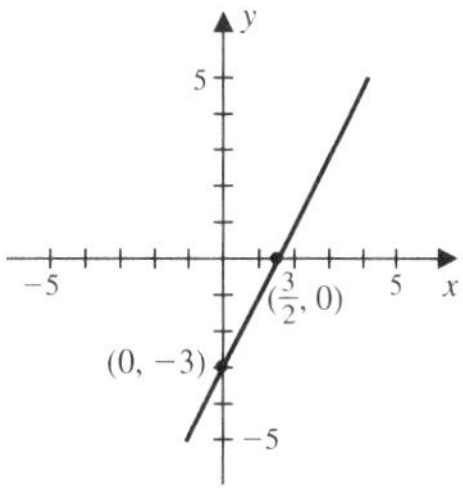

11. For $x + y = 1$, or $y = 1 - x$:
Intercepts: $(1, 0)$, $(0, 1)$;
Symmetry: No axis or origin symmetry.

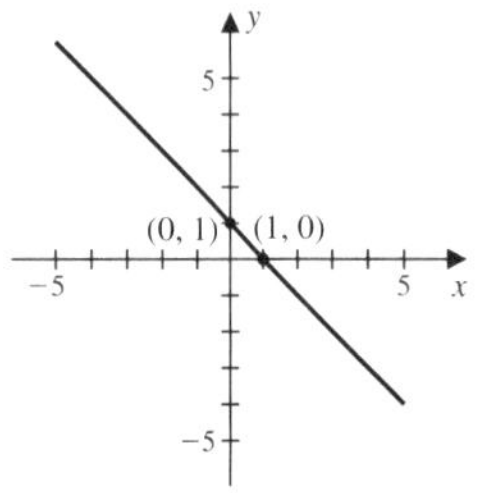

12. For $-2x - y = 2$, or $y = -2x - 2$:
Intercepts: $(-1, 0)$, $(0, -2)$;
Symmetry: No axis or origin symmetry.

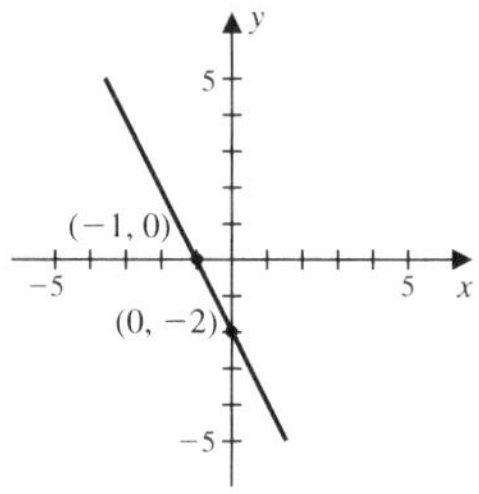

13. For $y = x^2 - 1$:
Intercepts: $(-1, 0)$, $(1, 0)$, $(0, -1)$;
Symmetry: y-axis.

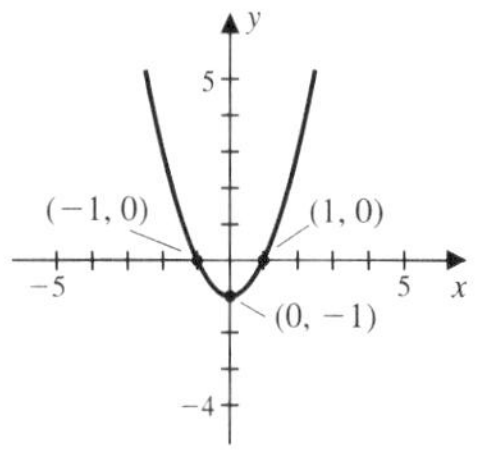

14. For $y = x^2 + 2$:
Intercepts: $(0, 2)$;
Symmetry: y-axis.

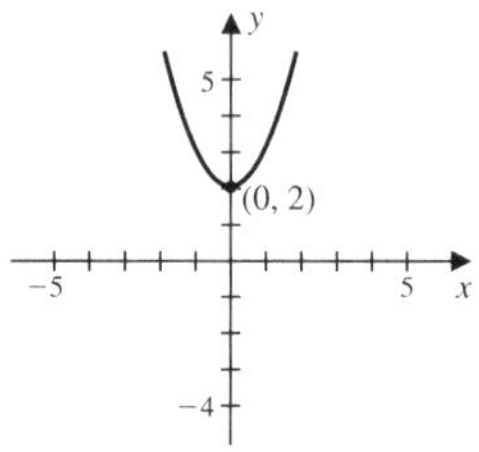

15. For $y = 1 - x^2$:
Intercepts: $(-1, 0)$, $(1, 0)$, $(0, 1)$;
Symmetry: y-axis.

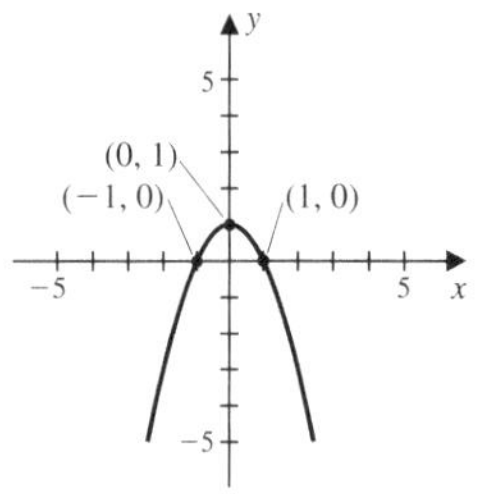

16. For $2y = x^2$, or $y = \frac{1}{2}x^2$:
Intercepts: $(0, 0)$;
Symmetry: y-axis.

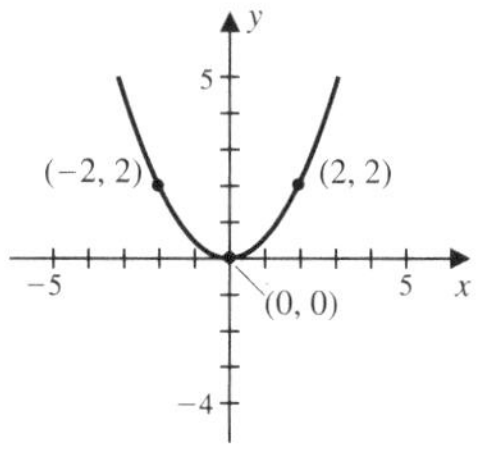

17. For $y = -2x^2$:
Intercepts: (0, 0);
Symmetry: y-axis.

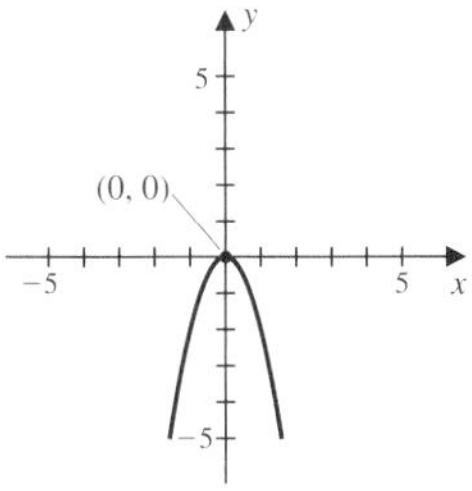

18. For $y = 2 - 2x^2$:
Intercepts: $(-1, 0)$, $(1, 0)$, $(0, 2)$;
Symmetry: y-axis.

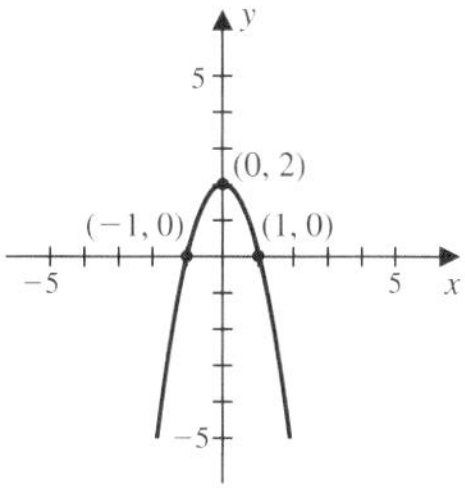

19. For $x = y^2 - 1$:
Intercepts: $(-1, 0)$, $(0, 1)$, $(0, -1)$;
Symmetry: x-axis.

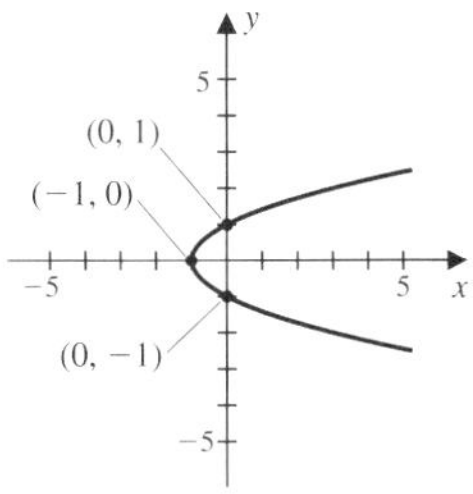

20. For $x = y^2 - 4$:
Intercepts: $(-4, 0)$, $(0, 2)$, $(0, -2)$;
Symmetry: x-axis.

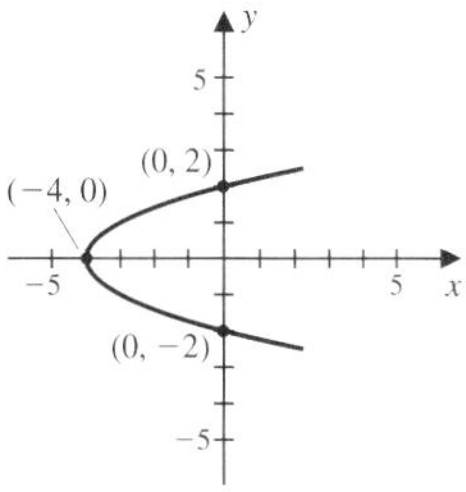

A

21. For $y = x^3 + 1$:
Intercepts: $(-1, 0)$, $(0, 1)$;
Symmetry: No axis or origin symmetry.

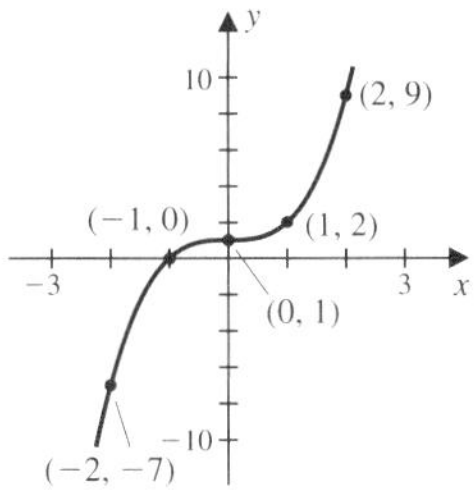

22. For $2y = x^3$, or $y = \frac{1}{2}x^3$:
Intercepts: (0, 0);
Symmetry: origin.

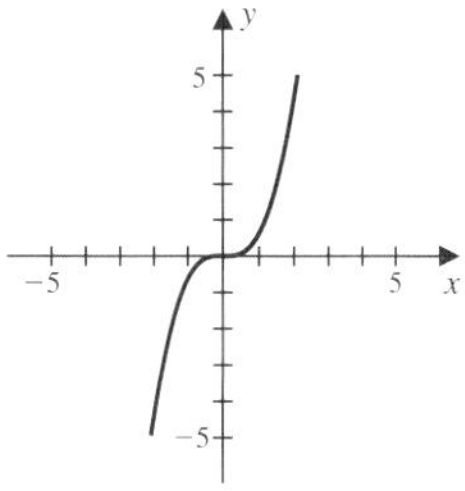

23. For $y = -x^3$:
Intercepts: $(0, 0)$;
Symmetry: origin.

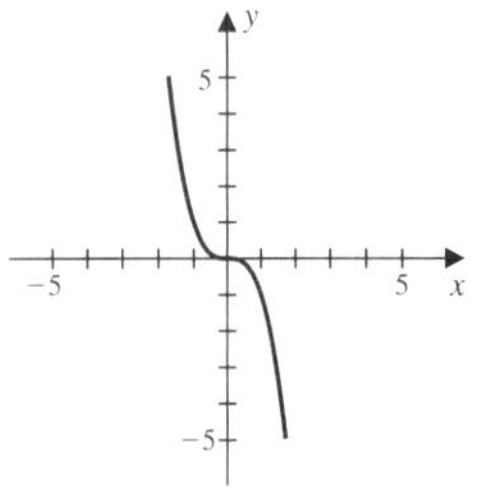

24. For $y = -x^3 + 1$:
Intercepts: $(1, 0)$, $(0, 1)$;
Symmetry: No axis or origin symmetry.

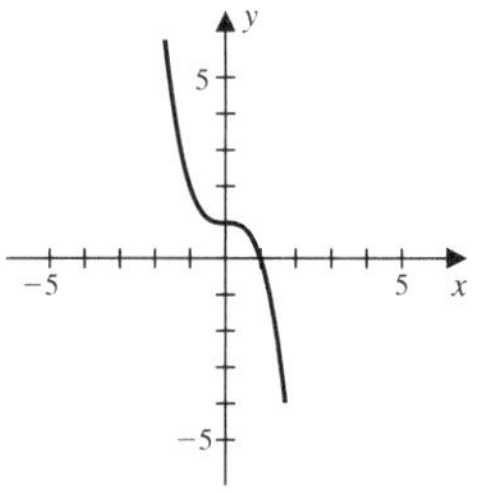

25. For $y = \frac{(x+2)(x-2)}{x-2} = x + 2$, when $x \neq 2$:
Intercepts: $(-2, 0)$, $(0, 2)$;
Symmetry: No axis or origin symmetry.

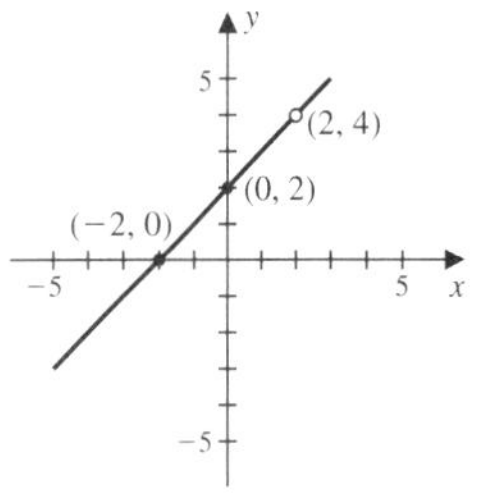

26. For $y = \frac{(x+3)(x+1)}{x+1} = x + 3$, when $x \neq -1$:
Intercepts: $(-3, 0)$, $(0, 3)$;
Symmetry: No axis or origin symmetry.

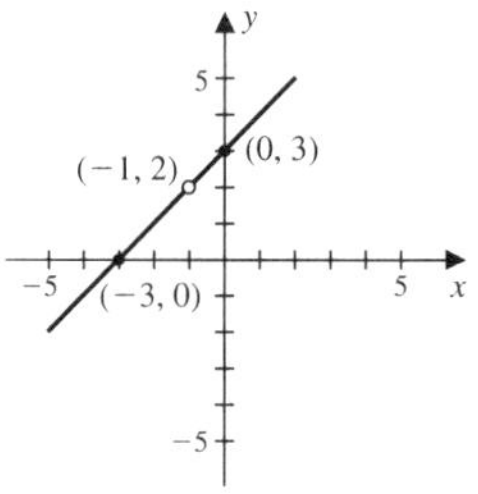

27. For $y = \frac{x^2-x-6}{x+2} = \frac{(x-3)(x+2)}{x+2} = x - 3$, when $x \neq -2$:
Intercepts: $(3, 0)$, $(0, -3)$;
Symmetry: No axis or origin symmetry.

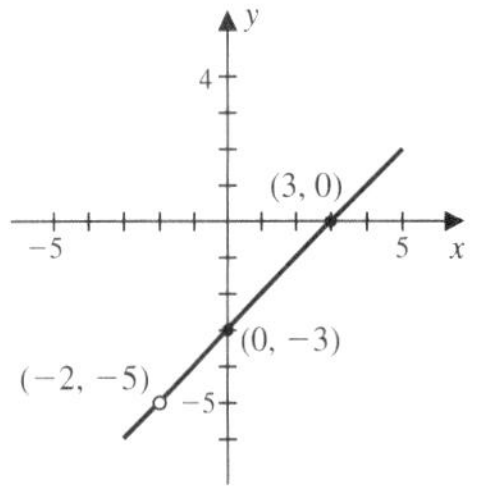

28. For $y = \frac{x^2+2x-3}{x+3} = \frac{(x+3)(x-1)}{x+3} = x - 1$, when $x \neq -3$:
Intercepts: $(1, 0)$, $(0, -1)$;
Symmetry: No axis or origin symmetry.

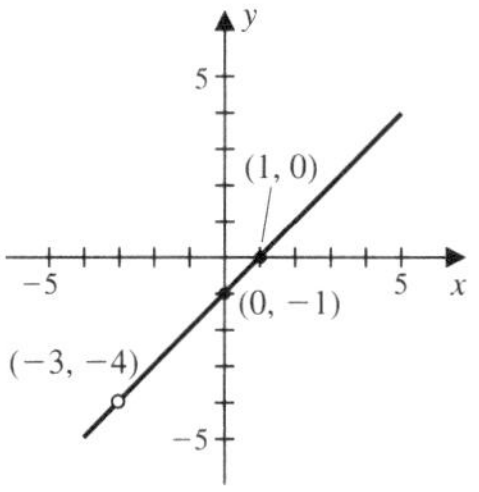

29. For $y = \sqrt{x} + 2$:
Intercepts: $(0, 2)$;
Symmetry: No axis or origin symmetry.

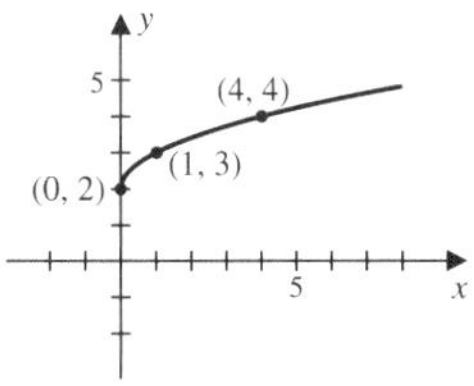

30. For $y = \sqrt{x - 1}$:
Intercepts: $(1, 0)$;
Symmetry: No axis or origin symmetry.

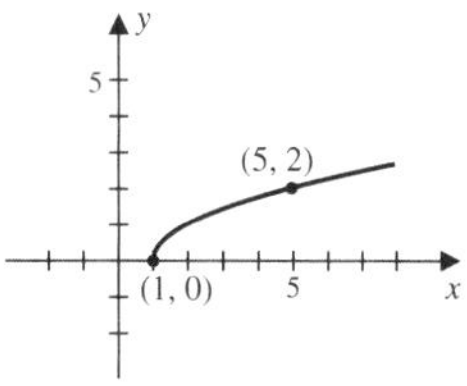

31. For $x^2 + y^2 = 4$:
Intercepts: $(-2, 0), (2, 0), (0, -2), (0, 2)$;
Symmetry: x-axis, y-axis, and origin.

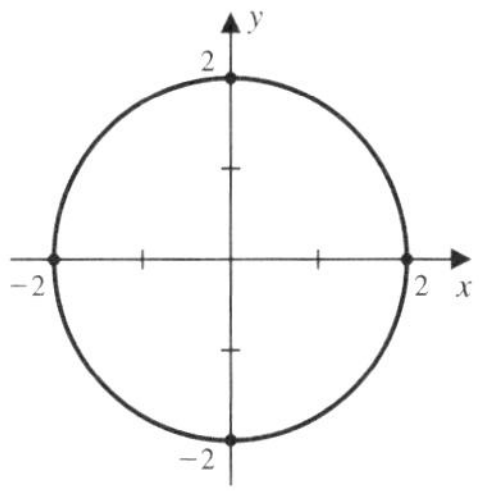

32. For $(x - 1)^2 + y^2 = 1$:
Intercepts: $(0, 0), (2, 0)$;
Symmetry: x-axis.

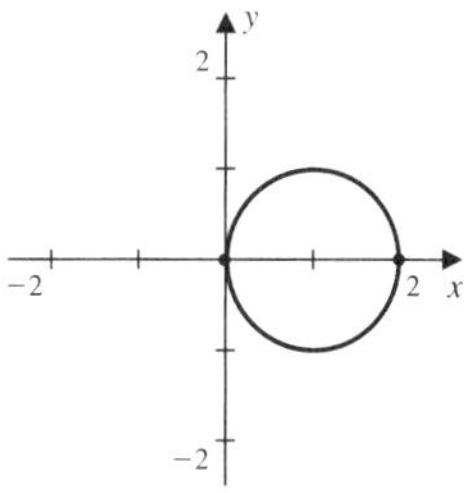

33. For $y = \sqrt{9 - x^2}$:
Intercepts: $(-3, 0), (3, 0), (0, 3)$;
Symmetry: y-axis.

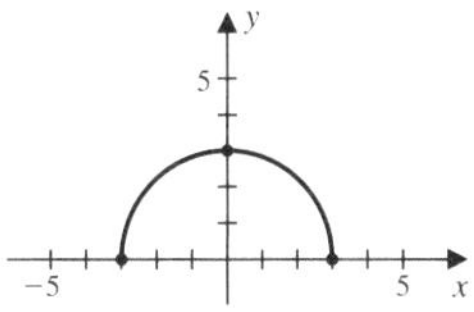

34. For $y = -\sqrt{9 - x^2}$:
Intercepts: $(-3, 0), (3, 0), (0, -3)$;
Symmetry: y-axis.

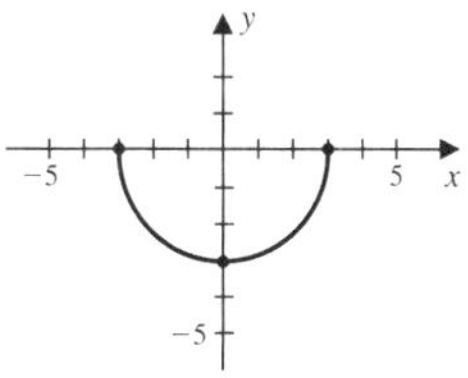

35. For $y = |x|$:
Intercepts: $(0, 0)$;
Symmetry: y-axis.

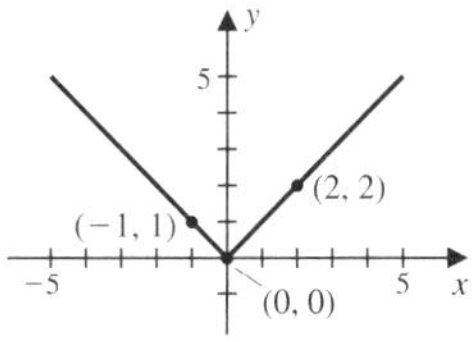

36. For $y = |x - 1|$:
Intercepts: $(1, 0), (0, 1)$;
Symmetry: No axis or origin symmetry.

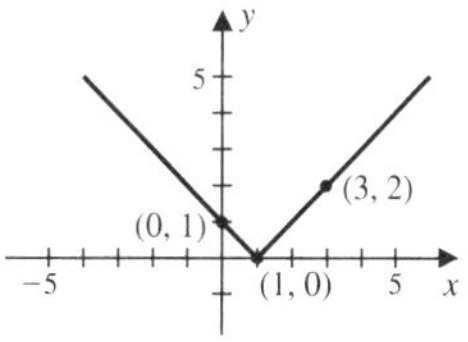

37. For $y = |x| - 1$:
Intercepts: $(-1, 0)$, $(1, 0)$, $(0, -1)$;
Symmetry: y-axis.

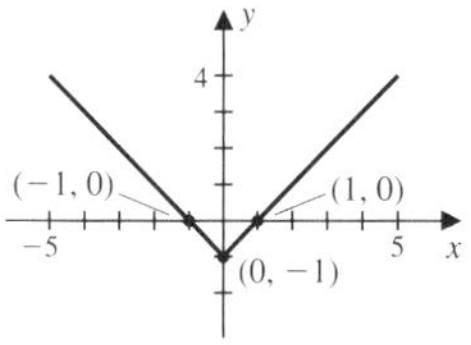

38. For $y = 2 - |x|$:
Intercepts: $(-2, 0)$, $(2, 0)$, $(0, 2)$;
Symmetry: y-axis.

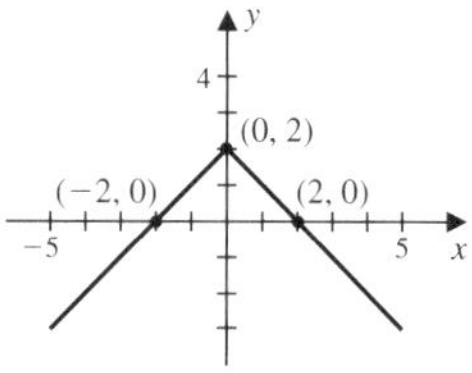

39. This graph has x-axis symmetry.

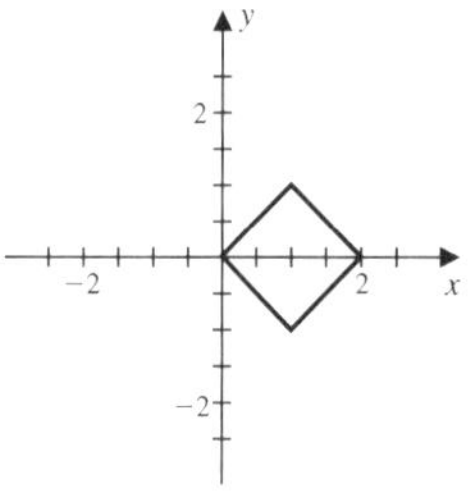

40. This graph has y-axis symmetry.

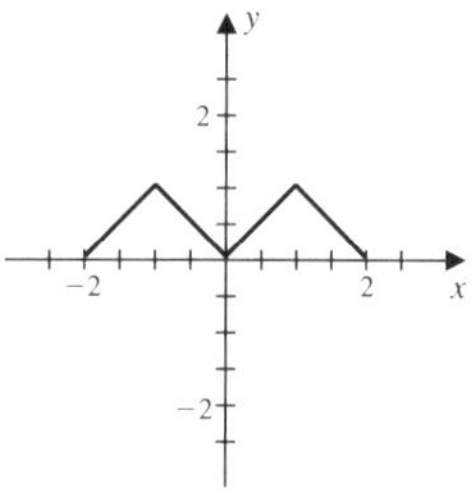

41. This graph has origin symmetry.

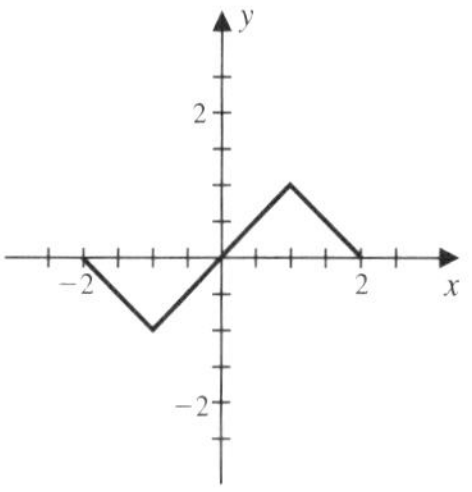

42. This graph has x-axis and y-axis symmetry.

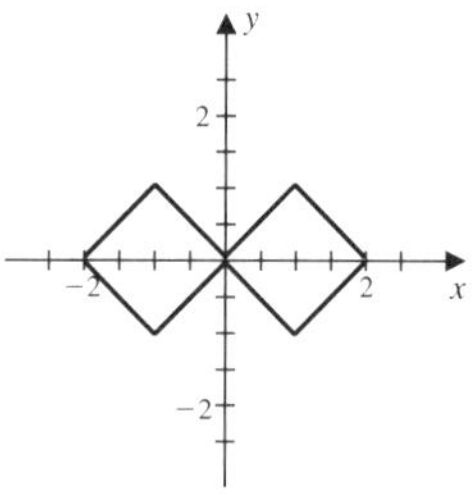

43. From the graph we can see the points of intersection of $y = x^2 + 1$ and $y = 2$ are $(1, 2)$ and $(-1, 2)$. Algebraically we have $x^2 + 1 = 2$ so $x^2 = 1$ and $x = \pm 1$. The distance between the two points of intersection is

$$d((1, 2), (-1, 2)) = \sqrt{(1 - (-1))^2 + (2 - 2)^2} = \sqrt{4} = 2.$$

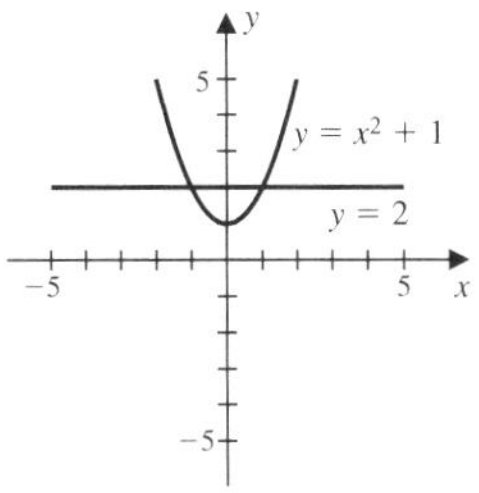

44. From the graph we can see the points of intersection of $y = x^2 - 3$ and $y = x + 3$ are $(3, 6)$ and $(-2, 1)$. Algebraically we have $x^2 - 3 = x + 3$ so $x^2 - x - 6 = (x - 3)(x + 2) = 0$ and $x = 3$, $x = -2$. The distance between the two points of intersection is

$$d((3, 6), (-2, 1)) = \sqrt{(3 - (-2))^2 + (6 - 1)^2} = \sqrt{25 + 25} = \sqrt{50} = 5\sqrt{2}.$$

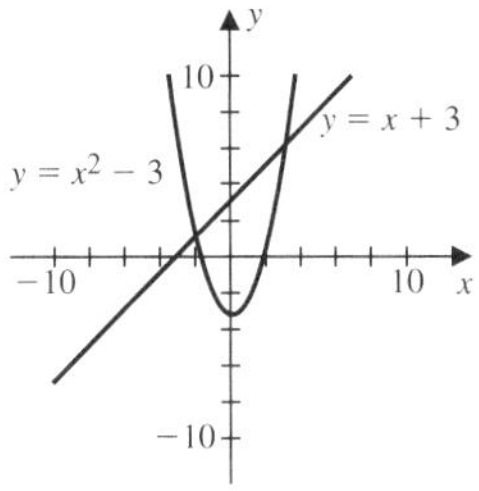

45. Assume the graph has symmetry with respect to both the x- and y-axes. To show the graph is symmetric with respect to the origin, we need to show that if (x, y) is on the graph, then $(-x, -y)$ is also on the graph. If (x, y) is on the graph and the graph is symmetric with respect to the x-axis, then $(x, -y)$ is also on the graph. But if $(x, -y)$ is on the graph and the graph is symmetric with respect to the y-axis, then $(-x, -y)$ is also on the graph. Hence the graph is symmetric with respect to the origin.

It is also true that if the graph has symmetry with respect to the origin and to one of the axes, then it has symmetry with respect to the other axis. Consider, for example, the situation of symmetry with respect to the origin and the x-axis. To show the graph is symmetric with respect to the y-axis, we need to show that if (x, y) is on the graph then $(-x, y)$ is also on the graph. If (x, y) is on the graph and the graph is symmetric with respect to the x-axis, then $(x, -y)$ is also on the graph. But if $(x, -y)$ is on the graph and the graph is symmetric with respect to the origin, then $(-x, -(-y)) = (-x, y)$ is also on the graph. Hence the graph is symmetric with respect to the y-axis.

Exercise Set 1.5 (Page 32)

1. Part (d) gives the best representation.

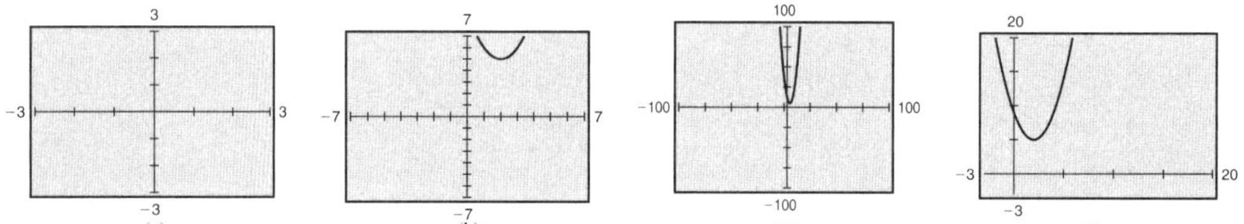

(a) (b) (c) (d)

2. Part (b) gives the best representation.

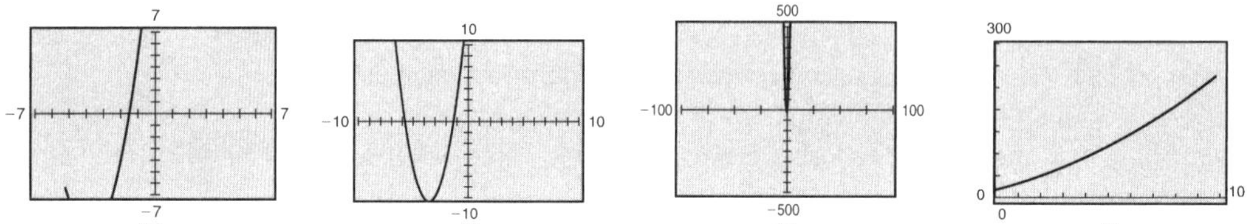

(a) (b) (c) (d)

3. Part (c) gives the best representation.

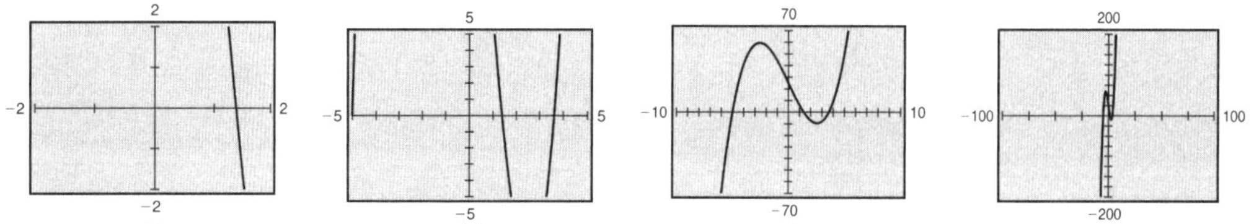

(a) (b) (c) (d)

4. Part (d) gives the best representation.

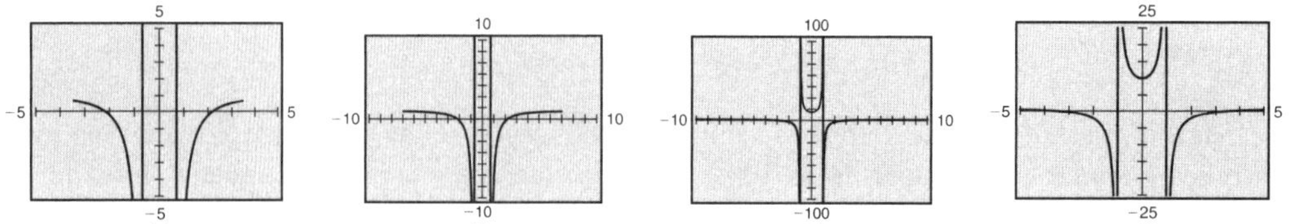

(a) (b) (c) (d)

5. $[-10, 10] \times [-10, 10]$

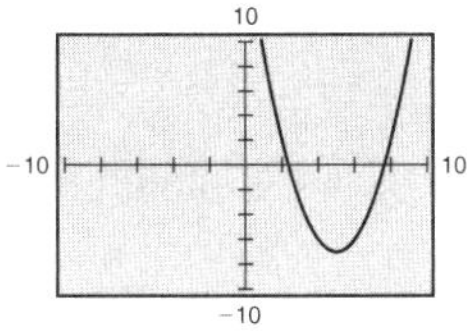

6. $[-20, 10] \times [-20, 70]$

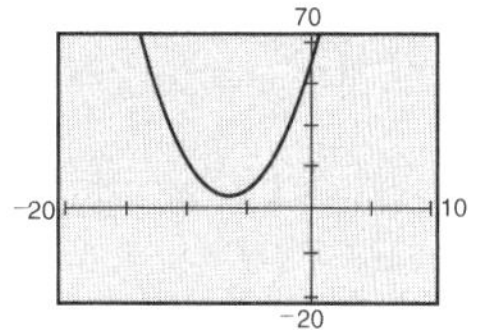

7. $[0, 20] \times [0, 10]$

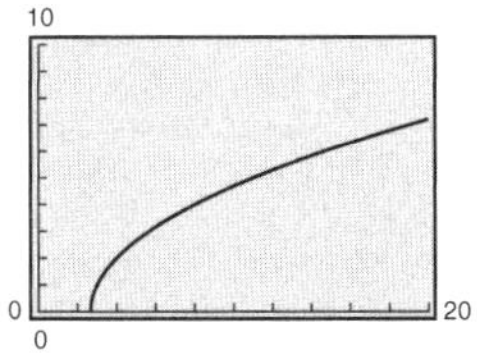

8. $[-20, 20] \times [0, 30]$

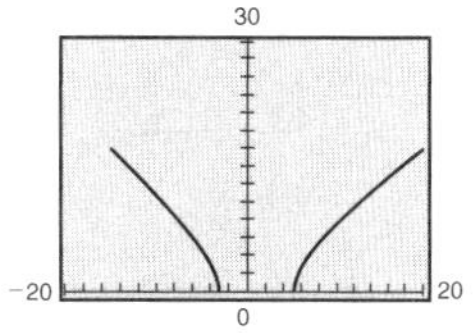

9. $[-10, 10] \times [-10, 10]$

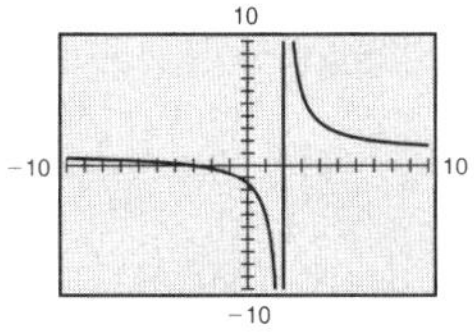

10. $[-5, 5] \times [-10, 10]$

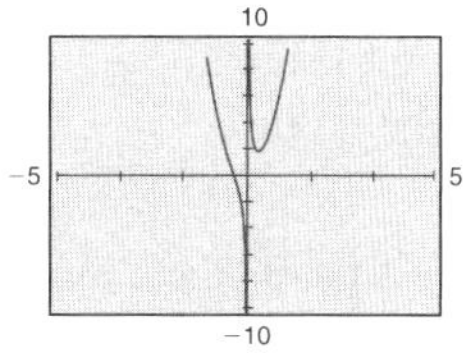

11. The graph of $y = x^3 + x^2 + 3x - 4$ crosses the x-axis at $x \approx 0.86$.

12. The graph of $y = x^3 - x^2 - 4x - 2$ crosses the x-axis at $x \approx -1$, $x \approx -0.7$, and $x \approx 2.7$.

13. The graph of $y = x^4 - 3x^3 + x^2 - 4$ crosses the x-axis at $x \approx -0.93$ and $x \approx 2.82$.

14. The graph of $y = x^4 + x^3 - 2x^2 - x + 1$ crosses the x-axis at $x \approx -1.6$, $x \approx -1$, $x \approx 0.6$ and $x \approx 1$.

15. The points of intersection are approximately $(1, 6)$ and $(-1, 0)$.

16. The points of intersection are approximately $(-2.73, -15.35)$, $(0.73, 5.39)$ and $(2.5, 20.63)$.

17. The points of intersection are approximately $(0, -1)$ and $(0.46, -0.79)$.

18. The points of intersection are approximately $(-2.3, -4.9)$ and $(0.95, -1.6)$.

19. **a.** We have $x \le -3.6$ or $x \ge 0.6$, and **b.** $x < -2.3$ or $1.3 < x < 3$.

20. The inequality $-x^2 + 2x + 1 \ge x^3 - 2x^2 - x + 2$ is satisfied for $0.3 \le x \le 2.2$ or $x \le -1.5$.

21. For x very large the graphs are almost identical. That is, as x grows without bound $x^4 - 4x^3 + 3x^2$ is approximately x^4.

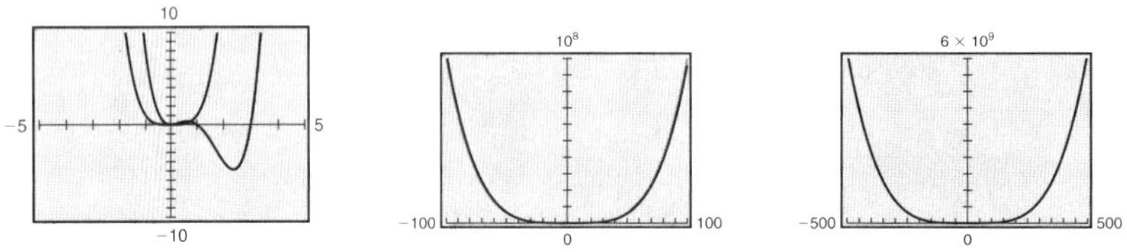

(a) (b) (c)

22. If $c > 0$, the graph has an appearance similar to that of $y = x^2$. If $c < 0$, the graph crosses the x-axis three times, just touching the point $(0, 0)$. As c increases in magnitude the other two points where the graph crosses the x-axis move further from the origin and are always symmetric on either side of the origin. The graph has a local high point at the origin and two local low points symmetric about the origin.

23. The graph appears to approach the horizontal line $y = \frac{a}{b}$ and the vertical line $x = \frac{1}{b}$.

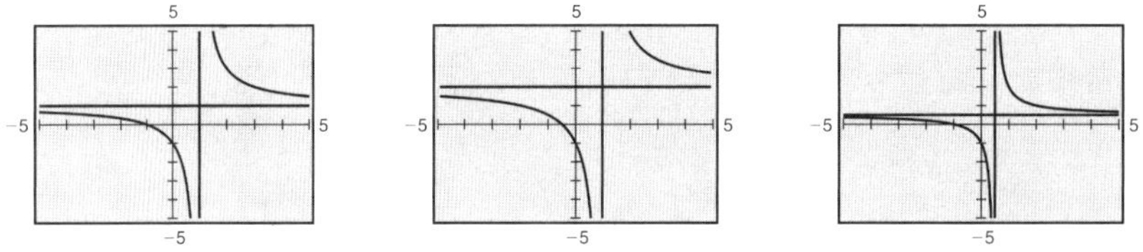

(a) (b) (c)

24. The value of a affects the inclination of the line. If $a > 0$, the line is increasing from left to right, and if $a < 0$ the line is decreasing. The larger the magnitude of a, the steeper the inclination. The constant b determines where the line crosses the y-axis.

25. If $a = b = 0$, the graph is the standard parabola $y = x^2$. The constants a and b shift $y = x^2$ either horizontally or vertically. If $a > 0$, the shift is to the right a units. If $a < 0$, the shift is to the left $|a|$ units. If $b > 0$, the shift is upward b units, and if $b < 0$, the shift is downward $|b|$ units.

26. **a.** When $0 < x < 1$: $\frac{1}{x^n} < \frac{1}{x^{n+2}}$. **b.** When $x > 1$: $\frac{1}{x^n} > \frac{1}{x^{n+2}}$.

c. When $-1 < x < 0$ and n is even we have $\frac{1}{x^n} < \frac{1}{x^{n+2}}$, but when n is odd we have $\frac{1}{x^n} > \frac{1}{x^{n+2}}$.

d. When $x < -1$ and n is even we have $\frac{1}{x^n} > \frac{1}{x^{n+2}}$, but when n is odd we have $\frac{1}{x^n} < \frac{1}{x^{n+2}}$.

Exercise Set 1.6 (Page 46)

1. For $f(x) = 4x^2 + 1$ we have:

a. $f(2) = 4(2)^2 + 1 = 17$

b. $f\left(\sqrt{3}\right) = 4\left(\sqrt{3}\right)^2 + 1 = 13$

c. $f\left(2+\sqrt{3}\right) = 4\left(2+\sqrt{3}\right)^2 + 1 = 4\left(4+4\sqrt{3}+3\right) + 1 = 29 + 16\sqrt{3}$

d. $f(2) + f\left(\sqrt{3}\right) = 17 + 13 = 30$

e. $f(2x) = 4(2x)^2 + 1 = 4(4x^2) + 1 = 16x^2 + 1$

f. $f(1-x) = 4(1-x)^2 + 1 = 4(1-2x+x^2) + 1 = 4x^2 - 8x + 5$

g. $f(x+h) = 4(x+h)^2 + 1 = 4(x^2 + 2hx + h^2) + 1 = 4x^2 + 8hx + 4h^2 + 1$

h. $f(x+h) - f(x) = 4x^2 + 8hx + 4h^2 + 1 - (4x^2 + 1) = 8hx + 4h^2$

2. For $f(x) = \sqrt{x+2}$ we have:

a. $f(-1) = \sqrt{-1+2} = 1$

b. $f(0) = \sqrt{2}$

c. $f(4) = \sqrt{6}$

d. $f(7) = \sqrt{9} = 3$

e. $f(a) = \sqrt{a+2}$

f. $f(2a-1) = \sqrt{2a-1+2} = \sqrt{2a+1}$

g. $f(x+h) = \sqrt{x+h+2}$

h. $f(x+h) - f(x) = \sqrt{x+h+2} - \sqrt{x+2}$

3. For $f(t) = |t-2|$ we have:

a. $f(4) = |4-2| = |2| = 2$

b. $f(1) = |1-2| = |-1| = 1$

c. $f(0) = |0-2| = |-2| = 2$

d. $f(t+2) = |t+2-2| = |t|$

e. $f(2-t^2) = \left|2-t^2-2\right| = \left|-t^2\right| = \left|t^2\right| = t^2$

f. $f(-t) = |-t-2| = |-(t+2)| = |t+2|$

4. For $f(x) = \begin{cases} x^2, & \text{if } x \geq 0 \\ 2x - 1, & \text{if } x < 0 \end{cases}$ we have:

a. $f(1) = 1^2 = 1$

b. $f(-1) = 2(-1) - 1 = -3$

c. $f(-2) = 2(-2) - 1 = -5$

d. $f(3) = 3^2 = 9$

e. For $a > 1$, $1 - a < 0$ and $f(1 - a) = 2(1 - a) - 1 = 1 - 2a$. If $a = 1$, $1 - a = 0$ and $f(0) = 0$.

f. For $0 < a < 1$, $1 < 1 + a < 2$ and $f(1 + a) = (1 + a)^2 = 1 + 2a + a^2$.

5. Reading from the graph given in the exercise we have:

a. $f(-2) = -3$, $f(0) = \frac{1}{2}$, $f(2) = 0$, and $f(3) = 2$

b. Domain$(f) = [-2, 3]$ and Range$(f) = [-3, 2]$

6. Reading from the graph given in the exercise we have:

a. $f(-2) = 2.5$, $f(0) = 0.5$, $f(2) = -2$ and $f(3) = 0$

b. Domain$(f) = [-2, 3]$ and Range$(f) = [-2.25, 0) \cup [0.5, 2.5]$

7. The graph satisfies the vertical line test, that is, each vertical line crosses the curve in at most one place, so it is the graph of a function.

8. There are vertical lines that cross the curve twice, so it is not the graph of a function.

9. The graph satisfies the vertical line test, so it is the graph of a function.

10. There are vertical lines that cross the curve twice, so it is not the graph of a function.

11. **a.** Since the only vertical line that does not intersect the curve is $x = 0$, the domain is $(-\infty, 0) \cup (0, \infty)$.

b. Since the only horizontal line that does not intersect the curve in at least one place is $y = 0$, the range is $(-\infty, 0) \cup (0, \infty)$.

12. **a.** Since the only vertical lines that do not intersect the curve are $x = -4$ and $x = 4$, the domain is all real numbers except $x = 4$ and $x = -4$, that is, $(-\infty, -4) \cup (-4, 4) \cup (4, \infty)$.

b. Since any horizontal line above $y = 0$ and below and including the line $y = 2$, fails to cross the curve, the range is $(-\infty, 0] \cup (2, \infty)$.

13. **a.** Domain$(f) = (-\infty, \infty)$
b. Range$(f) = (1, \infty) \cup \{-1\}$

14. **a.** Domain$(f) = (-\infty, 2)$
b. Range$(f) = (0, \infty)$

15. For $f(x) = x^2 - 1$ we have:
a. Domain$(f) = (-\infty, \infty)$.
b. Range$(f) = [-1, \infty)$.

16. For $f(x) = x^2 + 1$ we have:
a. Domain$(f) = (-\infty, \infty)$.
b. Range$(f) = [1, \infty)$.

17. For $f(x) = \sqrt{x} + 3$ we have:
a. Domain$(f) = [0, \infty)$.
b. Range$(f) = [3, \infty)$.

18. For $f(x) = \sqrt{x} - 2$ we have:
a. Domain$(f) = [0, \infty)$.
b. Range$(f) = [-2, \infty)$.

19. For $f(x) = x^2 - 2x + 1 = (x-1)^2$ we have:
a. Domain$(f) = (-\infty, \infty)$.
b. Range$(f) = [0, \infty)$.

20. For $f(x) = x^2 - 2x + 2 = (x-1)^2 + 1$ we have:
a. Domain$(f) = (-\infty, \infty)$.
b. Range$(f) = [1, \infty)$.

21. For $f(x) = \begin{cases} 1, & \text{if } x \geq 0 \\ -1, & \text{if } x < 0 \end{cases}$ we have:
a. Domain$(f) = (-\infty, \infty)$.
b. Range$(f) = \{-1, 1\}$.

22. For $f(x) = \begin{cases} x^2, & \text{if } x \geq 0 \\ -x, & \text{if } x < 0 \end{cases}$ we have:
a. Domain$(f) = (-\infty, \infty)$.
b. Range$(f) = [0, \infty)$.

23. For $f(x) = x^2 - 1$, $b = 0$, and $f(a) = a^2 - 1$

$$f(a) = 0 \quad \text{implies} \quad a^2 - 1 = 0$$

so $a^2 = 1$ and $a = \pm 1$.

24. For $f(x) = x^2 - 1$, $b = 2$, and $f(a) = a^2 - 1$

$$f(a) = 2 \quad \text{implies} \quad a^2 - 1 = 2$$

so $a^2 = 3$ and $a = \pm\sqrt{3}$.

25. For $f(x) = \sqrt{x-1}$, $b = \frac{1}{2}$, and $f(a) = \sqrt{a-1}$

$$f(a) = \frac{1}{2} \quad \text{implies} \quad \sqrt{a-1} = \frac{1}{2}$$

so $a - 1 = \frac{1}{4}$ and $a = \frac{5}{4}$.

26. For $f(x) = \sqrt{x} - 1$, $b = \frac{1}{2}$, and $f(a) = \sqrt{a} - 1$

$$f(a) = \frac{1}{2} \quad \text{implies} \quad \sqrt{a} - 1 = \frac{1}{2}$$

so $\sqrt{a} = \frac{3}{2}$ and $a = \frac{9}{4}$.

27. **a.** Domain$(f) = (-\infty, \infty)$
b. Domain$(f) = \{x : 2 - x \neq 0\} = \{x : x \neq 2\} = (-\infty, 2) \cup (2, \infty)$
c. Domain$(f) = \{x : 2 - x \geq 0\} = \{x : x \leq 2\} = (-\infty, 2]$
d. Domain$(f) = \{x : 2 - x > 0\} = (-\infty, 2)$

28. **a.** Domain$(f) = (-\infty, \infty)$
b. Domain$(f) = \{x : 3x + 1 \neq 0\} = \left(-\infty, -\frac{1}{3}\right) \cup \left(-\frac{1}{3}, \infty\right)$
c. Domain$(f) = \{x : 3x + 1 \geq 0\} = \left[-\frac{1}{3}, \infty\right)$
d. Domain$(f) = \{x : 3x + 1 > 0\} = \left(-\frac{1}{3}, \infty\right)$

29. **a.** Domain$(f) = \{x : x^2 - 1 \neq 0\} = \{x : x \neq \pm 1\} = (-\infty, -1) \cup (-1, 1) \cup (1, \infty)$

b. For $f(x) = \frac{x+1}{x^2-1} = \frac{x+1}{(x-1)(x+1)} = \frac{1}{x-1}$, the domain is still $(-\infty, -1) \cup (-1, 1) \cup (1, \infty)$.

c. We have

$$\begin{aligned}\text{Domain}(f) = &\left\{x : \frac{x^2}{x^2-1} \geq 0\right\} = \{0\} \cup \left\{x : \frac{x^2}{x^2-1} > 0\right\}\\ =&\{0\} \cup \{x : x^2 - 1 > 0\}\\ =&\{0\} \cup (-\infty, -1) \cup (1, \infty).\end{aligned}$$

30. **a.** For $f(x) = \frac{x+2}{x^2-5x+6} = \frac{x+2}{(x-3)(x-2)}$,
Domain$(f) = \{x : (x-3)(x-2) \neq 0\} = \{x : x \neq 3, x \neq 2\} = (-\infty, 2) \cup (2, 3) \cup (3, \infty)$.

b. For $\frac{x-2}{x^2-5x+6} = \frac{x-2}{(x-3)(x-2)} = \frac{1}{x-3}$,
Domain$(f) = \{x : (x-3)(x-2) \neq 0\} = \{x : x \neq 3, x \neq 2\} = (-\infty, 2) \cup (2, 3) \cup (3, \infty)$.

c. For $f(x) = \sqrt{\frac{(x+2)^2}{x^2-5x+6}}$, the domain is all x with

$$\frac{(x+2)^2}{x^2-5x+6} \geq 0, \text{ so } x = -2 \text{ or } x^2 - 5x + 6 > 0$$

which implies that $x = -2$, $x < 2$, or $x > 3$, and the domain is $(-\infty, 2) \cup (3, \infty)$.

31. **a.** For $f(x) = \sqrt{x(x-2)}$, Domain$(f) = \{x : x(x-2) \geq 0\} = (-\infty, 0] \cup [2, \infty)$.

b. For $f(x) = \sqrt{x(2-x)}$, Domain$(f) = \{x : x(2-x) \geq 0\} = \{x : x(x-2) \leq 0\} = [0, 2]$.

c. For $f(x) = \frac{x^2}{x^2-2x} = \frac{x^2}{x(x-2)}$,
Domain$(f) = \{x : x(x-2) \neq 0\} = \{x : x \neq 0, x \neq 2\} = (-\infty, 0) \cup (0, 2) \cup (2, \infty)$.

d. For $f(x) = \sqrt{\frac{x^2}{x^2-2x}} = \sqrt{\frac{x^2}{x(x-2)}}$, Domain$(f) = \{x : x(x-2) > 0\} = (-\infty, 0) \cup (2, \infty)$.

32. **a.** For $f(x) = \sqrt{(x+1)(x-1)}$, Domain$(f) = \{x : (x+1)(x-1) \geq 0\} = (-\infty, -1] \cup [1, \infty)$.

b. For $f(x) = \sqrt{(x+1)(1-x)}$,
Domain$(f) = \{x : (x+1)(1-x) \geq 0\} = \{x : (x+1)(x-1) \leq 0\} = [-1, 1]$.

c. For $f(x) = \frac{x^4-x^2}{x^2-1}$,
Domain$(f) = \{x : x^2 - 1 \neq 0\} = \{x : x \neq \pm 1\} = (-\infty, -1) \cup (-1, 1) \cup (1, \infty)$.

d. For $f(x) = \sqrt{\frac{x^4-x^2}{x^2-1}}$,
Domain$(f) = \left\{x : \frac{x^4-x^2}{x^2-1} = \frac{x^2(x^2-1)}{x^2-1} \geq 0\right\} = \{x : x \neq \pm 1\} = (-\infty, -1) \cup (-1, 1) \cup (1, \infty)$.

33. For $f(x) = x^2 + 2$ we have
$f(-x) = (-x)^2 + 2 = x^2 + 2$;
$-f(x) = -(x^2 + 2) = -x^2 - 2$;
$f\left(\frac{1}{x}\right) = \left(\frac{1}{x}\right)^2 + 2 = \frac{1}{x^2} + 2$;
$\frac{1}{f(x)} = \frac{1}{x^2+2}$;
$f(\sqrt{x}) = (\sqrt{x})^2 + 2 = x + 2$;
$\sqrt{f(x)} = \sqrt{x^2 + 2}$.

34. For $f(x) = x^2 + 2x$ we have
$f(-x) = (-x)^2 + 2(-x) = x^2 - 2x$;
$-f(x) = -(x^2 + 2x) = -x^2 - 2x$;
$f\left(\frac{1}{x}\right) = \left(\frac{1}{x}\right)^2 + 2\left(\frac{1}{x}\right) = \frac{1}{x^2} + \frac{2}{x}$;
$\frac{1}{f(x)} = \frac{1}{x^2+2x}$;
$f(\sqrt{x}) = (\sqrt{x})^2 + 2\sqrt{x} = x + 2\sqrt{x}$;
$\sqrt{f(x)} = \sqrt{x^2 + 2x}$.

35. For $f(x) = \frac{1}{x}$ we have
$f(-x) = \frac{1}{-x} = -\frac{1}{x}$;
$-f(x) = -\frac{1}{x}$;
$f\left(\frac{1}{x}\right) = \frac{1}{\frac{1}{x}} = x$;
$\frac{1}{f(x)} = \frac{1}{\frac{1}{x}} = x$;
$f(\sqrt{x}) = \frac{1}{\sqrt{x}} = \frac{1}{\sqrt{x}} \cdot \frac{\sqrt{x}}{\sqrt{x}} = \frac{\sqrt{x}}{x}$;
$\sqrt{f(x)} = \sqrt{\frac{1}{x}} = \frac{1}{\sqrt{x}} = \frac{\sqrt{x}}{x}$.

36. For $f(x) = \sqrt{x}$ we have
$f(-x) = \sqrt{-x}$;
$-f(x) = -\sqrt{x}$;
$f\left(\frac{1}{x}\right) = \sqrt{\frac{1}{x}} = \frac{1}{\sqrt{x}} = \frac{1}{\sqrt{x}} \cdot \frac{\sqrt{x}}{\sqrt{x}} = \frac{\sqrt{x}}{x}$;
$\frac{1}{f(x)} = \frac{1}{\sqrt{x}} = \frac{\sqrt{x}}{x}$;
$f(\sqrt{x}) = \sqrt{\sqrt{x}} = \left(x^{1/2}\right)^{1/2} = x^{\frac{1}{4}}$;
$\sqrt{f(x)} = \sqrt{\sqrt{x}} = x^{\frac{1}{4}}$.

37. For $f(x) = 2x - 4$:

a. $f(x+h) = 2(x+h) - 4 = 2x + 2h - 4$

b. $f(x+h) - f(x) = 2x + 2h - 4 - 2x + 4 = 2h$

c. The difference quotient is

$$\frac{f(x+h) - f(x)}{h} = \frac{2h}{h} = 2.$$

d. As h approaches 0, $\frac{f(x+h)-f(x)}{h}$ approaches 2.

38. For $f(x) = \frac{3}{2}x + \frac{1}{4}$:

a. $f(x+h) = \frac{3}{2}(x+h) + \frac{1}{4} = \frac{3}{2}x + \frac{3}{2}h + \frac{1}{4}$

b. $f(x+h) - f(x) = \frac{3}{2}x + \frac{3}{2}h + \frac{1}{4} - \frac{3}{2}x - \frac{1}{4} = \frac{3}{2}h$

c. The difference quotient is

$$\frac{f(x+h) - f(x)}{h} = \frac{\frac{3}{2}h}{h} = \frac{3}{2}.$$

d. As h approaches 0, $\frac{f(x+h)-f(x)}{h}$ approaches $\frac{3}{2}$.

39. For $f(x) = x^2$:

a. $f(x+h) = (x+h)^2 = x^2 + 2hx + h^2$

b. $f(x+h) - f(x) = x^2 + 2xh + h^2 - x^2 = 2xh + h^2$

c. The difference quotient is

$$\frac{f(x+h) - f(x)}{h} = \frac{2hx + h^2}{h} = \frac{h(2x+h)}{h} = 2x + h.$$

d. As h approaches 0, $\frac{f(x+h)-f(x)}{h}$ approaches $2x$.

40. For $f(x) = -x^2$:

a. $f(x+h) = -(x+h)^2 = -(x^2+2hx+h^2) = -x^2-2hx-h^2$

b. $f(x+h)-f(x) = -x^2-2hx-h^2+x^2 = -2hx-h^2$

c. The difference quotient is

$$\frac{f(x+h)-f(x)}{h} = \frac{-2hx-h^2}{h} = \frac{h(-2x-h)}{h} = -2x-h.$$

d. As h approaches 0, $\frac{f(x+h)-f(x)}{h}$ approaches $-2x$.

41. For $f(x) = 2-x-x^2$:

a. $f(x+h) = 2-(x+h)-(x+h)^2 = 2-x-h-x^2-2hx-h^2$

b. $f(x+h)-f(x) = 2-x-h-x^2-2hx-h^2-2+x+x^2 = -h-2hx-h^2$

c. The difference quotient is

$$\frac{f(x+h)-f(x)}{h} = \frac{-h-2hx-h^2}{h} = \frac{h(-1-2x-h)}{h} = -1-2x-h.$$

d. As h approaches 0, $\frac{f(x+h)-f(x)}{h}$ approaches $-1-2x$.

42. For $f(x) = 4x^2+3x+1$:

a.

$$\begin{aligned} f(x+h) =& 4(x+h)^2+3(x+h)+1 \\ =& 4(x^2+2hx+h^2)+3x+3h+1 \\ =& 4x^2+8hx+4h^2+3x+3h+1 \end{aligned}$$

b. $f(x+h)-f(x) = 4x^2+8hx+4h^2+3x+3h+1-4x^2-3x-1 = 8hx+4h^2+3h$

c. The difference quotient is

$$\frac{f(x+h)-f(x)}{h} = \frac{8hx+4h^2+3h}{h} = \frac{h(8x+4h+3)}{h} = 8x+4h+3.$$

d. As h approaches 0, $\frac{f(x+h)-f(x)}{h}$ approaches $8x+3$.

43. For $f(x) = \frac{1}{x}$:

a. $f(x+h) = \frac{1}{x+h}$

b. $f(x+h)-f(x) = \frac{1}{x+h} - \frac{1}{x} = \frac{x-(x+h)}{x(x+h)} = -\frac{h}{x(x+h)}$

c. The difference quotient is

$$\frac{f(x+h)-f(x)}{h} = \frac{-\frac{h}{x(x+h)}}{h} = \frac{-h}{hx(x+h)} = -\frac{1}{x(x+h)}.$$

d. As h approaches 0, $\frac{f(x+h)-f(x)}{h}$ approaches $-\frac{1}{x^2}$.

44. For $f(x) = x + \frac{1}{x}$:

a. $f(x+h) = x + h + \frac{1}{x+h}$

b. $f(x+h) - f(x) = x + h + \frac{1}{x+h} - x - \frac{1}{x} = h - \frac{h}{x(x+h)}$

c. The difference quotient is

$$\frac{f(x+h)-f(x)}{h} = \frac{h + \frac{1}{x+h} - \frac{1}{x}}{h} = \frac{h}{h} + \frac{-\frac{h}{x(x+h)}}{h} = 1 + \frac{-h}{hx(x+h)} = 1 - \frac{1}{x(x+h)}.$$

d. As h approaches 0, $\frac{f(x+h)-f(x)}{h}$ approaches $1 - \frac{1}{x^2}$.

45. For $f(x) = \frac{x}{x-3}$:

a. $f(x+h) = \frac{x+h}{x+h-3}$

b.

$$f(x+h) - f(x) = \frac{x+h}{x+h-3} - \frac{x}{x-3} = \frac{x^2 - 3x + hx - 3h - x^2 - xh + 3x}{(x+h-3)(x-3)}$$
$$= -\frac{3h}{(x+h-3)(x-3)}$$

c. The difference quotient is

$$\frac{f(x+h)-f(x)}{h} = \frac{-\frac{3h}{(x+h-3)(x-3)}}{h} = \frac{-3h}{h(x+h-3)(x-3)} = -\frac{3}{(x+h-3)(x-3)}.$$

d. As h approaches 0, $\frac{f(x+h)-f(x)}{h}$ approaches $-\frac{3}{(x-3)^2}$.

46. For $f(x) = \frac{3-5x}{2x}$:

a. $f(x+h) = \frac{3-5(x+h)}{2(x+h)} = \frac{3-5x-5h}{2x+2h}$

b.

$$f(x+h) - f(x) = \frac{3-5x-5h}{2x+2h} - \frac{3-5x}{2x} = \frac{3x - 5x^2 - 5x - 3x - 3h + 5x^2 + 5hx}{2x(x+h)}$$
$$= -\frac{3h}{2x(x+h)}$$

c. The difference quotient is

$$\frac{f(x+h)-f(x)}{h} = \frac{-\frac{3h}{2x(x+h)}}{h} = \frac{-3h}{2hx(x+h)} = \frac{-3}{2x(x+h)}.$$

d. As h approaches 0, $\frac{f(x+h)-f(x)}{h}$ approaches $-\frac{3}{2x^2}$.

47. For $f(x) = x^3$:

a. $f(x+h) = (x+h)^3 = x^3 + 3x^2h + 3xh^2 + h^3$

b. $f(x+h) - f(x) = x^3 + 3x^2h + 3xh^2 + h^3 - x^3 = 3x^2h + 3xh^2 + h^3$

c. The difference quotient is

$$\frac{f(x+h)-f(x)}{h} = \frac{3x^2h + 3xh^2 + h^3}{h} = \frac{h(3x^2 + 3xh + h^2)}{h}$$
$$= 3x^2 + 3xh + h^2.$$

d. As h approaches 0, $\frac{f(x+h)-f(x)}{h}$ approaches $3x^2$.

48. For $f(x) = 2x - x^3$:

a. $f(x+h) = 2(x+h) - (x+h)^3 = 2x + 2h - x^3 - 3x^2h - 3xh^2 - h^3$

b. $f(x+h) - f(x) = 2x + 2h - x^3 - 3x^2h - 3xh^2 - h^3 - 2x + x^3 = 2h - 3x^2h - 3xh^2 - h^3$

c. The difference quotient is

$$\frac{f(x+h)-f(x)}{h} = \frac{2h - 3x^2h - 3xh^2 - h^3}{h} = \frac{h(2 - 3x^2 - 3xh - h^2)}{h} = 2 - 3x^2 - 3xh - h^2.$$

d. As h approaches 0, $\frac{f(x+h)-f(x)}{h}$ approaches $2 - 3x^2$.

49. a. For $f(x) = \sqrt{x}$, we have $f(x+h) = \sqrt{x+h}$ and

$$\frac{f(x+h)-f(x)}{h} = \frac{\sqrt{x+h}-\sqrt{x}}{h}$$
$$= \frac{\sqrt{x+h}-\sqrt{x}}{h} \cdot \frac{\sqrt{x+h}+\sqrt{x}}{\sqrt{x+h}+\sqrt{x}} = \frac{(x+h)-x}{h(\sqrt{x+h}+\sqrt{x})}$$
$$= \frac{h}{h(\sqrt{x+h}+\sqrt{x})} = \frac{1}{\sqrt{x+h}+\sqrt{x}}.$$

b. As h approaches 0, the value of the difference quotient $\frac{f(x+h)-f(x)}{h}$ approaches

$$\frac{1}{\sqrt{x}+\sqrt{x}} = \frac{1}{2\sqrt{x}} = \frac{\sqrt{x}}{2x}.$$

50. a. For $f(x) = \frac{1}{\sqrt{x}}$ we have

$$\frac{f(x+h)-f(x)}{h} = \frac{\frac{1}{\sqrt{x+h}} - \frac{1}{\sqrt{x}}}{h} = \frac{\frac{\sqrt{x}-\sqrt{x+h}}{\sqrt{x+h}\sqrt{x}}}{h}$$

$$= \frac{\sqrt{x}-\sqrt{x+h}}{h\sqrt{x+h}\sqrt{x}} \cdot \frac{\sqrt{x}+\sqrt{x+h}}{\sqrt{x}+\sqrt{x+h}}$$

$$= \frac{x-(x+h)}{h\sqrt{x+h}\sqrt{x}(\sqrt{x}+\sqrt{x+h})}$$

$$= \frac{-1}{\sqrt{x+h}\sqrt{x}(\sqrt{x}+\sqrt{x+h})}.$$

b. As h approaches 0,

$$\frac{f(x+h)-f(x)}{h} \to \frac{-1}{\sqrt{x}\sqrt{x}(\sqrt{x}+\sqrt{x})} = -\frac{1}{2x\sqrt{x}} = -\frac{\sqrt{x}}{2x^2}.$$

51. The function is even, since the graph is symmetric with respect to the y-axis.

52. The function is even, since the graph is symmetric with respect to the y-axis.

53. The function is neither even nor odd, since the graph is not symmetric with respect to either the x-axis or the origin.

54. The function is odd, since the graph is symmetric with respect to the origin.

55. The left figure is even; the right figure is odd.

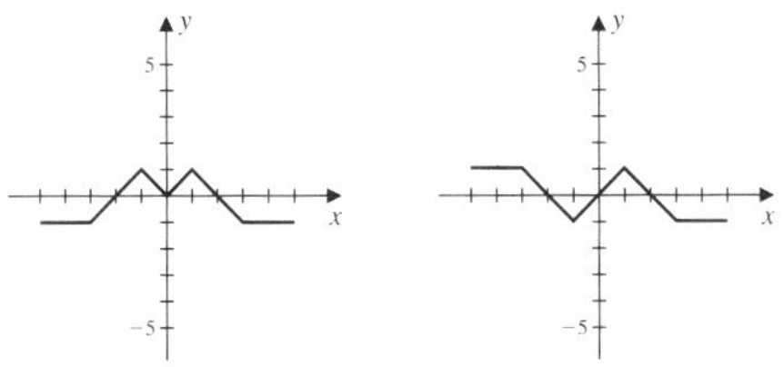

56. The left figure is even; the right figure is odd.

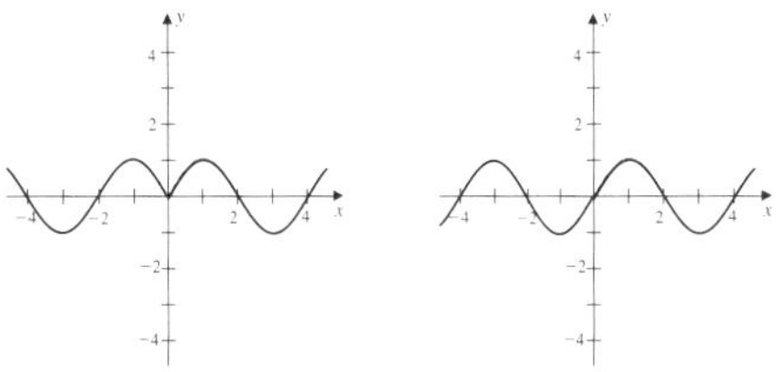

57. The transformations are shown in the figures.

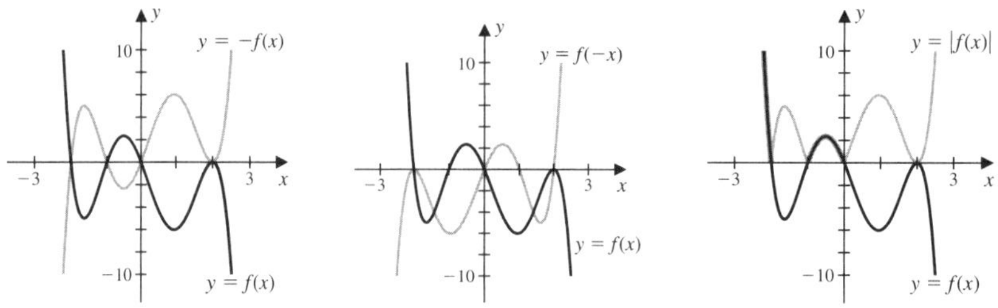

58. The following graph satisfies all the conditions.

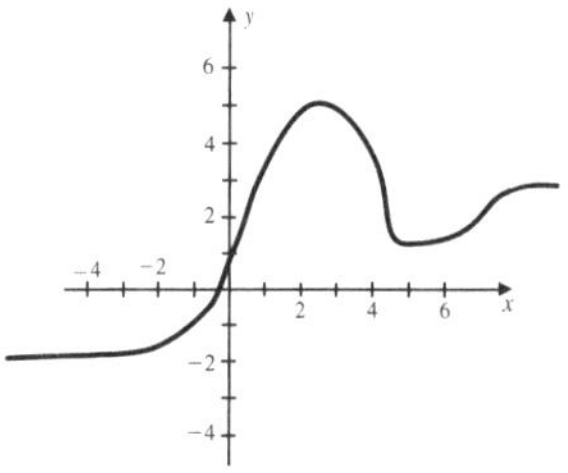

59. The distance between the ships is

$$d = \sqrt{(10(3+t))^2 + (15t)^2}$$
$$= \sqrt{325t^2 + 600t + 900}.$$

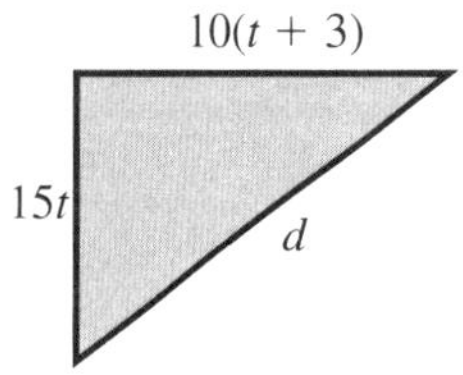

60. The volume of a cylinder is $V = \pi r^2 h$. So when $h = 2r$, $V(r) = 2\pi r^3$.

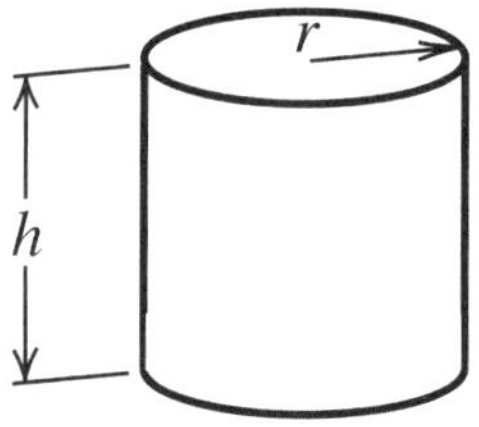

61. A rectangle with sides of length s and r has area $A = rs$ and perimeter $P = 2s + 2r$. Since $A = 64$, $r = 64/s$ and $P(s) = 2s + 128/s$.

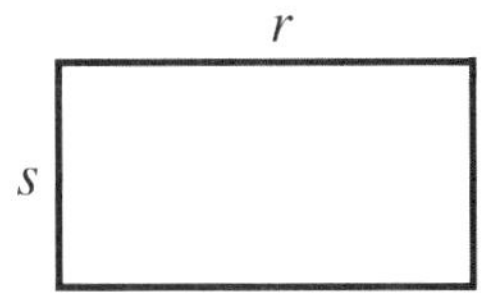

62. The area of a triangle with base b and altitude a is $A = \frac{1}{2}ab$. Since the triangle is equilateral, the base is x and the Pythagorean Theorem implies that the altitude is

$$\sqrt{x^2 + \left(\frac{x}{2}\right)^2} = \frac{\sqrt{3}}{2}x,$$

so the area is

$$A(x) = \frac{\sqrt{3}}{4}x^2.$$

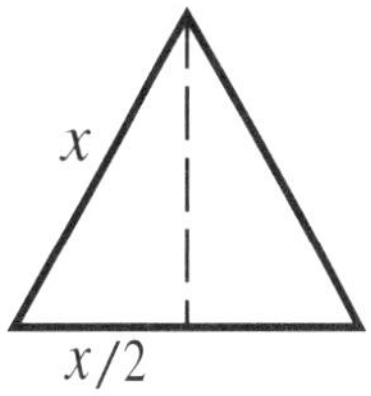

63. Let the length of the side of the cube be x. Since there are six faces of a cube, and each face of the cube has area x^2, the total surface area is $S = 6x^2$. The cube has volume $V = x^3$, so $x = \sqrt[3]{V}$. Hence $S(V) = 6(\sqrt[3]{V})^2 = 6V^{\frac{2}{3}}$.

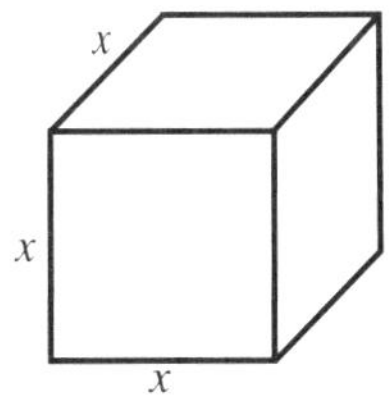

64. Let the radius of the circle be r. Then the area is $A = \pi r^2$ and the circumference is $C = 2\pi r$, or $r = C/2\pi$. Hence $A(C) = \pi(C/2\pi)^2 = C^2/(4\pi)$.

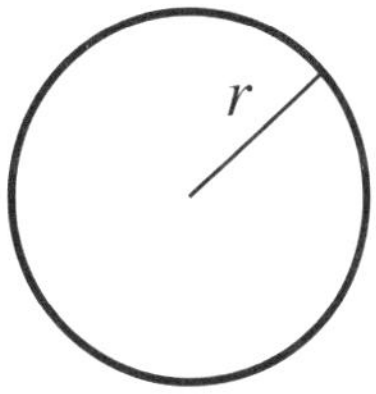

65. Let r be the radius of the circle and x and y be the lengths of the sides of the rectangle. Then $x^2 + y^2 = (2r)^2 = 4r^2$, so $y = \sqrt{4r^2 - x^2}$. Hence $A = xy = x\sqrt{4r^2 - x^2}$.

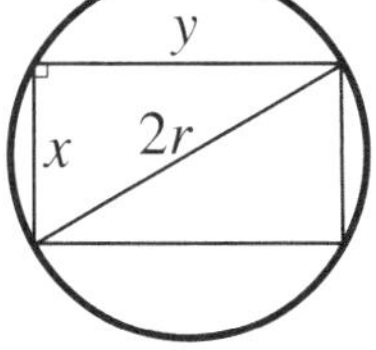

66. **a.** Let ℓ denote the length, w the width, P the perimeter, and A the area of the plot of land. Then

$$P = 2\ell + 2w, \quad \text{and} \quad A = \ell w = 432, \text{ so } \quad \ell = \frac{432}{w}$$

and

$$P(w) = 2\left(\frac{432}{w}\right) + 2w = 2w + \frac{864}{w}.$$

The domain is $(0, \infty)$.

b. $w \approx 20.8$ feet, $\ell \approx 20.8$ feet.

67. For $P(x) = 300x - 2x^2$:

a. Average rate of change:

$$\begin{aligned}\frac{P(x+h) - P(x)}{h} &= \frac{300(x+h) - 2(x+h)^2 - (300x - 2x^2)}{h} \\ &= \frac{300x + 300h - 2(x^2 + 2hx + h^2) - 300x + 2x^2}{h} \\ &= \frac{300h - 4hx - 2h^2}{h} = \frac{h(300 - 4x - 2h)}{h} = 300 - 4x - 2h\end{aligned}$$

b. If $h = 25$ and $x = 25$, then $x + h = 50$, and the average rate of change in the profit as the number of units changes from 25 to 50 is

$$300 - 4(25) - 2(25) = 150.$$

c. As h approaches 0, $\frac{P(x+h)-P(x)}{h}$ approaches $300 - 4x$, which is the instantaneous rate of change. And when $x = 25$, the instantaneous rate of change is $300 - 4(25) = 200$.

d. The graph shows $P(x)$ and the line joining $(25, P(25))$ and $(50, P(50))$.

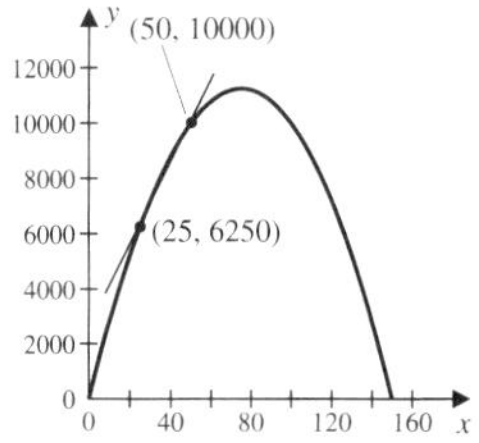

68. a. Let l, w, and h be the length, width, and height, respectively, of the box. Then $l = 8 - 2x$, $w = 8 - 2x$, and $h = x$, so the volume is $V(x) = x(8 - 2x)^2$, for $0 < x < 4$.

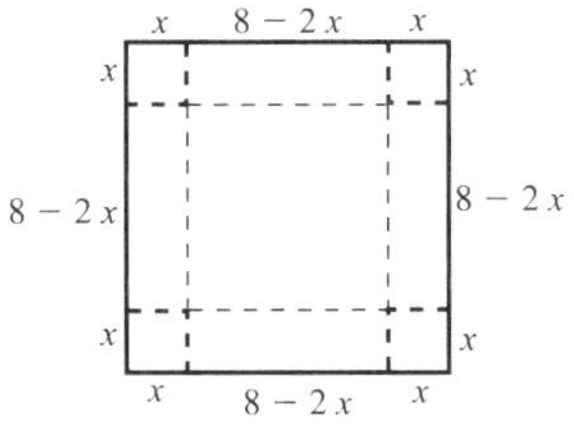

b. A graphing device shows that the maximum volume occurs when $x \approx 1.3$ and $V(1.3) \approx 37.9$.

69. a. Let r and h be the radius and height of the cylinder, respectively. Since the volume is 900 and $V = \pi r^2 h$, we have $900 = \pi r^2 h$ and $h = 900/(\pi r^2)$. The amount of material needed to construct the can is $M = 2(\pi r^2) + 2\pi r h$, so

$$M(r) = 2\pi r^2 + 2\pi r(900/\pi r^2) = 2\pi r^2 + 1800/r.$$

b. A graphing device shows that the minimum material occurs when $r \approx 5.2$, $h \approx 10.6$, and in this case $M \approx 516.1$.

70. a. The situation here is similar to that in Exercise 69(a), except that now $M = 2(2r)^2 + 2\pi r h$. Since $h = 900/(\pi r^2)$, we now have

$$M(r) = 8r^2 + 1800/r.$$

b. A graphing device shows that the minimum material used occurs when $r \approx 4.8$, $h \approx 12.4$, and in this case $M \approx 559.3$. It is reasonable that when the top and bottom of the can become more relatively inexpensive, the optimal solution should result in a decreased radius, an increased height, and the amount of material increases.

71. a. The combined area of the semi-circle regions is πr^2, and the area of the rectangle making up the remainder is $2rl$, so the total area is $A = \pi r^2 + 2rl$. Since the perimeter is 1 mile and is also $2\pi r + 2l$, we have $1 = 2\pi r + 2l$ and $l = (1 - 2\pi r)/2$. Hence

$$A(r) = \pi r^2 + 2r(1 - 2\pi r)/2 = \pi r^2 + r - 2\pi r^2 = r - \pi r^2.$$

b. A graphing device shows that the maximum area occurs when $r \approx 0.16$ and $A(0.16) \approx 0.08$.

72. a. Let x be the number of tickets exceeding 10. Then for $0 \le x$, the revenue $R(x)$ is

$$R(x) = (10 + x)(10 - 0.25x) = 100 + 7.5x - 0.25x^2.$$

b. The maximum revenue occurs when $x \approx 15$ and $R(15) = 25(6.25) = 156.25$.

73. **a.** Let x be the number of tickets exceeding 10. Then for $0 \le x$, the revenue $R(x)$ is

$$R(x) = (10 + x)(80 - 2x) = 800 + 60x - 2x^2.$$

b. The maximum revenue occurs when $x = 15$, and the maximum revenue is $R(15) = 25 \cdot 50 = \$1350$.

74. **a.** The conditions imply that $F = \frac{Gm_1m_2}{r^2}$, where G is the constant of proportionality.

b. The physical situation requires the objects be some positive distance apart, so the domain is $(0, \infty)$.

c. Depending on the values of G, m_1, and m_2, the graph appears as shown.

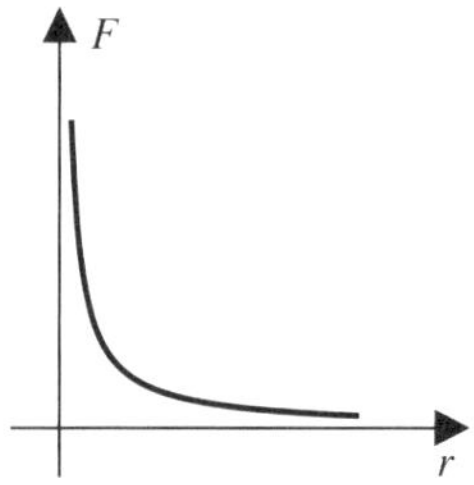

Exercise Set 1.7 (Page 58)

1. Slope: $m = \frac{2-0}{1-0} = 2$;
Equation: $y = 2x$

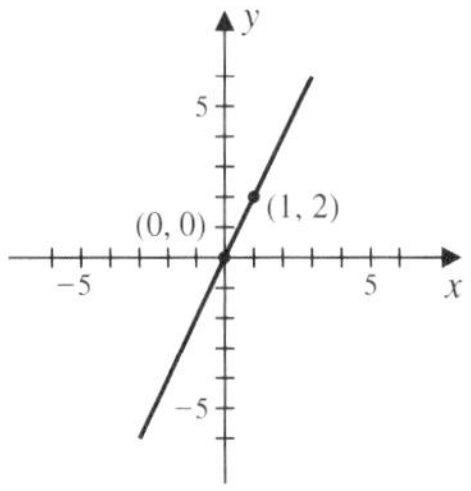

2. Slope: $m = \frac{-2-2}{3-1} = -2$;
Equation: $y - 2 = -2(x - 1)$ or $y = -2x + 4$

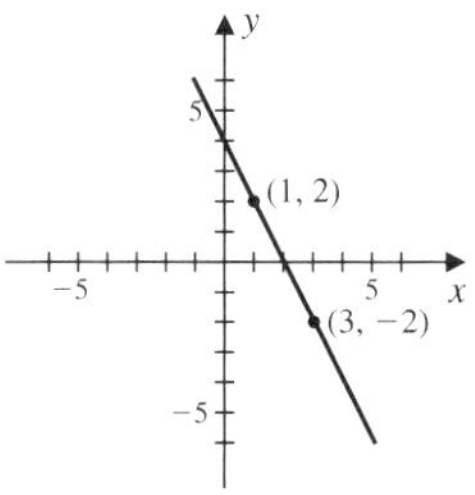

3. Slope: $m = \frac{-4-1}{2-(-2)} = -\frac{5}{4}$;
Equation: $y - 1 = -\frac{5}{4}(x + 2)$ or $y = -\frac{5}{4}x - \frac{3}{2}$

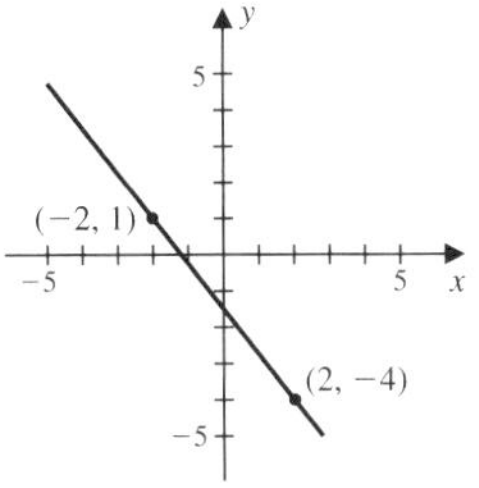

4. Slope: $m = \frac{1-(-3)}{5-1} = 1$;
Equation: $y - (-3) = x - 1$ or $y = x - 4$

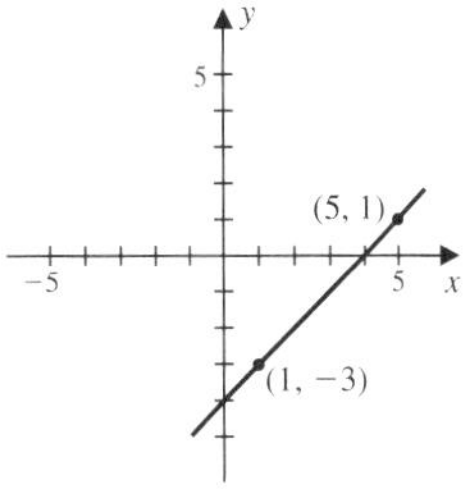

5. Slope: $m = \frac{3-3}{-1-2} = 0$;
Equation: $y = 3$

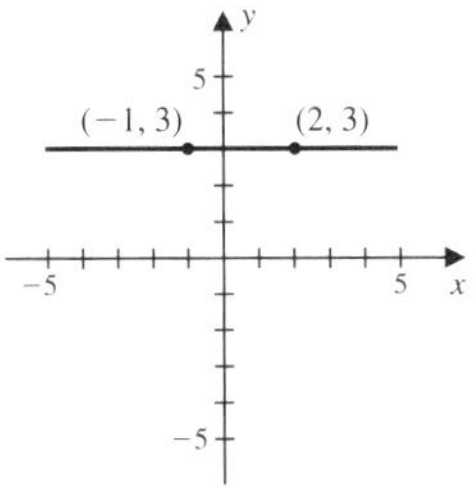

6. Slope: $m = \frac{5-3}{1-1}$ which is undefined;
Equation: $x = 1$

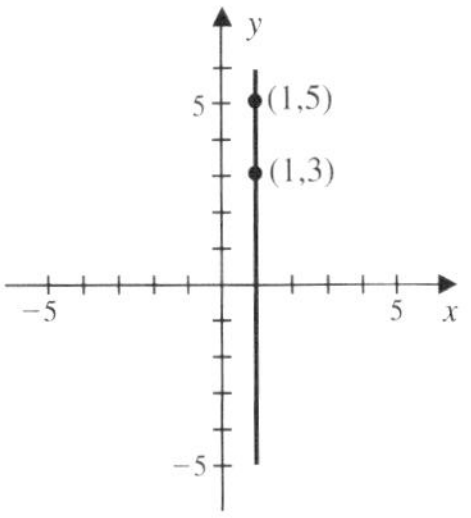

7. Slope: $m = \frac{6-(-3)}{-1-(-1)}$ which is undefined;
Equation: $x = -1$

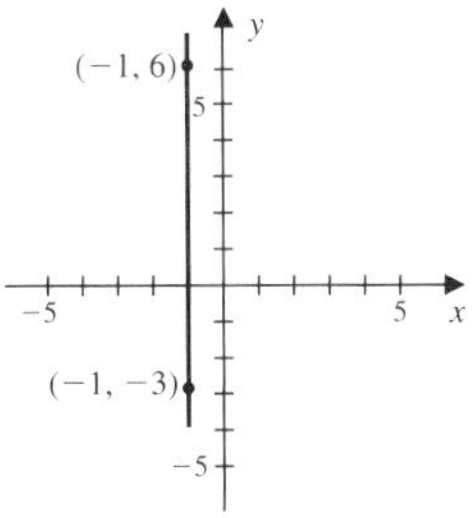

8. Slope: $m = \frac{-4-(-4)}{-2-2} = 0$;
Equation: $y = -4$

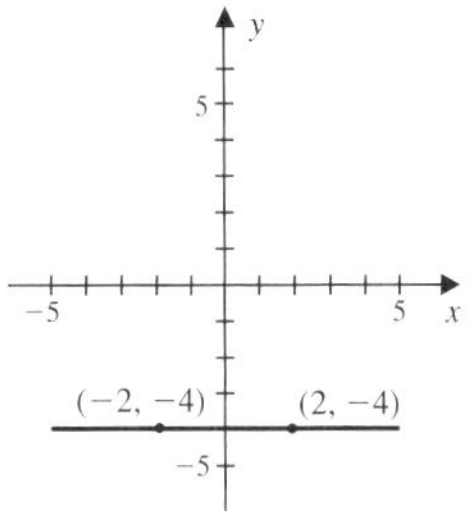

9. For $y = x + 2$, the slope is $m = 1$.

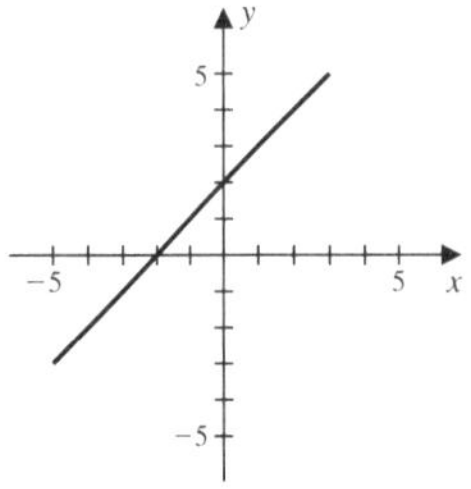

10. For $x + y = 3$ or $y = -x + 3$, the slope is $m = -1$.

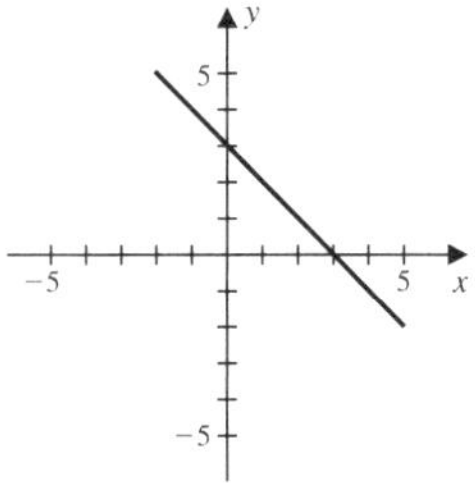

11. For $3x - y = 2$ or $y = 3x - 2$, the slope is $m = 3$.

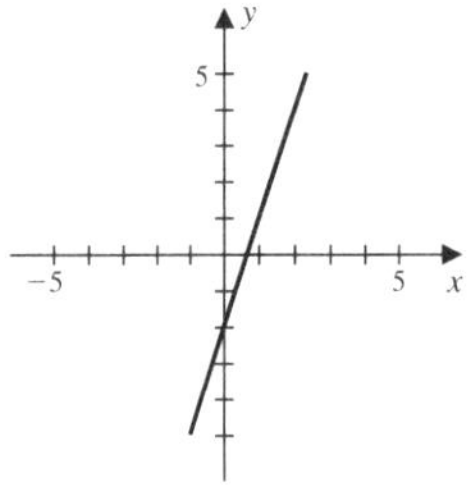

12. For $2x - y = -1$ or $-y = -2x - 1$ so $y = 2x + 1$, the slope is $m = 2$.

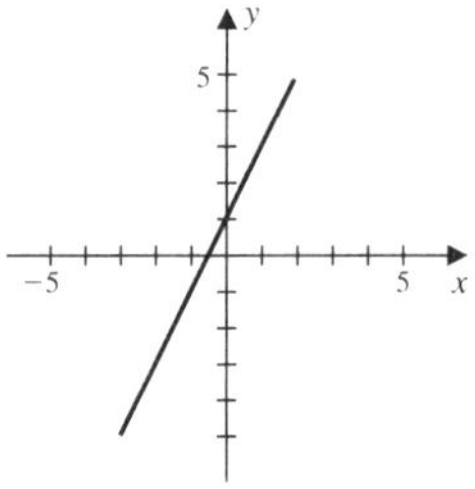

13. For $-2x - 3y = 1$ or $-3y = 2x + 1$ so $y = -\frac{2}{3}x - \frac{1}{3}$, the slope is $m = -\frac{2}{3}$.

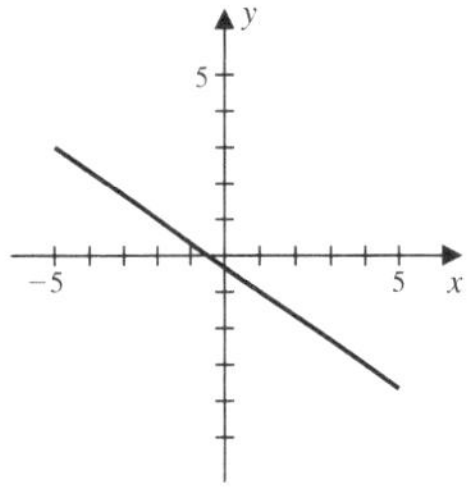

14. For $-3x - 2y = 1$ or $y = -\frac{3}{2}x - \frac{1}{2}$, the slope is $m = -\frac{3}{2}$.

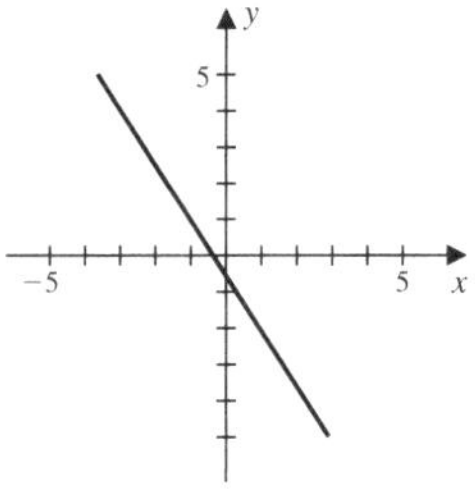

15. For $y = 2$, the slope is 0.

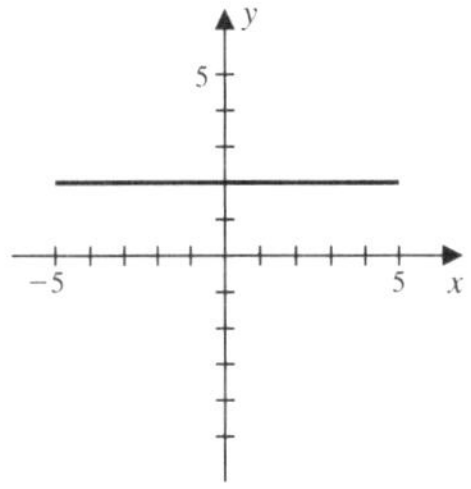

16. For $x = 2$, the slope is undefined.

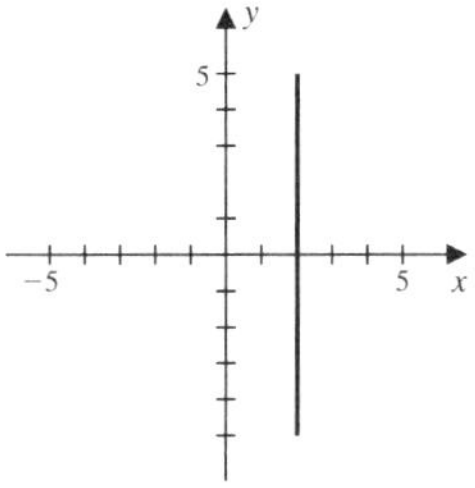

17. **a.** $y - 2 = x - 3$ implies $y = x - 1$

b. $y - 2 = -(x - 3)$ implies $y = -x + 5$

c. $y - 2 = 2(x - 3)$ implies $y = 2x - 4$

d. $y - 2 = \frac{1}{2}(x - 3)$ implies $y = \frac{1}{2}x + \frac{1}{2}$

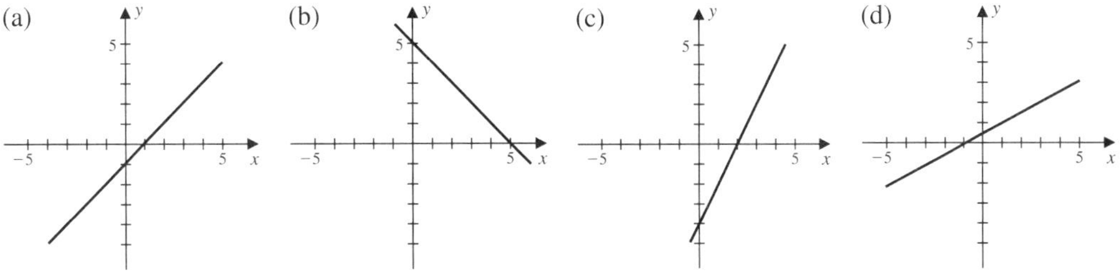

18. **a.** $y - (-2) = 2(x - (-1))$ implies $y = 2x$

b. $y + 2 = -2(x + 1)$ implies $y = -2x - 4$

c. $y + 2 = 3(x + 1)$ implies $y = 3x + 1$

d. $y + 2 = \frac{1}{3}(x + 1)$ implies $y = \frac{1}{3}x - \frac{5}{3}$

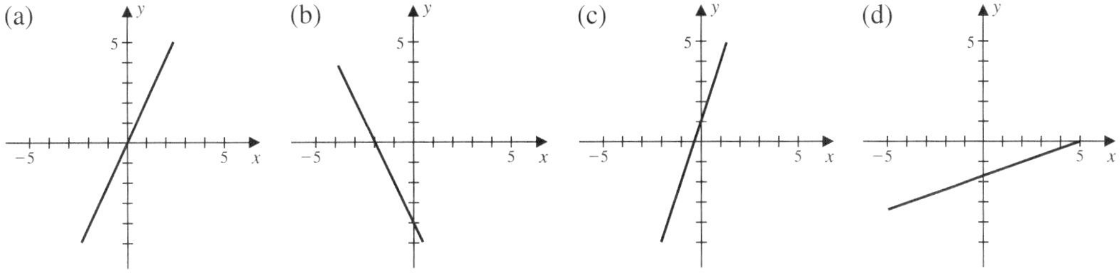

19. The slopes of the lines are:
(a) 0 (b) 1 (c) -1 (d) 1 (e) 2 (f) 0 (g) -1 (h) 2 (i) 1 (j) 1
Parallel Lines: (a) and (f); (c) and (g); (b), (d), (i), and (j); (e) and (h).

20. The slopes of the lines are:
(a) 0 (b) undefined (c) 2 (d) 3 (e) $\frac{1}{3}$ (f) 1 (g) -1 (h) -3 (i) $-\frac{1}{3}$ (j) $-\frac{1}{2}$
Perpendicular Lines: (a) and (b); (c) and (j); (d) and (i); (e) and (h); (f) and (g).

21. For $y = 2x + 1$:

a. The line parallel through $(0, 0)$ has same slope $m = 2$, so the equation is $y = 2x$.

b. The line perpendicular through $(0, 0)$ has slope negative reciprocal $m = -\frac{1}{2}$, so the equation is $y = -\frac{1}{2}x$.

22. For $x + y = -1$, that is, $y = -x - 1$:

a. The line parallel through $(0, 0)$ has same slope $m = -1$, so the equation is $y = -x$.

b. The line perpendicular through $(0, 0)$ has slope negative reciprocal $m = 1$, so the equation is $y = x$.

23. For $y = -2x + 3$:

a. The line parallel through $(-1, 2)$ has same slope $m = -2$, so the equation is $y - 2 = -2(x + 1)$ or $y = -2x$.

b. The line perpendicular through $(-1, 2)$ has slope negative reciprocal $m = \frac{1}{2}$, so the equation is $y - 2 = \frac{1}{2}(x + 1)$ or $y = \frac{1}{2}x + \frac{5}{2}$.

24. For $y = 3x - 2$:

a. The line parallel through $(1, 2)$ has same slope $m = 3$, so the equation is $y - 2 = 3(x - 1)$ or $y = 3x - 1$.

b. The line perpendicular through $(1, 2)$ has slope negative reciprocal $m = -\frac{1}{3}$, so the equation is $y - 2 = -\frac{1}{3}(x - 1)$ or $y = -\frac{1}{3}x + \frac{7}{3}$.

25. Since the y-intercept is 2, and the slope is -1, the equation is $y = -x + 2$.

26. We have $y = 3x - 2$.

27. We have $y - (-2) = 3(x - 1)$ or $y = 3x - 5$.

28. We have $y - (-3) = -2(x - (-1))$, which implies $y + 3 = -2(x + 1)$, so $y = -2x - 5$.

29. The line passes through $(2, 0)$ and $(0, 4)$, so the slope is $m = \frac{4-0}{0-2} = -2$, and the equation is $y - 0 = -2(x - 2)$ or $y = -2x + 4$.

30. The line passes through $(-2, 0)$ and $(0, -3)$, so the slope is $m = \frac{-3-0}{0-(-2)} = -\frac{3}{2}$, and the equation is $y + 3 = -\frac{3}{2}x$ or $y = -\frac{3}{2}x - 3$.

31. Since the line is parallel to the x-axis, its slope is 0, so the line is horizontal with equation $y = -1$.

32. Since the line is parallel to the y-axis, it is vertical with undefined slope and equation $x = -2$.

33. We have $2x - 3y = 2$, which implies $-3y = -2x + 2$, so $y = \frac{2}{3}x - \frac{2}{3}$. The parallel line also has slope $\frac{2}{3}$ and equation $y - 3 = \frac{2}{3}(x - 4)$ or $y = \frac{2}{3}x + \frac{1}{3}$.

34. We have $3x + 2y = 1$, which implies $2y = -3x + 1$, so $y = -\frac{3}{2}x + \frac{1}{2}$. The parallel line also has slope $-\frac{3}{2}$ and equation $y - 2 = -\frac{3}{2}(x + 1)$ or $y = -\frac{3}{2}x + \frac{1}{2}$.

35. A line perpendicular to $y = 2 - x$ has slope 1, and if it passes through $(1, 1)$, has equation $y - 1 = x - 1$ or $y = x$.

36. We have $x - 2y = 4$, which implies $-2y = -x + 4$, so $y = \frac{1}{2}x - 2$. A line perpendicular to this line has slope -2, and if it passes through $(-3, 5)$ it has equation $y - 5 = -2(x + 3)$ or $y = -2x - 1$.

37. a. The line tangent to the circle $x^2 + y^2 = 3$ at the point $(1, \sqrt{2})$ is perpendicular to the radius line passing through $(0, 0)$ and $(1, \sqrt{2})$. The slope of the radius line is $m = \frac{\sqrt{2}-0}{1-0} = \sqrt{2}$. A line tangent to the circle is perpendicular to this radius, so it has slope $-\frac{1}{\sqrt{2}} = -\frac{\sqrt{2}}{2}$, and if it passes through $(1, \sqrt{2})$ has equation

$$y - \sqrt{2} = -\frac{\sqrt{2}}{2}(x - 1) \quad \text{or} \quad y = -\frac{\sqrt{2}}{2}x + \frac{3\sqrt{2}}{2}.$$

b. The other point on the circle where the tangent line is parallel to this line is at the point diametrically opposite the point $(1, \sqrt{2})$, which is the point $(-1, -\sqrt{2})$. The tangent line at this point has equation

$$y + \sqrt{2} = -\frac{\sqrt{2}}{2}(x + 1) \quad \text{or} \quad y = -\frac{\sqrt{2}}{2}x - \frac{3\sqrt{2}}{2}.$$

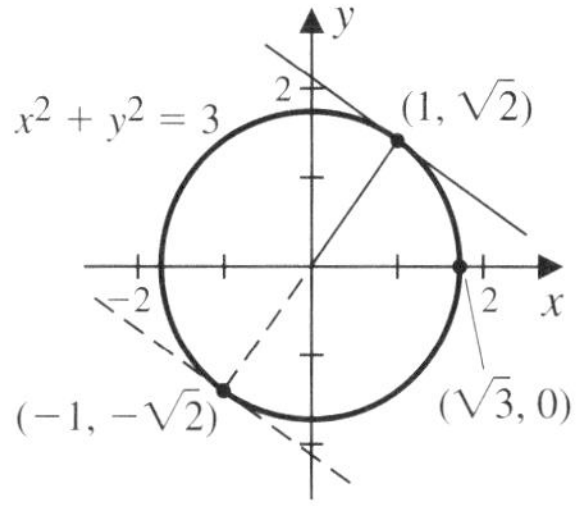

38. The line passes through $(a, 0)$ and $(0, b)$, so it has slope $m = \frac{b-0}{0-a} = -\frac{b}{a}$ and equation

$$y - 0 = -\frac{b}{a}(x - a) \quad \text{so} \quad y = -\frac{b}{a}x + b, \text{ that is, } \frac{b}{a}x + y = b \quad \text{and} \quad \frac{x}{a} + \frac{y}{b} = 1.$$

39. Since $v(t) = -32t$, $v(7) = -224$, and the rock is traveling downward at a speed of 224 feet/second.

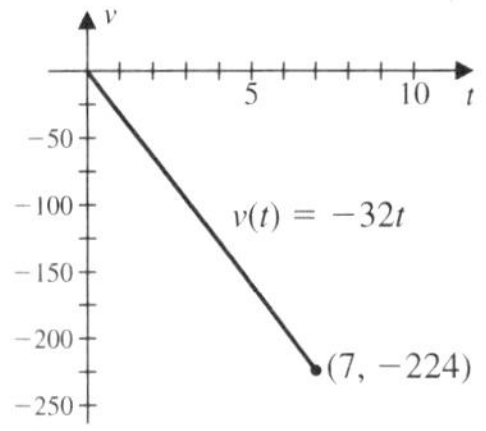

40. At sea level, water freezes at 32°F(0°C) and boils at 212°F(100°C), so the line plotting Fahrenheit against Celsius passes through the points (0, 32) and (100, 212). The slope of the line is $m = \frac{212-32}{100-0} = \frac{9}{5}$ and the equation is $F - 32 = \frac{9}{5}(C - 0)$ or $F = \frac{9}{5}C + 32$. The Fahrenheit temperature corresponding to 30°C is $F = \frac{9}{5}30 + 32 = 86°$.

41. a. For $W(n) = 500 - 0.5n$ we have the following graph.

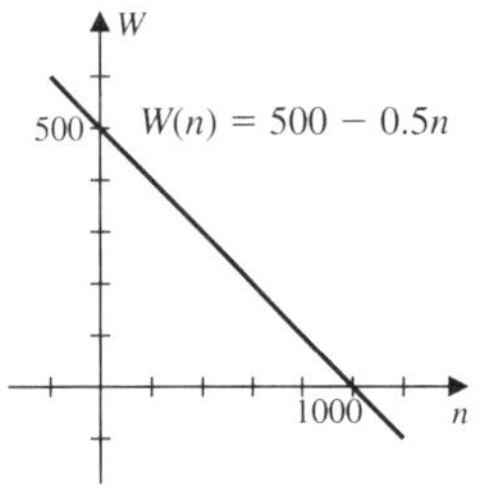

b. The total fish weight is

$$T(n) = n \cdot W(n) = n(500 - 0.5n) = 500n - 0.5n^2.$$

c. At $n = 1000$ the total weight is 0, so the fish population has disappeared. This is the predicted limiting value of the number of fish in the pond. For $n > 1000$, the total weight given by the equation is negative, which is physically unreasonable.

42. a. The linear model passes through the points (0, 10000) and (5, 2000). The slope of the line is

$$m = \frac{10000 - 2000}{0 - 5} = -1600$$

and the equation for the linear depreciation is

$$y - 10000 = -1600t \quad \text{or} \quad y = -1600t + 10000.$$

So the value of the equipment at time t is $V(t) = -1600t + 10000$.

b. After 2.5 years the equipment is worth $V(2.5) = -1600(2.5) + 10000 = 6000$.

c. We have

$$\frac{V(3) - V(1)}{3 - 1} = \frac{5200 - 8400}{2} = -1600,$$

and the negative sign indicates that the value is depreciating.

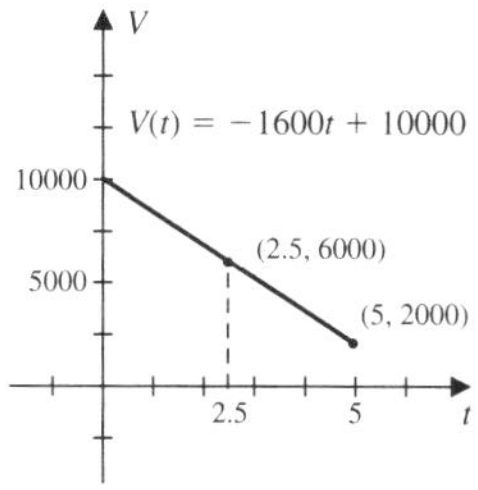

Exercise Set 1.8 (Page 69)

1. $y = x^2 + 1$

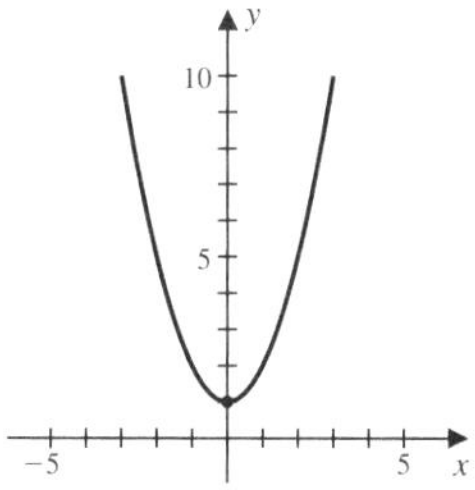

2. $y = x^2 - 2$

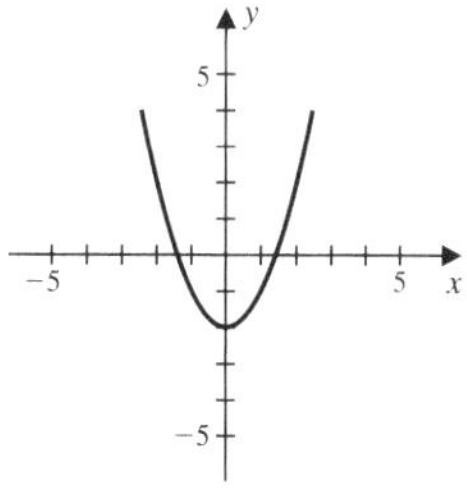

3. $y = -x^2 + 1$

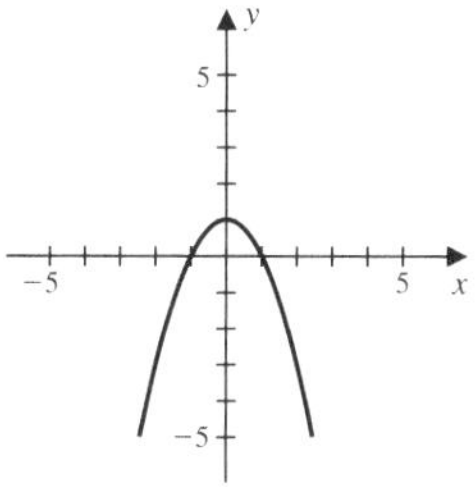

4. $y = -x^2 - 1$

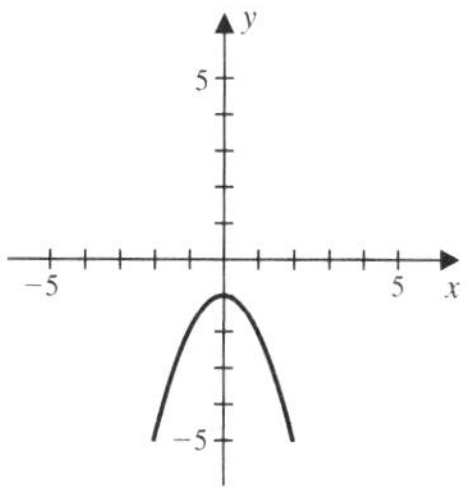

5. $y = (x-2)^2$

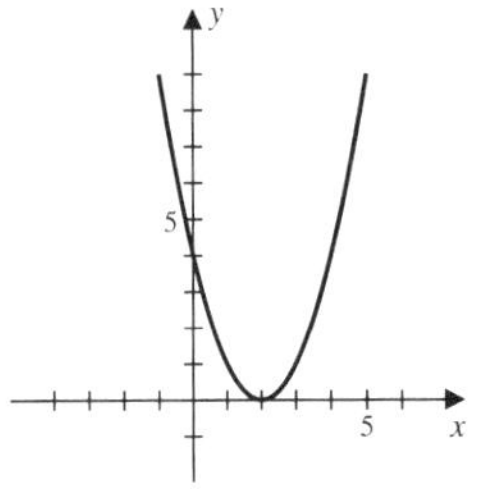

6. $y = (x+2)^2$

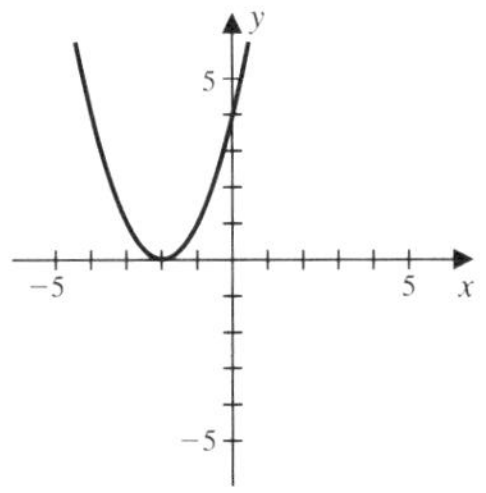

7. $y = (x+1)^2 - 1$

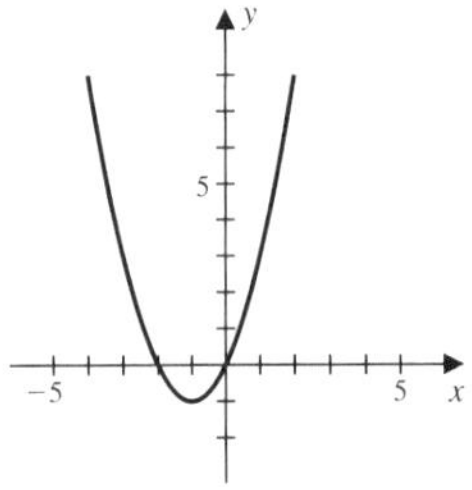

8. $y = -(x-1)^2 - 1$

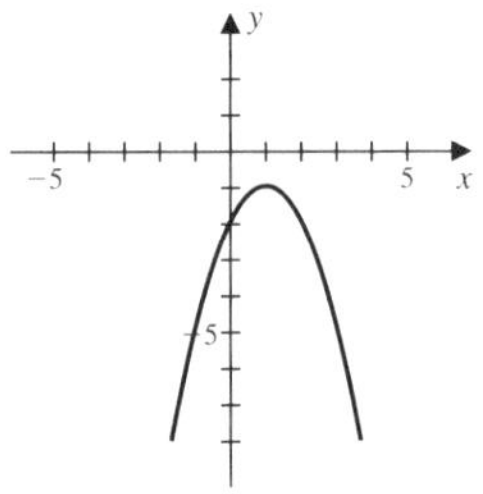

9. Completing the square gives

$$y = x^2 - 4x + 3 = x^2 - 4x + 4 - 4 + 3$$
$$= (x-2)^2 - 1.$$

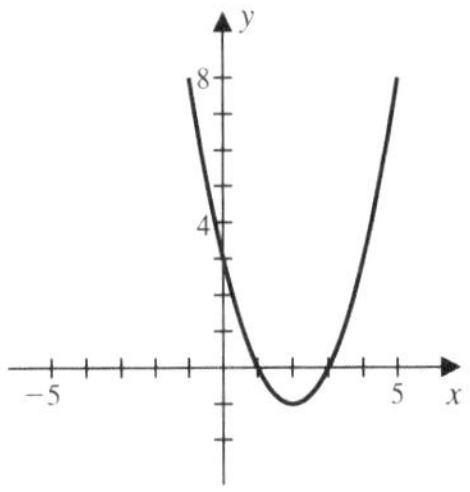

10. Completing the square gives

$$y = x^2 + 4x + 1$$
$$= x^2 + 4x + 4 - 4 + 1$$
$$= (x+2)^2 - 3.$$

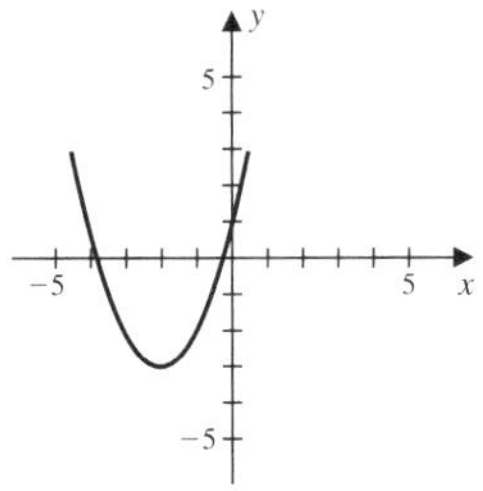

11. Completing the square gives

$$\begin{aligned} y &= -x^2 - 2x \\ &= -(x^2 + 2x) \\ &= -(x^2 + 2x + 1 - 1) \\ &= -(x+1)^2 + 1. \end{aligned}$$

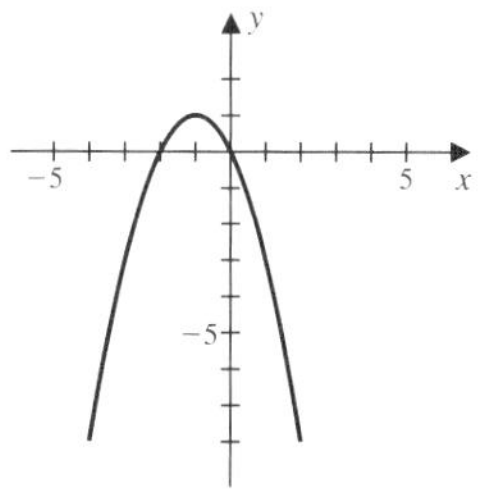

12. Completing the square gives

$$\begin{aligned} y &= -x^2 + 2x \\ &= -(x^2 - 2x) \\ &= -(x^2 - 2x + 1 - 1) \\ &= -(x-1)^2 + 1. \end{aligned}$$

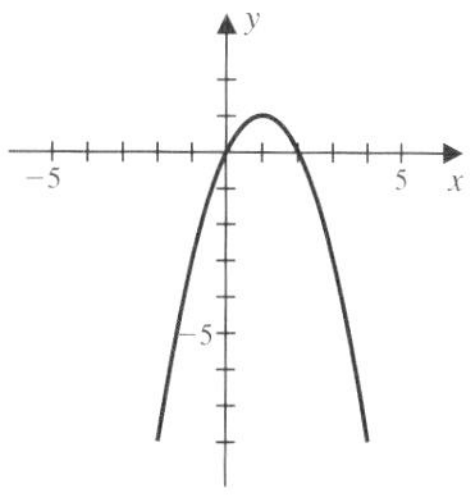

13. Completing the square gives

$$\begin{aligned} y &=3x^2 + 6x \\ &=3(x^2 + 2x) \\ &=3(x^2 + 2x + 1 - 1) \\ &=3(x+1)^2 - 3. \end{aligned}$$

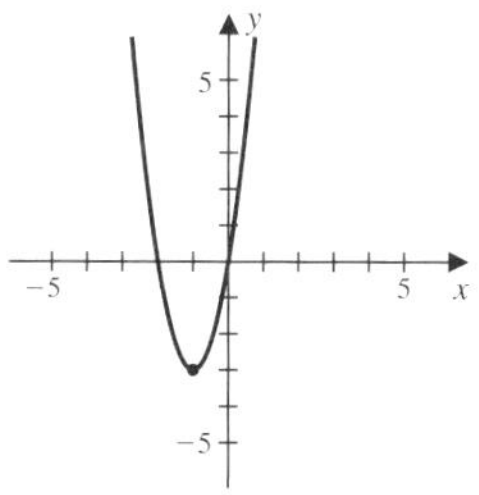

14. Completing the square gives

$$\begin{aligned} y &=2x^2 - 8x + 6 \\ &=2(x^2 - 4x + 4 - 4) + 6 \\ &=2(x-2)^2 - 2. \end{aligned}$$

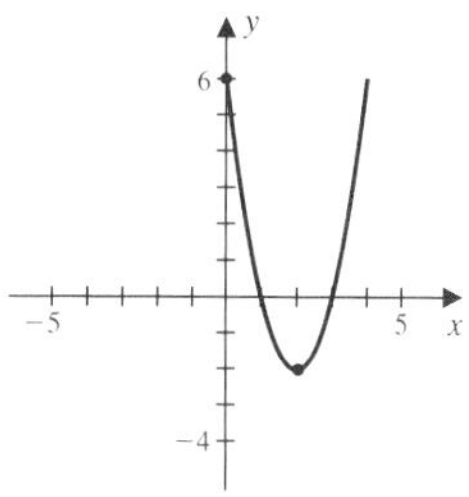

15. $y = \frac{1}{2}x^2 - 1$

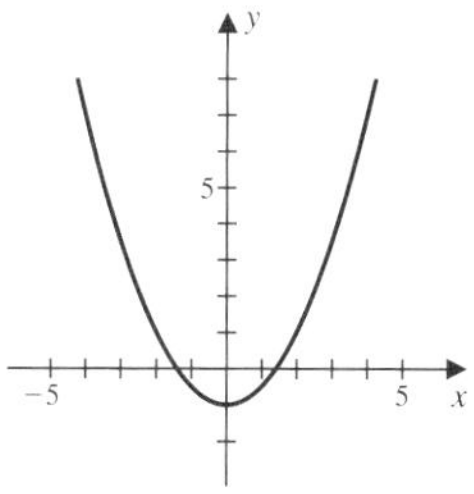

16. $y = \frac{1}{2}x^2 + 2$

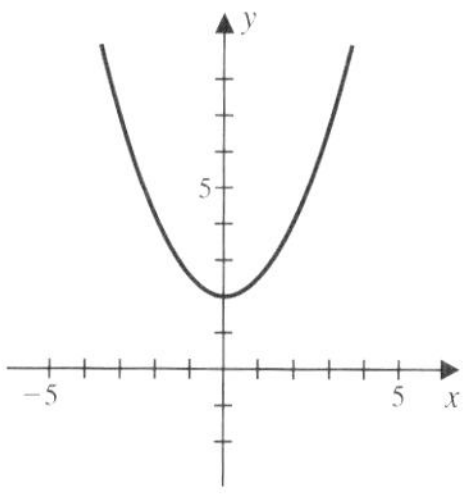

17. Completing the square gives

$$\begin{aligned} y &= \frac{1}{2}x^2 - 2x + 2 \\ &= \frac{1}{2}\left(x^2 - 4x + 4\right) \\ &= \frac{1}{2}(x-2)^2. \end{aligned}$$

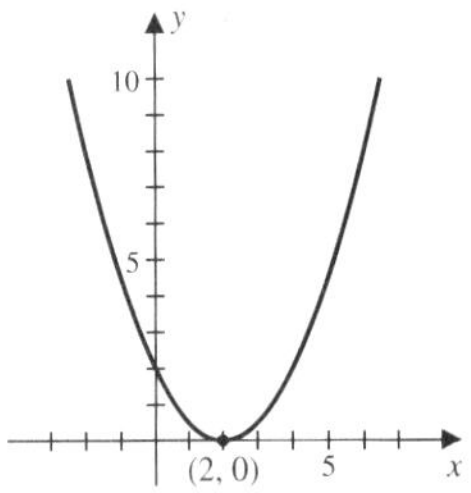

18. Completing the square gives

$$\begin{aligned} y &= \frac{1}{3}x^2 - 2x + 1 \\ &= \frac{1}{3}\left(x^2 - 6x + 9 - 9\right) + 1 \\ &= \frac{1}{3}(x-3)^2 - 2. \end{aligned}$$

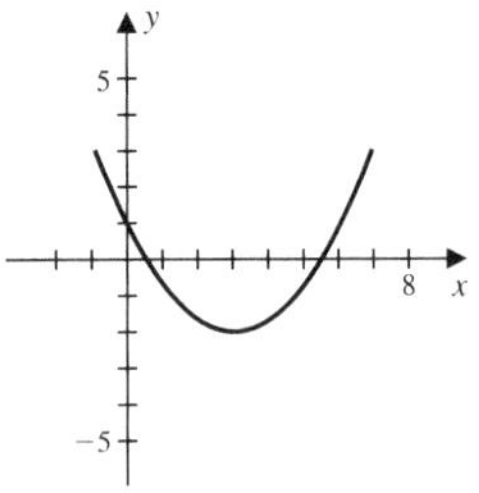

19. a. $f(x) = x^2 - 6x + 7 = (x^2 - 6x + 9 - 9) + 7 = (x-3)^2 - 2$

b. To find the x-intercepts solve

$$(x-3)^2 - 2 = 0, \quad \text{that is} \quad (x-3)^2 = 2, \quad \text{so} \quad x = \pm\sqrt{2} + 3,$$

and the intercepts are $(3 + \sqrt{2}, 0)$ and $(3 - \sqrt{2}, 0)$. The y-intercept is $(0, 7)$.

c. Since the parabola opens upward the vertex $(3, -2)$ is a minimum point so the minimum value of the function is $f(3) = -2$.

20. a. $f(x) = x^2 + 4x + 5 = (x^2 + 4x + 4 - 4) + 5 = (x+2)^2 + 1$

b. The graph does not cross the x-axis and so there are no x-intercepts. The y-intercept is $(0, 5)$.

c. Since the parabola opens upward, the vertex $(-2, 1)$ is a minimum point so the minimum value of the function is $f(-2) = 1$.

21. a. We have $f(x) = -x^2 + 6x - 8 = -(x^2 - 6x + 9 - 9) - 8 = -(x-3)^2 + 1$.

b. To find the x-intercepts solve

$$-(x-3)^2 + 1 = 0, \quad \text{that is} \quad (x-3)^2 = 1 \quad \text{so} \quad x = \pm 1 + 3,$$

and the intercepts are $(4, 0)$ and $(2, 0)$. The y-intercept is $(0, -8)$.

c. Since the parabola opens downward the vertex $(3, 1)$ is a maximum point so the maximum value of the function is $f(3) = 1$.

22. a. $f(x) = -x^2 - 4x - 4 = -(x^2 + 4x + 4) = -(x+2)^2$

b. To find the x-intercepts solve

$$-(x+2)^2 = 0, \quad \text{that is} \quad (x+2)^2 = 0 \quad \text{so} \quad x = -2,$$

and the intercept is $(-2, 0)$. The y-intercept is $(0, -4)$.

c. Since the parabola opens downward the vertex $(-2, 0)$ is a maximum point so the maximum value of the function is $f(-2) = 0$.

23. For $6x^2 - 5x + 1 = 0$,

$$x = \frac{-(-5) \pm \sqrt{(-5)^2 - 4(6)(1)}}{2(6)} = \frac{5 \pm \sqrt{25 - 24}}{12} = \frac{5 \pm 1}{12}$$

so

$$x = \frac{1}{2} \text{ or } x = \frac{1}{3}.$$

24. For $12x^2 + x - 1 = 0$,

$$x = \frac{-1 \pm \sqrt{(1)^2 - 4(12)(-1)}}{2(12)} = \frac{-1 \pm \sqrt{1 + 48}}{24} = \frac{-1 \pm 7}{24}$$

so

$$x = \frac{1}{4} \text{ or } x = -\frac{1}{3}.$$

25. For $2x^2 - 4x + 1 = 0$,

$$x = \frac{-(-4) \pm \sqrt{(-4)^2 - 4(2)(1)}}{2(2)} = \frac{4 \pm \sqrt{16 - 8}}{4} = \frac{4 \pm \sqrt{8}}{4} = \frac{4 \pm 2\sqrt{2}}{4}$$

so

$$x = 1 + \frac{\sqrt{2}}{2} \text{ or } x = 1 - \frac{\sqrt{2}}{2}.$$

26. For $4x^2 - 8x + 1 = 0$,

$$x = \frac{-(-8) \pm \sqrt{(-8)^2 - 4(4)(1)}}{2(4)} = \frac{8 \pm \sqrt{64 - 16}}{8} = \frac{8 \pm \sqrt{48}}{8} = \frac{8 \pm 4\sqrt{3}}{8}$$

so

$$x = 1 + \frac{\sqrt{3}}{2} \text{ or } x = 1 - \frac{\sqrt{3}}{2}.$$

27. For $2x^2 + 4x + 3 = 0$,

$$x = \frac{-4 \pm \sqrt{(4)^2 - 4(2)(3)}}{2(2)} = \frac{-4 \pm \sqrt{16 - 24}}{4} = \frac{-4 \pm \sqrt{-8}}{4}.$$

Since the square root of a negative number is not a real number, the quadratic has no solutions and there are no x-intercepts.

28. For $4x^2 - 8x + 7 = 0$,

$$x = \frac{-(-8) \pm \sqrt{(-8)^2 - 4(4)(7)}}{2(4)} = \frac{8 \pm \sqrt{64 - 112}}{8} = \frac{8 \pm \sqrt{-48}}{8}.$$

Since the square root of a negative number is not a real number, the quadratic has no solutions and there are no x-intercepts.

29. The various graphs are shown.

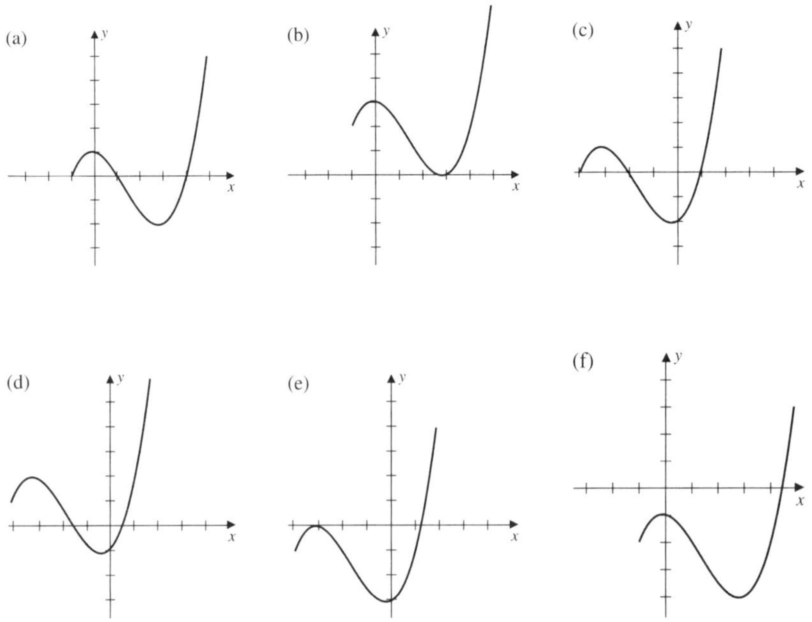

30. **a.** An upward shift by 2 units of $y = x^2$, so is (v).

b. Since $y = x^2 - 4x + 4 = (x - 2)^2$, the curve is a right shift by 2 units of $y = x^2$, so is (ii).

c. A left shift by 2 units of $y = x^2$, so is (iv).

d. A left shift by 1 unit of $y = x^2$, followed by a downward shift of 2 units so is (iii).

e. Since $y = x^2 - 2x + 3 = (x - 1)^2 + 2$, the curve is a right shift by 1 unit of $y = x^2$, followed by an upward shift of 2 units so is (vi).

f. A reflection of $y = x^2$ about the x-axis followed by a downward shift of 1 unit so is (i).

31. The function will have the form $f(x) = a(x - 1)^2 + 3$ and if the graph passes through the point $(-2, 5)$ we have

$$5 = a(-2 - 1)^2 + 3, \quad \text{that is} \quad 5 = 9a + 3 \quad \text{so} \quad a = \frac{2}{9},$$

and $f(x) = \frac{2}{9}(x - 1)^2 + 3$.

32. The function will have the form $f(x) = a(x + 2)^2 + 2$ and if the graph passes through the point $(1, -4)$ we have

$$-4 = a(1 + 2)^2 + 2, \quad \text{that is} \quad -4 = 9a + 2 \quad \text{so} \quad a = -\frac{6}{9} = -\frac{2}{3},$$

and $f(x) = -\frac{2}{3}(x+2)^2 + 2$.

33. **a.** If x is in the domain of $f(x) = \sqrt{x^2-3}$, then

$$x^2 - 3 = \left(x - \sqrt{3}\right)\left(x + \sqrt{3}\right) \geq 0 \quad \text{so} \quad x \leq -\sqrt{3} \quad \text{or} \quad x \geq \sqrt{3},$$

which can be written $\left(-\infty, -\sqrt{3}\right] \cup \left[\sqrt{3}, \infty\right)$.

b. The domain of $f(x) = \sqrt{x^2 - \frac{1}{2}x} = \sqrt{x\left(x - \frac{1}{2}\right)}$ is all x with $x\left(x - \frac{1}{2}\right) \geq 0$ so $x \leq 0$ or $x \geq \frac{1}{2}$, which can be written $(-\infty, 0] \cup \left[\frac{1}{2}, \infty\right)$.

34. For $f(x) = 100x^2$ and $g(x) = 0.1x^3$ we have

$$f(x) \geq g(x), \quad \text{that is} \quad 100x^2 \geq 0.1x^3 \quad \text{so} \quad \frac{1}{10}x^3 - 100x^2 \leq 0 \quad \text{and} \quad \frac{1}{10}x^2(x-1000) \leq 0$$

if and only if

$$x - 1000 \leq 0 \quad \text{or} \quad x \leq 1000\,, \quad \text{that is } x \text{ is in } (-\infty, 1000].$$

35. Let $f(x) = ax^2 + bx + c$.

a. Since $(1, 1)$ is on the graph, we have $1 = f(1) = a + b + c$. So $a + b + c = 1$.

b. The y-intercept is 6, which implies $6 = f(0) = c$. So $c = 6$.

c. The vertex of a parabola is $\left(-\frac{b}{2a}, \frac{4ac-b^2}{4a}\right)$. The vertex is $(1, 1)$, which implies $-\frac{b}{2a} = 1$ and $\frac{4ac-b^2}{4a} = 1$, so $b = -2a$ and

$$1 = \frac{4ac - b^2}{4a} = \frac{4ac - (-2a)^2}{4a} = \frac{4ac - 4a^2}{4a} = \frac{4a(c-a)}{4a} = c - a, \text{ when } a \neq 0.$$

Since $c - a = 1$, we have $c = a + 1$.

d. $a + b + c = 1$ and $c = 6$ so $a + b = -5$. But $c = a + 1$ so $a = 5$ and $b = -10$. So $a = 5, b = -10, c = 6$ and $f(x) = 5x^2 - 10x + 6$.

36. a. Let $s(t) = 576 + 144t - 16t^2$. Completing the square gives

$$\begin{aligned} -16t^2 + 144t + 576 &= -16(t^2 - 9t) + 576 \\ &= -16\left(t^2 - 9t + \frac{81}{4} - \frac{81}{4}\right) + 576 \\ &= -16\left(t - \frac{9}{2}\right)^2 + 324 + 576 \\ &= -16\left(t - \frac{9}{2}\right)^2 + 900. \end{aligned}$$

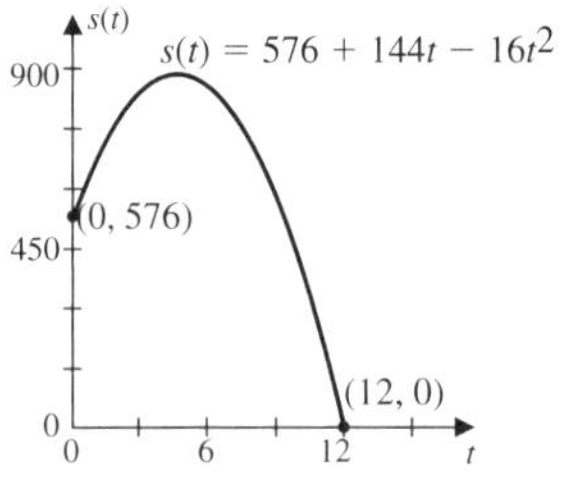

b. When the rock is on the ground we have

$$s(t) = 576 + 144t - 16t^2 = 0 \quad \text{implies} \quad 16t^2 - 144t - 576 = 16(t^2 - 9t - 36) = 0$$

if and only if

$$16(t - 12)(t + 3) = 0 \quad \text{so} \quad t = 12 \quad \text{or} \quad t = -3.$$

It takes the rock 12 seconds to reach the ground.

c. The rock reaches its maximum height at the vertex of the parabolic path. From part (a) the vertex of the parabola is $\left(\frac{9}{2}, 900\right)$, so the rock reaches a maximum height of 900 feet $\frac{9}{2}$ seconds after it has been thrown.

d. The domain consists of only real numbers greater than or equal to 0 since t represents time, and should not exceed the value of t for which the rock strikes the ground. This implies that the domain is [0, 12]. The range is all real numbers greater than or equal to 0, and less than the maximum height the rock reaches. Hence the range is [0, 900].

37. **a.** Let $v(t) = 144 - 32t$.

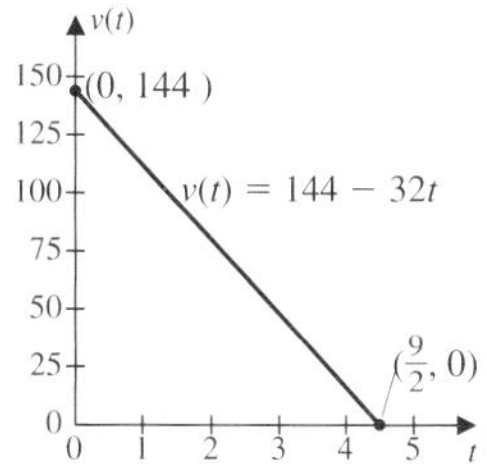

b. From Exercise 36(b), the rock strikes the ground at $t = 12$, so the velocity of the rock is $v(12) = 144 - 32(12) = -240$ feet/second. The minus sign indicates the rock is falling to the earth.

c. $\text{Domain}(v) = [0, 12]$; $\text{Range}(v) = [-240, 144]$

38. Let $C(x) = x^2 - 120x + 4000$.

The parabola opens upward, so the vertex is the minimum point on the curve. To find the number of items to produce in order to minimize the unit cost, complete the square on the quadratic equation to find the vertex.

$$x^2 - 120x + 4000 = x^2 - 120x + 3600 - 3600 + 4000 = (x - 60)^2 + 400$$

The vertex is at $(60, 400)$, so the number of items to produce is 60 per day with a minimum cost of \$400.00.

39. Let $P(x) = -0.1x^2 + 160x - 20000$.

a. The parabola opens downward, so the vertex is the maximum point on the curve. To find the number of terminals to produce in order to maximize the profit, complete the square and find the vertex.

$$\begin{aligned} -0.1x^2 + 160x - 20000 &= -0.1(x^2 - 1600x) - 20000 \\ &= -0.1(x^2 - 1600x + 640000 - 640000) - 20000 \\ &= -0.1(x - 800)^2 + 64000 - 20000 \\ &= -0.1(x - 800)^2 + 44000 \end{aligned}$$

The vertex is at $(800, 44000)$, so the company should produce 800 terminals to maximize the profit.

b. The maximum profit is \$44,000.00.

40. **a.** The distance from (x, y) to $(5, 1)$ is

$$d = \sqrt{(x-5)^2 + (y-1)^2} = \sqrt{(x-5)^2 + (x^2+2-3)^2} = \sqrt{(x-5)^2 + (x^2-1)^2}.$$

b. A graphing device indicates that the point on the graph of $y = x^2 + 2$ that is closet to (5, 1) is approximately (1, 3).

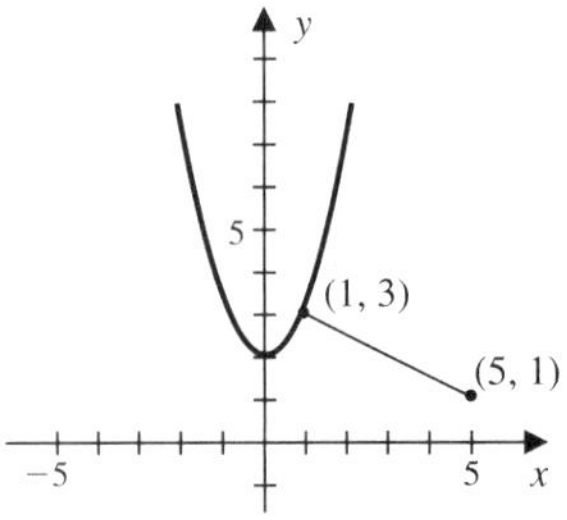

41. a. The area of the rectangle, for $x > 0$, is $A(x) = 2x(4 - x^2) = 8x - 2x^3$.

b. A graphing device indicates that the maximum area occurs when the width, w, and the height, h, satisfy $w \approx 2.3$, $h \approx 2.7$, and $A(w/2) \approx 6.2$.

42. Let $y = a(x - 1970)^2 + b(x - 1970) + c$.

a. Substituting the values 1970, 1990, and 2000 into the equation gives the three equations

$$c = 80, \quad 400a + 20b + c = 690, \quad \text{and} \quad 900a + 30b + c = 1299$$

and solving the system of two equations in two unknowns

$$400a + 20b = 610 \quad \text{and} \quad 900a + 30b = 1219,$$

gives $a = \frac{76}{75}$ and $b = \frac{307}{30}$. The parabola fitting the data is

$$y = \frac{76}{75}(x - 1970)^2 + \frac{307}{30}(x - 1970) + 80.$$

b. Substituting 2010 into the equation gives

$$\frac{76}{75}(40)^2 + \frac{307}{30}(40) + 80 \approx 2111 \text{ billion.}$$

43. **a.** Let $P(x) = 200x - x^2$. Completing the square gives

$$P(x) = -x^2 + 200x = -(x^2 - 200x)$$
$$= -(x^2 - 200x + 10000 - 10000) = -(x - 100)^2 + 10000.$$

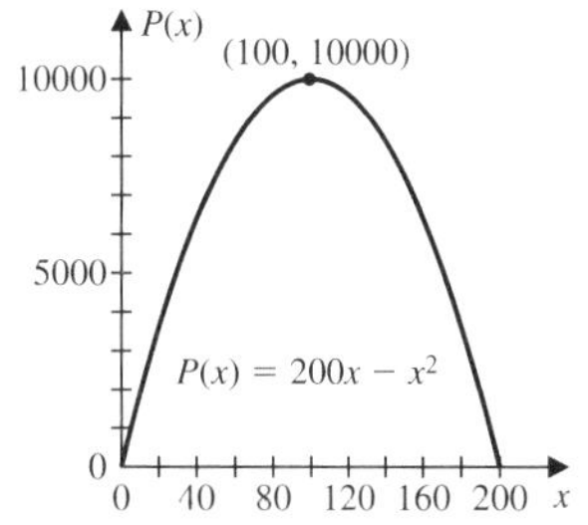

b. The vertex of the parabola will give the maximum profit, so from part (a) the vertex is at (100, 10000) and hence 100 units should be produced giving a maximum profit is \$10,000.00.

c. We have

$$\frac{P(x+h) - P(x)}{h} = \frac{200(x+h) - (x+h)^2 - (200x - x^2)}{h}$$
$$= \frac{200x + 200h - (x^2 + 2hx + h^2) - 200x + x^2}{h}$$
$$= \frac{200h - 2hx - h^2}{h} = \frac{h(200 - 2x - h)}{h}$$
$$= 200 - 2x - h.$$

As h approaches 0, the difference quotient approaches $200 - 2x$.

44. **a.** Let $v(t) = -4t^2 - 4t + 80$. Completing the square gives

$$-4t^2 - 4t + 80 = -4(t^2 + t) + 80 = -4\left(t^2 + t + \frac{1}{4} - \frac{1}{4}\right) + 80 = -4\left(t + \frac{1}{2}\right)^2 + 81.$$

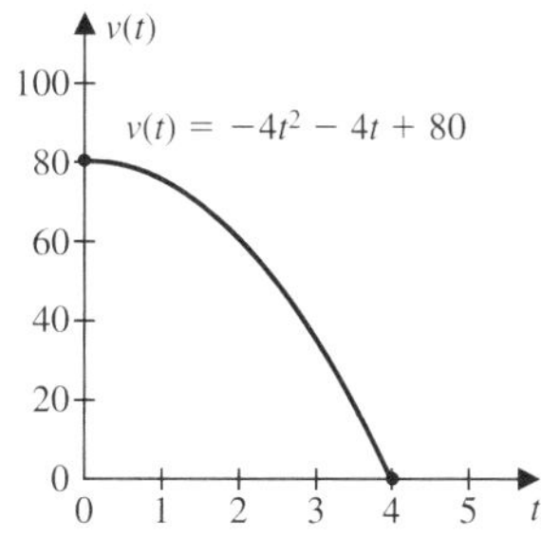

b. The car is at rest when

$$v(t) = -4t^2 - 4t + 80 = -4(t^2 + t - 20) = -4(t+5)(t-4) = 0 \quad \text{so} \quad t = -5, t = 4.$$

Since t is time, $t = -5$ is not a solution, and the car comes to rest after $t = 4$ seconds.

c.

t	0	$\frac{1}{2}$	1	$\frac{3}{2}$	2	$\frac{5}{2}$	3	$\frac{7}{2}$	4
$v(t)$	80	77	72	65	56	45	32	17	0

d. slowest: 77 feet/second; fastest: 80 feet/second

e.

	slowest	fastest
second 1/2 sec	72	77
third 1/2 sec	65	72

f.

	minimum distance	maximum distance
first 1/2 sec	$77 \times 1/2 = 38.5$ ft	$80 \times 1/2 = 40$ ft
second 1/2 sec	$72 \times 1/2 = 36$ ft	$77 \times 1/2 = 38.5$ ft

g. Lower bound:

$$\frac{1}{2}[77 + 72 + 65 + 56 + 45 + 32 + 17] = \frac{1}{2} \cdot 364 = 182$$

Upper bound:

$$\frac{1}{2}[80 + 77 + 72 + 65 + 56 + 45 + 32 + 17] = \frac{1}{2} \cdot 444 = 222$$

Review Exercises For Chapter 1 (Page 71)

1. a. $-1 \le x \le 7$ **b.**

2. a. $0 < x < \sqrt{5}$ **b.**

3. a. $x < 7$ **b.**

4. a. $-5 \le x$ **b.**

5. a. $(-\infty, -5)$ **b.**
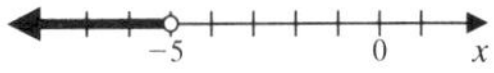

6. a. $(-3, 3]$ **b.**
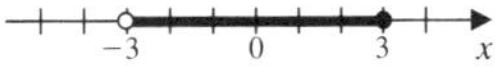

7. a. $[2, 10)$ **b.**

8. a. $(-\infty, 3]$ **b.**

9. $2x + 3 \geq 4$ implies $2x \geq 1$ so $x \geq \frac{1}{2}$

10. $-(3x + 4) \geq 5$, which implies $3x + 4 \leq -5$ implies $3x \leq -9$ so $x \leq -3$

11. $x^2 + 2x + 1 \geq 1$ implies $x^2 + 2x \geq 0$, so $x(x + 2) \geq 0$. The solution set is $(-\infty, -2] \cup [0, \infty)$.

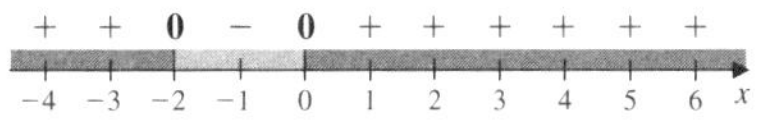

12. $x^2 - 4x > -3$ implies $x^2 - 4x + 3 > 0$ so $(x - 1)(x - 3) > 0$. The solution set is $(-\infty, 1) \cup (3, \infty)$.

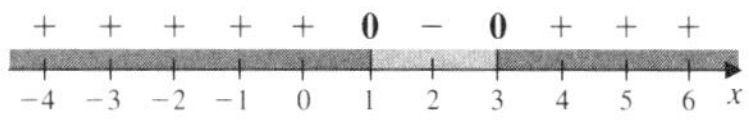

13. $(x - 2)(x + 2)(x + 3) \geq 0$
The solution set is $[-3, -2] \cup [2, \infty)$.

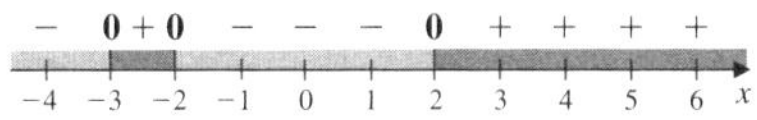

14. $(x - 1)(x - 4)(x + 2) < 0$
The solution set is $(-\infty, -2) \cup (1, 4)$.

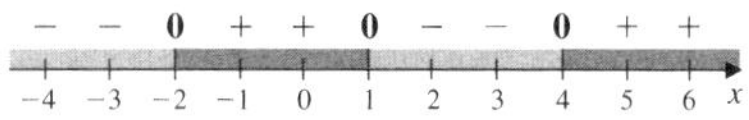

15. $x^2 + 3x > 0$ implies $x(x + 3) > 0$. The solution set is $(-\infty, -3) \cup (0, \infty)$.

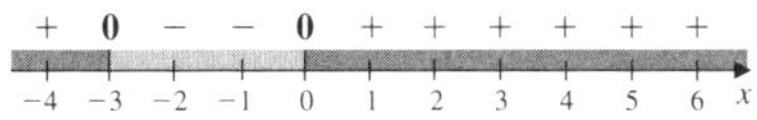

16. $x^3 - 4x^2 \leq 0$ implies $x^2(x - 4) \leq 0$, so $x - 4 \leq 0$ and $x \leq 4$.

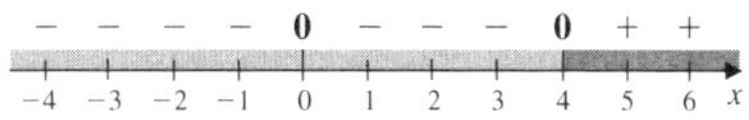

17. $\frac{2x-1}{x+1} \leq -2$ implies $\frac{2x-1}{x+1} + 2 \leq 0$ so $\frac{2x-1+2(x+1)}{x+1} \leq 0$ and $\frac{4x+1}{x+1} \leq 0$. The solution set is $\left(-1, -\frac{1}{4}\right]$.

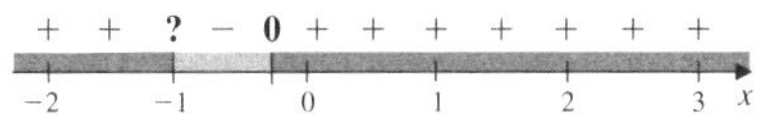

18. $\frac{x^2-9}{x^2-4} \leq 0$ implies $\frac{(x+3)(x-3)}{(x+2)(x-2)} \leq 0$. The solution set is $[-3, -2) \cup (2, 3]$.

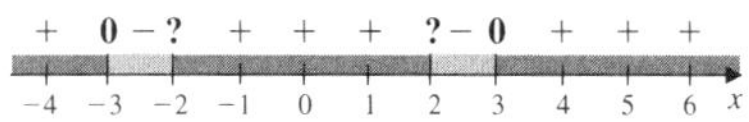

19. $|2x - 3| < 5$ implies $-5 < 2x - 3 < 5$ so $-2 < 2x < 8$ and $-1 < x < 4$

20. $|4x - 2| < 0.01$ implies $-0.01 < 4x - 2 < 0.01$ so $1.99 < 4x < 2.01$ and $0.4975 < x < 0.5025$

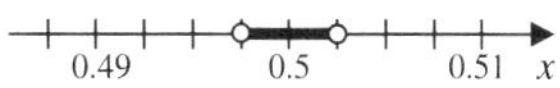

21. $|2 - x| \geq 2$ implies $2 - x \geq 2$ or $2 - x \leq -2$ so $-x \geq 0$ or $-x \leq -4$ and $x \leq 0$ or $x \geq 4$. That is, $(-\infty, 0] \cup [4, \infty)$.

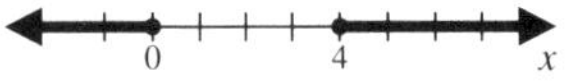

22. $|x^2 - 4| \leq 1$ so $-1 \leq x^2 - 4 \leq 1$. That is, $x^2 - 3 \geq 0$ and $x^2 - 5 \leq 0$, which implies $\left(x - \sqrt{3}\right)\left(x + \sqrt{3}\right) \geq 0$ and $\left(x - \sqrt{5}\right)\left(x + \sqrt{5}\right) \leq 0$. Hence $x \geq \sqrt{3}$ or $x \leq -\sqrt{3}$ and $-\sqrt{5} \leq x \leq \sqrt{5}$. In interval notation the solution set is $\left[-\sqrt{5}, -\sqrt{3}\right] \cup \left[\sqrt{3}, \sqrt{5}\right]$.

23. $2 < y \leq 3$

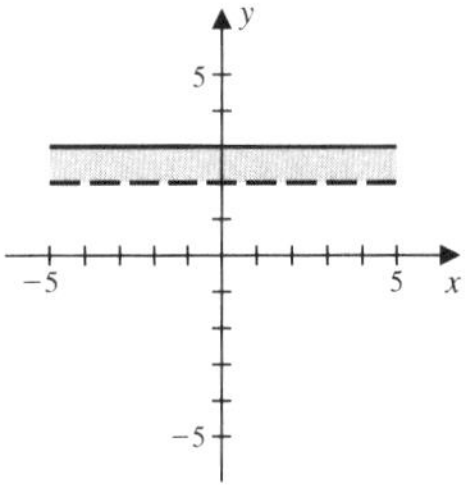

24. $|y| < 2$ implies $-2 < y < 2$

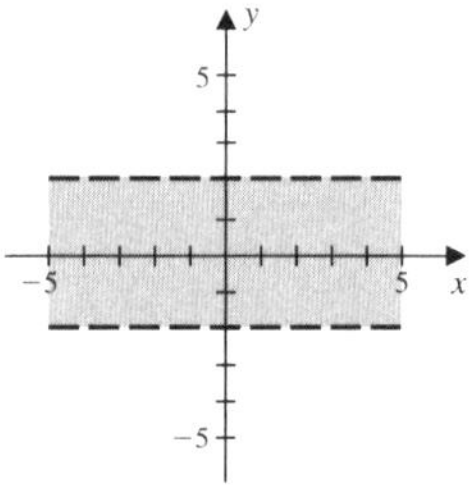

25. $|x - 1| < 2$ implies $-2 < x - 1 < 2$ so $-1 < x < 3$

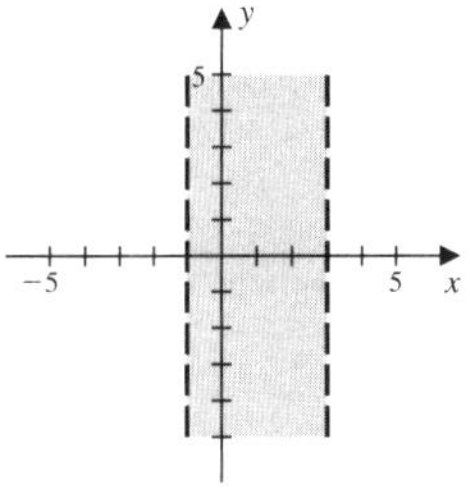

26. $-1 \leq x \leq 1$ and $-2 \leq y \leq 2$

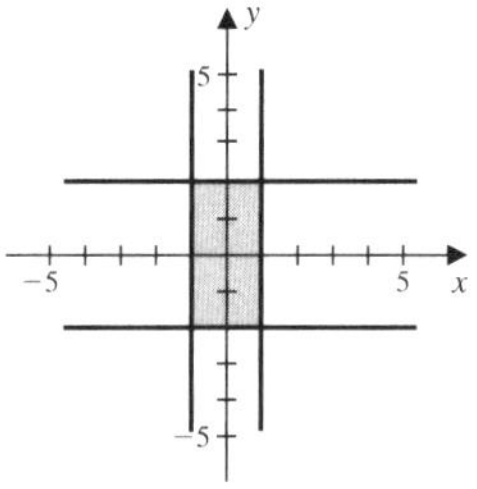

27. $|x| + |y| = 1$

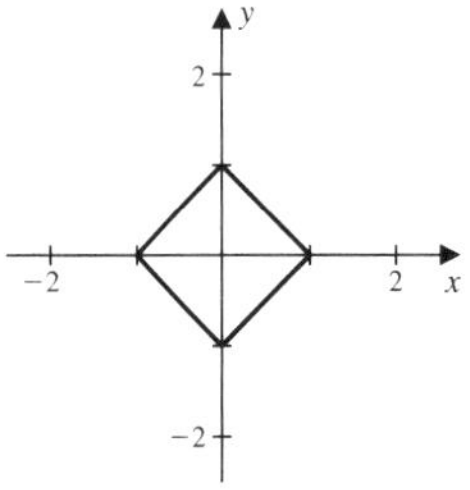

28. $1 \leq |x| + |y| \leq 4$

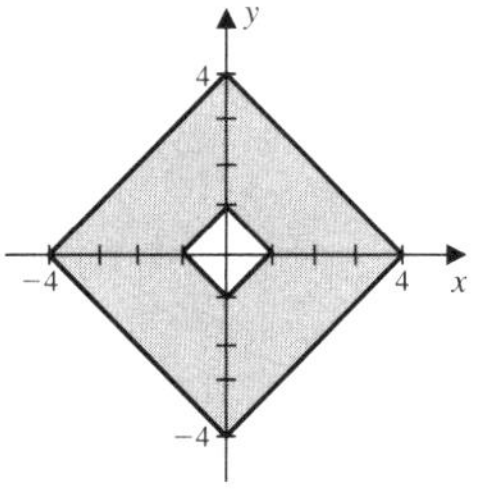

29. $|x| > |y|$

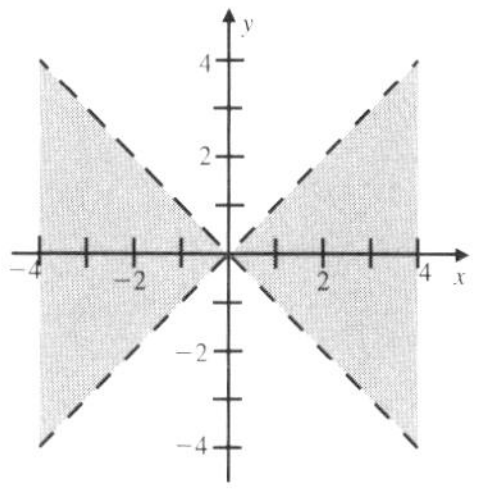

30. $|x + 1| \le |y|$

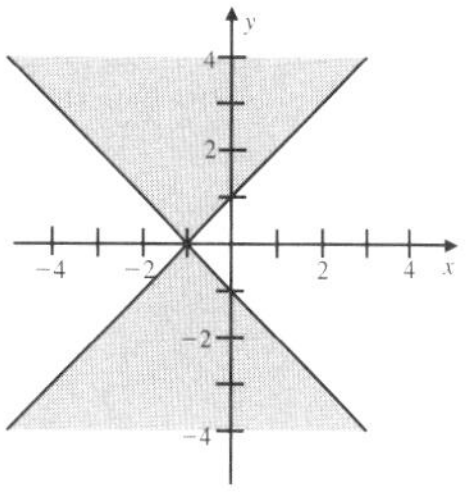

31. a. For $f(x) = x^2 - 3$ the domain is the set of all real numbers.

b. Since the graph is obtained from the basic graph of $y = x^2$, shifted 3 units downward, the vertex is $(0, -3)$, so the range is $[-3, \infty)$.

32. a. For $f(x) = \frac{1}{2x-5}$ the domain is the set of all real numbers for which the denominator is not zero, so the domain is $\{x : x \neq 5/2\} = \left(-\infty, \frac{5}{2}\right) \cup \left(\frac{5}{2}, \infty\right)$.

b. The range is all real numbers except $x = 0$, so the range is $(-\infty, 0) \cup (0, \infty)$.

33. a. For $f(x) = \sqrt{x - 2} + 2$ the domain is the set of all real numbers for which the expression under the radical is nonnegative, so the domain is $\{x : x - 2 \ge 0\} = [2, \infty)$.

b. The range is $[2, \infty)$.

34. a. Since the function $f(x) = \begin{cases} -2x + 1, & \text{if } x < 0 \\ -x^2, & \text{if } x \ge 0 \end{cases}$ is defined at every real number, the domain is the set of all real numbers.

b. The y-intercept of the graph of $y = -x^2$ is 0 and of $y = -2x + 1$ is 1, although the point $(0, 1)$ is not on the curve. So the range is $(-\infty, 0] \cup (1, \infty)$.

35. a. For

$$f(x) = \frac{1}{x^2 - 6x + 8}$$

the domain is

$$\{x \mid x^2 - 6x + 8 \neq 0\} = \{x : (x-4)(x-2) \neq 0\} = (-\infty, 2) \cup (2, 4) \cup (4, \infty).$$

b. For

$$f(x) = \frac{x-2}{x^2 - 6x + 8} = \frac{x-2}{(x-4)(x-2)} = \frac{1}{x-4}$$

the domain is still $(-\infty, 2) \cup (2, 4) \cup (4, \infty)$, since $x = 2$ can not be substituted into the original equation.

c. For

$$f(x) = \sqrt{\frac{x^2}{x^2 - 6x + 8}}$$

the domain is

$$\{x \mid (x-4)(x-2) > 0 \text{ or } x = 0\} = (-\infty, 2) \cup (4, \infty).$$

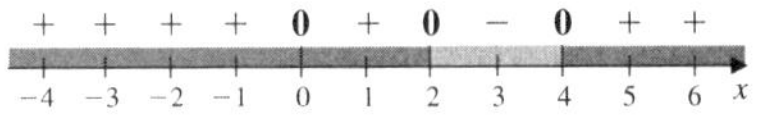

36. a. For $f(x) = \frac{x-1}{x^2+3x+2} = \frac{x-1}{(x+2)(x+1)}$ the domain is $\{x : (x+2)(x+1) \neq 0\} = (-\infty, -2) \cup (-2, -1) \cup (-1, \infty)$.

b. For $f(x) = \frac{x+1}{x^2+3x+2} = \frac{x+1}{(x+2)(x+1)} = \frac{1}{x+2}$, the domain is still $(-\infty, -2) \cup (-2, -1) \cup (-1, \infty)$, since $x = -1$ can not be substituted into the original equation.

c. For $f(x) = \sqrt{\frac{x-1}{x^2+3x+2}}$ the domain is $\left\{x \mid \frac{x-1}{(x+2)(x+1)} \geq 0\right\} = (-2, -1) \cup [1, \infty)$.

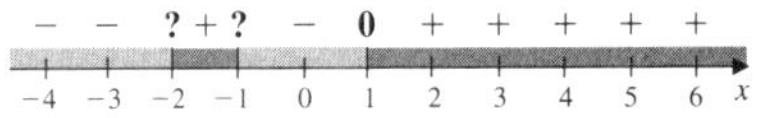

37. For $f(x) = 7x + 4$ we have:

a. $f(x+h) = 7(x+h) + 4 = 7x + 7h + 4$

b.

$$\frac{f(x+h) - f(x)}{h} = \frac{7x + 7h + 4 - (7x+4)}{h} = \frac{7h}{h} = 7.$$

38. For $f(x) = -\frac{3}{2}x + 5$ we have:

a. $f(x+h) = -\frac{3}{2}(x+h) + 5 = -\frac{3}{2}x - \frac{3}{2}h + 5$

b.

$$\frac{f(x+h) - f(x)}{h} = \frac{-\frac{3}{2}x - \frac{3}{2}h + 5 - \left(-\frac{3}{2}x + 5\right)}{h} = \frac{-\frac{3}{2}h}{h} = -\frac{3}{2}.$$

39. For $f(x) = x^2 - 1$ we have:

a. $f(x+h) = (x+h)^2 - 1 = x^2 + 2hx + h^2 - 1$

b.

$$\frac{f(x+h) - f(x)}{h} = \frac{x^2 + 2hx + h^2 - 1 - (x^2 - 1)}{h} = \frac{2hx + h^2}{h} = \frac{h(2x+h)}{h} = 2x + h.$$

40. For $f(x) = 2x^2 - x$ we have:

a. $f(x+h) = 2(x+h)^2 - (x+h) = 2x^2 + 4hx + 2h^2 - x - h$

b.

$$\begin{aligned}\frac{f(x+h) - f(x)}{h} &= \frac{2x^2 + 4hx + 2h^2 - x - h - (2x^2 - x)}{h} \\ &= \frac{4hx + 2h^2 - h}{h} = \frac{h(4x + 2h - 1)}{h} = 4x + 2h - 1.\end{aligned}$$

41. For $f(x) = \frac{1}{x-1}$ we have:

a. $f(x+h) = \frac{1}{x+h-1}$

b.

$$\begin{aligned}\frac{f(x+h) - f(x)}{h} &= \frac{\frac{1}{x+h-1} - \frac{1}{x-1}}{h} = \frac{\frac{x-1-(x+h-1)}{(x+h-1)(x-1)}}{h} \\ &= \frac{-h}{h(x+h-1)(x-1)} = \frac{-1}{(x+h-1)(x-1)}.\end{aligned}$$

42. For $f(x) = \frac{1}{2-x}$ we have:

a. $f(x+h) = \frac{1}{2-x-h}$

b.

$$\begin{aligned}\frac{f(x+h) - f(x)}{h} &= \frac{\frac{1}{2-x-h} - \frac{1}{2-x}}{h} = \frac{\frac{2-x-(2-x-h)}{(2-x-h)(2-x)}}{h} \\ &= \frac{h}{h(2-x-h)(2-x)} = \frac{1}{(2-x-h)(2-x)}.\end{aligned}$$

43. **a.** (ii) **b.** (v) **c.** (vi) **d.** (iv) **e.** (i) **f.** (iii)

44. **a.** (ii) **b.** (iii) $x^2 + x = x^2 + x + \frac{1}{4} - \frac{1}{4} = \left(x + \frac{1}{2}\right)^2 - \frac{1}{4}$

c. (iv) $x^2 - 2x - 1 = x^2 - 2x + 1 - 1 - 1 = (x-1)^2 - 2$

d. (i) $x^2 + 2x + 3 = x^2 + 2x + 1 - 1 + 3 = (x+1)^2 + 2$

45. **a., d.**

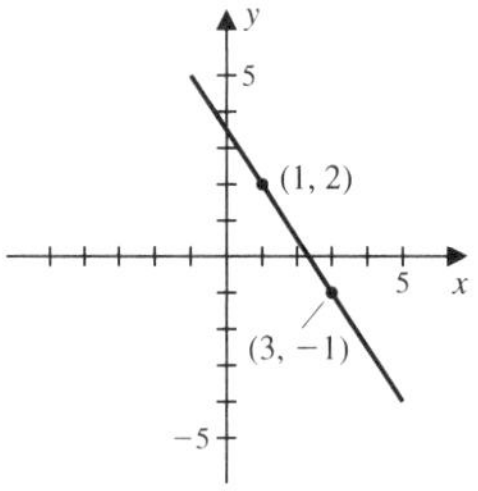

b. $d = \sqrt{(3-1)^2 + (-1-2)^2} = \sqrt{13}$ **c.** $\left(\frac{3+1}{2}, \frac{-1+2}{2}\right) = \left(2, \frac{1}{2}\right)$

e. $m = \frac{2-(-1)}{1-3} = -\frac{3}{2}$; $y - (-1) = -\frac{3}{2}(x-3)$ implies $y = -\frac{3}{2}x + \frac{7}{2}$

46. **a., d.**

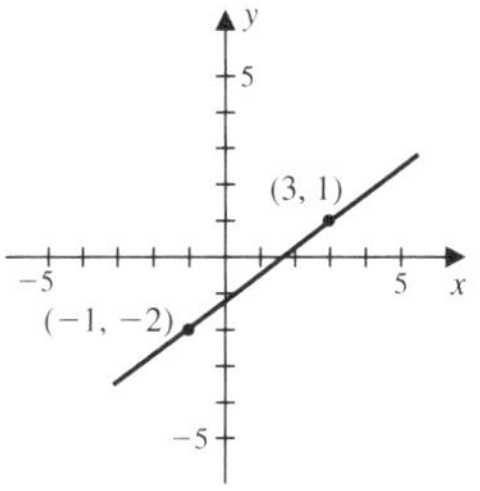

b. $d = \sqrt{(3-(-1))^2 + (1-(-2))^2} = \sqrt{25} = 5$ **c.** $\left(\frac{3-1}{2}, \frac{1-2}{2}\right) = \left(1, -\frac{1}{2}\right)$

e. $m = \frac{-2-1}{-1-3} = \frac{3}{4}$; $y - 1 = \frac{3}{4}(x-3)$implies $y = \frac{3}{4}x - \frac{5}{4}$

47. **a., d.**

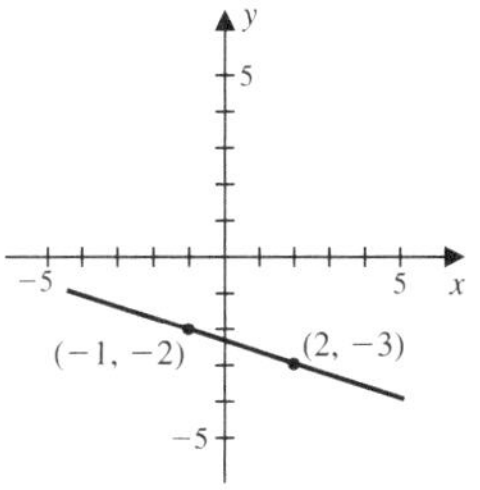

b. $d = \sqrt{(-1-2)^2 + (-2-(-3))^2} = \sqrt{10}$ **c.** $\left(\frac{-1+2}{2}, \frac{-2-3}{2}\right) = \left(\frac{1}{2}, -\frac{5}{2}\right)$

e. $m = \frac{-3-(-2)}{2-(-1)} = -\frac{1}{3}$; $y-(-2) = -\frac{1}{3}(x-(-1))$ implies $y = -\frac{1}{3}x - \frac{7}{3}$

48. a., d.

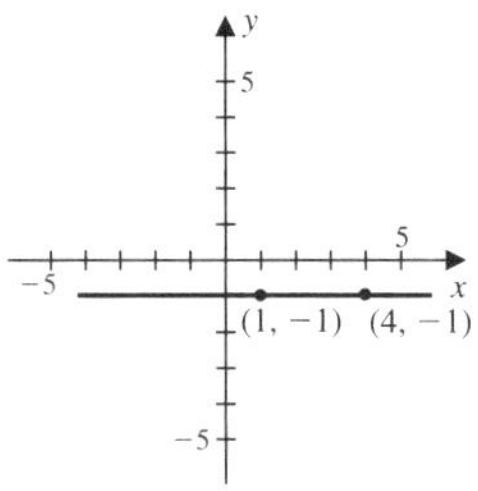

b. $d = \sqrt{(1-4)^2 + (-1-(-1))^2} = \sqrt{9} = 3$ **c.** $\left(\frac{1+4}{2}, \frac{-1-1}{2}\right) = \left(\frac{5}{2}, -1\right)$

e. $m = \frac{-1-(-1)}{4-1} = 0$; $y = -1$

49. a. $m = 4$; $y - 0 = 4(x-0)$ implies $y = 4x$ **b.** $m = -\frac{1}{4}$; $y = -\frac{1}{4}x$

50. a. $m = \frac{1}{2}$; $y - 1 = \frac{1}{2}(x-1)$ implies $y = \frac{1}{2}x + \frac{1}{2}$ **b.** $m = -2$; $y - 1 = -2(x-1)$ implies $y = -2x + 3$

51. $-7x - 5y = -1$ implies $y = -\frac{7}{5}x + \frac{1}{5}$

a. $m = -\frac{7}{5}$; $y - (-3) = -\frac{7}{5}(x-(-1))$ implies $y = -\frac{7}{5}x - \frac{22}{5}$

b. $m = \frac{5}{7}$; $y + 3 = \frac{5}{7}(x+1)$ implies $y = \frac{5}{7}x - \frac{16}{7}$

52. $2x - 3y = 4$ implies $y = \frac{2}{3}x - \frac{4}{3}$

a. $m = \frac{2}{3}$; $y - (-1) = \frac{2}{3}(x-1)$ implies $y = \frac{2}{3}x - \frac{5}{3}$

b. $m = -\frac{3}{2}$; $y - (-1) = -\frac{3}{2}(x-1)$ implies $y = -\frac{3}{2}x + \frac{1}{2}$

53. **a.** For $y = (x - 1)^2 - 2$, we have the following graph.

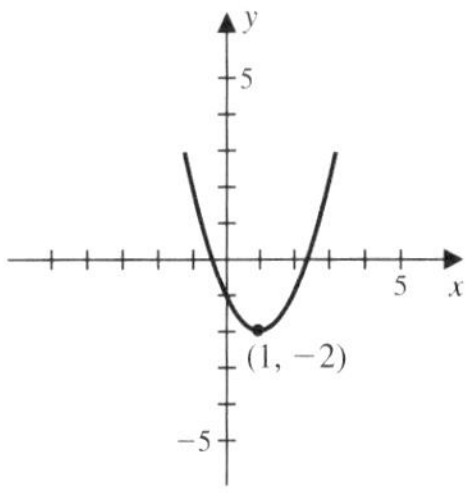

b. The range is $[-2, \infty)$.
c. The x-intercepts occur when $0 = (x - 1)^2 - 2$ so $x = 1 \pm \sqrt{2}$. The y-intercept is $(0, -1)$.
d. The minimum value of y is -2 when $x = 1$.

54. **a.** For $y = -(x + 1)^2 + 1$, we have the following graph.

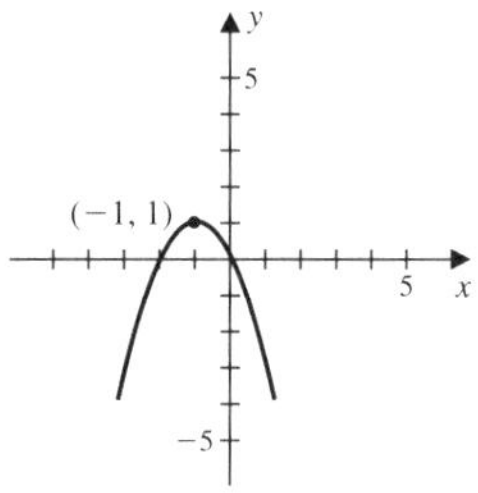

b. The range is $(-\infty, 1]$.
c. The x-intercepts occur when $0 = -(x + 1)^2 + 1$ so $x = -2$ or $x = 0$. The y-intercept is $(0, 0)$.
d. The maximum value of y is 1 when $x = -1$.

55. **a.** For $y = -(x + 1)^2 - 1$, we have the following graph.

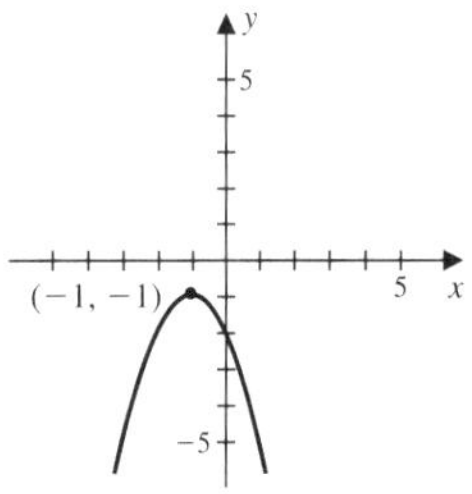

b. The range is $(-\infty, -1]$.
c. There are no x-intercepts. The y-intercept is $(0, -2)$.
d. The maximum value of y is -1 when $x = -1$.

56. **a.** For $y = (x - 1)^2 + 1$, we have the following graph.

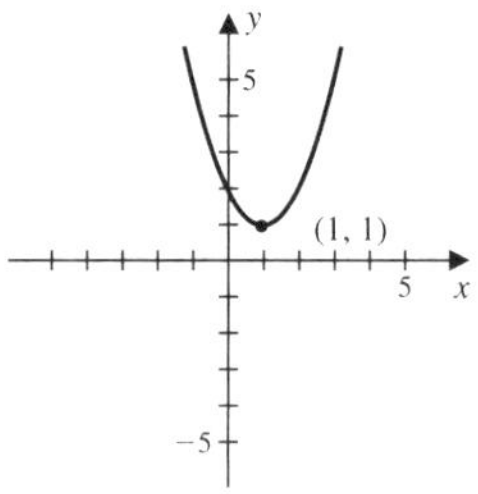

b. The range is $[1, \infty)$.
c. There are no x-intercepts. The y-intercept is $(0, 2)$.
d. The minimum value of y is 1 when $x = 1$.

57. a. We have $y = x^2 - 4x = x^2 - 4x + 4 - 4 = (x-2)^2 - 4$.

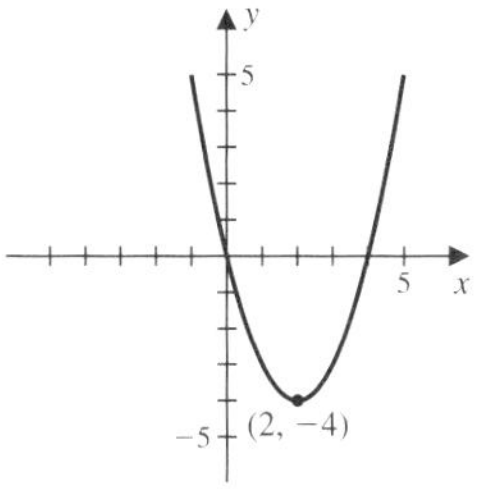

b. The range is $[-4, \infty)$.

c. x-intercepts: $(0, 0)$, $(4, 0)$ since $0 = x^2 - 4x = x(x-4)$ implies $x = 0$, $x = 4$;

y-intercept: $(0, 0)$;

d. A minimum occurs at the point $(2, -4)$.

58. a. We have $y = -x^2 + 4x - 5 = -(x^2 - 4x) - 5 = -(x^2 - 4x + 4 - 4) - 5 = -(x-2)^2 - 1$.

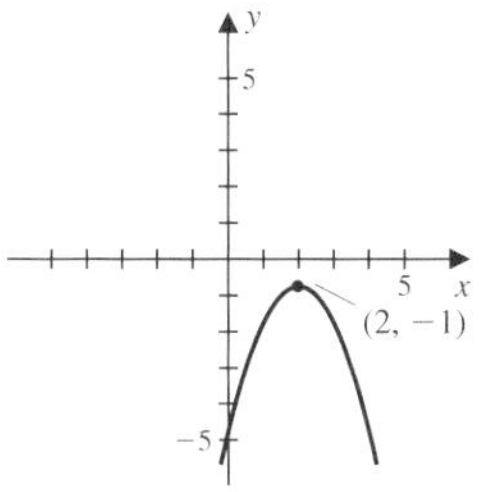

b. The range is $(-\infty, -1]$.

c. x-intercepts: none; y-intercept: $(0, -5)$;

d. A maximum occurs at the point $(2, -1)$.

59. a. We have $y = 2x^2 - 12x + 18 = 2(x^2 - 6x) + 18 = 2(x^2 - 6x + 9 - 9) + 18 = 2(x-3)^2$.

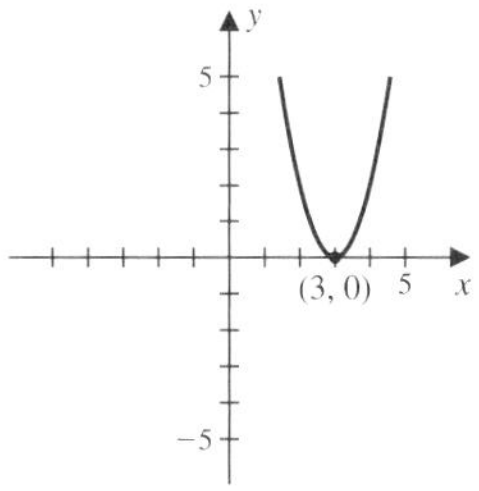

b. The range is $[0, \infty)$.

c. x-intercepts: $(3, 0)$; y-intercept: $(0, 18)$;

d. A minimum occurs at the point (3, 0).

60. a. We have

$$y = 2x^2 + 5x + 10 = 2\left(x^2 + \frac{5}{2}x\right) + 10 = 2\left(x^2 + \frac{5}{2}x + \frac{25}{16} - \frac{25}{16}\right) + 10 = 2\left(x + \frac{5}{4}\right)^2 + \frac{55}{8}.$$

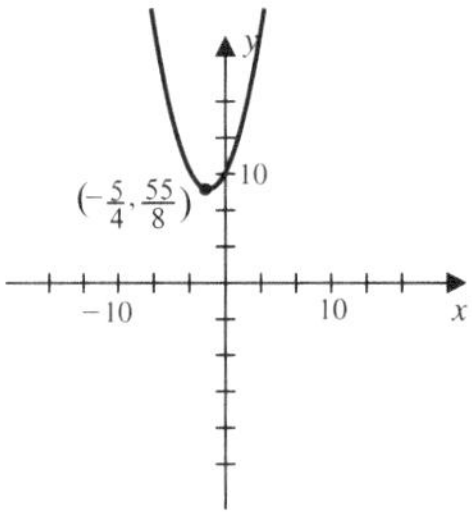

b. The range is $\left[\frac{55}{8}, \infty\right)$.

c. x-intercepts: none; y-intercept: (0, 10);

d. A minimum occurs at the point $\left(-\frac{5}{4}, \frac{55}{8}\right)$.

61. a. We have

$$y = -\frac{1}{2}x^2 + 3x - 3 = -\frac{1}{2}(x^2 - 6x) - 3 = -\frac{1}{2}\left(x^2 - 6x + 9 - 9\right) - 3 = -\frac{1}{2}(x-3)^2 + \frac{3}{2}.$$

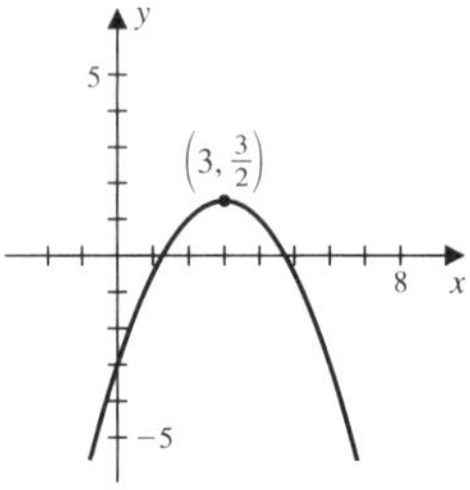

b. The range is $\left(-\infty, \frac{3}{2}\right]$.

c. x-intercepts: $\left(3 + \sqrt{3}, 0\right)$, $\left(3 - \sqrt{3}, 0\right)$, since
$-\frac{1}{2}(x-3)^2 = -\frac{3}{2}$ implies $(x-3)^2 = 3$ and $x = 3 \pm \sqrt{3}$.
y-intercept: (0, −3);

d. A maximum occurs at the point $\left(3, \frac{3}{2}\right)$.

62. **a.** We have

$$y = -\frac{1}{3}x^2 + 2x - 1 = -\frac{1}{3}(x^2 - 6x) - 1 = -\frac{1}{3}\left(x^2 - 6x + 9 - 9\right) - 1 = -\frac{1}{3}(x-3)^2 + 2.$$

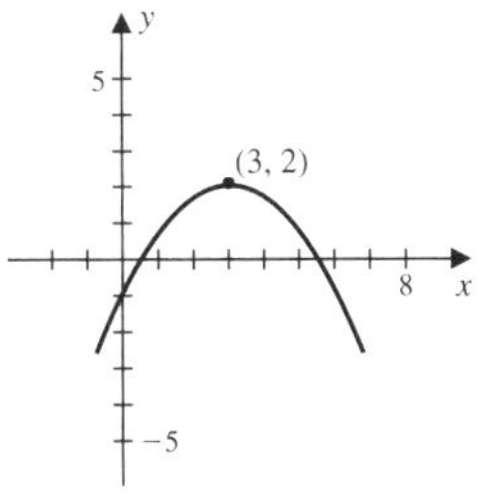

b. The range is $(-\infty, 2]$.

c. x-intercepts: $\left(3+\sqrt{6}, 0\right)$, $\left(3-\sqrt{6}, 0\right)$, since
$-\frac{1}{3}(x-3)^2 = -2$ implies $(x-3)^2 = 6$ so $x = 3 \pm \sqrt{6}$ y-intercept: $(0, -1)$;

d. A maximum occurs at the point $(3, 2)$.

63. center: $(0, 0)$; radius: 4

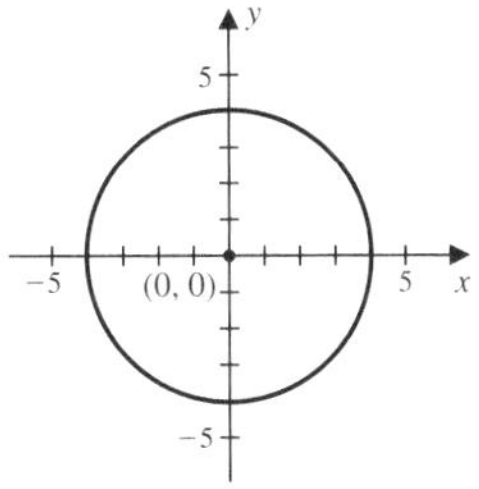

64. center: $(1, 0)$; radius: 1

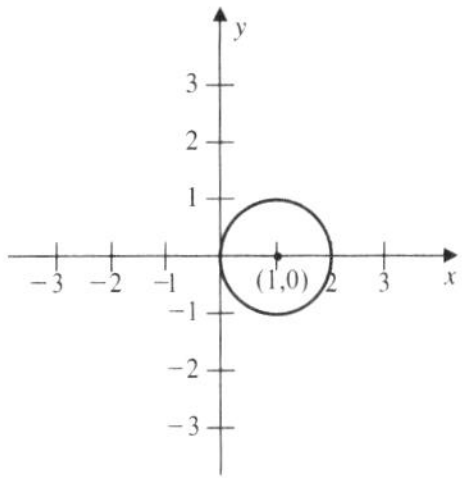

65. center: $(-2, 1)$; radius: 3

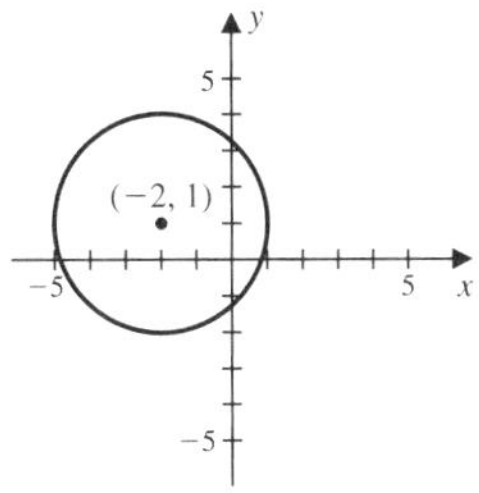

66. center: $(2, 3)$; radius 2

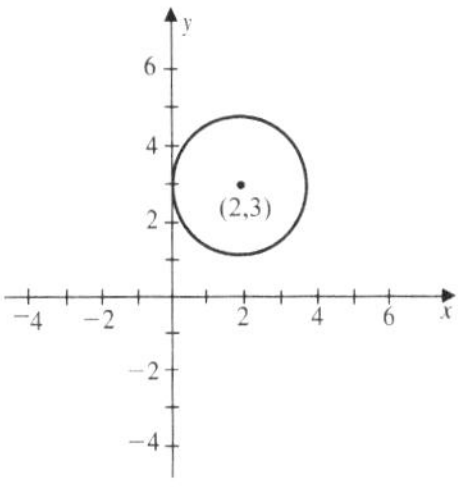

67. Completing the square on the x- and y- terms gives

$$4 = x^2 + 4x + 4 - 4 + y^2 + 2y + 1 - 1, \text{ which implies } (x+2)^2 + (y+1)^2 = 9.$$

center: $(-2, -1)$; radius: 3

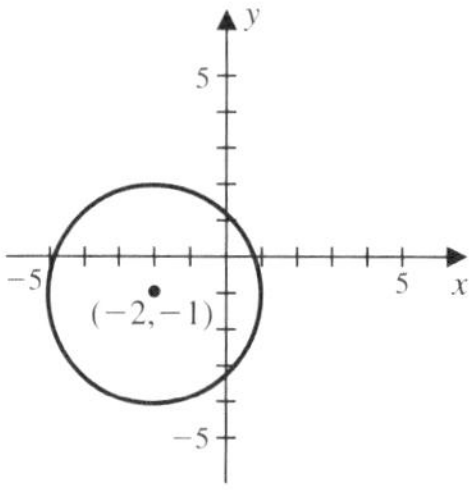

68. Completing the square on the x- and y- terms gives

$$0 = x^2 + 6x + 9 - 9 + y^2 - 2y + 1 - 1 - 6, \text{ which implies } (x+3)^2 + (y-1)^2 = 16.$$

center: $(-3, 1)$; radius: 4

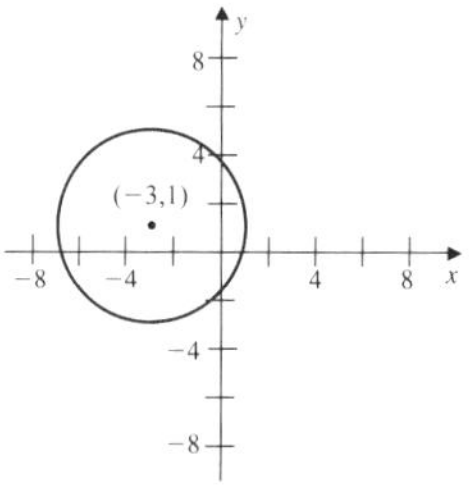

69. To show that the points $(-3, 3)$, $(3, -5)$, and $(7, -2)$ are vertices of a rectangle, show that the line passing through $(-3, 3)$ and $(3, -5)$ is perpendicular to the line passing through $(3, -5)$ and $(7, -2)$. Computing the slopes of the lines gives

$$m_1 = \frac{3-(-5)}{-3-3} = -\frac{4}{3} \quad \text{and} \quad m_2 = \frac{-5-(-2)}{3-7} = \frac{3}{4}.$$

Since the slopes are negative reciprocals of one another, the lines are perpendicular. To find the fourth vertex, first note that $(7, -2)$ is 4 to the right and 3 units upward from $(3, -5)$. Moving this distance from $(-3, 3)$ places the other vertex of the rectangle at $(-3+4, 3+3) = (1, 6)$.

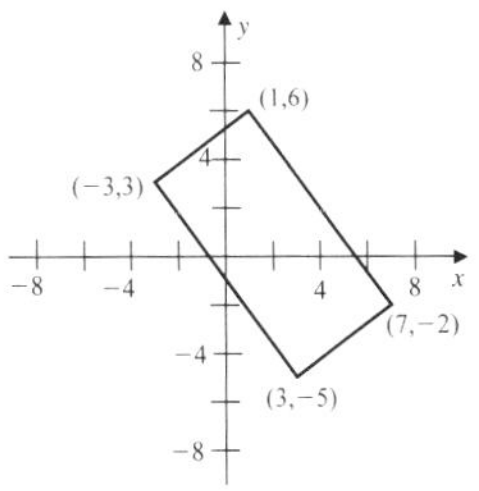

70. The line passes through $(3, 0)$ and $(0, 2)$, so the slope is $m = \frac{2-0}{0-3} = -\frac{2}{3}$ and the equation is $y = -\frac{2}{3}x + 2$.

71. The line perpendicular to $y = 2x - 2$ and passing through the point $(-2, -1)$ has slope $m = -\frac{1}{2}$ and equation $y + 1 = -\frac{1}{2}(x+2)$, or $y = -\frac{1}{2}x - 2$. The point of intersection of $y = 2x - 2$ and $y = -\frac{1}{2}x - 2$ is found from the equation

$$2x - 2 = -\frac{1}{2}x - 2, \text{ so } \frac{5}{2}x = 0 \text{ and } x = 0,\ y = -2.$$

Then $d((-2, -1), (0, -2)) = \sqrt{(-2-0)^2 + (-1-(-2))^2} = \sqrt{5}$.

72. A radius for the circle is the line segment between $(-3, -1)$ and $(-5, -3)$, so the radius is

$$r = \sqrt{(-3-(-5))^2 + (-1-(-3))^2} = \sqrt{4+4} = \sqrt{8}.$$

The equation of the circle is $(x+3)^2 + (y+1)^2 = 8$.

73. If the circle has center $(4, 2)$ and is tangent to the x-axis, then the circle also passes through $(2, 0)$. So the radius is 2 and the equation is $(x-4)^2 + (y-2)^2 = 4$.

74. We have

$$0 = x^2 - 4x + y^2 - 2y - 1$$

$$= x^2 - 4x + 4 - 4 + y^2 - 2y + 1 - 1 - 1$$

so $(x-2)^2 + (y-1)^2 = 6$.

The circle has center $(2, 1)$ so the line passing through $(2, 1)$ and $(3, 4)$ has slope $m = \frac{4-1}{3-2} = 3$ and equation $y - 1 = 3(x-2)$ implying $y = 3x - 5$.

75. We need to find $(x, 0)$ so that

$$\sqrt{(x-(-2))^2 + (0-(-3))^2} = \sqrt{(x-3)^2 + (0-5)^2}$$

so $\sqrt{(x+2)^2 + 9} = \sqrt{(x-3)^2 + 25}$ and $x^2 + 4x + 13 = x^2 - 6x + 34$.

This reduces to $10x = 21$, so $x = \frac{21}{10}$, and the point is $\left(\frac{21}{10}, 0\right)$.

76. The graph of $y = g(x)$ is shown.

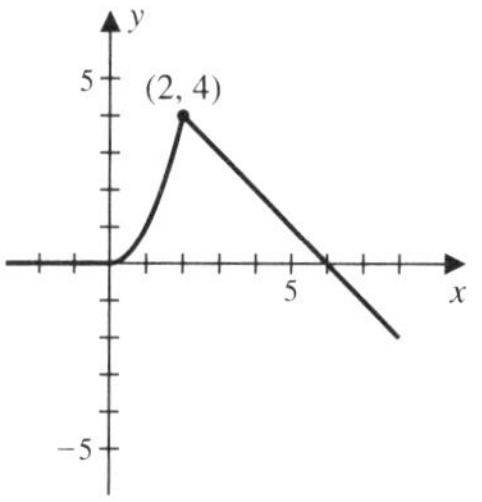

77. **a.** The data points appear to lie on a parabola with vertex $(3, 10)$ that opens downward and has been vertically elongated by a factor of 4. This implies that the equation of the parabola is $y = -4(x-3)^2 + 10$.

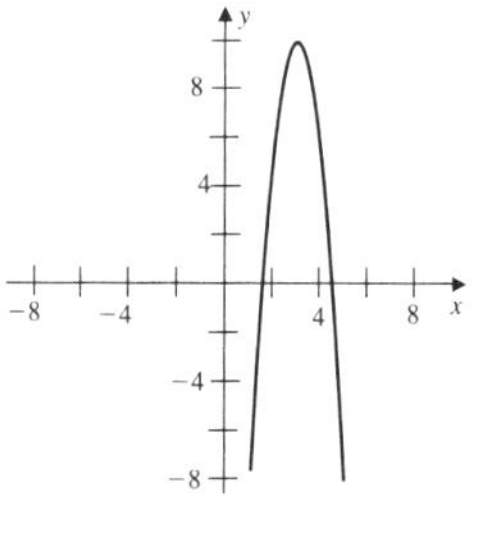

b. For each single unit change in x, the change in y is -2, so the data points represent a line with slope -2. The line passes through the point $(0, 1)$, so the equation is $y - 1 = -2(x - 0)$, that is, $y = -2x + 1$.

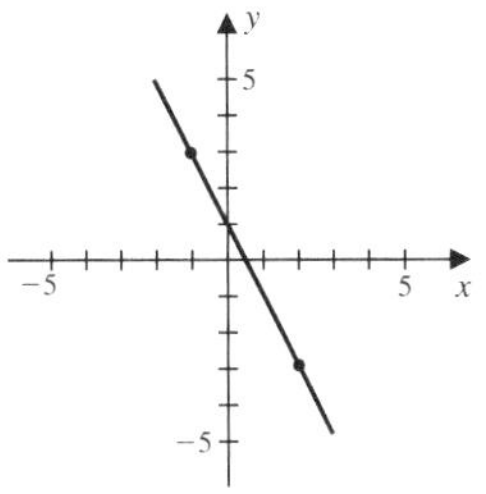

78. a. From the graph of $s(t) = -4.8t^2 + 57.6t$ the maximum height is approximately 172.8 meters, occurring when $t = 6$ seconds.

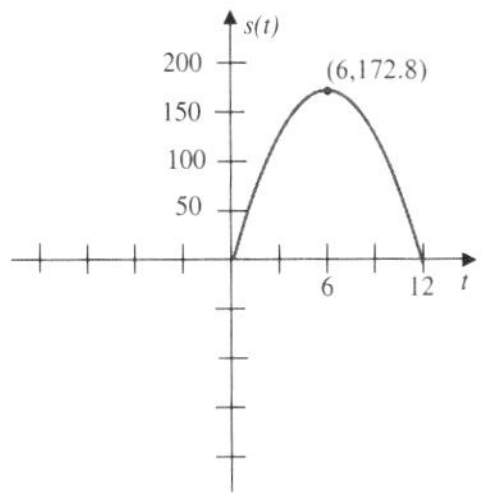

b. We need to solve $-4.8t^2 + 57.6t \geq 120$ so

$$0 \leq -4.8t^2 + 57.6t - 120 = -4.8(t^2 - 12t + 25),$$

which implies that $t^2 - 12t + 25 \leq 0$. Solving $t^2 - 12t + 25 = 0$ gives

$$t = \frac{12 \pm \sqrt{144 - 4(1)(25)}}{2} = \frac{12 \pm \sqrt{44}}{2} = \frac{12 \pm 2\sqrt{11}}{2} = 6 \pm \sqrt{11}.$$

So $2.7 \approx 6 - \sqrt{11} \leq t \leq 6 + \sqrt{11} \approx 9.3$.

c. The ball strikes the ground when

$$0 = -4.8t^2 + 57.6t = -4.8t(t - 12) \text{ so } t = 0, t = 12.$$

Since $t > 0$, we have $t = 12$.

79. Let x denote the price over \$36.00 for a complete dinner and let $P(x)$ denote the amount taken in by the restaurant, so that

$$P(x) = (36 + x)(200 - 4x) = 7200 + 56x - 4x^2.$$

From the graph of $P(x) = 7200+56x-4x^2$ we see that the maximum return occurs at approximately $x = 7$. So the owner should charge $36 + 7 =$ \$43.00 per dinner. By completing the square on the quadratic we have

$$\begin{aligned} -4x^2 + 56x + 7200 &= -4(x^2 - 14x) + 7200 \\ &= -4(x^2 - 14x + 49 - 49) + 7200 = -4(x - 7)^2 + 7396, \end{aligned}$$

so the exact maximum return is when $x = 7$.

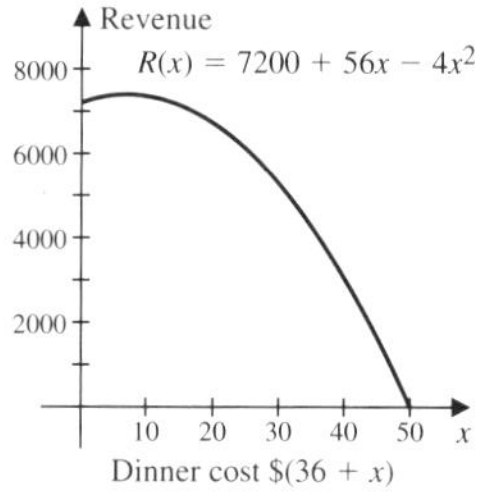

80. The cost per day to rent from agency A is $21 + 0.21x$ dollars and from agency B is $32 + 0.18x$ dollars, where x is the number of miles the car is driven.

a. If the car is driven 320 miles, the cost from agency A is $21 + 0.21(320) =$ \$88.20 and the cost from agency B is $32 + 0.18(320) =$ \$89.60. This implies that a driver should rent from agency A.

b. Since the slope of the line describing agency A's cost is greater than the slope of the line describing agency B's cost, it will eventually overtake the line with the smaller positive slope. The point of intersection determines the number of miles driven where the cost to both agencies is the same and for any number of miles greater agency B's cost is less. So solve $21 + 0.21x = 32 + 0.18x$ which implies that $0.03x = 11$ and $x = \frac{11}{0.03} \approx 367$ miles.

81. a. The amount of fence required is the perimeter of the plot plus the length of fence down the middle. If ℓ and w are the length and width of the plot, then the amount of fence, F, is $F = 3\ell + 2w$, where the assumption is the length of the middle section of fence is l. Since the area of the plot is 432, we have $432 = \ell w$. This implies that $w = \frac{432}{\ell}$ and that

$$F(\ell) = 3\ell + 2 \cdot \frac{432}{\ell} = 3\ell + \frac{864}{\ell}.$$

b. The domain of the function is $(0, \infty)$.

82. The volume of the box is

$$V(x) = x(5-2x)(9-2x) = 45x - 28x^2 + 4x^3.$$

The graph indicates that the box of maximum volume is constructed when a square of about 1 centimeter is removed from each corner of the sheet of cardboard.

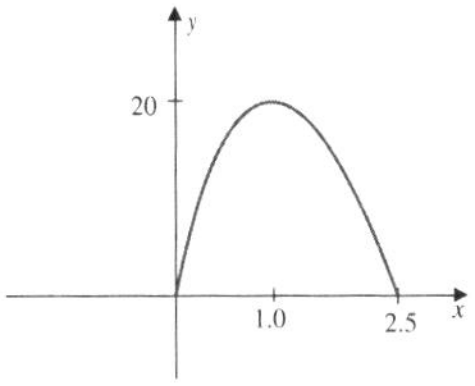

83. **a.** If the relationship between cost and number of units produced per day is linear, then the points (100, 3200) and (500, 9600) are on the line. The slope of the line is

$$m = \frac{9600-3200}{500-100} = \frac{6400}{400} = 16$$

and the equation of the line is $y - 3200 = 16(x - 100) = 16x - 1600$.
Hence $y = 16x + 1600$.

b. The slope of line indicates that for each one unit increase in the number of units produced, results in an increase of $16.00 in the cost.

c. The y-intercept is 1600 and represents the fixed costs of production.

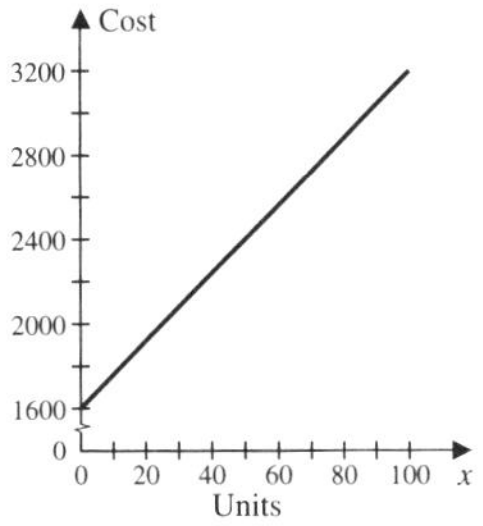

84. Part (d) gives the best representation.

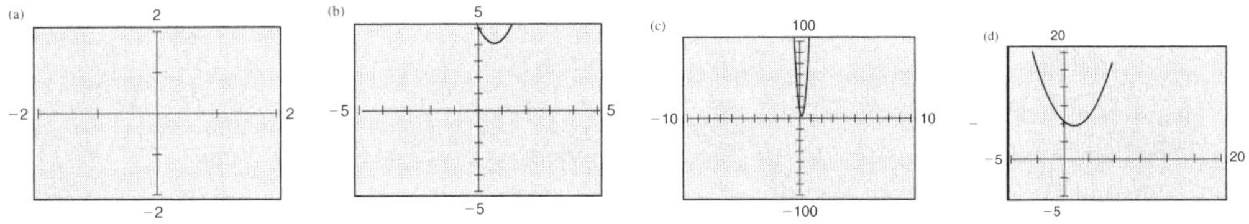

85. Part (d) gives the best representation.

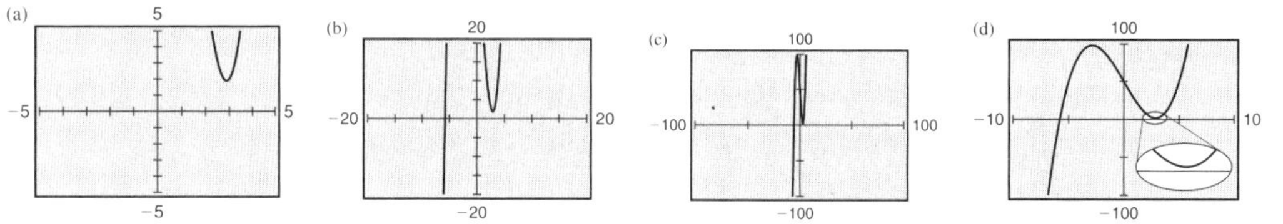

86. **a.** $y = x^2 - 16x + 61$ **b.** $y = x^2 + 10x + 38$ **c.** $y = \sqrt{x^2 - 10x + 29}$ **d.** $y = \dfrac{x+4}{x-3}$

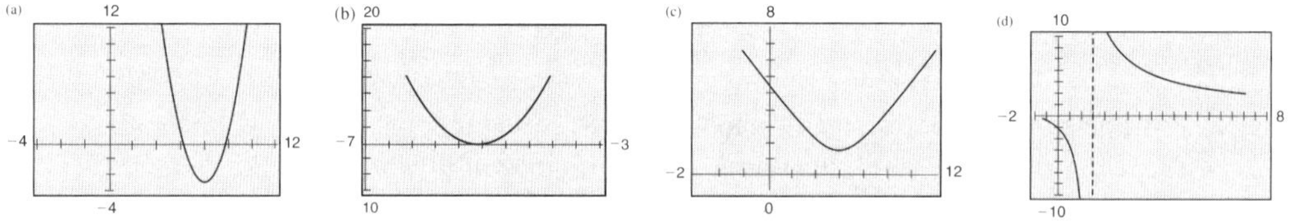

87. **a.** Between A and B and to the right of E. **b.** Between C and D. **c.** Between B and C.

d. Between D and E.

88. This graph has the required properties.

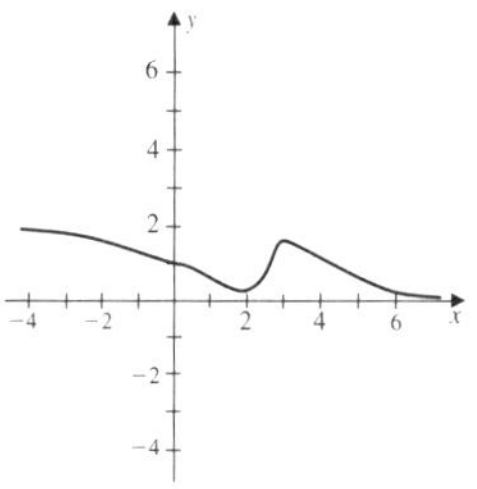

Chapter 1 Exercises for Calculus (Page 75)

1. For each increase in x of 4 units, the y-coordinate $y = g(x)$ always increases by 12 units, so g is linear and f is quadratic. The slope of the line $y = g(x)$ is $m = \frac{12}{4} = 3$ and the line passes through $(-2, 2)$, so

$$y - 2 = 3(x + 2) \text{ implies } y = g(x) = 3x + 8.$$

The quadratic $y = f(x) = ax^2$ and passes through $(-2, 2)$, so $2 = f(-2) = 4a$, which implies $a = \frac{1}{2}$ and $f(x) = \frac{1}{2}x^2$.

2. Let $3x + ay = -2a$.

a. $ay = -3x - 2a$ implies $y = -\frac{3}{a}x - 2$. If the slope is 2, then $-\frac{3}{a} = 2$ so $a = -\frac{3}{2}$.

b. A line is horizontal provided the slope is 0, but a can not be zero, so there are no values of a.

c. The line is vertical provided the slope is undefined, which holds when $a = 0$.

d. If the line passes through $(0, -2)$, then $3(0) + a(-2) = -2a$ implies $-2a = -2a$, which is a true statement for all values of a. The line passes through $(0, -2)$ for all $a \neq 0$.

3. a. The table shows that the solution is $14 \leq n \leq 16$.

n	12	13	14	15	16	17
$\frac{n(n+1)}{2}$	78	91	105	120	136	153

b. The table shows that the solution is $8 \leq n \leq 9$.

n	5	6	7	8	9	10
$\frac{n(n+1)(2n+1)}{6}$	55	91	140	204	285	385

4. a. The graph of $y = ax^2 + bx$ has x-intercepts at $x = 0$ and $x = -b/a$, and a vertex at $(-b/(2a), -b^2/(4a))$.

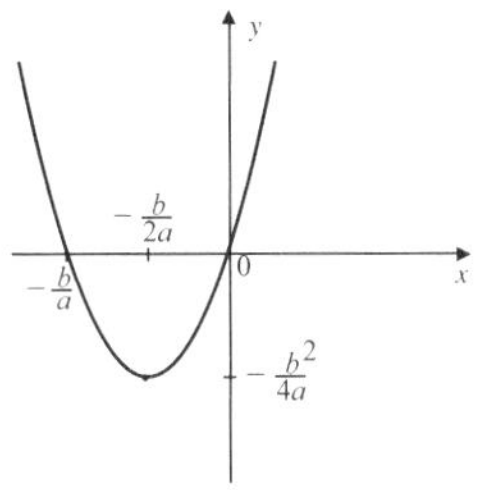

b. To show that $y = ax^2 + bx$ is symmetric about the line $x = -\frac{b}{2a}$, select values equidistant on either side of the line and show they yield the same y-coordinate for the corresponding point on the curve. Let $z > 0$. Then

$$a\left(-\frac{b}{2a} + z\right)^2 + b\left(-\frac{b}{2a} + z\right) = a\left(\frac{b^2}{4a^2} - \frac{b}{a}z + z^2\right) - \frac{b^2}{2a} + bz = az^2 + \frac{b^2}{4a} - \frac{b^2}{2a} = az^2 - \frac{b^2}{4a}$$

and

$$a\left(-\frac{b}{2a}-z\right)^2+b\left(-\frac{b}{2a}-z\right)=a\left(\frac{b^2}{4a^2}+\frac{b}{a}z+z^2\right)-\frac{b^2}{2a}-bz=az^2+\frac{b^2}{4a}-\frac{b^2}{2a}=az^2-\frac{b^2}{4a},$$

so the graph is symmetric about the vertical line $y=-\frac{b}{2a}$.

c. Since the graph of $y=ax^2+bx+c$ is simply a vertical shift of the graph of $y=ax^2+bx$, the result in part (b) implies that it also is symmetric about the vertical line $x=-\frac{b}{2a}$.

5. The graph of the temperature might be as follows.

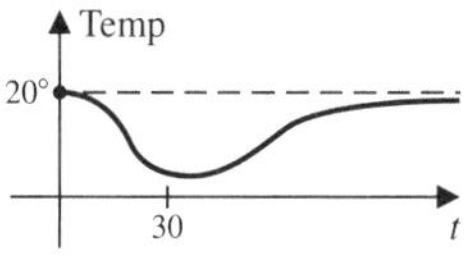

6. The graph of the child's temperature might be as follows.

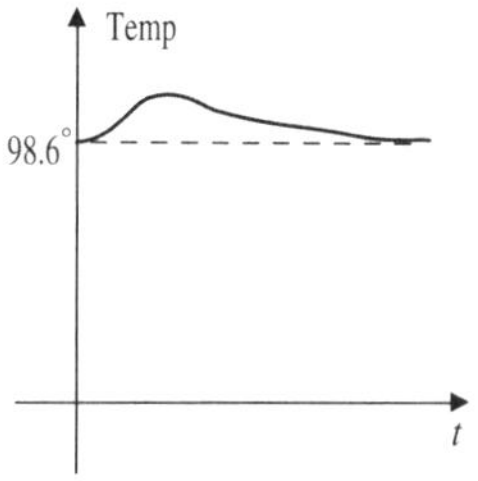

7. Graphs for the airplane are shown below.

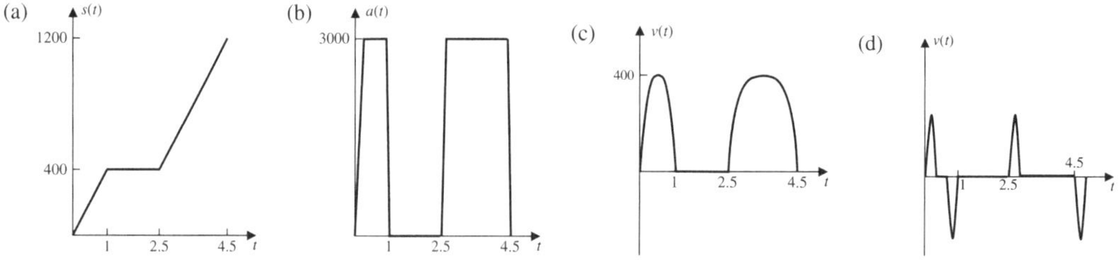

8. a. The surface area of the can is the sum of the area of the top and bottom plus the area of the walls. Hence the surface area is $S = 2\pi r^2 + 2\pi rh$. Since the volume of the can is 900 centimeters^3, we have $900 = V = \pi r^2 h \Rightarrow h = \frac{900}{\pi r^2}$. So

$$S = 2\pi r^2 + 2\pi r\left(\frac{900}{\pi r^2}\right) = 2\pi r^2 + \frac{1800}{r}.$$

b. From the graph we see that the radius of the can that yields the minimum surface area is approximately $r = 5.2$ centimeters. The height is then $h = \frac{900}{\pi(5.2)^2} \approx 10.6$ centimeters.

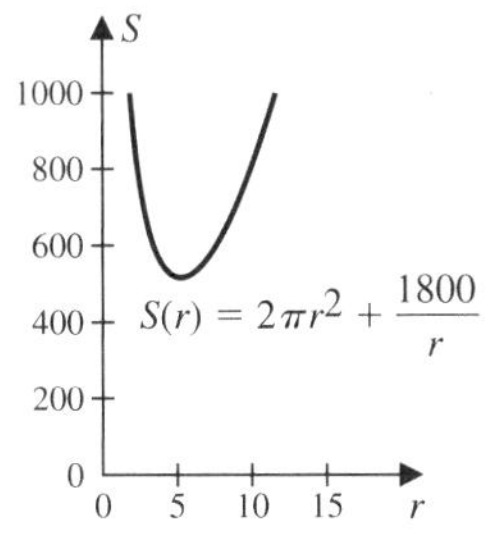

9. a. Let x be the number purchased. Then the price per item is

$$P(x) = \begin{cases} 300, & \text{if } 0 \le x \le 100, \\ 300 - (x - 100) = 400 - x, & \text{if } 100 < x \le 150, \\ 225, & \text{if } 150 < x. \end{cases}$$

b. The graph of the price per item is given below.

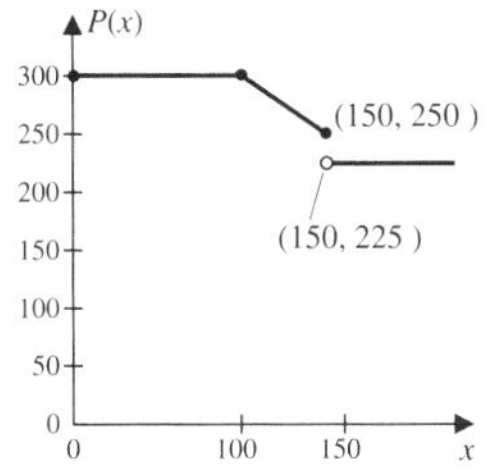

10. a. Let x be the number of trees planted per acre. Then the yield of a one acre plot is

$$Y(x) = \begin{cases} 300x, & \text{if } 0 \le x \le 30 \\ (300 - 5(x - 30))x, & \text{if } 30 < x \le 90 \end{cases} = \begin{cases} 300x, & \text{if } 0 \le x \le 30 \\ 450x - 5x^2, & \text{if } 30 < x \le 90 \end{cases}$$

The bound of 90 is given since if 90 trees are planted, there will be no fruit per tree.

b. The maximum yield occurs when $x = 45$ and $Y(45) = 450(45) - 5(45)^2 = 10,125$lb.

11. **a.** The length of the rod at a temperature t above 0° C, is $L(t) = \ell + a\ell t$, where ℓ is the original length of the rod and $a = 11 \times 10^{-6}$ is the coefficient of linear expansion. Since the length of the rod is 2 meters at 0°C, the length of the rod is $L(t) = 2 + (22 \times 10^{-6})t$.

b. $L(1000) = 2 + (22 \times 10^{-6})(1000) = 2 + 22 \times 10^{-3}$.

12. **a.** The slope of the radius is $\frac{\sqrt{3}/2}{1/2} = \sqrt{3}$. The tangent line to the point is perpendicular to the radius through the point, so the slope of the tangent line is $-\frac{1}{\sqrt{3}} = -\frac{\sqrt{3}}{3}$. The equation of the tangent line is

$$y - \frac{\sqrt{3}}{2} = -\frac{\sqrt{3}}{3}\left(x - \frac{1}{2}\right) \quad \text{so} \quad y = -\frac{\sqrt{3}}{3}x + \frac{\sqrt{3}}{6} + \frac{\sqrt{3}}{2} = -\frac{\sqrt{3}}{3}x + \frac{2\sqrt{3}}{3}.$$

b. For $y = f(x) = \sqrt{1 - x^2}$, we have

$$\begin{aligned}\frac{f(x+h) - f(x)}{h} &= \frac{\sqrt{1-(x+h)^2} - \sqrt{1-x^2}}{h} \\ &= \frac{\sqrt{1-(x+h)^2} - \sqrt{1-x^2}}{h}\left(\frac{\sqrt{1-(x+h)^2} + \sqrt{1-x^2}}{\sqrt{1-(x+h)^2} + \sqrt{1-x^2}}\right) \\ &= \frac{-h(2x+h)}{h(\sqrt{1-(x+h)^2} + \sqrt{1-x^2})}.\end{aligned}$$

As $h \to 0$ this reduces to

$$\frac{-2x}{2\sqrt{1-x^2}} = -\frac{x}{\sqrt{1-x^2}} \quad \text{and when } x = 1/2 \quad -\frac{1/2}{\sqrt{1-1/4}} = -\frac{1}{2\sqrt{3/4}} = -\frac{1}{\sqrt{3}} = -\frac{\sqrt{3}}{3}.$$

This value agrees, of course, with the slope of the tangent line found in part (a).

Chapter 1 Chapter Test (Page 76)

1. True.

2. False. Since $x^2 - 3x + 2 = (x-1)(x-2)$, the solutions are $x = 2$ and $x = 1$.

3. False. By the Quadratic Formula, the zeros of $f(x) = x^2 + 2x - 4$ are

$$\frac{-2}{2} \pm \frac{\sqrt{4+16}}{2} = -1 \pm \sqrt{5}.$$

Since $\sqrt{5}$ is not a rational number, the zeros are not rational.

4. True.

5. True.

6. False. Since $-3x + 4 \geq 10$ implies that $-3x \geq 6$, which implies that $x \leq -2$, the solution set of the inequality is the interval $(-\infty, -2]$.

7. False. Another solution is $x = 2/3$.

8. True.

9. True.

10. False. If we rewrite $|2x - 5| = 3$ in its equivalent form, $|x - 5/2| = 3/2$, we see that the equation implies that the distance from x to $5/2$ is $3/2$.

11. True.

12. False. Since the denominator cannot be 0, $x = 3$ must be excluded, resulting in the domain $(3, \infty)$.

13. True.

14. False. Since

$$\sqrt{x^2 + x - 2} = \sqrt{(x-1)(x+2)},$$

the domain is $(-\infty, -2] \cup [1, \infty)$.

15. True.

16. False. Completing the square on x and on y gives

$$x^2 + 2x + 1 + y^2 - 4y + 4 = 4 + 1 + 4$$

or

$$(x+1)^2 + (y-2)^2 = 9.$$

This is the equation of a circle with radius 3, but the center is at $(-1, 2)$.

17. False. Since $y = \frac{2}{3}x - \frac{4}{3}$, the slope of the line is $2/3$.

18. True.

19. True.

20. False. Since $x - 3y = 3$ has slope $1/3$ and $4x - 6y = 5$ has slope $2/3$, the slopes differ and the lines are not parallel.

21. False. Since $2x + y = 2$ has slope -2 and $2y + x = -1$ has slope $-1/2$, the products of the slopes $(-2)(-1/2) = 1 \neq -1$, so the lines are not perpendicular.

22. True.

23. True.

24. False. The equation of the line that passes through the two points $(2, 1)$ and $(-3, 2)$ has slope $(2-1)/(-3-2) = -1/5$ and equation

$$-\frac{1}{5} = \frac{y-1}{x-2}$$

which simplifies to $5y + x = 7$.

25. False. Since $y = 1$ is not included, the shaded region is described by $y < 1$.

26. True.

27. False. Since $x = 0$ is included and $x = 2$ is not, the shaded region is described by $0 \leq x < 2$.

28. True.

29. True.

30. False. The region outside the shaded area is described by $|x - 1| > 3$ or $|y + 1| \geq 2$. This implies that either

$$x < -2 \text{ or } 4 < x$$

or that

$$y \leq -3 \text{ or } 1 \leq y.$$

31. False. Since

$$x^2 + 4x + 3 = x^2 + 4x + 4 - 1 = (x+2)^2 - 1,$$

the vertex of the parabola $y = x^2 + 4x + 3$ is at $(-2, -1)$.

32. False. The parabola $y = -(x+1)^2 - 2$ has a maximum point at $(-1, -2)$.

33. True.

34. False. Since $y = 2$ is not in the range, the range of the function is $(-\infty, 0] \cup (2, \infty)$.

35. False. The difference quotient for the function $f(x) = -2x + 3$ reduces to

$$\frac{-2x - 2h + 3 + 2x - 3}{h} = -2.$$

36. True.

37. False. The graph of an even function is symmetric with respect to the y-axis and the graph of an odd function is symmetric with respect to the origin.

38. True.

39. True.

40. False. The graph of $y = (x-1)^2 + 2$ is obtained by shifting the graph of $y = x^2$ to the right 1 unit and upward 2 units.

41. True.

42. False. The leading coefficient of the red parabola is more than the leading coefficient of the green parabola.

43. False. The leading coefficient of the yellow parabola is positive.

44. True.

45. False. Since the vertex of the yellow parabola is $(-3, -2)$ the equation has the form

$$y = a(x+3)^2 - 2.$$

The parabola also passes through $(0, 7)$, so

$$7 = a(9) - 2 \quad \text{and} \quad a = 1.$$

Its equation is $y = (x+3)^2 - 2 = x^2 + 6x + 7$.

Exercise Set 2.2 (Page 86)

1. $f(x) = |x - 3|$

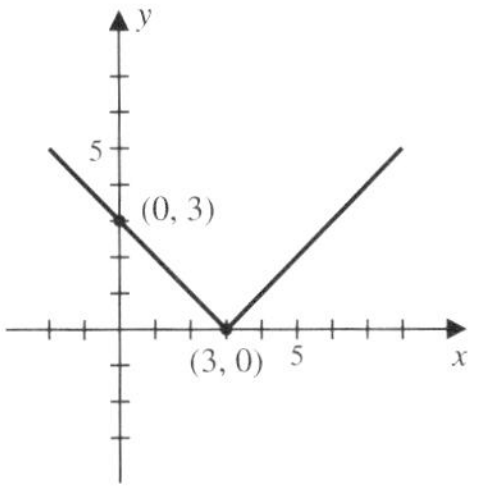

2. $f(x) = |x + 1| - 1$

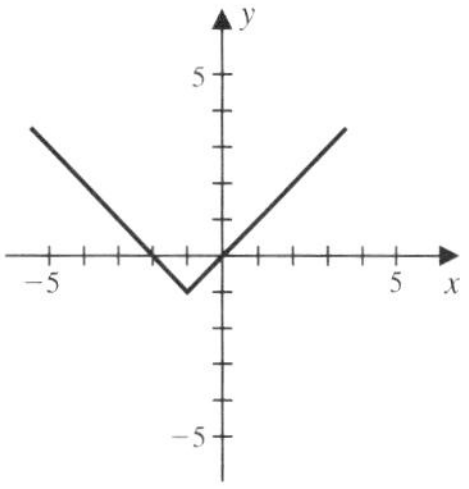

3. $f(x) = |x + 2| - 2$

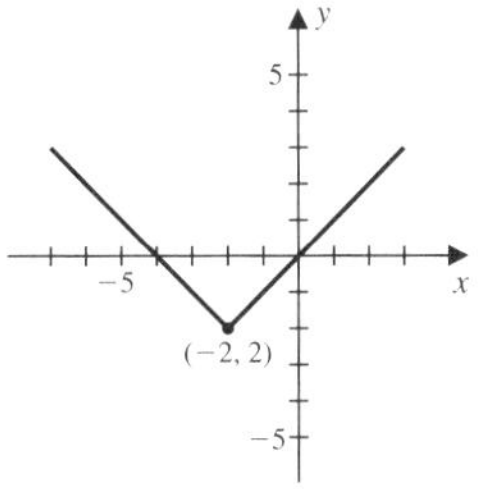

4. $f(x) = |x - 2| + 2$

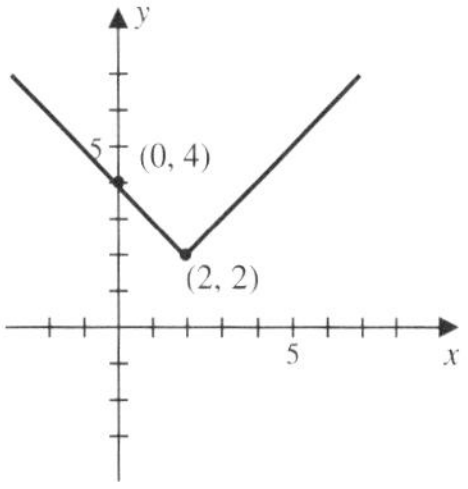

5. $f(x) = -2|x|$

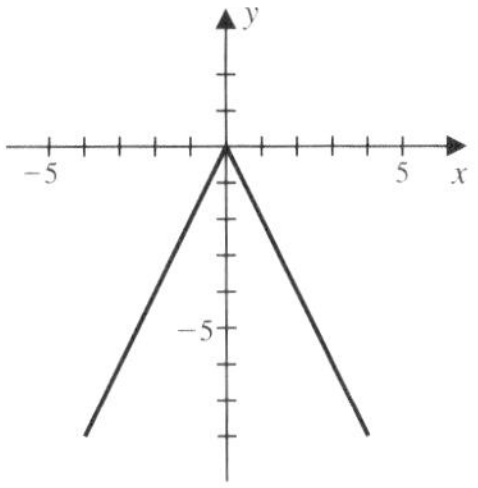

6. $f(x) = |2x - 5| = \left|2\left(x - \frac{5}{2}\right)\right| = 2\left|x - \frac{5}{2}\right|$

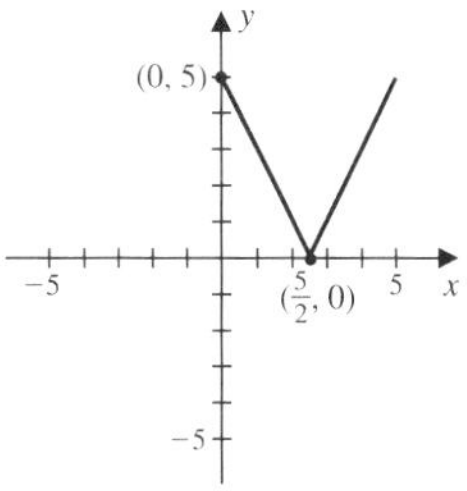

7. $f(x) = -3|x - 1| + 1$

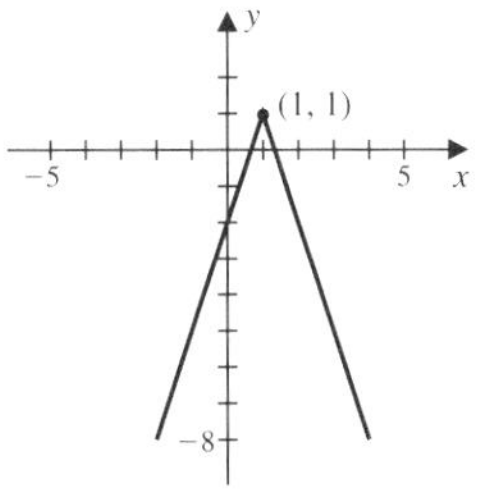

8. $f(x) = -2|x + 1| - 3$

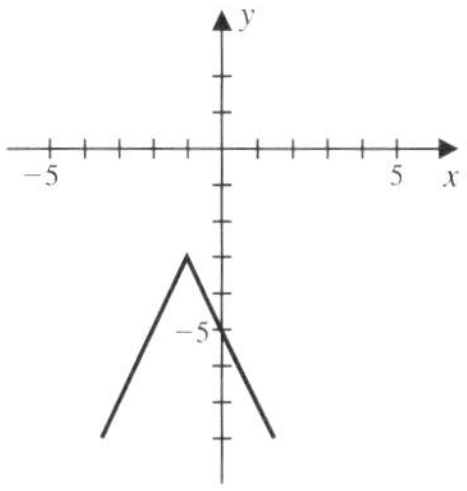

9. **a.** $g(x) = \sqrt{x} + 2$

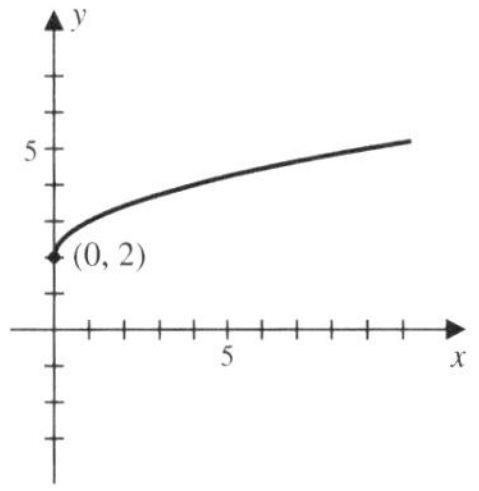

b. Domain: $g(x) = \sqrt{x} + 2$ is defined when $x \geq 0$, so the domain is $[0, \infty)$.
Range: Since $\sqrt{x} + 2 \geq 2$, the range is $[2, \infty)$.

10. **a.** $g(x) = \sqrt{x} - 1$

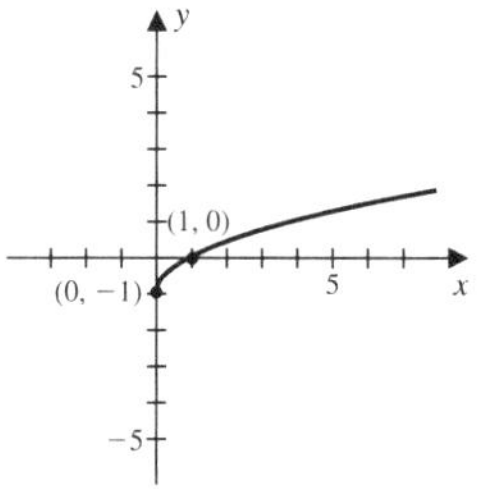

b. Domain: $g(x) = \sqrt{x} - 1$ is defined when $x \geq 0$, so the domain is $[0, \infty)$.
Range: Since $\sqrt{x} - 1 \geq -1$, the range is $[-1, \infty)$.

11. **a.** $g(x) = \sqrt{x + 2} - 2$

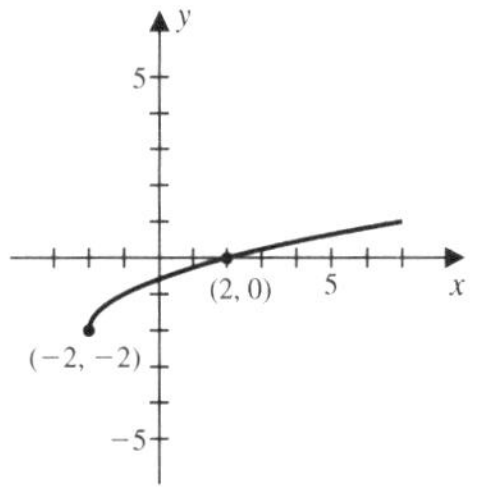

b. Domain: $g(x) = \sqrt{x + 2} - 2$ is defined when $x + 2 \geq 0$, so $x \geq -2$ and the domain is $[-2, \infty)$.
Range: Since $\sqrt{x + 2} - 2 \geq -2$, the range is $[-2, \infty)$.

12. **a.** $g(x) = \sqrt{x - 1} + 2$

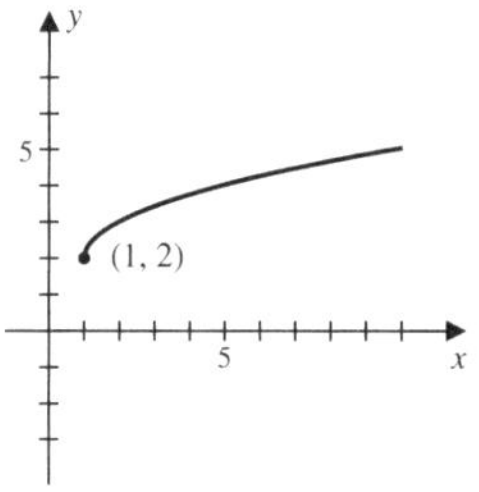

b. Domain: $g(x) = \sqrt{x - 1} + 2$ is defined when $x - 1 \geq 0$, so $x \geq 1$ and the domain is $[1, \infty)$.
Range: Since $\sqrt{x - 1} + 2 \geq 2$, the range is $[2, \infty)$.

13. a. $g(x) = -\sqrt{x+2}$

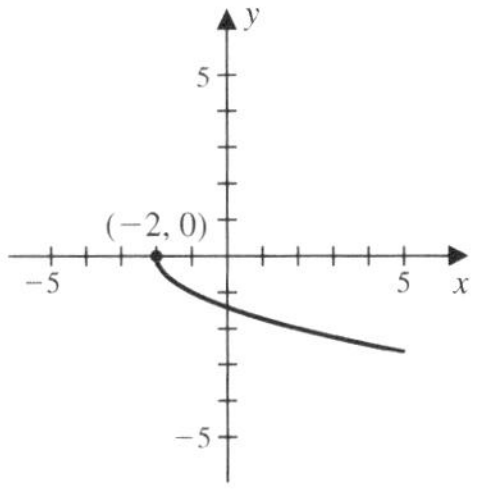

b. Domain: $g(x) = -\sqrt{x+2}$ is defined when $x + 2 \geq 0$, so $x \geq -2$ and the domain is $[-2, \infty)$.
Range: Since $-\sqrt{x+2} \leq 0$, the range is $(-\infty, 0]$.

14. a. $g(x) = 2 - \sqrt{x+2}$

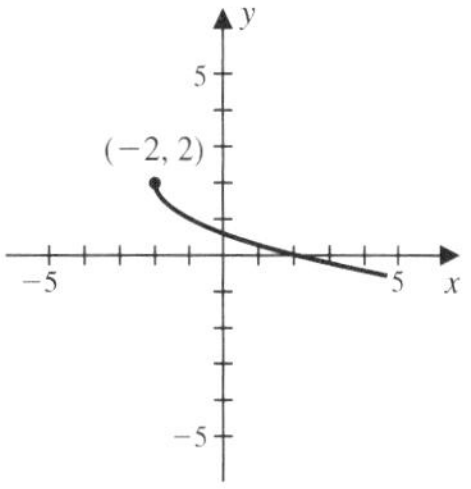

b. Domain: $g(x) = 2 - \sqrt{x+2}$ is defined when $x + 2 \geq 0$, so $x \geq -2$ and the domain is $[-2, \infty)$.
Range: Since $2 - \sqrt{x+2} \leq 2$, the range is $(-\infty, 2]$.

15. a. $g(x) = \sqrt{2-x} - 1 = \sqrt{-(x-2)} - 1$

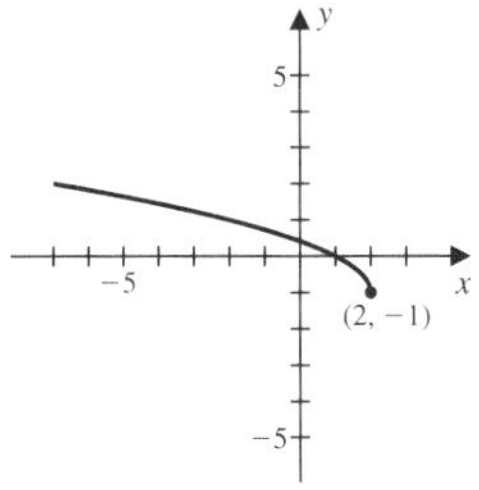

b. Domain: $g(x) = \sqrt{2-x} - 1$ is defined when $2 - x \geq 0$, that is, $x - 2 \leq 0$, so $x \leq 2$ and the domain is $(-\infty, 2]$.
Range: Since $\sqrt{2-x} - 1 \geq -1$, the range is $[-1, \infty)$.

16. a. $g(x) = \sqrt{3-x} + 2 = \sqrt{-(x-3)} + 2$

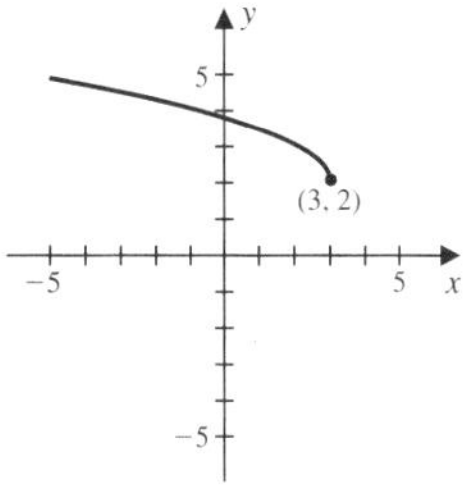

b. Domain: $g(x) = \sqrt{3-x} + 2$ is defined when $3 - x \geq 0$, that is, $x - 3 \leq 0$, so $x \leq 3$ and the domain is $(-\infty, 3]$.
Range: Since $\sqrt{3-x} + 2 \geq 2$, the range is $[2, \infty)$.

17. a. $g(x) = x^3 + 1$ **b.** $g(x) = x^3 - 1$ **c.** $g(x) = (x+1)^3$ **d.** $g(x) = (x-1)^3$
e. $g(x) = 2x^3$ **f.** $g(x) = -2x^3$

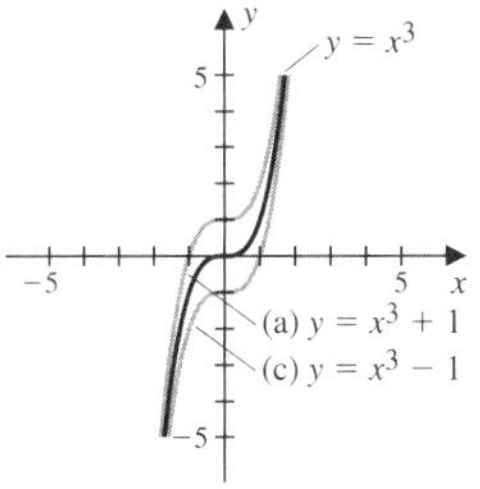

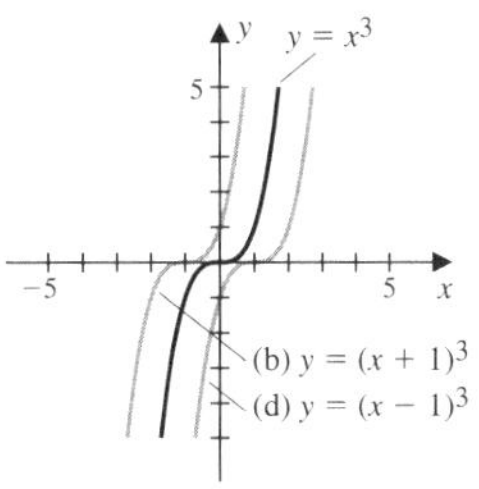

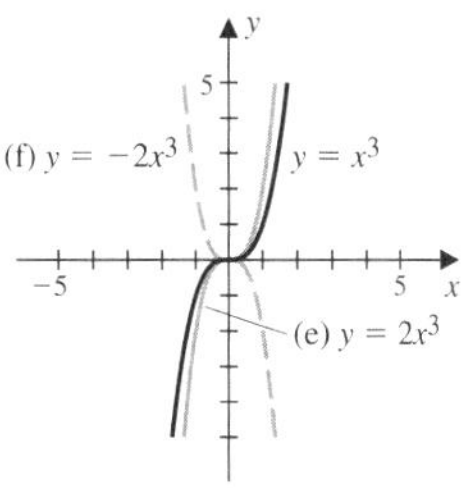

18. **a.** $g(x) = \sqrt[3]{x} + 1$ **b.** $g(x) = \sqrt[3]{x+1}$ **c.** $g(x) = \sqrt[3]{x} - 1$ **d.** $g(x) = \sqrt[3]{x-1}$
e. $g(x) = 2\sqrt[3]{x}$ **f.** $g(x) = -2\sqrt[3]{x}$

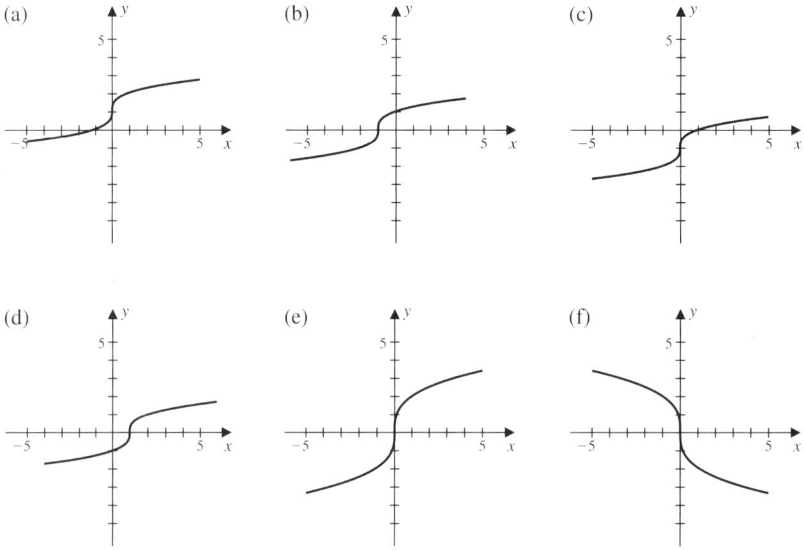

19. $f(x) = 2x - 3$

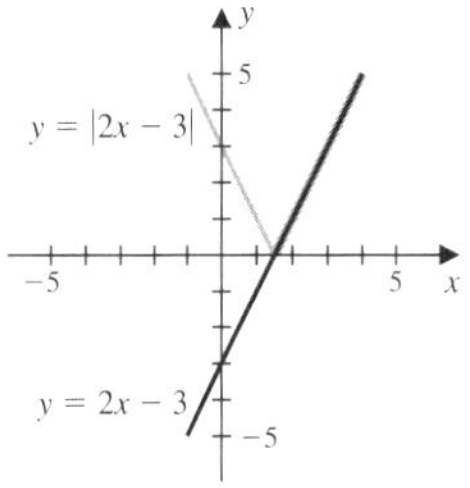

20. $f(x) = -x + 2$

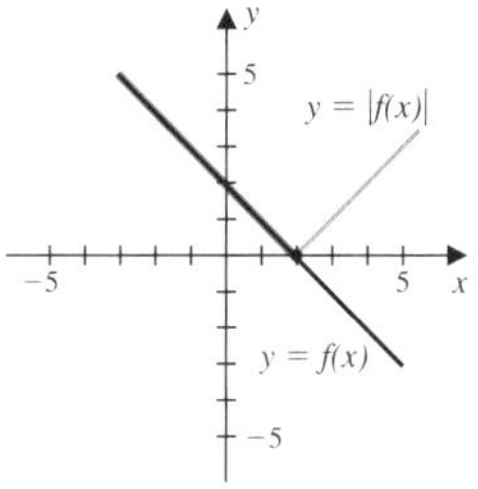

21. $f(x) = -x^2 - 2$

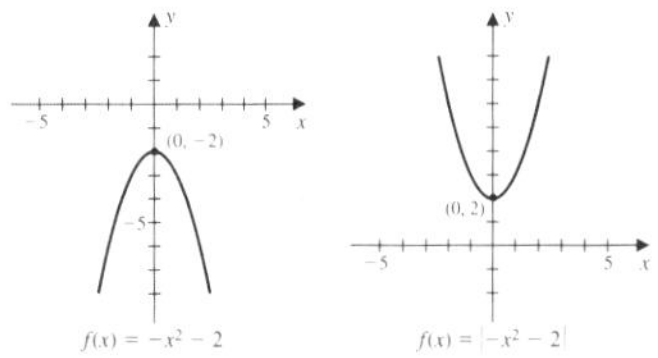

22. $f(x) = -x^2 + 2$

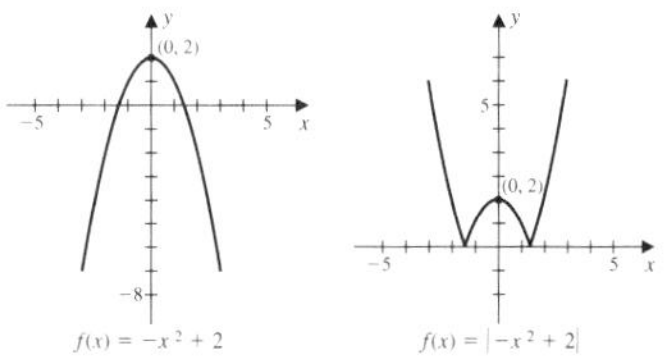

23. $f(x) = -(x-1)^2 + 1$

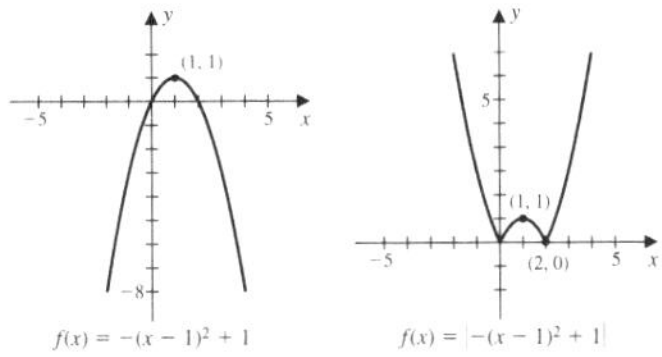

24. $f(x) = (x+2)^2 - 3$

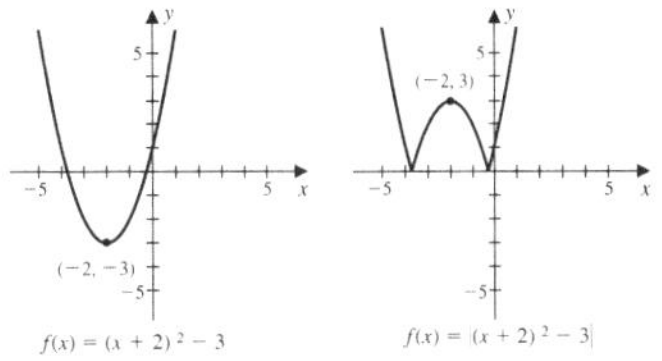

25.

$$\begin{aligned} f(x) &= x^2 - 4x + 2 \\ &= x^2 - 4x + 4 - 4 + 2 \\ &= (x-2)^2 - 2 \end{aligned}$$

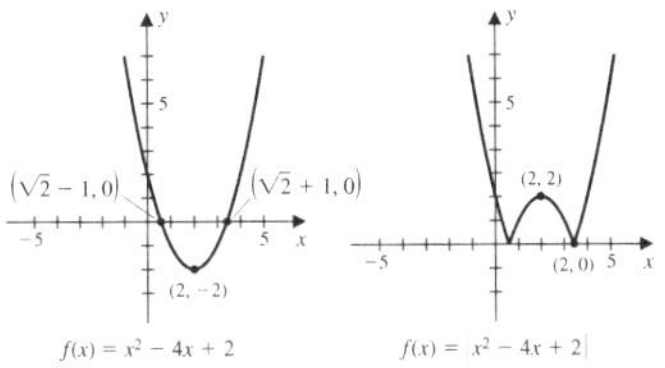

26.

$$\begin{aligned} f(x) &= x^2 - 5x + 6 \\ &= x^2 - 5x + \frac{25}{4} - \frac{25}{4} + 6 \\ &= \left(x - \frac{5}{2}\right)^2 - \frac{1}{4} \end{aligned}$$

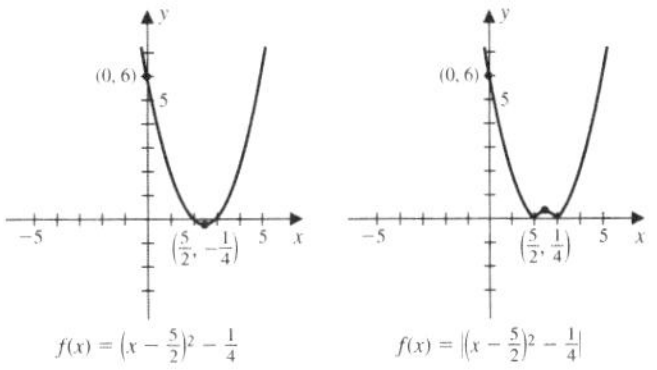

27. $f(x) = \lfloor x - 2 \rfloor$

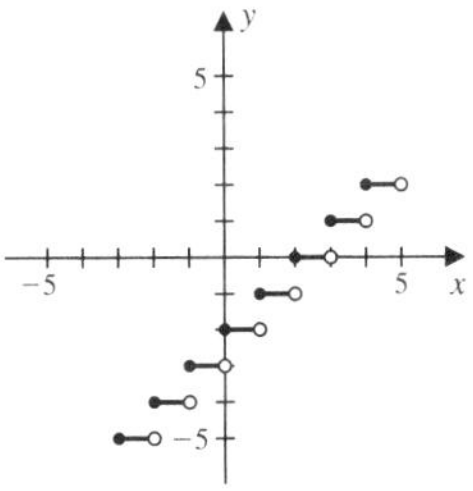

28. $f(x) = \lfloor x \rfloor - 2$

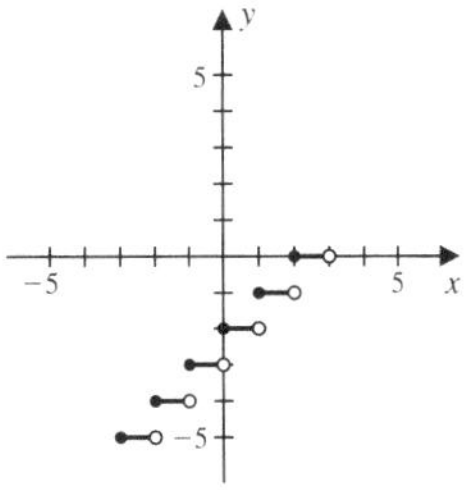

29. $f(x) = \lfloor x + 1 \rfloor - 2$

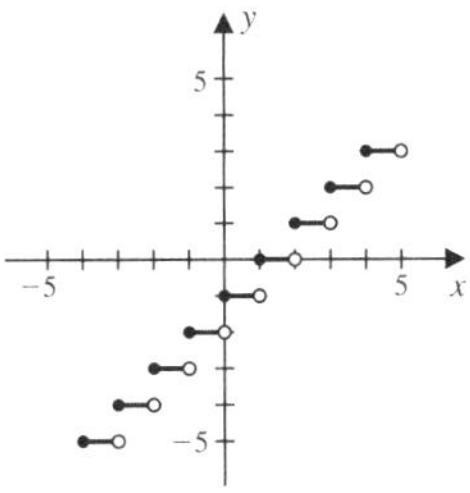

30. $f(x) = \lfloor x - 2 \rfloor + 1$

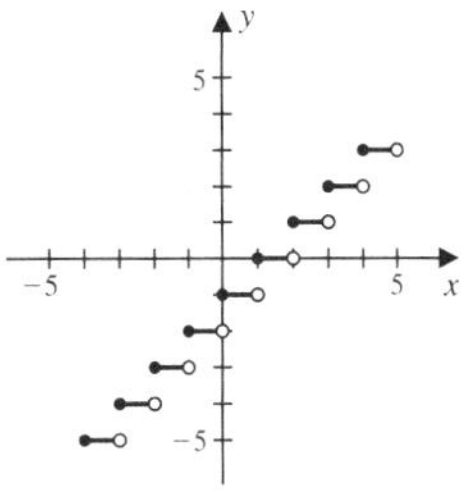

31. $f(x) = 2 - \lfloor x \rfloor$

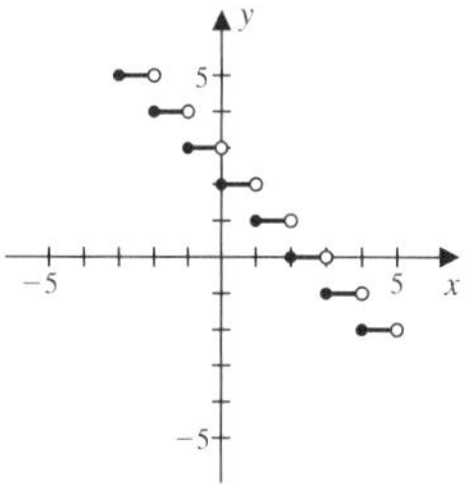

32. $f(x) = 2 - \lfloor x - 1 \rfloor$

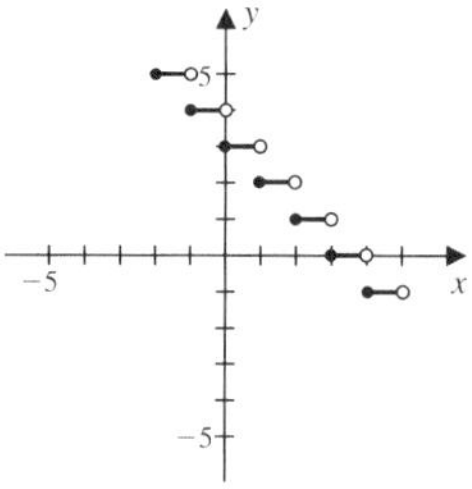

33. **a.** $y = f(x) + 1$ **b.** $y = f(x - 2)$ **c.** $y = 2f(x)$ **d.** $y = f(2x)$
e. $y = -f(x)$ **f.** $y = f(-x)$ **g.** $y = |f(x)|$ **h.** $y = f(|x|)$

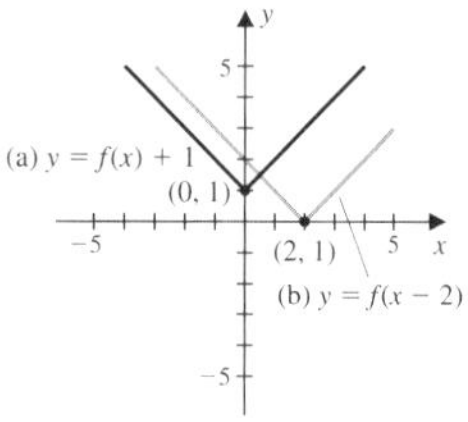

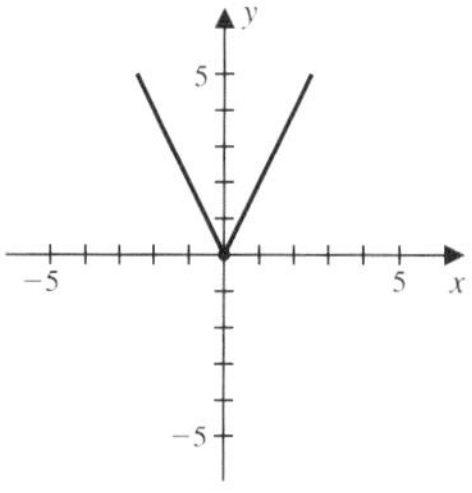

(c) and (d) $y = 2f(x) = f(2x)$

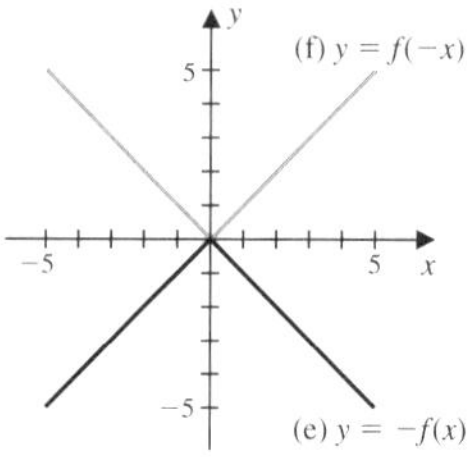

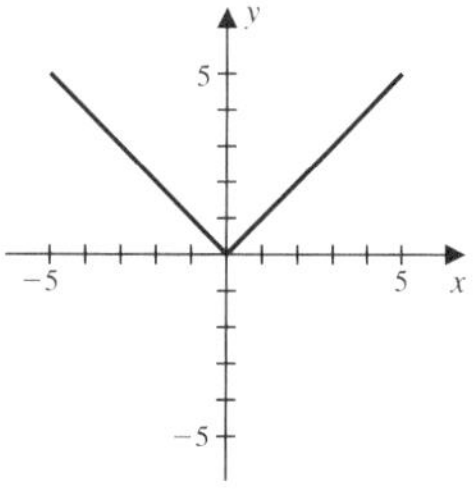

(g) and (h) $y = |f(x)| = f(|x|)$

34. **a.** $y = f(x) + 1$ **b.** $y = f(x - 2)$ **c.** $y = 2f(x)$ **d.** $y = f(2x)$
e. $y = -f(x)$ **f.** $y = f(-x)$ **g.** $y = |f(x)|$ **h.** $y = f(|x|)$

(a)

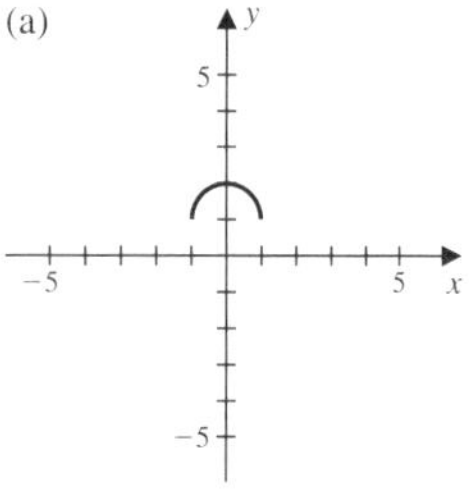

(b)

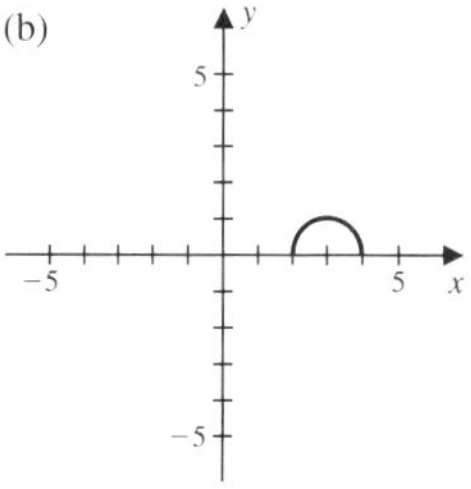

(c)

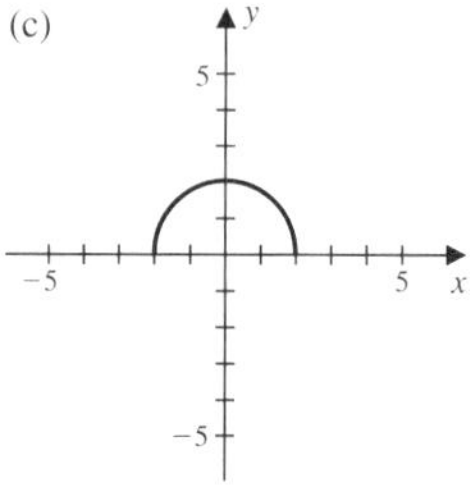

(d)

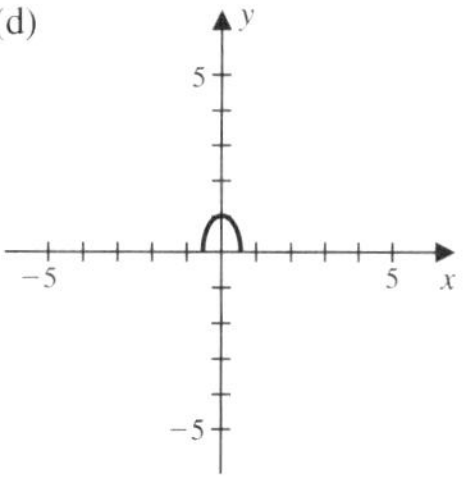

(e)

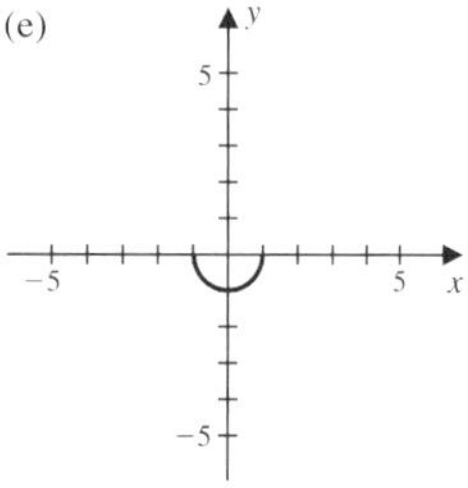

(f)

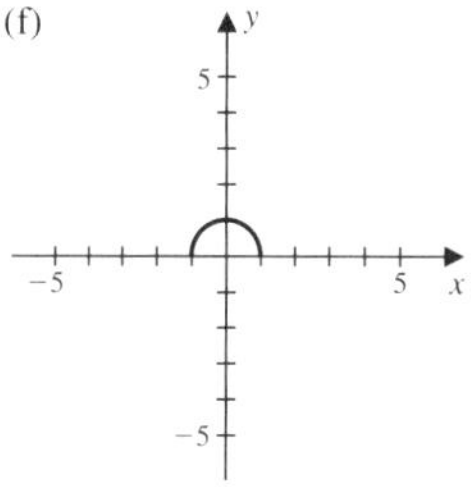

(g)

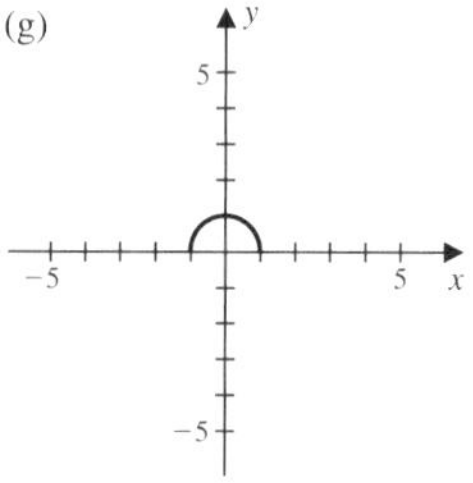

(h)

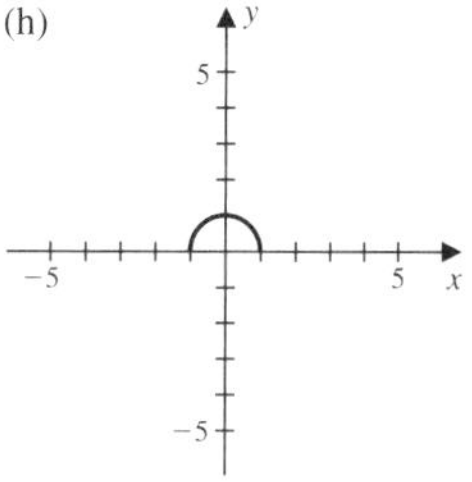

35. a. The graphs of parts (a), (b), and (c) are shown in left, center, and right, respectively.

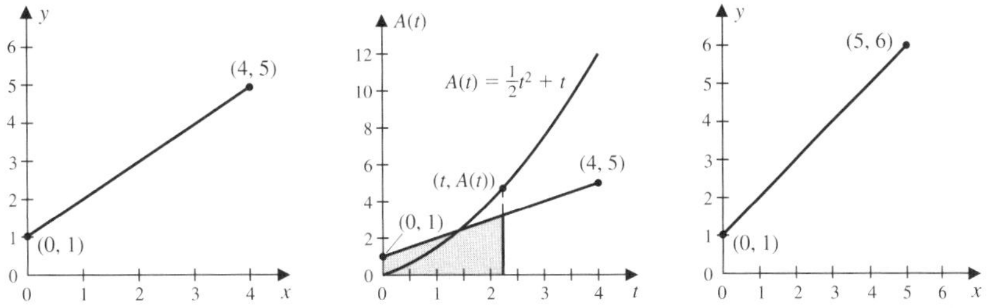

b. The area under the curve is the area of a trapezoid, which implies

$$A(t) = \frac{1}{2}(1 + (t + 1)) \cdot t = t + \frac{1}{2}t^2.$$

c.
$$\begin{aligned} d(x) &= d\left((0,0),(x, f(x)\right) = d((0,0),(x, x+1)) \\ &= \sqrt{(x-0)^2 + (x+1-0)^2} = \sqrt{x^2 + x^2 + 2x + 1} \\ &= \sqrt{2x^2 + 2x + 1} \end{aligned}$$

36. a. The graph of the distance from C is shown in the figure.

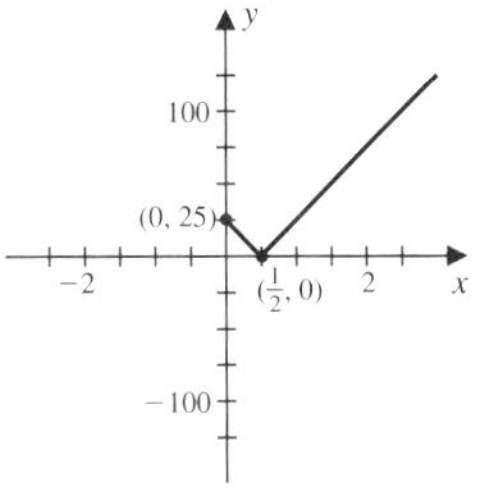

b. The line on the interval $\left[0, \frac{1}{2}\right]$ passes through the points (0, 25) and $\left(\frac{1}{2}, 0\right)$ so the slope is $m = \frac{25}{-\frac{1}{2}} = -50$. The equation of the line is $y = -50x + 25$. On the interval $\left[\frac{1}{2}, 2\right]$ the line passes through $\left(\frac{1}{2}, 0\right)$ and (2, 75) so the slope is $m = \frac{75}{\frac{3}{2}} = 50$. The equation of the line is $y = 50\left(x - \frac{1}{2}\right) = 50x - 25$. The distance can then be expressed as $|50x - 25|$.

37. a. Domain$(f) = [0, 13]$; Range$(f) = \{.37, .60, .83, .., 3.36\} = \{.37 + 0.23n, n = 0, 1, ..., 13\}$

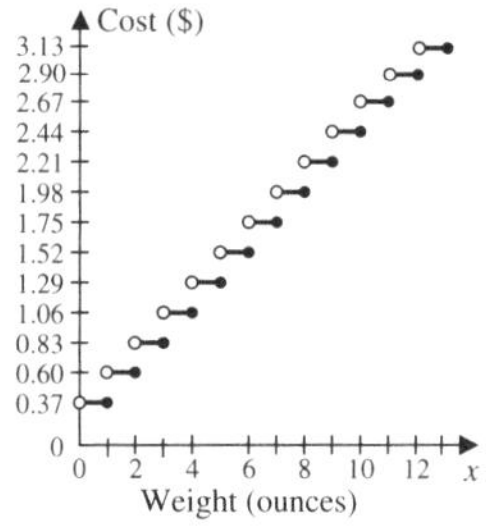

b. We have

$$P(w) = \begin{cases} 0.37 + 0.23\lfloor w \rfloor, & \text{if } w \text{ is not a positive integer} \\ 0.37 + 0.23(w-1) = 0.14 + 0.23w, & \text{if } w \text{ is a positive integer} \end{cases}$$

Also,

$$0.14 - 0.23\lfloor -w \rfloor = \begin{cases} 0.14 - 0.23(-\lfloor w \rfloor - 1), & \text{if } w \text{ is not a positive integer} \\ 0.14 - 0.23(-\lfloor w \rfloor), & \text{if } w \text{ is a positive integer} \end{cases}$$

$$= \begin{cases} 0.37 + 0.23\lfloor w \rfloor, & \text{if } w \text{ is not a positive integer} \\ 0.37 + 0.23\lfloor w \rfloor, & \text{if } w \text{ is a positive integer} \end{cases}$$

so, in any case, $P(w) = 0.14 - 0.23\lfloor -w \rfloor$.

38. $$f(x) = \begin{cases} -x + 75, & \text{if } 0 \le x \le 75, \\ x - 75, & \text{if } 75 \le x \le 117.5, \\ -x + 160, & \text{if } 117.5 \le x \le 160, \\ x - 160, & \text{if } 160 \le x \le 241. \end{cases} = \begin{cases} |x - 75|, & \text{if } 0 \le x \le 117.5, \\ |x - 160|, & \text{if } 117.5 \le x \le 214. \end{cases}$$

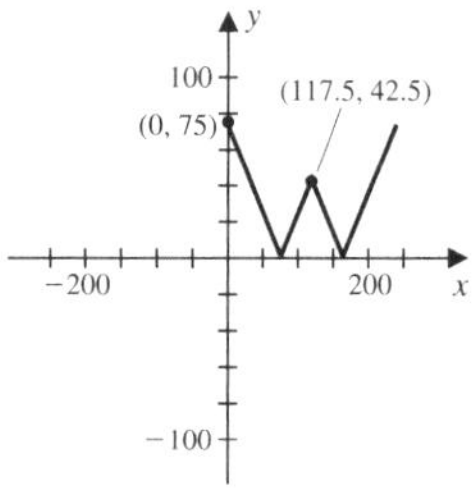

39. **a.** For approximating $\sqrt{2}$ start with $a = 1.4$. The table gives four iterations.

a	$b = \frac{1}{2}\left(a + \frac{2}{a}\right)$
1.4	1.414285714
1.414285714	1.414213564
1.414213564	1.414213562
1.414213562	1.414213562

So $\sqrt{2} \approx 1.414313562 \approx 1.414$.

b. For approximating $\sqrt{13}$ start with $a = 3.5$. The table gives four iterations.

a	$b = \frac{1}{2}\left(a + \frac{13}{a}\right)$
3.5	3.607142857
3.607142857	3.605551627
3.605551627	3.605551276
3.605551276	3.605551276

So $\sqrt{3} \approx 3.605551276 \approx 3.606$.

Notice that the number of decimal places of accuracy approximately doubles with each iteration of the technique. This provides an extremely efficient technique for determining square roots.

40. Let n be the unique integer satisfying $n \le x < n + 1$. Then $n + 1 \le x + 1 < n + 2$ and $n + 3 \le x + 3 < n + 4$, so $\lfloor x + 3 \rfloor - 2 = n + 3 - 2 = n + 1 = \lfloor x + 1 \rfloor$.

Exercise Set 2.3 (Page 95)

1. For $f(x) = x$ and $g(x) = 3x - 2$ we have

$(f + g)(x) = f(x) + g(x) = x + (3x - 2) = 4x - 2;$

$(f - g)(x) = f(x) - g(x) = x - (3x - 2) = x - 3x + 2 = -2x + 2;$

$(f \cdot g)(x) = f(x)g(x) = x(3x-2) = 3x^2 - 2x$;

$(f/g)(x) = \frac{f(x)}{g(x)} = \frac{x}{3x-2}$.

The domain of $f+g$, $f-g$, and $f \cdot g$ is Domain(f) $\cap$ Domain(g) $= (-\infty, \infty) \cap (-\infty, \infty) = (-\infty, \infty)$.

The domain of f/g is Domain(f) $\cap$ Domain(g) excluding any values that make the denominator 0. The denominator is 0 when $3x-2=0$ implies $x = \frac{2}{3}$, so Domain$(f/g) = \left(-\infty, \frac{2}{3}\right) \cup \left(\frac{2}{3}, \infty\right)$.

2. For $f(x) = x^2$ and $g(x) = x+1$ we have

$(f+g)(x) = f(x) + g(x) = x^2 + x + 1$;

$(f-g)(x) = f(x) - g(x) = x^2 - (x+1) = x^2 - x - 1$;

$(f \cdot g)(x) = f(x)g(x) = x^2(x+1) = x^3 + x^2$;

$(f/g)(x) = \frac{f(x)}{g(x)} = \frac{x^2}{x+1}$.

The domain of $f+g$, $f-g$, and $f \cdot g$ is
Domain(f) $\cap$ Domain(g) $= (-\infty, \infty) \cap (-\infty, \infty) = (-\infty, \infty)$.

The domain of f/g is Domain(f) $\cap$ Domain(g) excluding any values that make the denominator 0. The denominator is 0 when $x+1=0$ implies $x=-1$, so $D(f/g) = (-\infty, -1) \cup (-1, \infty)$.

3. For $f(x) = \frac{1}{x}$ and $g(x) = \frac{x}{x-2}$ we have

$(f+g)(x) = f(x) + g(x) = \frac{1}{x} + \frac{x}{x-2} = \frac{(x-2)+x^2}{x(x-2)} = \frac{x^2+x-2}{x^2-2x}$;

$(f-g)(x) = f(x) - g(x) = \frac{1}{x} - \frac{x}{x-2} = \frac{(x-2)-x^2}{x(x-2)} = \frac{-x^2+x-2}{x^2-x}$;

$(f \cdot g)(x) = f(x)g(x) = \frac{x}{x(x-2)} = \frac{1}{x-2}$;

$(f/g)(x) = \frac{f(x)}{g(x)} = \frac{\frac{1}{x}}{\frac{x}{x-2}} = \frac{1}{x} \cdot \frac{x-2}{x} = \frac{x-2}{x^2}$.

Domain of $f : x \neq 0$; Domain of $g : x - 2 \neq 0$ implies $x \neq 2$

So Domain(f) $\cap$ Domain(g) $= \{x \mid x \neq 0 \text{ and } x \neq 2\} = (-\infty, 0) \cup (0, 2) \cup (2, \infty)$.

The domain of $f+g$, $f-g$, and $f \cdot g$ is Domain(f) $\cap$ Domain(g) $= (-\infty, 0) \cup (0, 2) \cup (2, \infty)$.

The domain of f/g is Domain(f) $\cap$ Domain(g) excluding any values that make the denominator 0, which is $x=0$, which has already been excluded. So
Domain$(f/g) = (-\infty, 0) \cup (0, 2) \cup (2, \infty)$.

4. For $f(x) = \frac{1}{x-1}$ and $g(x) = \frac{1}{x+1}$ we have

$(f+g)(x) = f(x) + g(x) = \frac{1}{x-1} + \frac{1}{x+1} = \frac{(x+1)+(x-1)}{(x-1)(x+1)} = \frac{2x}{x^2-1}$;

$(f-g)(x) = f(x) - g(x) = \frac{1}{x-1} - \frac{1}{x+1} = \frac{(x+1)-(x-1)}{(x-1)(x+1)} = \frac{2}{x^2-1}$;

$(f \cdot g)(x) = f(x)g(x) = \frac{1}{x-1} \cdot \frac{1}{x+1} = \frac{1}{x^2-1}$;

$(f/g)(x) = \frac{f(x)}{g(x)} = \frac{\frac{1}{x-1}}{\frac{1}{x+1}} = \frac{1}{x-1} \cdot \frac{x+1}{1} = \frac{x+1}{x-1}$.

The domain of f is all real numbers so the denominator is not 0, that is, all $x \neq 1$, and the domain of g is all x so the denominator is not 0, so is all $x \neq -1$. Then
Domain(f) $\cap$ Domain(g) $= \{x \mid x \neq 1 \text{ and } x \neq -1\} = (-\infty, -1) \cup (-1, 1) \cup (1, \infty)$.

The domain of $f + g$, $f - g$, and $f \cdot g$ is
Domain(f) $\cap$ Domain(g) $= (-\infty, -1) \cup (-1, 1) \cup (1, \infty)$.

The domain of f/g is Domain(f) $\cap$ Domain(g) excluding any values that make the denominator 0. The denominator is 0 when $x = 1$, which has already been excluded from the intersection, so
Domain(f/g) $= (-\infty, -1) \cup (-1, 1) \cup (1, \infty)$.

5. For $f(x) = \sqrt{x+2}$ and $g(x) = \sqrt{2-x}$ we have

$(f + g)(x) = f(x) + g(x) = \sqrt{x+2} + \sqrt{2-x}$;

$(f - g)(x) = f(x) - g(x) = \sqrt{x+2} - \sqrt{2-x}$;

$(f \cdot g)(x) = f(x)g(x) = \sqrt{x+2}\sqrt{2-x} = \sqrt{(x+2)(2-x)} = \sqrt{4-x^2}$;

$(f/g)(x) = \frac{f(x)}{g(x)} = \frac{\sqrt{x+2}}{\sqrt{2-x}}$.

Domain of $f : x + 2 \geq 0$ implies $x \geq -2$; Domain of $g : 2 - x \geq 0$ implies $x \leq 2$

So Domain(f) $\cap$ Domain(g) $= [-2, 2]$.

The domain of $f + g$, $f - g$, and $f \cdot g$ is Domain(f) $\cap$ Domain(g) $= [-2, 2]$.

The domain of f/g is Domain(f) $\cap$ Domain(g) , excluding any values that make the denominator 0, which is $x = 2$, so Domain(f/g) $= [-2, 2)$.

6. For $f(x) = \frac{1}{x}$ and $g(x) = \sqrt{x-1}$ we have

$(f + g)(x) = f(x) + g(x) = \frac{1}{x} + \sqrt{x-1} = \frac{1+x\sqrt{x-1}}{x}$;

$(f - g)(x) = f(x) - g(x) = \frac{1}{x} - \sqrt{x-1} = \frac{1-x\sqrt{x-1}}{x}$;

$(f \cdot g)(x) = f(x)g(x) = \frac{1}{x} \cdot \sqrt{x-1} = \frac{\sqrt{x-1}}{x}$;

$(f/g)(x) = \frac{f(x)}{g(x)} = \frac{\frac{1}{x}}{\sqrt{x-1}} = \frac{1}{x\sqrt{x-1}}$.

The domain of f is all real numbers so the denominator is not 0, that is, all $x \neq 0$. The domain of g is all x so the radical is defined, that is, all x satisfying $x - 1 \geq 0$ implies $x \geq 1$. Since 0 is not in the domain of g, Domain(f) $\cap$ Domain(g) $=$ Domain(g) $= [1, \infty)$.

The domain of $f + g$, $f - g$, and $f \cdot g$ is Domain(f) $\cap$ Domain(g) $= [1, \infty)$.

The domain of f/g is Domain(f) $\cap$ Domain(g) excluding any values that make the denominator 0. The denominator is 0 when $x = 0$, or $x - 1 = 0$ implies $x = 0$, or $x = 1$, so
Domain(f/g) $= (1, \infty)$.

7. For $f(x) = \begin{cases} -1, & \text{if } x < 0 \\ 1, & \text{if } x \geq 0 \end{cases}$ and $g(x) = \begin{cases} 1, & \text{if } x < 0 \\ 0, & \text{if } x \geq 0 \end{cases}$ we perform the operations separately for $x < 0$ and for $x \geq 0$.

$(f + g)(x) = f(x) + g(x) = \begin{cases} 0, & \text{if } x < 0 \\ 1, & \text{if } x \geq 0 \end{cases}$;

$(f - g)(x) = f(x) - g(x) = \begin{cases} -2, & \text{if } x < 0 \\ 1, & \text{if } x \geq 0 \end{cases}$;

$(f \cdot g)(x) = f(x)g(x) = \begin{cases} -1, & \text{if } x < 0 \\ 0, & \text{if } x \geq 0 \end{cases}$;

The quotient is defined only for $x < 0$, since $g(x) = 0$ for $x \geq 0$. So $(f/g)(x) = -1$, for $x < 0$.

The domain of $f + g$, $f - g$, $f \cdot g$ is $(-\infty, \infty)$ and the domain of f/g is $(-\infty, 0)$.

8. For $f(x) = \begin{cases} 0, & \text{if } x < 0 \\ x, & \text{if } x \geq 0 \end{cases}$ and $g(x) = \begin{cases} -x, & \text{if } x < 0 \\ 0, & \text{if } x \geq 0 \end{cases}$ we perform the operations separately for $x < 0$ and for $x \geq 0$.

$(f + g)(x) = f(x) + g(x) = \begin{cases} -x, & \text{if } x < 0 \\ x, & \text{if } x \geq 0 \end{cases} = |x|$;

$(f - g)(x) = f(x) - g(x) = \begin{cases} x, & \text{if } x < 0 \\ x, & \text{if } x \geq 0 \end{cases} = x$;

$(f \cdot g)(x) = f(x)g(x) = \begin{cases} 0, & \text{if } x < 0 \\ 0, & \text{if } x \geq 0 \end{cases} = 0$;

The quotient is defined only for $x < 0$ since $g(x) = 0$ for $x \geq 0$. So $(f/g)(x) = 0$ for $x < 0$.

The domain of $f + g$, $f - g$, $f \cdot g$ is $(-\infty, \infty)$ and the domain of f/g is $(-\infty, 0)$.

9. $g(x) = x - 1$; $h(x) = \frac{1}{x-1}$

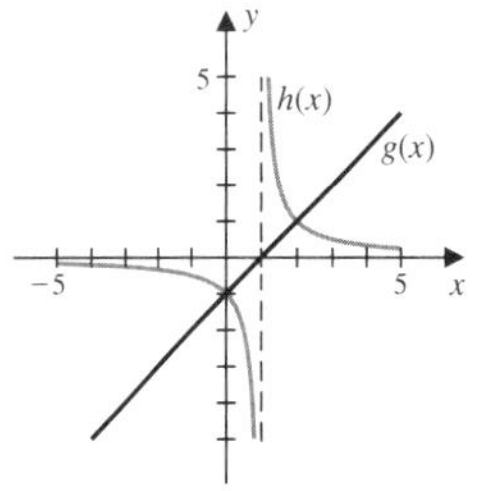

10. $g(x) = 2 - x$; $h(x) = \frac{1}{2-x}$

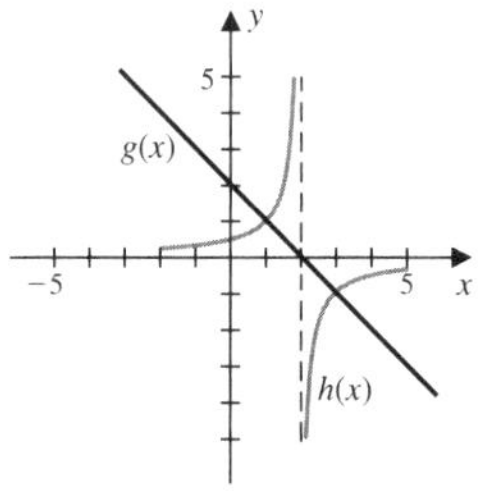

11. $g(x) = |x|; h(x) = \frac{1}{|x|}$

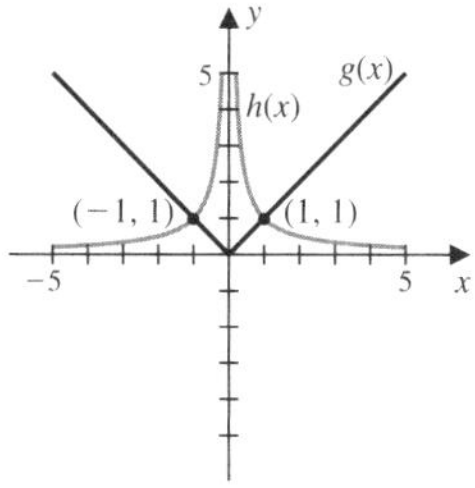

12. $g(x) = |x - 1|; h(x) = \frac{1}{|x-1|}$

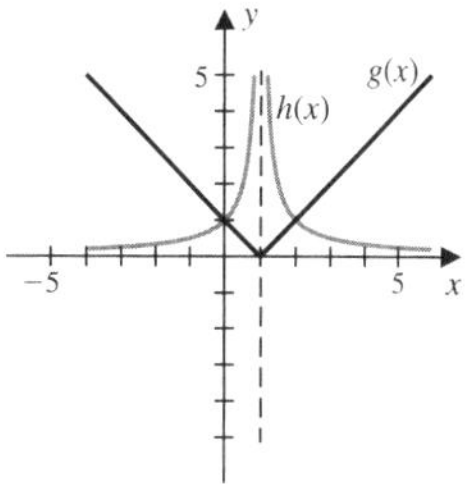

13. $g(x) = x^2 - 1; h(x) = \frac{1}{x^2-1}$

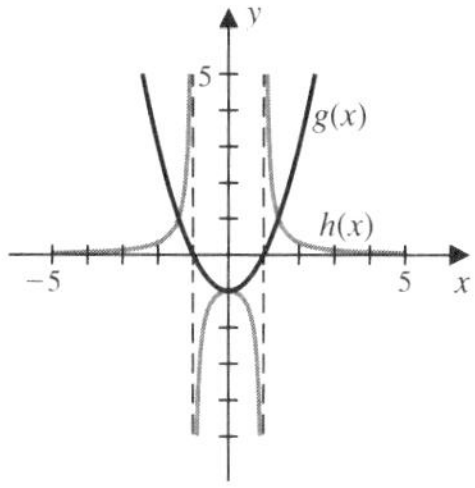

14. $g(x) = x^2 + 1; h(x) = \frac{1}{x^2+1}$

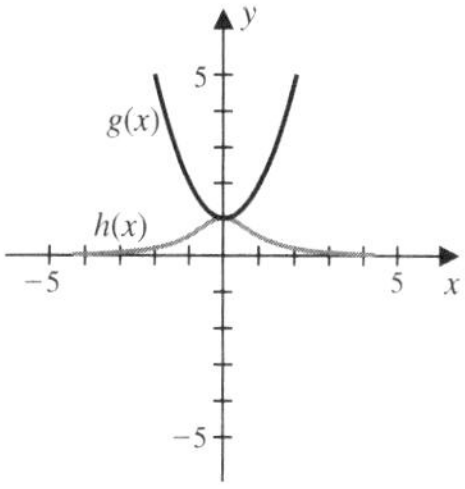

15.

$$\begin{aligned} g(x) =& x^2 - 4x + 3 \\ =& x^2 - 4x + 4 - 4 + 3 \\ =& (x - 2)^2 - 1 \end{aligned}$$

and

$$h(x) = \frac{1}{(x - 2)^2 - 1}$$

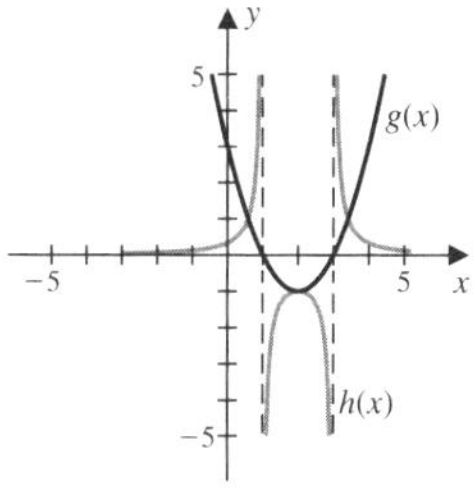

16.

$$\begin{aligned} g(x) =& x^2 - x - 2 \\ =& x^2 - x + \frac{1}{4} - \frac{1}{4} - 2 \\ =& \left(x - \frac{1}{2}\right)^2 - \frac{9}{4} \end{aligned}$$

and

$$h(x) = \frac{1}{\left(x - \frac{1}{2}\right)^2 - \frac{9}{4}}$$

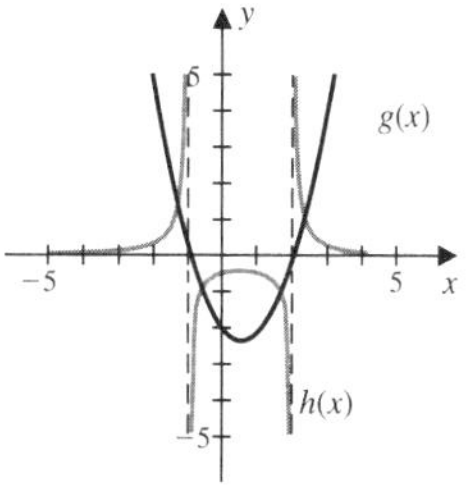

17. The figure shows the graphs of $f + g$ and $f - g$.

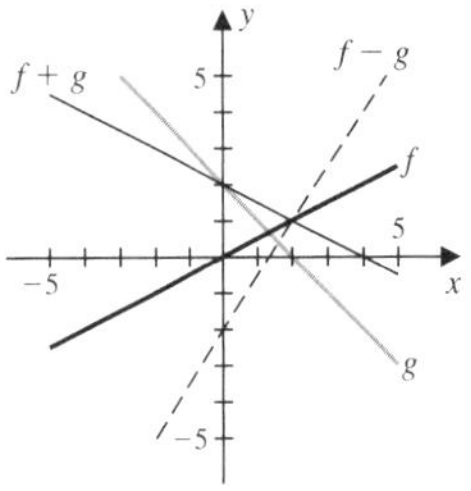

18. The figure shows the graphs of $f + g$ and $f - g$.

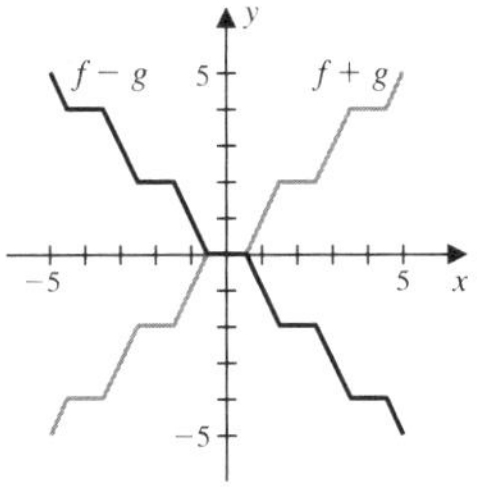

19. The figure shows the graphs of $f + g$ and $f - g$.

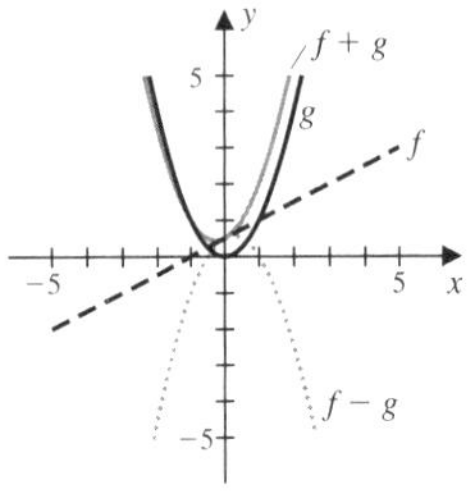

20. The figure shows the graphs of $f + g$ and $f - g$.

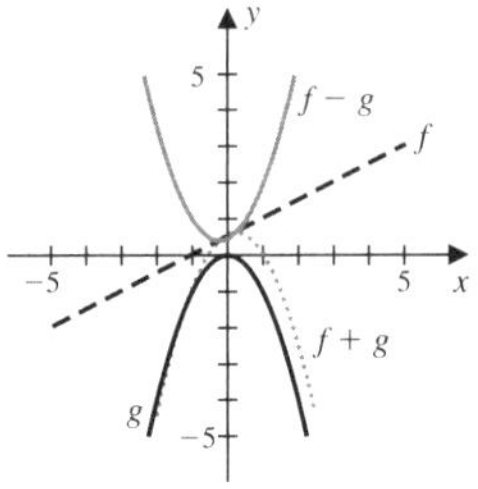

21. We have $f(x) = \frac{x^2-4}{x-2} = \frac{(x-2)(x+2)}{x-2} = x + 2$, for $x \neq 2$. The denominator of the original expression is 0 at $x = 2$, making the quotient undefined. To graph $y = f(x) = \frac{x^2-4}{x-2}$, simply graph $y = x + 2$ and remove the point with x-coordinate 2. That is, remove the point $(2, 4)$. So the graphs of $f(x) = \frac{x^2-4}{x-2}$ and $g(x) = x + 2$ are identical except for the point $(2, 4)$ removed from the graph of $y = f(x)$.

22. We have $f(x) = \frac{x^3-2x^2}{x} = \frac{x^2(x-2)}{x} = x(x - 2)$, for $x \neq 0$. The denominator of the original expression is 0 at $x = 0$, making the quotient undefined. To graph $y = f(x) = \frac{x^3-2x^2}{x}$, simply graph $y = x^2 - 2x$ and remove the point with x-coordinate 0. That is, remove the point $(0, 0)$. So the graphs of $f(x) = \frac{x^3-2x^2}{x}$ and $g(x) = x^2 - 2x$ are identical except for the point $(0, 0)$ removed from the graph of $y = f(x)$.

23. The figure shows the graphs of $y = \sqrt{x-1}$, $y = \sqrt{x+2}$, and $y = \sqrt{x-1} + \sqrt{x+2}$.

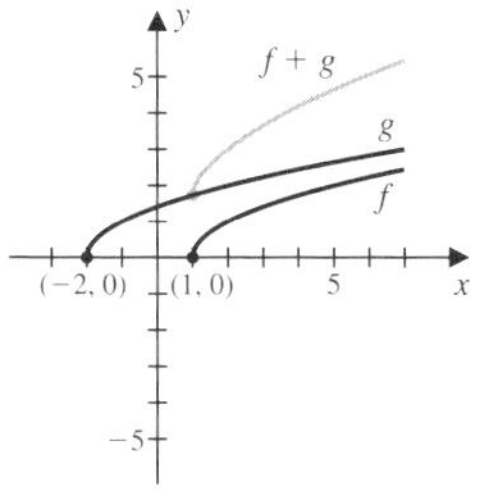

24. The figure shows the graphs of $y = x^2 - 1$, $y = x^3$, and $y = x^3 + x^2 - 1$.

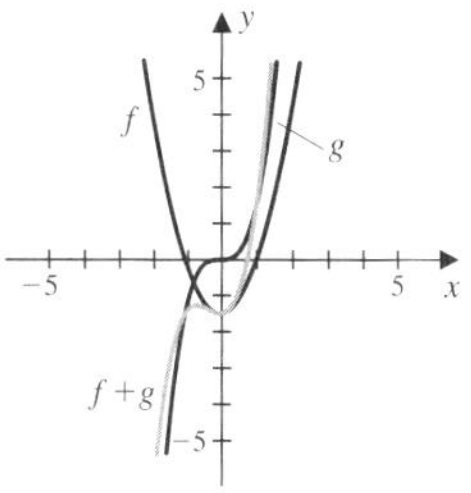

25. The figure shows the graphs of $y = x^3 - 2x^2 - x + 2$, $y = x^3 - 7x + 6$, and $y = 2x^3 - 2x^2 - 8x + 8$.

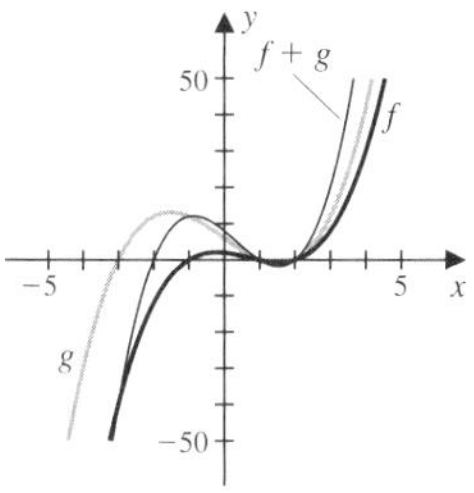

26. The figure shows the graphs of $y = -x^3 - x^2 + 2x$, $y = x^3 - x^2 - 2x$, and $y = -2x^2$.

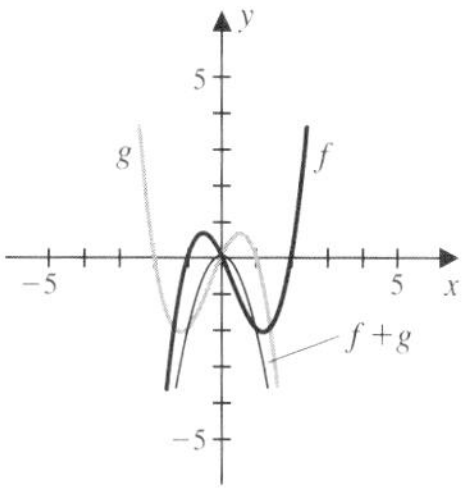

Exercise Set 2.4 (Page 104)

1. For $f(x) = 2x - 3$ and $g(x) = x^2 + 2$ we have

$$(f \circ g)(2) = f(g(2)) = f(2^2 + 2) = f(6) = 2(6) - 3 = 9.$$

2. For $f(x) = 2x - 3$ and $g(x) = x^2 + 2$ we have

$$(g \circ f)(-1) = g(f(-1)) = g(2(-1) - 3) = g(-5) = (-5)^2 + 2 = 27.$$

3. For $f(x) = 2x - 3$ and $g(x) = x^2 + 2$ we have

$$(f \circ g)(-3) = f(g(-3)) = f((-3)^2 + 2) = f(11) = 2(11) - 3 = 19.$$

4. For $f(x) = 2x - 3$ and $g(x) = x^2 + 2$ we have

$$(g \circ f)(5) = g(f(5)) = g(2(5) - 3) = g(7) = (7)^2 + 2 = 51.$$

5. For $f(x) = 2x - 3$ and $g(x) = x^2 + 2$ we have

$$(f \circ f)(-2) = f(f(-2)) = f(2(-2) - 3) = f(-7) = 2(-7) - 3 = -17.$$

6. For $f(x) = 2x - 3$ and $g(x) = x^2 + 2$ we have

$$(g \circ g)(-1/2) = g(g(-1/2)) = g\left(\left(-\frac{1}{2}\right)^2 + 2\right) = g\left(\frac{9}{4}\right) = \left(\frac{9}{4}\right)^2 + 2 = \frac{113}{16}.$$

7. For $f(x) = 2x + 1$ and $g(x) = 3x - 1$ we have

$$(f \circ g)(x) = f(g(x)) = f(3x - 1) = 2(3x - 1) + 1 = 6x - 1;$$
$$(g \circ f)(x) = g(f(x)) = g(2x + 1) = 3(2x + 1) - 1 = 6x + 2.$$

Since the domain of f and g is the set of all real numbers, the domain of each of the compositions is also the set of all real numbers.

8. For $f(x) = x^2 + 1$ and $g(x) = x - 1$ we have

$$(f \circ g)(x) = f(g(x)) = f(x - 1) = (x - 1)^2 + 1 = x^2 - 2x + 2;$$
$$(g \circ f)(x) = g(f(x)) = g(x^2 + 1) = x^2 + 1 - 1 = x^2.$$

Since the domain of f and g is the set of all real numbers, the domain of each of the compositions is also the set of all real numbers.

9. For $f(x) = \frac{1}{x}$ and $g(x) = x^2 + 2x$ we have

$$(f \circ g)(x) = f(g(x)) = f(x^2 + 2x) = \frac{1}{x^2 + 2x}$$

and

$$(g \circ f)(x) = g(f(x)) = g\left(\frac{1}{x}\right) = \left(\frac{1}{x}\right)^2 + 2\left(\frac{1}{x}\right) = \frac{1}{x^2} + \frac{2}{x} = \frac{1 + 2x}{x^2}.$$

Domain $f : \{x \mid x \neq 0\} = (-\infty, 0) \cup (0, \infty)$; Domain $g : (-\infty, \infty)$; Domain $f \circ g : \{x \mid x^2 + 2x = x(x + 2) \neq 0\} = \{x \mid x \neq 0, x \neq -2\} = (-\infty, -2) \cup (-2, 0) \cup (0, \infty)$; Domain $g \circ f : \{x \neq 0 \mid 1/x \text{ is defined }\} = \{x \mid x \neq 0\} = (-\infty, 0) \cup (0, \infty)$.

10. For $f(x) = \frac{3}{x-6}$ and $g(x) = x^2 - x$ we have

$$(f \circ g)(x) = f(g(x)) = f(x^2 - x) = \frac{3}{x^2 - x - 6};$$

$$\begin{aligned}(g \circ f)(x) = g(f(x)) = g\left(\frac{3}{x-6}\right) &= \left(\frac{3}{x-6}\right)^2 - \frac{3}{x-6} \\ &= \frac{9}{(x-6)^2} - \frac{3}{x-6} = \frac{9 - 3(x-6)}{(x-6)^2} \\ &= \frac{27 - 3x}{(x-6)^2}.\end{aligned}$$

Domain $f : \{x \mid x \neq 6\} = (-\infty, 6) \cup (6, \infty)$; Domain $g : (-\infty, \infty)$;
Domain $f \circ g$:
$\{x \mid x^2 - x \neq 6\} = \{x \mid (x-3)(x+2) \neq 0\} = (-\infty, -2) \cup (-2, 3) \cup (3, \infty)$;
Domain $g \circ f : \{x \neq 6 \mid f(x) \text{ is defined}\} = (-\infty, 6) \cup (6, \infty)$.

11. For $f(x) = \sqrt{x-1}$ and $g(x) = x^2 - 3$ we have

$$(f \circ g)(x) = f(g(x)) = f(x^2 - 3) = \sqrt{x^2 - 3 - 1} = \sqrt{x^2 - 4};$$

$$(g \circ f)(x) = g(f(x)) = f(\sqrt{x-1}) = (\sqrt{x-1})^2 - 3 = x - 4.$$

Domain $f : \{x \mid x - 1 \geq 0\} = [1, \infty)$; Domain $g : (-\infty, \infty)$;
Domain $f \circ g$:
$\{x \mid x^2 - 3 \geq 1\} = \{x \mid (x-2)(x+2) \geq 0\} = (-\infty, -2] \cup [2, \infty)$;
Domain $g \circ f : \{x \mid x \geq 1 \text{ and } f(x) \text{ is defined}\} = [1, \infty)$.

12. For $f(x) = \sqrt{x-9}$ and $g(x) = x^2$ we have

$$(f \circ g)(x) = f(g(x)) = f(x^2) = \sqrt{x^2 - 9};$$

$$(g \circ f)(x) = g(f(x)) = g(\sqrt{x-9}) = (\sqrt{x-9})^2 = x - 9.$$

Domain $f : \{x \mid x - 9 \geq 0\} = [9, \infty)$; Domain $g : (-\infty, \infty)$;
Domain $f \circ g : \{x \mid x^2 \geq 9\} = \{x \mid (x-3)(x+3) \geq 0\} = (-\infty, -3] \cup [3, \infty)$;
Domain $g \circ f : \{x \mid x \geq 9 \text{ and } f(x) \text{ is defined}\} = [9, \infty)$.

13. For $f(x) = \frac{1}{x}$ and $g(x) = \frac{1}{x+1}$ we have

$$(f \circ g)(x) = f(g(x)) = f\left(\frac{1}{x+1}\right) = \frac{1}{\frac{1}{x+1}} = x + 1;$$

$$(g \circ f)(x) = g(f(x)) = g\left(\frac{1}{x}\right) = \frac{1}{\frac{1}{x} + 1} = \frac{1}{\frac{1+x}{x}} = \frac{x}{x+1}.$$

Domain $f : \{x \mid x \neq 0\} = (-\infty, 0) \cup (0, \infty)$;
Domain $g : \{x \mid x + 1 \neq 0\} = (-\infty, -1) \cup (-1, \infty)$;
Domain $f \circ g : \{x \neq -1 \mid \frac{1}{x+1} \neq 0\} = (-\infty, -1) \cup (-1, \infty)$;
Domain $g \circ f : \{x \neq 0 \mid \frac{1}{x} \neq -1\} = (-\infty, -1) \cup (-1, 0) \cup (0, \infty)$.

14. For $f(x) = \frac{x}{x-1}$ and $g(x) = \frac{x-1}{x+1}$ we have

$$(f \circ g)(x) = f(g(x)) = f\left(\frac{x-1}{x+1}\right) = \frac{\frac{x-1}{x+1}}{\frac{x-1}{x+1} - 1} = \frac{\frac{x-1}{x+1}}{\frac{x-1-(x+1)}{x+1}} = \frac{\frac{x-1}{x+1}}{\frac{-2}{x+1}} = \frac{x-1}{-2} = -\frac{1}{2}x + \frac{1}{2};$$

$$(g \circ f)(x) = g(f(x)) = g\left(\frac{x}{x-1}\right) = \frac{\frac{x}{x-1} - 1}{\frac{x}{x-1} + 1} = \frac{\frac{x-(x-1)}{x-1}}{\frac{x+(x-1)}{x-1}} = \frac{\frac{1}{x-1}}{\frac{2x-1}{x-1}} = \frac{1}{2x-1}.$$

Domain $f : \{x \mid x - 1 \neq 0\} = (-\infty, 1) \cup (1, \infty)$;
Domain $g : \{x \mid x + 1 \neq 0\} = (-\infty, -1) \cup (-1, \infty)$;
Domain $f \circ g : \{x \neq -1 \mid \frac{x-1}{x+1} \neq 1\} = (-\infty, -1) \cup (-1, \infty)$;
Domain
$g \circ f : \left\{x \neq 1 \mid \frac{x}{x-1} \neq -1\right\} = \{x \neq 1 : 2x \neq 1\} = \left\{x \neq 1 \mid x \neq \frac{1}{2}\right\} = \left(-\infty, \frac{1}{2}\right) \cup \left(\frac{1}{2}, 1\right) \cup (1, \infty)$.

15. The inside operation of $h(x) = \left(3x^2 - 2\right)^4$ is $3x^2 - 2$ and the outside operation is raising to the fourth power, x^4. So define $g(x) = 3x^2 - 2$ and $f(x) = x^4$. This gives

$$(f \circ g)(x) = f(g(x)) = f(3x^2 - 2) = (3x^2 - 2)^4 = h(x).$$

16. The inside operation of $h(x) = \left(x^2 - x + 1\right)^8$ is $x^2 - x + 1$ and the outside operation is raising to the eighth power, x^8. So define $g(x) = x^2 - x + 1$ and $f(x) = x^8$. This gives

$$(f \circ g)(x) = f(g(x)) = f(x^2 - x + 1) = (x^2 - x + 1)^8 = h(x).$$

17. The inside operation of $h(x) = \sqrt{x-2}$ is $x - 2$ and the outside operation is the square root of a number, $\sqrt{x}$. So define $g(x) = x - 2$ and $f(x) = \sqrt{x}$. This gives

$$(f \circ g)(x) = f(g(x)) = f(x - 2) = \sqrt{x - 2} = h(x).$$

18. The inside operation of $h(x) = \sqrt[3]{x + \sqrt{x}}$ is $x + \sqrt{x}$ and the outside operation is taking the cube root, $\sqrt[3]{x}$. So define $g(x) = x + \sqrt{x}$ and $f(x) = \sqrt[3]{x}$. This gives

$$(f \circ g)(x) = f(g(x)) = f(x + \sqrt{x}) = \sqrt[3]{x + \sqrt{x}} = h(x).$$

19. Define the first operation to be $x + 2$ and then take the reciprocal, so let $g(x) = x + 2$ and $f(x) = \frac{1}{x}$. Then

$$(f \circ g)(x) = f(g(x)) = f(x+2) = \frac{1}{x+2} = h(x).$$

20. The inside operation of $h(x) = \left|x^2 + x + 1\right|$ is $x^2 + x + 1$ and the outside operation is the absolute value of a number, $|x|$. So define $g(x) = x^2 + x + 1$ and $f(x) = |x|$. This gives

$$(f \circ g)(x) = f(g(x)) = f(x^2 + x + 1) = \left|x^2 + x + 1\right| = h(x).$$

21. **a.** (i) $(f \circ g)(-2) = f(g(-2)) = f(1) = 2$ (ii) $(g \circ f)(-2) = g(f(-2)) = g(0) = 0$
(iii) $(g \circ f)(2) = g(f(2)) = g(0) = 0$ (iv) $(f \circ g)(2) = f(g(2)) = f(-1) = -2$

b. (i) $(f \circ g)(x) = f(g(x)) = 0$ whenever $g(x) = -2, 0, 2,$ that is, $x = -4, 0, 4$
(ii) $(g \circ f)(x) = g(f(x)) = 0$ whenever $f(x) = 0,$ that is, $x = -2, 0, 2$

22. The completed table assumes f is even and g is odd.

x	$f(x)$	$g(x)$	$(f \circ g)(x)$	$(g \circ f)(x)$
-3	2	2	-1	1
-2	-1	-1	3	3
-1	3	3	2	-2
0	0	0	0	0
1	3	-3	2	-2
2	-1	1	3	3
3	2	-2	-1	1

23. The figure shows the graphs of $y = f(2x)$, $y = f(-2x)$, $y = f(\frac{1}{2}x)$, and $y = f(-\frac{1}{2}x)$.

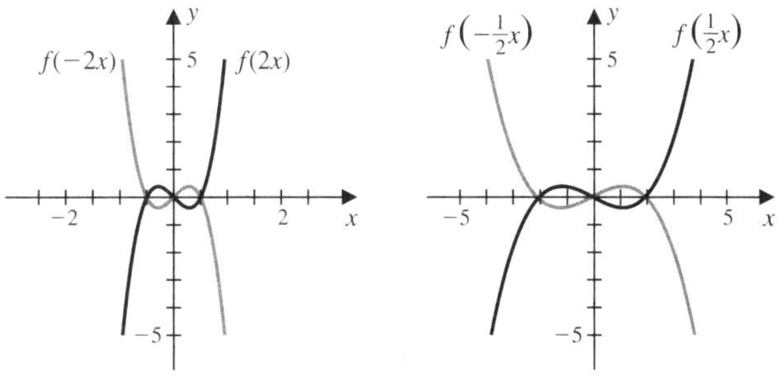

24. The figure shows the graphs of $y = f(2x)$, $y = f(-2x)$, $y = f(\frac{1}{2}x)$, and $y = f(-\frac{1}{2}x)$.

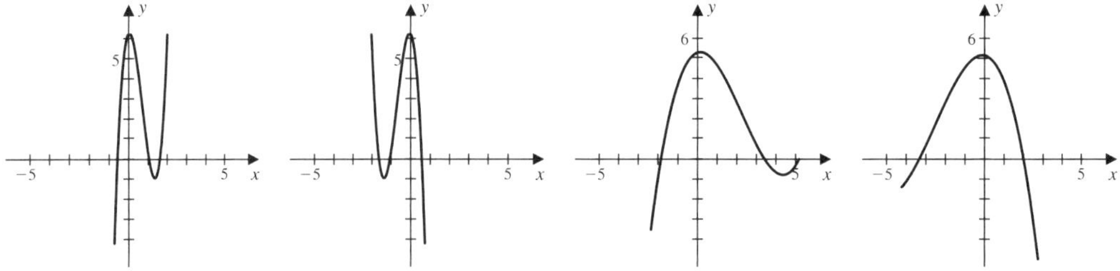

25. **a.** $y = x - 1$ **b.** $y = (x - 1)^2$ **c.** $y = x^2 - 2x = (x - 1)^2 - 1$ **d.** $y = |x^2 - 2x|$
e. $y = \frac{1}{|x^2 - 2x|}$ **f.** $y = -\frac{1}{|x^2 - 2x|}$

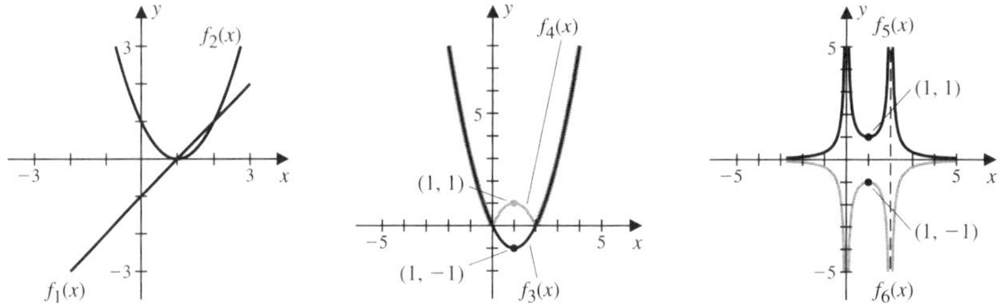

26. **a.** $y = x - 2$ **b.** $y = (x - 2)^2$ **c.** $y = x^2 - 4x + 2 = (x - 2)^2 - 2$ **d.** $y = |x^2 - 4x + 2|$
e. $y = \frac{1}{|x^2 - 4x + 2|}$ **f.** $y = \frac{2}{|x^2 - 4x + 2|}$

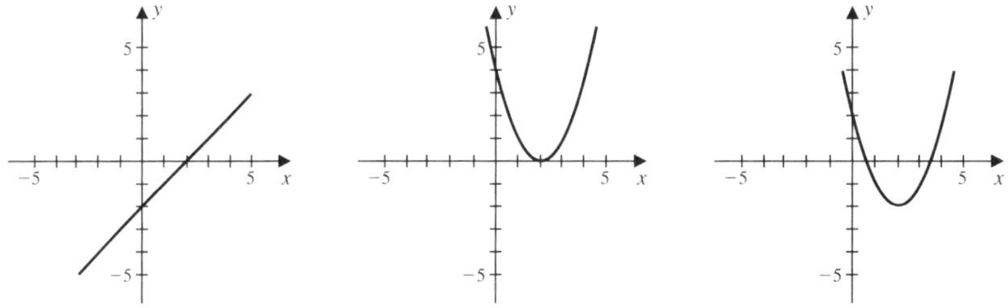

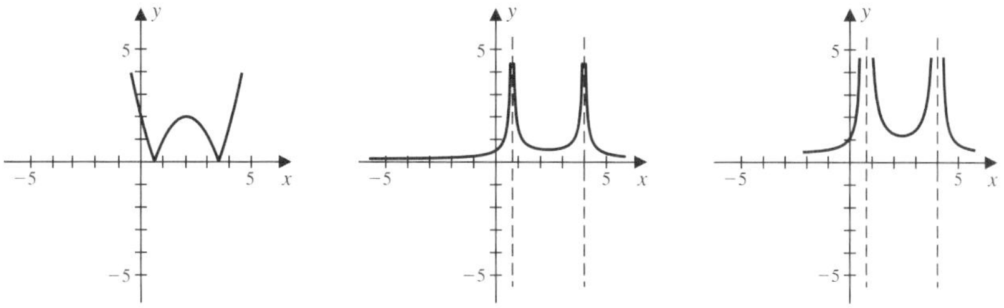

27. a. For $f(x) = \frac{1}{|x^2+2x-1|}$, completing the square on the quadratic in the denominator gives

$$f(x) = \frac{1}{|x^2+2x+1-1-1|} = \frac{1}{|(x+1)^2-2|}.$$

Then letting $g_1(x) = (x+1)^2$, $g_2(x) = x-2$, $g_3(x) = |x|$ and $g_4(x) = \frac{1}{x}$ we have

$$\begin{aligned} g_4(g_3(g_2(g_1(x)))) &= g_4(g_3(g_2((x+1)^2))) = g_4(g_3((x+1)^2-2)) \\ &= g_4\left(\left|(x+1)^2-2\right|\right) = \frac{1}{|(x+1)^2-2|} = f(x). \end{aligned}$$

b. The graphs of $y = g_1(x)$, $y = g_2(g_1(x))$, $y = g_3(g_2(g_1(x)))$, and $y = f(x)$ are shown in the figures.

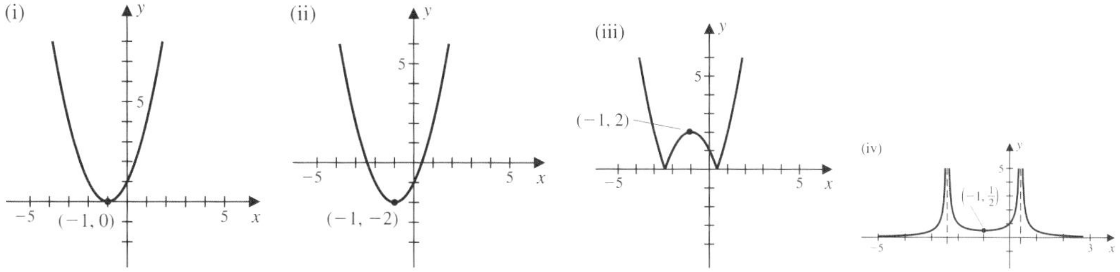

28. a. For $f(x) = \frac{1}{|x^2-6x+10|}$, completing the square on the quadratic in the denominator gives

$$f(x) = \frac{1}{|x^2-6x+9-9+10|} = \frac{1}{|(x-3)^2+1|}.$$

Then letting $g_1(x) = (x-3)^2$, $g_2(x) = x+1$, $g_3(x) = |x|$ and $g_4(x) = \frac{1}{x}$ we have

$$\begin{aligned} g_4(g_3(g_2(g_1(x)))) &= g_4(g_3(g_2((x-3)^2))) = g_4(g_3((x-3)^2+1)) \\ &= g_4\left(\left|(x-3)^2+1\right|\right) = \frac{1}{|(x-3)^2+1|} = f(x). \end{aligned}$$

b. The graphs of $y = g_1(x)$, $y = g_2(g_1(x))$, $y = g_3(g_2(g_1(x)))$, and $y = f(x)$ are shown in the figures.

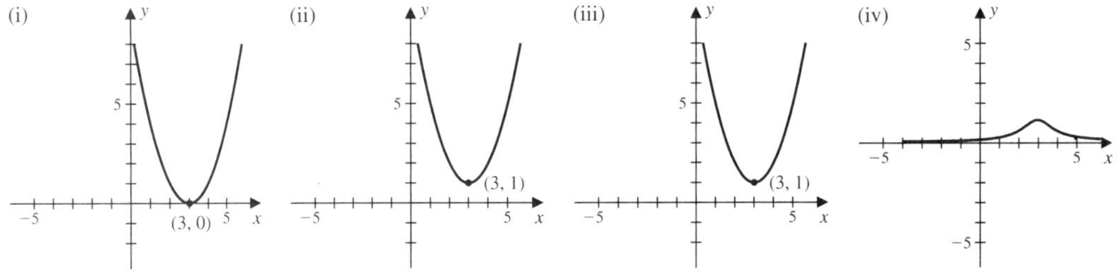

29. Let $f(x) = x$ and $g(x) = x + 1$. Then $(f \circ g)(x) = f(g(x)) = f(x+1) = x + 1$ and $(g \circ f)(x) = g(f(x)) = g(x) = x + 1$. There are many other examples.

30. If f and g are odd functions, then $f(-x) = -f(x)$ and $g(-x) = -g(x)$. So

$$(f \circ g)(-x) = f(g(-x)) = f(-g(x)) = -f(g(x)) = -(f \circ g)(x)$$

and

$$(g \circ f)(-x) = g(f(-x)) = g(-f(x)) = -g(f(x)) = -(g \circ f)(x).$$

31. If f is an odd function and g is an even function, then $f(-x) = -f(x)$ and $g(-x) = g(x)$. So

$$(f \circ g)(-x) = f(g(-x)) = f(g(x)) = (f \circ g)(x)$$

and

$$(g \circ f)(-x) = g(f(-x)) = g(-f(x)) = g(f(x)) = (g \circ f)(x).$$

32. Let $f(x) = ax + b$. Then

$$(f \circ f)(x) = f(f(x)) = a(ax + b) + b = a^2x + b(a + 1),$$

so $(f \circ f)(x) = 4x + 3$ which implies $a^2x + b(a + 1) = 4x + 3$. Then $a^2 = 4$, $b(a + 1) = 3$, so $a = 2, 3b = 3$ or $a = -2, -b = 3$ which means $a = 2, b = 1$ or $a = -2, b = -3$. So two possible linear functions are $f(x) = 2x + 1$ or $f(x) = -2x - 3$.

33. Let $f(x) = ax + b$ and $g(x) = cx + d$. Then

$$(f \circ g)(x) = a(cx + d) + b = acx + ad + b$$
$$(g \circ f)(x) = c(ax + b) + d = acx + bc + d.$$

a. $(f \circ g)(x) = (g \circ f)(x)$ implies $ad + b = bc + d$.

b. $(f \circ g)(x) = f(x)$ implies $ac = a$ and $ad + b = b$ so $c = 1$ and $d = 0$.

c. $(f \circ g)(x) = g(x)$ implies $ac = c$ and $ad + b = d$ so $a = 1, b = 0$.

34. **a.** A linear function. This is seen from the equations in Exercise 33.

b. Let $f(x) = ax + b$ and $g(x) = cx^2 + dx + e$. Then

$$f(g(x)) = a(cx^2 + dx + e) + b = acx^2 + adx + ae + b$$

and

$$g(f(x)) = c(ax + b)^2 + d(ax + b) + e = ca^2x^2 + (2abc + ad)x + cb^2 + bd + e$$

both of which are quadratic.

c. Let $f(x) = ax^2 + bx + c$ and $g(x) = dx^2 + ex + f$. Then

$$f(g(x)) = a(dx^2 + ex + f)^2 + b(dx^2 + ex + f) + c$$

has highest power x^4 with leading coefficient ad^2. Similarly $(g \circ f)(x)$ is a fourth degree polynomial.

35. $g(x) = 3f(x - 1) + 2$

36. $g(x) = -f\left(\frac{1}{2}(x + 2)\right) = -f\left(\frac{1}{2}x + 1\right)$

37. The volume of a sphere of radius r is $V(r) = \frac{4}{3}\pi r^3$. So $V(t) = \frac{4}{3}\pi(3 + 0.01t)^3$.

38. The volume of a cone of radius r and height h is $V = \frac{\pi}{3}r^2h$, and if the radius and height are equal we have

$$V(h) = \frac{\pi}{3}h^2h = \frac{\pi}{3}h^3 \text{ and } V(t) = \frac{\pi}{3}(10 + 0.25t)^3.$$

39. **a.** The volume of a sphere of radius r is $V(r) = \frac{4}{3}\pi r^3$ so

$$V(t) = V(r(t)) = \frac{4}{3}\pi\left(3\sqrt{t} + 5\right)^3 \text{ centimeters}^3.$$

b. The surface area is $S(r) = 4\pi r^2$ so

$$S(t) = S(r(t)) = 4\pi\left(3\sqrt{t} + 5\right)^2 \text{ centimeters}^2.$$

40. Since $s(t) = 576 + 144t - 16t^2$ we have

$$\begin{aligned}\bar{s}(t) &= s(t-2) \\ &= 576 + 144(t-2) - 16(t-2)^2 \\ &= 576 + 144t - 288 - 16(t^2 - 4t + 4) = 224 + 208t - 16t^2.\end{aligned}$$

To find the domain, determine the time t when the ball strikes the ground. Factoring we have

$$224 + 208t - 16t^2 = -16\,(t+1)\,(t-14)$$

and the rock strikes the ground at time $t = 14$ seconds and the domain is $[2, 14]$. Since the parabola is the parabola of Exercise 36 of Exercise Set 1.8, shifted 2 units to the right, the vertex has the same height of 900. So the range is $[0, 900]$.

Exercise Set 2.5 (Page 114)

1. The function is one-to-one since every horizontal line crosses the curve in only one point.

2. The function is not one-to-one since the horizontal line $y = 1$ crosses the curve in many places.

3. The function is not one-to-one since many horizontal lines cross the curve in more than one point.

4. The function is one-to-one since every horizontal line crosses the curve in only one point.

5. The function is not one-to-one since every horizontal line that crosses the curve intersects the curve in two points.

6. The function is one-to-one since every horizontal line that crosses the curve intersects the curve in only one point.

7. The function defined by $f(x) = 3x - 4$ is one-to-one, since

$$f(x_1) = f(x_2) \quad \text{implies} \quad 3x_1 - 4 = 3x_2 - 4 \quad \text{so} \quad 3x_1 = 3x_2 \quad \text{and} \quad x_1 = x_2.$$

8. The function defined by $f(x) = \sqrt{x-1}$ is one-to-one, since

$$f(x_1) = f(x_2) \quad \text{implies} \quad \sqrt{x_1 - 1} = \sqrt{x_2 - 1} \quad \text{so} \quad x_1 - 1 = x_2 - 1 \quad \text{and} \quad x_1 = x_2.$$

9. The function defined by $f(x) = |x - 1| + 1$ is not one-to-one. For example, if $x_1 = 0$ and $x_2 = 2$, then

$$f(x_1) = f(0) = |0 - 1| + 1 = 2$$

and

$$f(x_2) = f(2) = |2 - 1| + 1 = 2.$$

So we have found x_1 and x_2, with $x_1 \neq x_2$ and $f(x_1) = f(x_2)$.

10. The function defined by $f(x) = x^2 - 3x + 2$ is not one-to-one, since

$$f(x) = (x - 2)(x - 1) = 0 \text{ implies } x = 1 \text{ or } x = 2.$$

That is $f(1) = f(2) = 0$.

11. The function defined by $f(x) = x^4 + 1$ is not one-to-one, since, for example, $f(-1) = 2 = f(1)$.

12. The function defined by $f(x) = x^6 - 2$ is not one-to-one, since, for example, $f(-1) = -1 = f(1)$.

13. The function defined by $f(x) = \begin{cases} x^2, & \text{if } x \geq 0 \\ x - 1 & \text{if } x < 0 \end{cases}$ is one-to-one, since the graph satisfies the horizontal line test. That is, every horizontal line that crosses the graph does so in only one point.

14. The function defined by $f(x) = \begin{cases} \sqrt{x}, & \text{if } x \geq 0 \\ x + 1 & \text{if } x < 0 \end{cases}$ is not one-to-one, since the graph does not satisfy the horizontal line test.

15. The figure shows the graph of the function and its inverse.

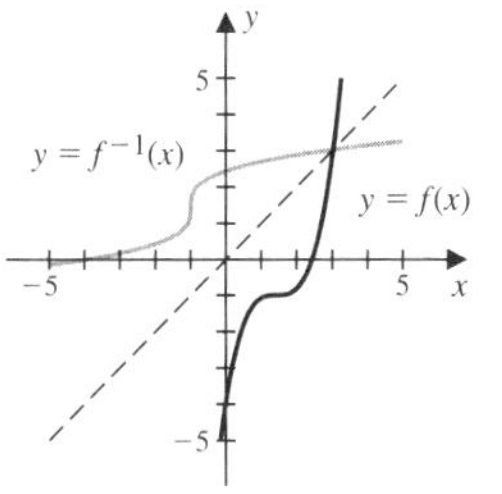

16. The figure shows the graph of the function and its inverse.

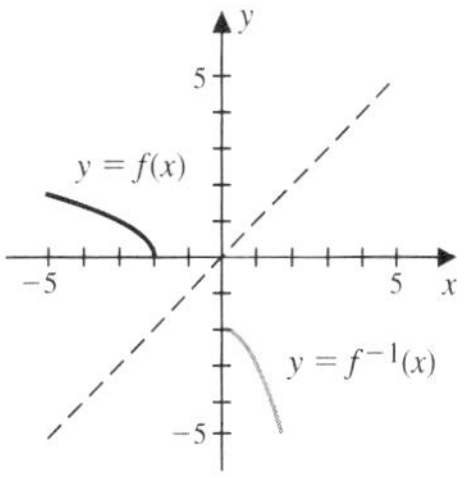

17. The figure shows the graph of the function and its inverse.

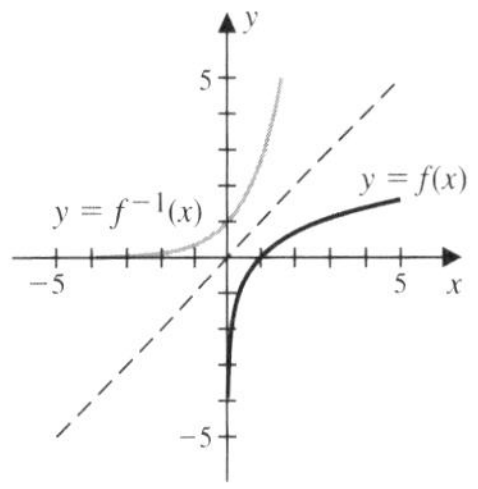

18. The figure shows the graph of the function and its inverse.

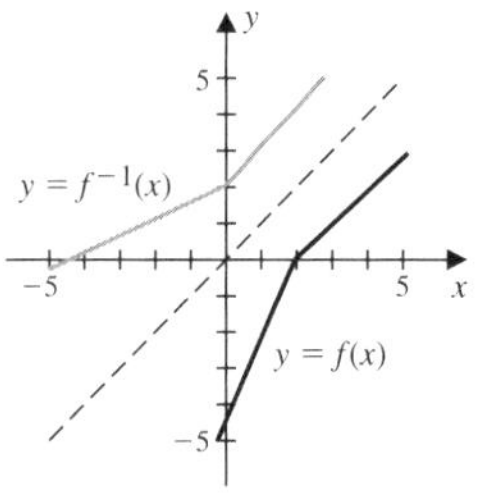

19. Since $f(x) = 2$, $f^{-1}(2) = x$. Solving $x + 1 = 2$ implies $x = 1$. So $f^{-1}(2) = 1$.

20. Since $f(x) = 0$, $f^{-1}(0) = x$. Solving $x - 2 = 0$ implies $x = 2$. So $f^{-1}(0) = 2$.

21. Since $f(x) = 1$, $f^{-1}(1) = x$. Solving $3x - 2 = 1$ implies $x = 1$. So $f^{-1}(1) = 1$.

22. Since $f(x) = 2$, $f^{-1}(2) = x$. Solving $3x - 2 = 2$ implies $x = \frac{4}{3}$. So $f^{-1}(2) = \frac{4}{3}$.

23. Since $f(x) = 2$, $f^{-1}(2) = x$. Solving $\frac{1}{x} = 2$ implies $x = \frac{1}{2}$. So $f^{-1}(2) = \frac{1}{2}$.

24. Since $f(x) = -8$, $f^{-1}(-8) = x$. Solving $x^3 = -8$ implies $x = -2$. So $f^{-1}(-8) = -2$.

25. Let $f(x) = 2x - 1$. Then f is one-to-one since

$$f(x_1) = f(x_2) \quad \text{implies} \quad 2x_1 - 1 = 2x_2 - 1 \quad \text{so} \quad x_1 = x_2.$$

To find f^{-1} : $y = 2x - 1$ implies $2x = y + 1$ so $x = \frac{y+1}{2}$, and $f^{-1}(x) = \frac{x+1}{2} = \frac{x}{2} + \frac{1}{2}$.

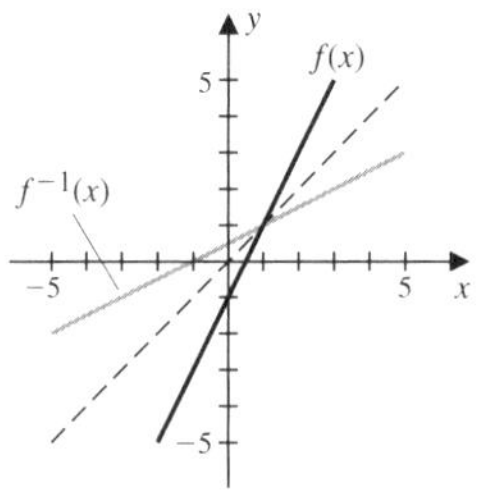

26. Let $f(x) = \frac{4x-3}{2}$. Then f is one-to-one since

$$f(x_1) = f(x_2) \quad \text{implies} \quad \frac{4x_1 - 3}{2} = \frac{4x_2 - 3}{2} \quad \text{so} \quad 4x_1 - 3 = 4x_2 - 3 \quad \text{and} \quad x_1 = x_2.$$

To find f^{-1} : $y = \frac{4x-3}{2}$ implies $2y = 4x - 3$ or $4x = 2y + 3$ so $x = \frac{2y+3}{4}$, and $f^{-1}(x) = \frac{2x+3}{4}$.

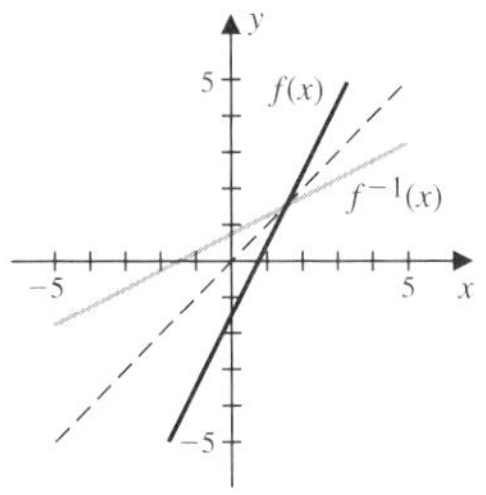

27. Let $f(x) = \sqrt{x-3}$. Then f is one-to-one since

$$f(x_1) = f(x_2) \quad \text{implies} \quad \sqrt{x_1 - 3} = \sqrt{x_2 - 3} \quad \text{so} \quad x_1 - 3 = x_2 - 3 \quad \text{and} \quad x_1 = x_2.$$

To find f^{-1} : $y = \sqrt{x-3}$ implies $x - 3 = y^2$ so $x = y^2 + 3$, and $f^{-1}(x) = x^2 + 3$.
Note: The domain of f is $[3, \infty)$, which is also the range of f^{-1} and the range of f is $[0, \infty)$, which is also the domain of f^{-1}.

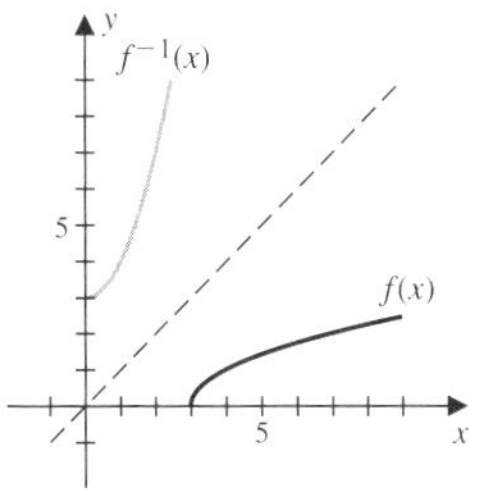

28. Let $f(x) = \sqrt{x+2}$. Then f is one-to-one since

$$f(x_1) = f(x_2) \quad \text{implies} \quad \sqrt{x_1+2} = \sqrt{x_2+2} \quad \text{so} \quad x_1 + 2 = x_2 + 2 \quad \text{and} \quad x_1 = x_2.$$

To find f^{-1} : $y = \sqrt{x+2}$ implies $x + 2 = y^2$ so $x = y^2 - 2$, and $f^{-1}(x) = x^2 - 2$.
Note: The domain of f is $[-2, \infty)$, which is also the range of f^{-1} and the range of f is $[0, \infty)$, which is also the domain of f^{-1}.

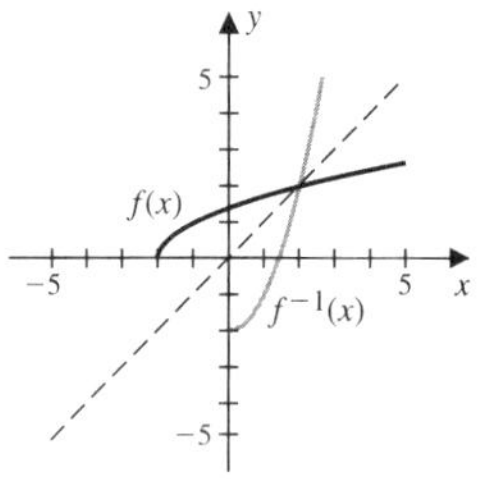

29. Let $f(x) = \frac{1}{x}$. Then f is one-to-one since

$$f(x_1) = f(x_2) \quad \text{implies} \quad \frac{1}{x_1} = \frac{1}{x_2} \quad \text{so} \quad x_1 = x_2.$$

To find f^{-1} : $y = \frac{1}{x}$ implies $xy = 1$ so $x = \frac{1}{y}$, and $f^{-1}(x) = \frac{1}{x}$.

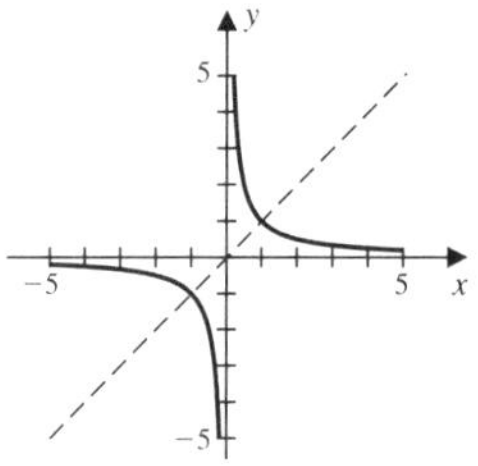

30. Let $f(x) = \frac{1}{x+1}$. Then f is one-to-one since

$$f(x_1) = f(x_2) \quad \text{implies} \quad \frac{1}{x_1+1} = \frac{1}{x_2+1} \quad \text{so} \quad x_2 + 1 = x_1 + 1 \quad \text{and} \quad x_1 = x_2.$$

To find f^{-1} : $y = \frac{1}{x+1}$ implies $(x+1)y = 1$ so $x + 1 = \frac{1}{y}$ or $x = \frac{1}{y} - 1 = \frac{1-y}{y}$, and $f^{-1}(x) = \frac{1-x}{x}$.

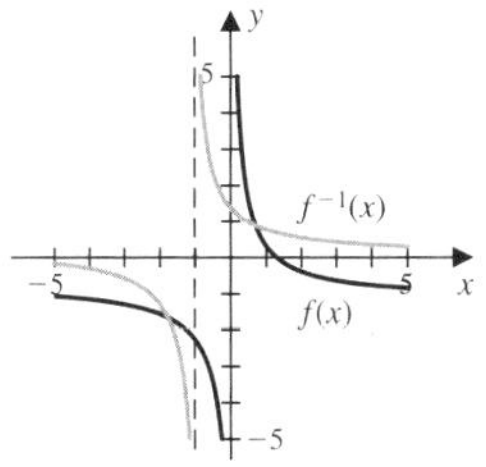

31. Let $f(x) = \frac{1}{\sqrt{x}}$. Then f is one-to-one since

$$f(x_1) = f(x_2) \quad \text{implies} \quad \frac{1}{\sqrt{x_1}} = \frac{1}{\sqrt{x_2}} \quad \text{so} \quad \sqrt{x_2} = \sqrt{x_1} \quad \text{and} \quad x_1 = x_2.$$

To find f^{-1} : $y = \frac{1}{\sqrt{x}}$ implies $\sqrt{x} = \frac{1}{y}$ so $x = \frac{1}{y^2}$, and $f^{-1}(x) = \frac{1}{x^2}$.
Note: The domain of f is $(0, \infty)$, which is also the range of f^{-1} and the range of f is $(0, \infty)$, which is also the domain of f^{-1}.

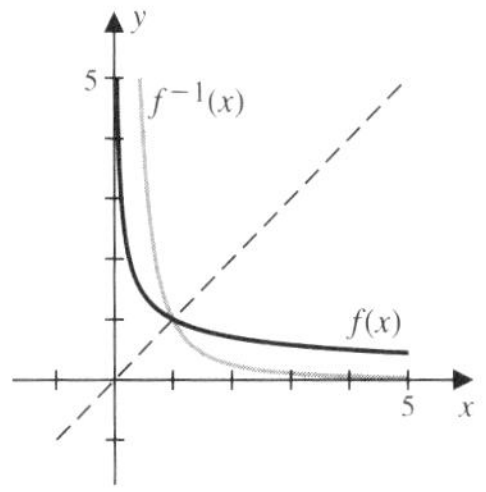

32. Let $f(x) = \frac{1}{\sqrt{x}} + 1$. Then f is one-to-one since

$$f(x_1) = f(x_2) \quad \text{implies} \quad \frac{1}{\sqrt{x_1}} + 1 = \frac{1}{\sqrt{x_2}} + 1 \quad \text{so} \quad \sqrt{x_2} = \sqrt{x_1} \quad \text{and} \quad x_1 = x_2.$$

To find f^{-1} : $y = \frac{1}{\sqrt{x}} + 1$ implies $\frac{1}{\sqrt{x}} = y - 1$ so $\sqrt{x} = \frac{1}{y-1}$ implies $x = \frac{1}{(y-1)^2}$, and $f^{-1}(x) = \frac{1}{(x-1)^2}$.

Note: The domain of f is $(0, \infty)$, which is also the range of f^{-1} and the range of f is $(1, \infty)$, which is also the domain of f^{-1}.

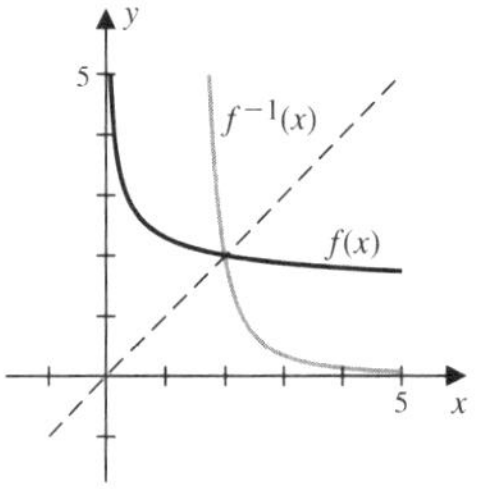

33. Let $f(x) = 1 + x^3$. Then f is one-to-one since

$$f(x_1) = f(x_2) \quad \text{implies} \quad 1 + x_1^3 = 1 + x_2^3 \quad \text{so} \quad x_1^3 = x_2^3 \quad \text{and} \quad x_1 = x_2.$$

To find f^{-1} : $y = 1 + x^3$ implies $x^3 = y - 1$ so $x = (y-1)^{\frac{1}{3}}$, and $f^{-1}(x) = (x-1)^{\frac{1}{3}}$.

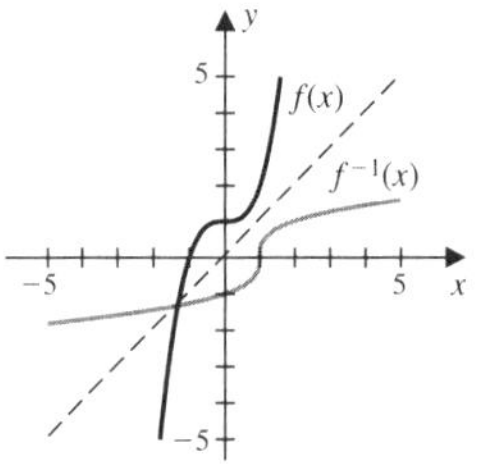

34. Let $f(x) = 2 - x^3$. Then f is one-to-one since

$$f(x_1) = f(x_2) \quad \text{implies} \quad 2 - x_1^3 = 2 - x_2^3 \quad \text{so} \quad x_1^3 = x_2^3 \quad \text{and} \quad x_1 = x_2.$$

To find f^{-1} : $y = 2 - x^3$ implies $x^3 = 2 - y$ so $x = (2 - y)^{\frac{1}{3}}$, and $f^{-1}(x) = (2 - x)^{\frac{1}{3}}$.

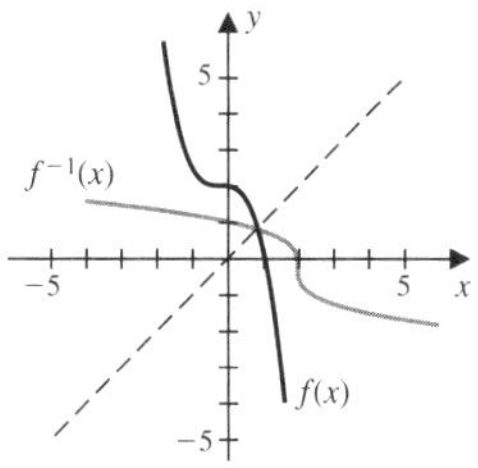

35. Let $f(x) = x^2 + 1$, $x \geq 0$. Then f is one-to-one since

$$f(x_1) = f(x_2) \quad \text{implies} \quad x_1^2 + 1 = x_2^2 + 1 \quad \text{so} \quad x_1^2 = x_2^2 \quad \text{and} \quad x_1 = x_2,$$

since $x \geq 0$ (otherwise we could only conclude that $|x_1| = |x_2|$).
To find f^{-1} : $y = x^2 + 1$ implies $y - 1 = x^2$ so $x = \sqrt{y - 1}$, and $f^{-1}(x) = \sqrt{x - 1}$.
Note: The domain of f is $[0, \infty)$, which is also the range of f^{-1} and the range of f is $[1, \infty)$, which is also the domain of f^{-1}.

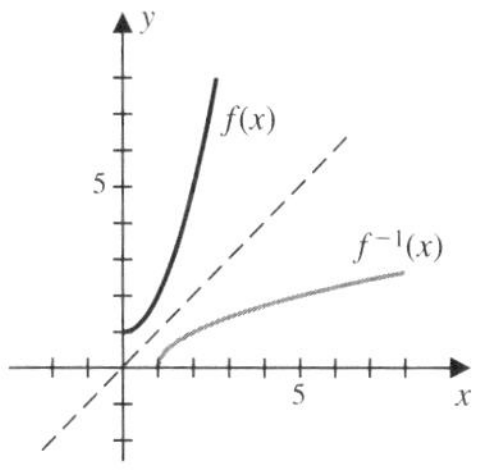

36. Let $f(x) = 1 - x^2, x \le 0$. Then f is one-to-one since

$$f(x_1) = f(x_2) \quad \text{implies} \quad 1 - x_1^2 = 1 - x_2^2 \quad \text{so} \quad x_1^2 = x_2^2 \quad \text{and} \quad x_1 = x_2,$$

since $x \le 0$ (otherwise we could only conclude that $|x_1| = |x_2|$).
To find f^{-1} : $y = 1 - x^2$ implies $y - 1 = -x^2$ so $x^2 = 1 - y$ implies $x = -\sqrt{1-y}$, and $f^{-1}(x) = -\sqrt{1-x}$.
Note: The domain of f is $(-\infty, 0]$, which is also the range of f^{-1} and the range of f is $(-\infty, 1]$, which is also the domain of f^{-1}.

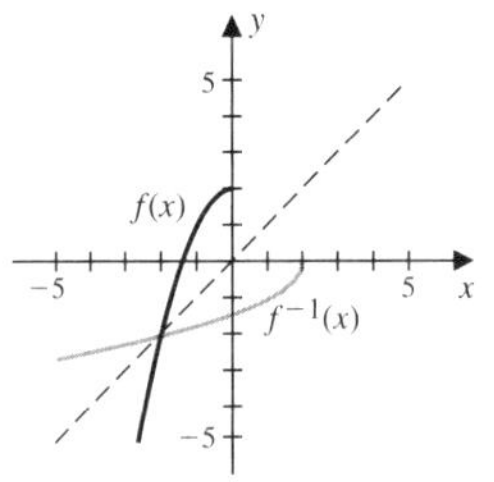

37. a. The graph of the function $f(x) = |2 - x|$ is V-shaped, opens upward and the point is at $(2, 0)$. The line $y = 1$, for example, must cross the curve two times, and $|2 - x| = 1$ implies $2 - x = 1$ or $2 - x = -1$ so $x = 1$ or $x = 3$. So the function is not one-to-one.

b. The function is one-to-one on the interval $[2, \infty)$.

f^{-1} for f restricted on $[2, \infty)$: Since $2 - x \le 0$ for $x \ge 2$,

$$y = -(2 - x) = x - 2 \text{ implies } x = y + 2, \text{ so } f^{-1}(x) = x + 2.$$

The domain of f^{-1} is the range of f, which equals $[0, \infty)$.

38. a. For $f(x) = -|x - 3| - 2$, $f(2) = -|-1| - 2 = -3$ and $f(4) = -|1| - 2 = -3$ so the function is not one-to-one.

b. The function is one-to-one on the interval $[3, \infty)$.

f^{-1} for f restricted on $[3, \infty)$: Since $x - 3 \ge 0$ for $x \ge 3$,

$$y = -(x - 3) - 2 = -x + 1 \text{ implies } x = 1 - y, \text{ so } f^{-1}(x) = 1 - x.$$

The domain of f^{-1} is the range of f, which equals $(-\infty, -2]$.

39. a. We have $f(x) = x^2 - 2x = x(x - 2)$, and $f(x) = 0$ for both $x = 0$ and $x = 2$, so f is not one-to-one. The graph is a parabola that opens upward and has vertex $(1, -1)$.

b. The function is one-to-one on the interval $[1, \infty)$.

f^{-1} for f restricted on $[1, \infty)$:

$$y = x^2 - 2x = x^2 - 2x + 1 - 1 = (x - 1)^2 - 1 \text{ implies } y + 1 = (x - 1)^2$$

if and only if

$$x - 1 = \sqrt{y+1} \text{ implies } x = \sqrt{y+1} + 1, \quad \text{so } f^{-1}(x) = \sqrt{x+1} + 1.$$

The domain of f^{-1} is $[-1, \infty)$ which is also the range of f.

40. **a.** We have $f(x) = x^2 - 2x + 3 = (x+1)(x-3)$, and $f(x) = 0$ for both $x = -1$ and $x = 3$, so f is not one-to-one.

b. Since

$$\begin{aligned} f(x) &= x^2 - 2x + 3 \\ &= x^2 - 2x + 1 - 1 + 3 \\ &= (x-1)^2 + 2 \end{aligned}$$

the graph is the parabola with vertex $(1, 2)$. The function is one-to-one on the interval $[1, \infty)$.

f^{-1} for f restricted on $[1, \infty)$:

$$y = x^2 - 2x + 3 = (x-1)^2 + 2 \text{ implies } (x-1)^2 = y - 2$$

if and only if

$$x - 1 = \sqrt{y-2} \text{ implies } x = \sqrt{y-2} + 1, \text{ so } f^{-1}(x) = 1 + \sqrt{x-2}.$$

The domain of f^{-1} is $[2, \infty)$ which is also the range of f.

41. For $f(x) = mx + b$

$$f(x_1) = f(x_2) \text{ implies } mx_1 + b = mx_2 + b \text{ so } mx_1 = mx_2$$

if and only if

$$x_1 = x_2 \text{ provided m} \neq 0.$$

So for all $m \neq 0$ the function $f(x) = mx + b$ is one-to-one. Solving for x gives

$$y = mx + b \text{ implies } y - b = mx \text{ implies } x = \frac{y-b}{m}, \quad \text{so } f^{-1}(x) = \frac{x-b}{m}.$$

42. Since $K = C + 273$ and $F = \frac{9}{5}C + 32$ we have $F = \frac{9}{5}(K - 273) + 32$. Since the function is linear, it is one-to-one. To find the inverse we solve

$$F - 32 = \frac{9}{5}(K - 273) \text{ implies } \frac{5}{9}(F - 32) = K - 273 \text{ implies } K = \frac{5}{9}(F - 32) + 273.$$

Review Exercises for Chapter 2 (Page 115)

1. **a.** The graph of $y = \sqrt{x-1} + 2$ is (ii).
b. The graph of $y = \sqrt{x} - 1$ is (i).
c. The graph of $y = -\sqrt{x+2} + 1$ is (iv).
d. The graph of $y = 2\sqrt{x}$ is (iii).

2. **a.** The graph of $y = -|x-1| + 2$ is (ii).
b. The graph of $y = 2|x| - 1$ is (iv).
c. The graph of $y = |x+1| - 1$ is (iii).
d. The graph of $y = |x| + 2$ is (i).

3. **a.** For $f(x) = 2x - 1$ and $g(x) = x + 2$,

$$(f+g)(x) = 3x + 1;\ (f-g)(x) = x - 3;$$
$$(fg)(x) = 2x^2 + 3x - 2;\ (f/g)(x) = \frac{2x-1}{x+2};$$
$$(f \circ g)(x) = 2x + 3;\ (g \circ f)(x) = 2x + 1;$$
$$(f \circ f)(x) = 4x - 3;\ (g \circ g)(x) = x + 4.$$

b. The domain of $f+g$, $f-g$, fg, $f \circ g$, $g \circ f$, $f \circ f$ and $g \circ g$ is $(-\infty, \infty)$. The domain of f/g is $(-\infty, -2) \cup (-2, \infty)$.

4. **a.** For $f(x) = x^2$ and $g(x) = x - 2$,

$$(f+g)(x) = x^2 + x - 2;\ (f-g)(x) = x^2 - x + 2;$$
$$(fg)(x) = x^3 - 2x^2;\ (f/g)(x) = \frac{x^2}{x-2};$$
$$(f \circ g)(x) = (x-2)^2;\ (g \circ f)(x) = x^2 - 2;$$
$$(f \circ f)(x) = x^4;\ (g \circ g)(x) = x - 4.$$

b. The domain of $f+g$, $f-g$, fg, $f \circ g$, $g \circ f$, $f \circ f$ and $g \circ g$ is $(-\infty, \infty)$. The domain of f/g is $(-\infty, 2) \cup (2, \infty)$.

5. **a.** For $f(x) = x + 1$ and $g(x) = \sqrt{x-1}$,

$$(f+g)(x) = x + 1 + \sqrt{x-1};\ (f-g)(x) = x + 1 - \sqrt{x-1};$$
$$(fg)(x) = (x+1)\sqrt{x-1};\ (f/g)(x) = \frac{x+1}{\sqrt{x-1}};$$
$$(f \circ g)(x) = \sqrt{x-1} + 1;\ (g \circ f)(x) = \sqrt{x};$$
$$(f \circ f)(x) = x + 2;\ (g \circ g)(x) = \sqrt{\sqrt{x-1} - 1};$$

b. The domain of $f+g$, $f-g$, fg and $f \circ g$ is $[1, \infty)$. The domain of f/g is $(1, \infty)$. The domain of $g \circ f$ is $[0, \infty)$. The domain of $f \circ f$ is $(-\infty, \infty)$. The domain of $g \circ g$ is $[2, \infty)$.

6. a. For $f(x) = 2x - 3$ and $g(x) = \frac{1}{x}$,

$$(f + g)(x) = 2x - 3 + \frac{1}{x}; \ (f - g)(x) = 2x - 3 - \frac{1}{x};$$
$$(fg)(x) = \frac{2x - 3}{x}; \ (f/g)(x) = (2x - 3)x;$$
$$(f \circ g)(x) = \frac{2}{x} - 3; \ (g \circ f)(x) = \frac{1}{2x - 3};$$
$$(f \circ f)(x) = 2(2x - 3) - 3 = 4x - 9; \ (g \circ g)(x) = x;$$

b. The domain of $f + g$, $f - g$, fg, f/g, $f \circ g$, and $g \circ g$ is $(-\infty, 0) \cup (0, \infty)$. The domain of $g \circ f$ is $(-\infty, \frac{3}{2}) \cup (\frac{3}{2}, \infty)$. The domain of $f \circ f$ is $(-\infty, \infty)$.

7. One choice is $f(x) = x^6$ and $g(x) = x^3 - 2x + 1$.

8. One choice is $f(x) = 1/x^3$ and $g(x) = x^4 - 2x + 1$.

9. One choice is $f(x) = \sqrt{x}$ and $g(x) = 2x + 2$.

10. One choice is $f(x) = x^{\frac{3}{2}}$ and $g(x) = x + 1$.

11. One choice is $f(x) = |x|$ and $g(x) = x^2 - 2x + 1$.

12. One choice is $f(x) = |x| + 1$ and $g(x) = x^2 - 2x$.

13. The graph of $f(x) = x + 2$ and its reciprocal are shown.

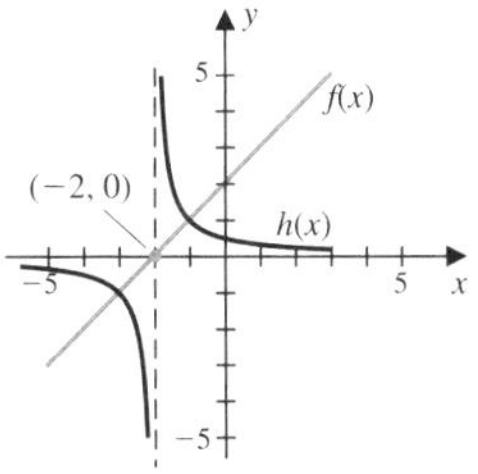

14. The graph of $f(x) = x^2 - 2$ and its reciprocal are shown.

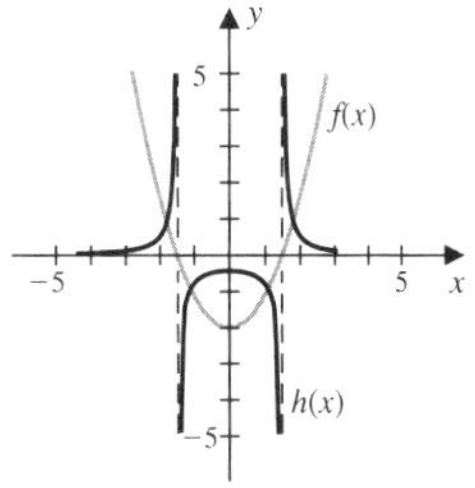

15. The graph of

$$f(x) = x^2 - 2x - 3 = (x - 1)^2 - 4$$

and its reciprocal are shown.

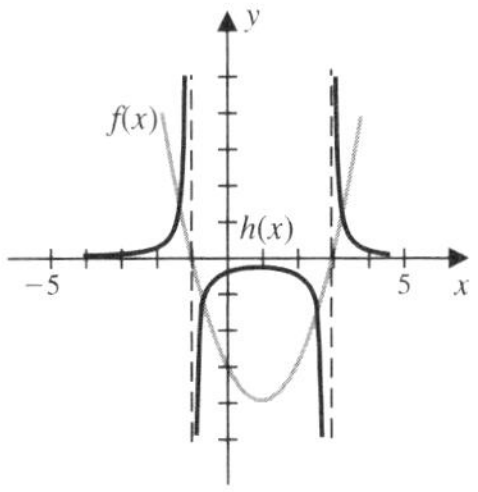

16. The graph of $f(x) = |x + 2|$ and its reciprocal are shown.

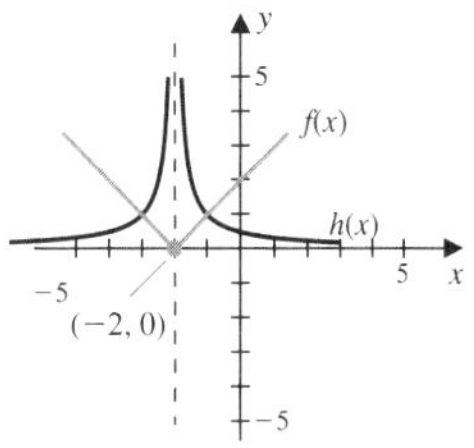

17. $f(x) = |x + 1|$

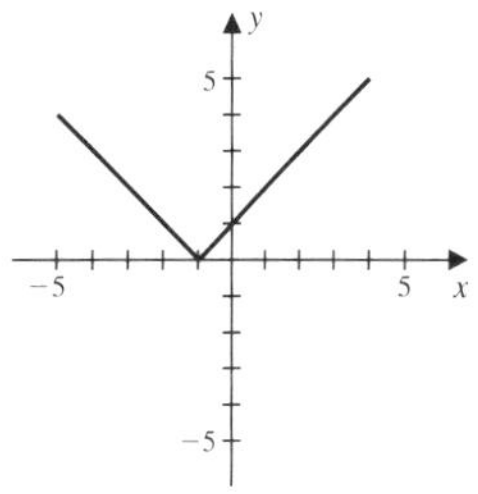

18. $f(x) = |x - 3| + 3$

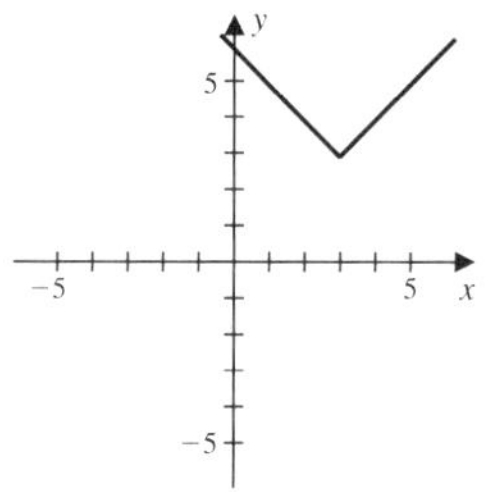

19. $f(x) = |x + 2| - 2$

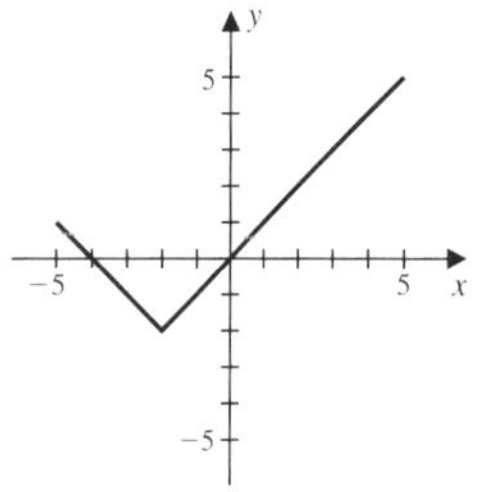

20. $f(x) = |2x + 4| - 3 = 2|x + 2| - 3$

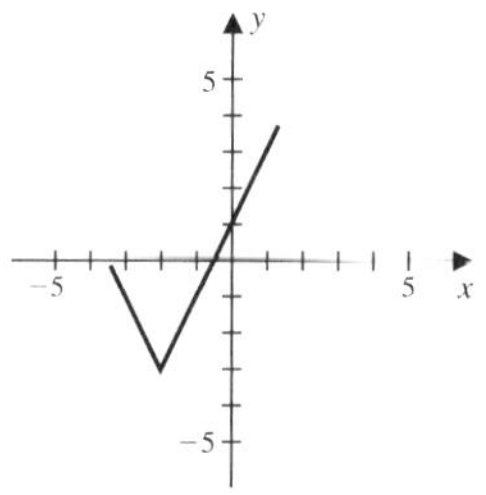

21. $g(x) = \sqrt{x + 2}$

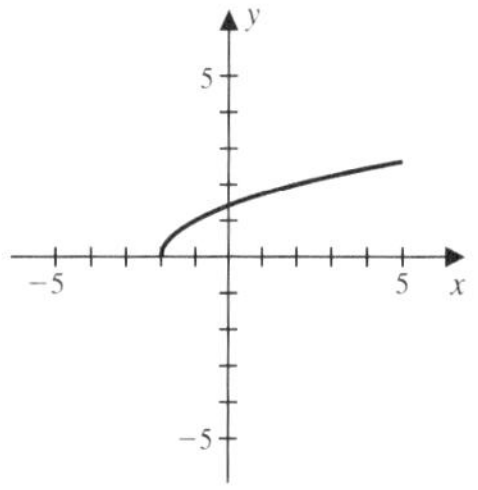

22. $g(x) = \sqrt{x - 2} + 2$

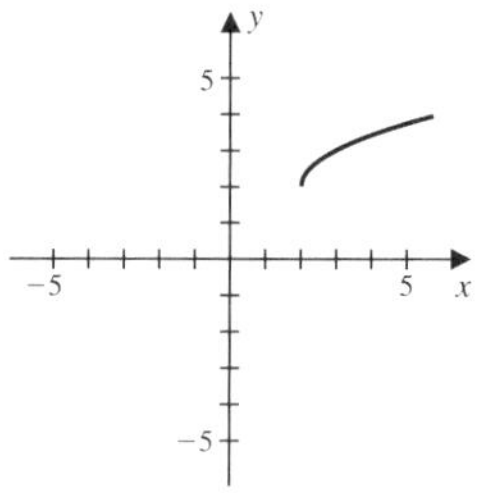

23. $g(x) = -\sqrt{x} - 2$

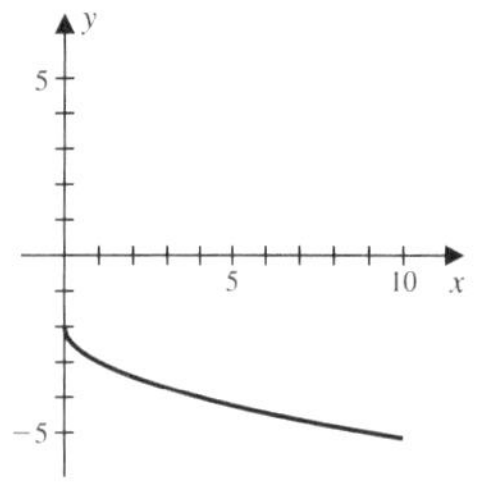

24. $g(x) = 2 - \sqrt{x - 2}$

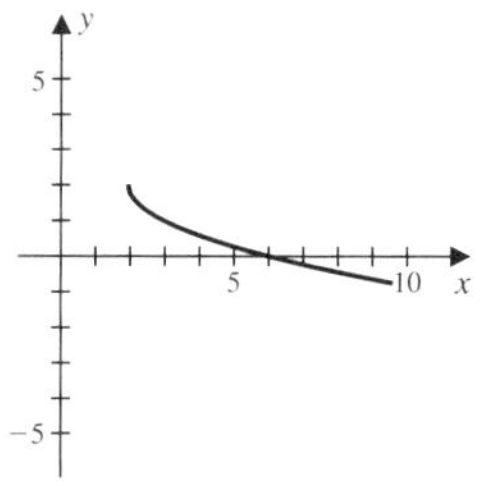

25. $f(x) = \lfloor x + 2 \rfloor$

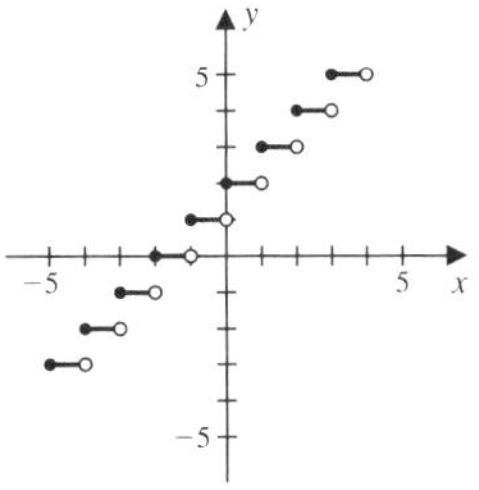

26. $f(x) = \lfloor x \rfloor + 2$

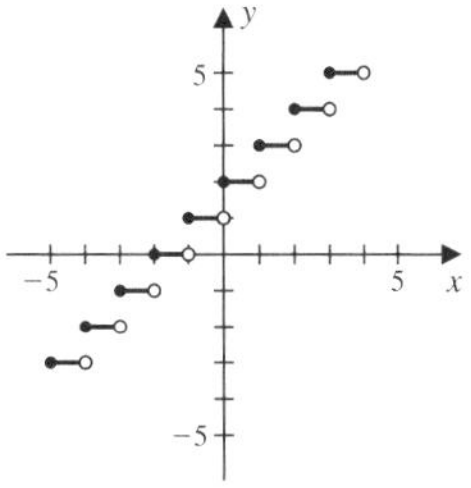

27. $f(x) = -\lfloor x \rfloor$

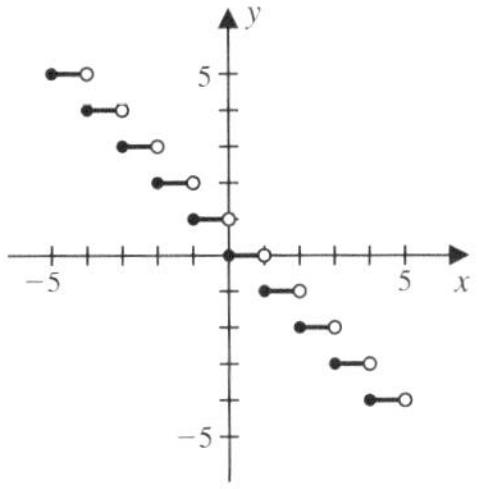

28. $f(x) = 2\lfloor x \rfloor + 1$

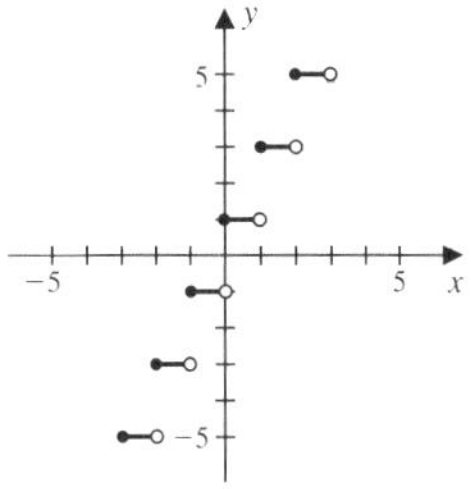

29. $f(x) = 3x - 2$

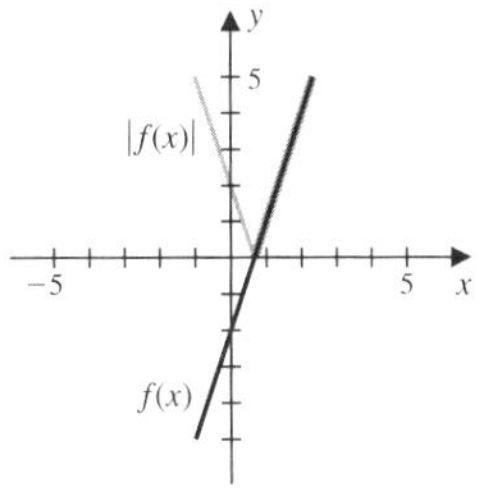

30. $f(x) = -2x + 1$

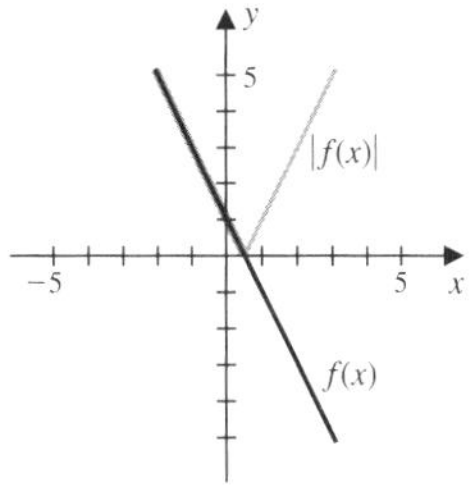

31. $f(x) = -x^2 + 3$

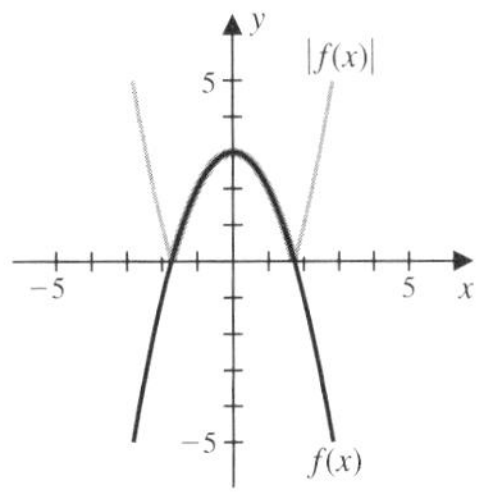

32. First complete the square to give

$$\begin{aligned} f(x) &= x^2 - 6x + 8 \\ &= x^2 - 6x + 9 - 9 + 8 \\ &= (x - 3)^2 - 1. \end{aligned}$$

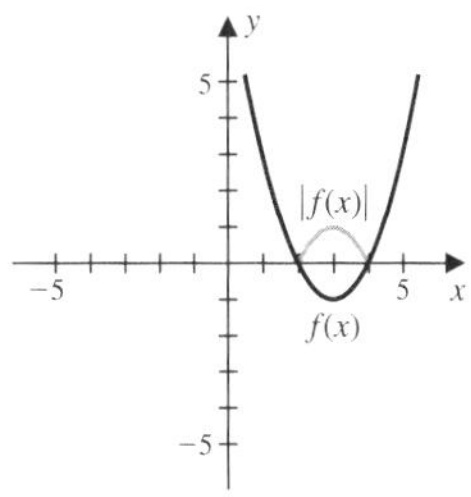

33. **a.** $y = f(x+1)$ **b.** $y = f(x+1)+1$ **c.** $y = f(x-1)$ **d.** $y = f(x-1)+2$
e. $y = f(x+1)-1$ **f.** $y = f(x-1)-2$ **e.** $y = f(x+1)-1$ **f.** $y = f(x-1)-2$
g. $y = f(2x)$ **h.** $y = f(x/2)$

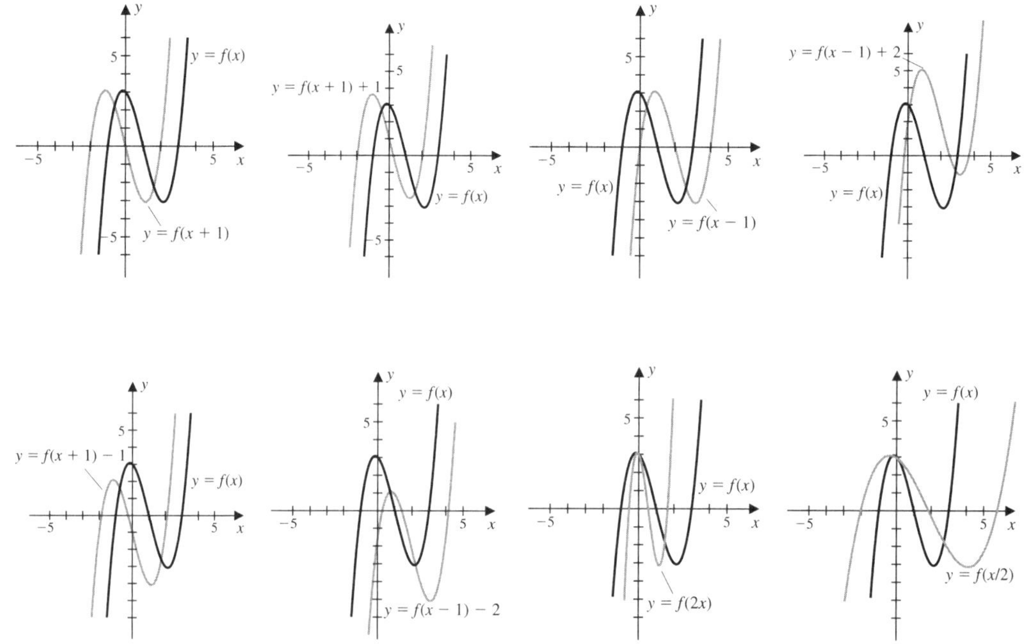

34. **i.** (a) x-axis symmetry (b) Not a function.
ii. y-axis symmetry (b) Is a function. (c) Domain: $(-\infty, -2) \cup (-2, 2) \cup (2, \infty)$; Range: $(-\infty, 0] \cup (1, \infty)$ (d) Even (e) Not one-to-one
iii. (a) origin symmetry (b) Is a function. (c) Domain: $(-\infty, \infty)$; Range: $(-\infty, \infty)$ (d) Odd (e) Not one-to-one
iv. (a) Origin symmetry (b) Is a function. (c) Domain: $(-\infty, \infty)$; Range: $(-\infty, \infty)$ (d) Odd (e) one-to-one (f)

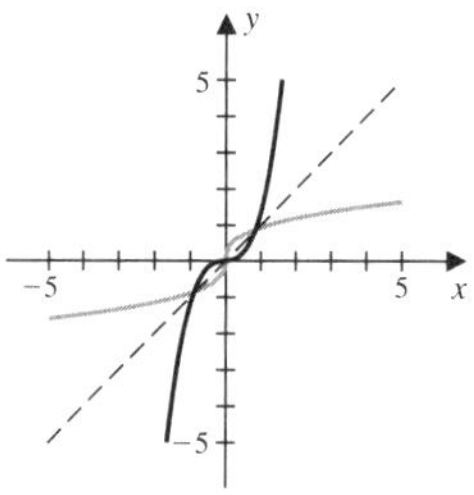

35. **a.** For $f(x) = 2x^2 - 3$ we have $f(-x) = 2(-x)^2 - 3 = 2x^2 - 3$;

$-f(x) = -(2x^2 - 3) = -2x^2 + 3$; $f\left(\frac{1}{x}\right) = 2\left(\frac{1}{x}\right)^2 - 3 = \frac{2}{x^2} - 3 = \frac{2-3x^2}{x^2}$;

$\frac{1}{f(x)} = \frac{1}{2x^2-3}$; $f\left(\sqrt{x}\right) = 2\left(\sqrt{x}\right)^2 - 3 = 2x - 3$; $\sqrt{f(x)} = \sqrt{2x^2 - 3}$.

b. For $f(x) = \frac{1}{x^2}$ we have $f(-x) = \frac{1}{(-x)^2} = \frac{1}{x^2}$; $-f(x) = -\frac{1}{x^2}$;

$f\left(\frac{1}{x}\right) = \frac{1}{\left(\frac{1}{x}\right)^2} = x^2$; $\frac{1}{f(x)} = \frac{1}{\frac{1}{x^2}} = x^2$;

$f\left(\sqrt{x}\right) = \frac{1}{\left(\sqrt{x}\right)^2} = \frac{1}{x}$; $\sqrt{f(x)} = \sqrt{\frac{1}{x^2}} = \frac{1}{x}$.

36. **a.** $f(x) = g(h_5(x)) = g(x^5) = \frac{1}{x^5}$

b. $f(x) = w(v(h_4(x))) = w(v(x^4)) = w(x^4 - 2) = \sqrt{x^4 - 2}$

c. $f(x) = h_6(v(h_2))) = h_6(v(x^2)) = h_6(x^2 - 2) = (x^2 - 2)^6$

d. $f(x) = g(w(v(h_3(x)))) = g(w(v(x^3))) = g(w(x^3 - 2)) = g(\sqrt{x^3 - 2}) = \frac{1}{\sqrt{x^3-2}}$

37. $g(x) = \frac{1}{2}f(x+2) - 3$

38. $g(x) = f(3(x-3)) + 1$

39. **a.** For $f(x) = 2x + 1$,

$$f(x_1) = f(x_2) \text{ implies } 2x_1 + 1 = 2x_2 + 1 \text{ so } 2x_1 = 2x_2 \text{ and } x_1 = x_2.$$

The function is one-to-one. To find the inverse we have

$$y = 2x + 1 \text{ which implies } 2x = y - 1 \text{ so } x = \frac{y-1}{2}$$

and $f^{-1}(x) = \frac{x-1}{2}$.

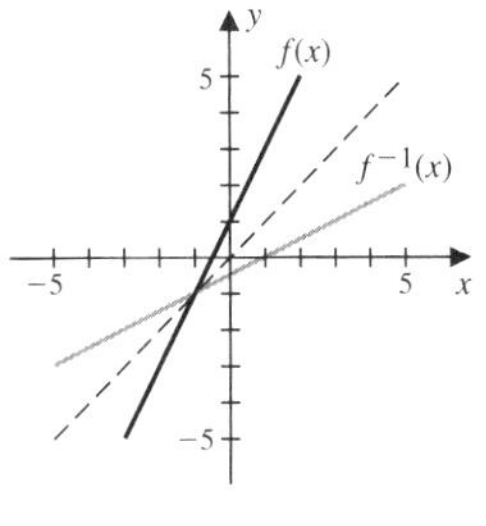

b. For $f(x) = |x - 1| + 1$, $f(0) = 2 = f(2)$, and hence the function is not one-to-one.

c. The function defined by $f(x) = 2 + x^3$ is one-to-one, since

$$f(x_1) = f(x_2) \text{ implies } 2 + x_1^3 = 2 + x_2^3 \text{ so } x_1^3 = x_2^3 \text{ and } x_1 = x_2.$$

To find the inverse we have

$$y = 2 + x^3 \text{ which implies } x^3 = y - 2 \text{ so } x = \sqrt[3]{y-2}$$

and $f^{-1}(x) = \sqrt[3]{x-2}$.

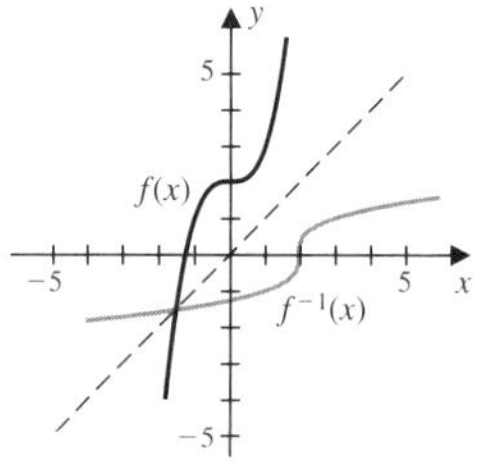

d. The graph of $f(x) = x^2 + 2x - 2$ is a parabola, so there are horizontal lines that cross the graph two times. Hence the function is not one-to-one. To show this algebraically, we have

$$f(x) = x^2 + 2x - 2 = x^2 + 2x + 1 - 1 - 2 = (x+1)^2 - 3.$$

So

$$f(x) = 0 \text{ implies } x + 1 = \pm\sqrt{3} \text{ implies } x = 1 \pm \sqrt{3}.$$

So both $f(1 + \sqrt{3}) = 0$ and $f(1 - \sqrt{3}) = 0$, which implies that the function is not one-to-one.

e. The function defined by $f(x) = \frac{1}{x^2}$ is not one-to-one, since $f(-1) = 1$ and $f(1) = 1$.

f. The function defined by $f(x) = \sqrt{x-1} + 2$ is one-to-one, since

$$f(x_1) = f(x_2) \text{ implies } \sqrt{x_1 - 1} + 2 = \sqrt{x_2 - 1} + 2 \text{ so } \sqrt{x_1 - 1} = \sqrt{x_2 - 1} \text{ implies } x_1 - 1 = x_2 - 1$$

and $x_1 = x_2$. To find the inverse we have

$$y = \sqrt{x-1} + 2 \text{ which implies } \sqrt{x-1} = y - 2 \text{ so } x - 1 = (y-2)^2 \text{ implies } x = (y-2)^2 + 1$$

and $f^{-1}(x) = (x-2)^2 + 1$.

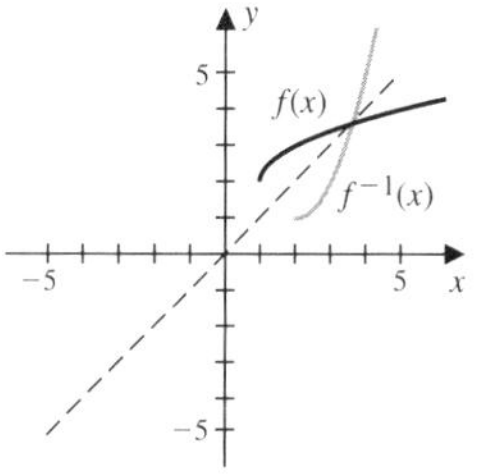

40. **a.** The graph of $y = f(x) = x^2 - 2$ for $x \geq 0$ is shown.

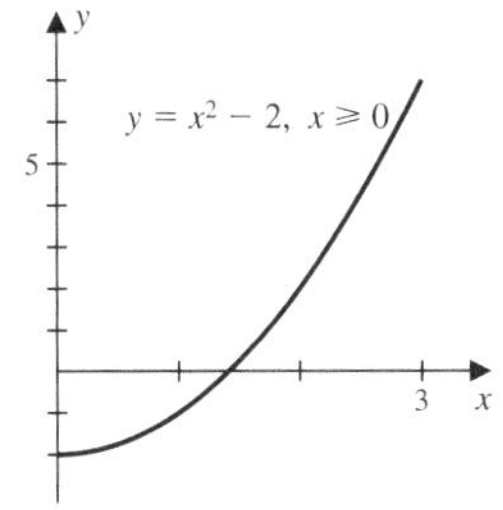

b. The graph of the inverse function is the reflection about $y = x$ of the graph of $f(x) = x^2 - 2$, when $x \geq 0$.

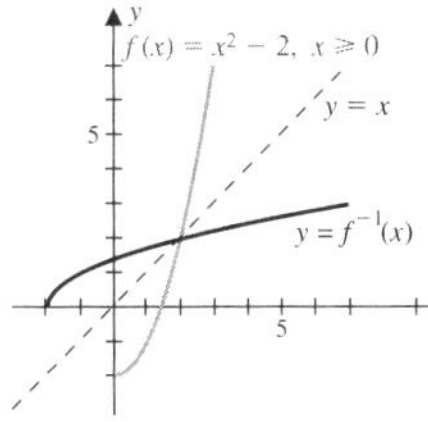

c. $y = x^2 - 2$ implies $x^2 = y + 2$ so $x = \sqrt{y+2}$, and $f^{-1}(x) = \sqrt{x+2}$.

41. **a.** The graph of the function restricted to the subset $[2, \infty)$ is shown.

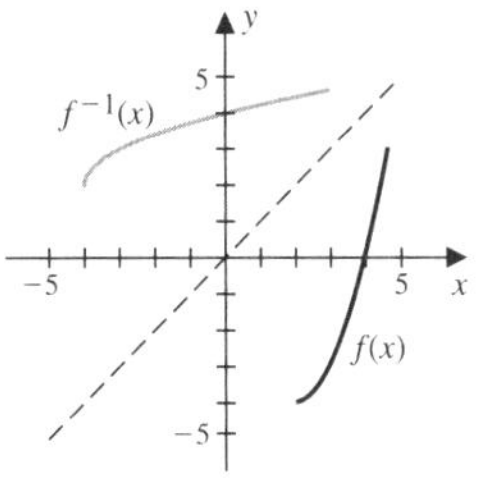

b. Since $f(x) = x^2 - 4x = x^2 - 4x + 4 - 4 = (x-2)^2 - 4,$ the vertex of the parabola is $(2, -4)$. If the domain is restricted to $[2, \infty)$, the new function is one-to-one.

c. $y = (x-2)^2 - 4$ implies $y + 4 = (x-2)^2$ so $x = \sqrt{y+4} + 2$, and $f^{-1}(x) = \sqrt{x+4} + 2$. The domain of f^{-1} is $[-4, \infty)$.

42. **a.** If the point, C, where the new construction to point B begins is x miles east of point A, we can apply the Pythagorean Theorem to the right triangle to determine that the cost of construction is

$$C(x) = 200000x + 400000\sqrt{(40-x)^2 + 400}.$$

b. Plot the total cost with respect to the distance x, where the new construction begins from point A. The minimum cost corresponds to the x-coordinate of the low point on the graph. Approximately 28.5 miles of the old road should be restored in order to minimize the cost of construction.

43. **a.** Since the cost is $C(x) = 30x + 1500$ and the selling price is $p(x) = 120 - 0.1x$, the profit function is given by

$$P(x) = xp(x) - C(x) = x(120 - 0.1x) - 30x - 1500 = -0.1x^2 + 90x - 1500.$$

b. To find the number of units sold that yields the maximum profit we find the vertex of the parabola. Then

$$\begin{aligned} -0.1x^2 + 90x - 1500 &= -0.1(x^2 - 900x) - 1500 \\ &= -0.1(x^2 + 900x + 202500 - 202500) - 1500 \\ &= -0.1(x - 450)^2 + 20250 - 1500 \\ &= -0.1(x - 450)^2 + 18750. \end{aligned}$$

So the vertex is (450, 18750) and 450 units should be produced to yield the maximum profit.

c. The maximum profit is \$18,750.00.

44. **a.** $x = 1, 2, 4$ **b.** $x \le 1$ or $2 \le x \le 4$ **c.** $x = 0, 3$ **d.** $0 \le x \le 1.5$ or $3 \le x \le 5$

e. $(f + g)(3) = f(3) + g(3) = 0 + 3 = 3$; $(f + g)(4) = f(4) + g(4) = 1.5 + 1.5 = 3$; $(f - g)(3) = f(3) - g(3) = 0 - 3 = -3$; $(f - g)(4) = f(4) - g(4) = 1.5 - 1.5 = 0$

f. $(f \circ g)(4) = f(g(4)) = f(1.5) = 2.25$; $(g \circ f)(5) = g(f(5)) = g(3) = 3$

g. The range of f and the range of g is $[0, 3]$. **h.** $f(5) = 3$

Chapter 2 Exercises for Calculus (Page 118)

1. The sign of $f(x)$ and the zeros of $f(x)$ give the graph of $1/f(x)$.

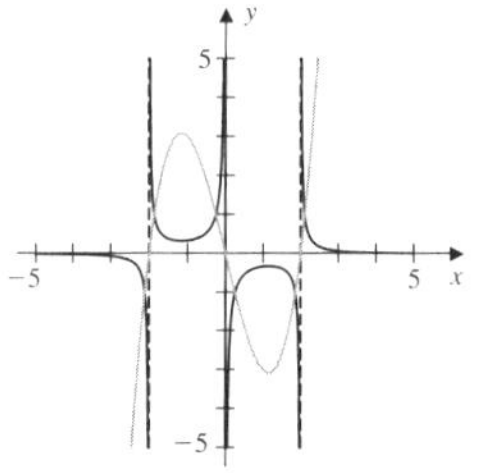

2. a. The Pythagorean Theorem implies that the distance of the car from C is $D(d) = \sqrt{4 + d^2}$ miles.

b. Since the car is traveling at 60 mi/hr,

$$d = 60t, \text{ for t in hours, and } D(t) = \sqrt{4 + (60t)^2} \text{ miles.}$$

c. With $d(t) = 60t$ and $D(d) = \sqrt{4 + d^2}$, we also have

$$D(t) = (D \circ d)(t) = \sqrt{4 + (60t)^2}.$$

3. a. The graphs of the World population, $W(x)$, and urban population, $U(x)$, are shown.

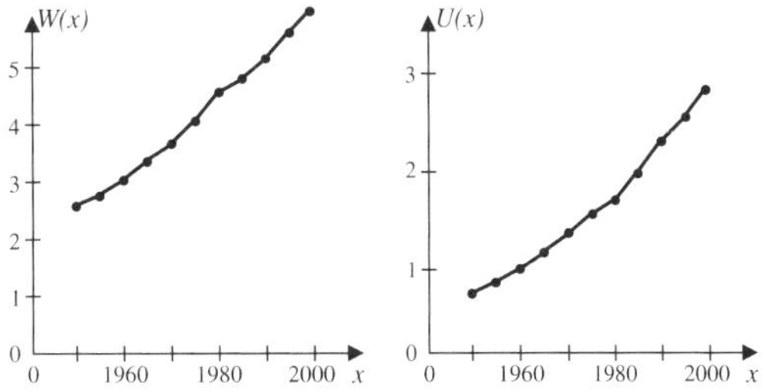

b. The graph of the non-urban population, $N(x) = W(x) - U(x)$, and the percentage of urban population, $S(x) = 100 \cdot U(x)/W(x)$, are shown.

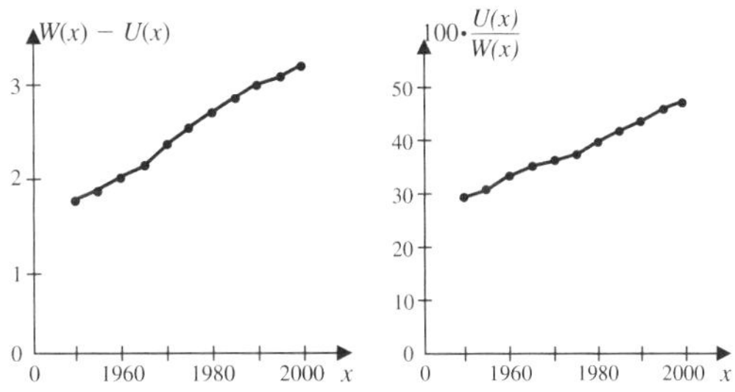

4. a. $R = kP(M - P)$

b. In the figure shown the constant of proportionality is $k = 2$ and the maximum sustainable population is $M = 10{,}000$.

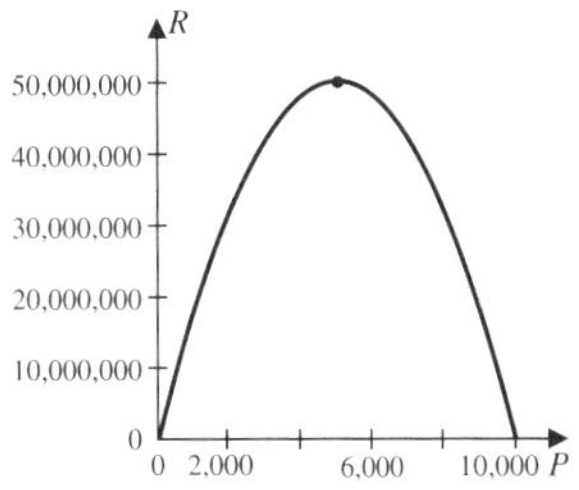

5. a. If x pairs of shoes are produced, the average cost for a pair of shoes is

$$\frac{C(x)}{x} = \frac{295 + 3.28x + 0.003x^2}{x} = \frac{295}{x} + 3.28 + 0.003x.$$

The graph of $y = \frac{295}{x} + 3.28 + 0.003x$ has a minimum when $x \approx 314$, so to minimize the average cost about 314 pairs of shoes should be produced.

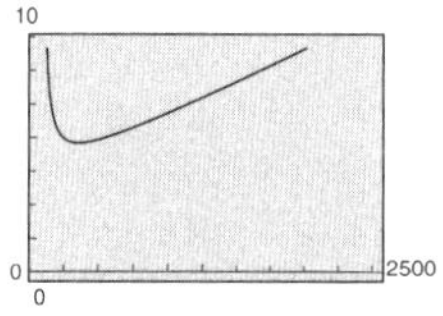

b. The revenue to the company if x pairs of shoes are sold is given by

$$R(x) = xp(x) = x\left(7.47 + \frac{321}{x}\right) = 7.47x + 321.$$

The profit is then given by

$$\begin{aligned} P(x) =& R(x) - C(x) = 7.47x + 321 - (295 + 3.28x + 0.003x^2) \\ =& 26 + 4.19x - 0.003x^2. \end{aligned}$$

The graph of $y = 26 + 4.19x - 0.003x^2$ has a maximum when $x \approx 698$, so to maximize the profit about 698 pairs of shoes should be produced.

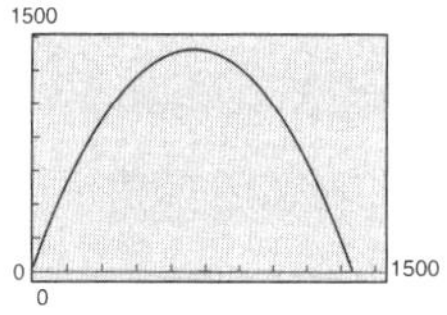

6. a. Since $PV = 8000$ we have $P = \frac{8000}{V}$. The average rate of change of P as V increases from 200 to 250 is

$$\begin{aligned} \frac{P(250) - P(200)}{250 - 200} =& \frac{\frac{8000}{250} - \frac{8000}{200}}{50} \\ =& \frac{32 - 40}{50} = -\frac{8}{50} = -\frac{4}{25} \text{ pounds/inch}^3. \end{aligned}$$

b. We have

$$\frac{P(200 + h) - P(200)}{h} = \frac{\frac{8000}{200+h} - \frac{8000}{200}}{h}$$

$$= \frac{8000(200) - 8000(200+h)}{h(200)(200+h)}$$
$$= \frac{-8000h}{h(200)(200+h)} = -\frac{40}{200+h}$$

So as h approaches 0, $-\frac{40}{200+h}$ approaches $-\frac{4}{200} = -\frac{1}{25}$, which is the instantaneous rate of change.

7. a. The first ten terms of the Fibonacci sequence are 1, 1, 2, 3, 5, 8, 13, 21, 34, and 55.

b. If d is an integer that divides both F_{n+1} and F_n, then d also divides

$$F_{n-1} = F_{n+1} - F_n.$$

Similarly, d must divide

$$F_{n-2} = F_n - F_{n-1}.$$

Proceeding in this manner, we eventually have the conclusion that d must divide $F_1 = 1$, hence $d = 1$.

8. For $f(x) = -x + b$, we have

$$\begin{aligned} f^1(x) &= -x + b \\ f^2(x) &= f(f(x)) = -(-x+b) + b = x - b + b = x \\ f^3(x) &= f(f^2(x)) = -x + b = f(x) \\ f^4(x) &= f(f^3(x)) = f(f(x)) = f^2(x) = x. \end{aligned}$$

So the powers cycle between $f(x) = -x + b$, and $f^2(x) = x$, which is the orbit.

9. a. We have

$$(f \circ f)(x) = f(f(x)) = \begin{cases} f(2x), & 0 \le x \le 1/2 \\ f(2-2x), & 1/2 < x \le 1 \end{cases}$$
$$= \begin{cases} 2(2x), & 0 \le x \le 1/4 \\ 2 - 2(2x), & 1/4 < x \le 1/2 \\ 2 - 2(2-2x), & 1/2 < x \le 3/4 \\ 2(2-2x), & 3/4 < x \le 1 \end{cases}$$
$$= \begin{cases} 4x, & 0 \le x \le 1/4 \\ 2 - 4x, & 1/4 < x \le 1/2 \\ -2 + 4x, & 1/2 < x \le 3/4 \\ 4 - 4x, & 3/4 < x \le 1 \end{cases}$$

b. The graphs of $y = f(x)$ and $y = (f \circ f)(x) = f(f(x))$ are shown.

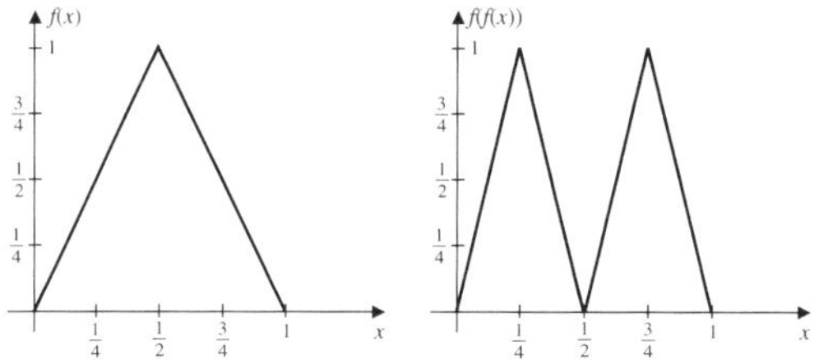

10. For all values of x we have

$$(f \circ g)(x) = f(bx + a) = a(bx + a) + b = abx + a^2 + b$$

and

$$(g \circ f)(x) = g(ax + b) = b(ax + b) + a = abx + b^2 + a.$$

If the two compositions are the same, then

$$a^2 + b = b^2 + a, \quad \text{so} \quad b - a = b^2 - a^2 = (b - a)(b + a).$$

So $f \circ g = g \circ f$ if and only if $b = a$ or $a + b = 1$.

11. Let $f(x) = \frac{x+a}{x+b}$, for $a \neq b$.

a. <u>f^{-1} exists:</u> First we need to show f is one-to-one. Since $a \neq b$, if $f(x_1) = f(x_2)$ then

$$\frac{x_1 + a}{x_1 + b} = \frac{x_2 + a}{x_2 + b} \text{ or } (x_1 + a)(x_2 + b) = (x_2 + a)(x_1 + b).$$

So

$$x_1x_2 + bx_1 + ax_2 + ab = x_1x_2 + bx_2 + ax_1 + ab$$

which implies $(b - a)x_1 = (b - a)x_2$ so $x_1 = x_2$, since $a \neq b$.

<u>$f^{-1}(x)$:</u> Solving for x in the equation $y = \dfrac{x + a}{x + b}$ gives

$$y(x + b) = x + a, \quad yx - x = a - yb, \quad x(y - 1) = a - yb \quad \text{and} \quad x = \frac{a - yb}{y - 1}.$$

So

$$f^{-1}(x) = \frac{a - bx}{x - 1}.$$

b. <u>Domain of f :</u> $\{x \mid x \neq -b\}$; <u>Range of f :</u> $\{x \mid x \neq 1\}$

Domain of f^{-1} : $\{x \mid x \neq 1\}$; Range of f^{-1} : $\{x \mid x \neq -b\}$

c.

$$(f^{-1} \circ f)(x) = f^{-1}(f(x)) = f^{-1}\left(\frac{x+a}{x+b}\right) = \frac{a - b\left(\frac{x+a}{x+b}\right)}{\left(\frac{x+a}{x+b} - 1\right)}$$

$$= \frac{\frac{a(x+b)-b(x+a)}{x+b}}{\frac{x+a-(x+b)}{x+b}} = \frac{\frac{ax-bx}{x+b}}{\frac{a-b}{x+b}} = \frac{ax - bx}{a - b} = \frac{x(a-b)}{a-b} = x$$

$$(f \circ f^{-1})(x) = f(f^{-1}(x)) = f\left(\frac{a - bx}{x - 1}\right) = \frac{\frac{a-bx}{x-1} + a}{\frac{a-bx}{x-1} + b}$$

$$= \frac{\frac{a-bx+a(x-1)}{x-1}}{\frac{a-bx+b(x-1)}{x-1}} = \frac{\frac{ax-bx}{x-1}}{\frac{a-b}{x-1}} = \frac{ax - bx}{a - b} = \frac{x(a-b)}{a-b} = x$$

Chapter 2 Chapter Test (Page 119)

1. True.

2. False. The range is $[1, \infty)$.

3. False. The function is increasing on $(-\infty, -1)$ and decreasing on $(-1, \infty)$.

4. False. The domain is $(-\infty, \infty)$ and the range is $(-\infty, 2]$.

5. True.

6. False. $f(-2) = -3$

7. True.

8. False. The function is constantly 2.

9. False. The function is constantly -1.

10. True.

11. False. Shift to the right 1 unit and upward 2 units.

12. True.

13. True.

14. False. $(g - f)(1) = -2$

15. False. $(f/g)(3) = 1/3$

16. False. The domain is $\{x | x \neq -3 \text{ and } x \neq 1\}$.

17. False. $(f \circ g)(-1) = -3$

18. False. $(g \circ f)(-1) = -3$

19. True.

20. True.

21. True.

22. False. It is sometimes the case that $(f \circ g)(x) = (g \circ f)(x)$.

23. True.

24. True.

25. False. $(f + g)(2) = 6$

26. False. $(f - g)(3) = 0$

27. False. $(f \circ g)(2) = 4$

28. True.

29. True.

30. False. $f^{-1}(x) = \frac{1}{2}\sqrt[3]{x+1}$

31. False. The scaling is by a factor greater than 1.

32. False. The black curve is reflected about the x-axis, and then the translations as stated are performed.

33. False. $y = -|x + 2| + 1$

34. True.

35. True.

36. True.

37. True.

38. False. $y = 2f(x)$

39. False. As $x \to 2^+$, $f(x) \to \infty$, but as $x \to 2^-$, $f(x) \to -\infty$.

40. True.

41. True.

42. True.

Exercise Set 3.2 (Page 133)

1. **a.** The factor $(x + 1)^2$ implies the function has a zero of multiplicity 2 at $x = -1$ so the curve just touches at $x = -1$ without crossing through $x = -1$ as in (iii).

b. The factor $(x - 1)^2$ implies the function has a zero of multiplicity 2 at $x = 1$ so the curve just touches at $x = 1$ without crossing through $x = 1$ as in (iv).

c. Curve (i) is the only curve that does not exhibit the behavior of zeros of multiplicity greater than 1, and has zeros at $x = 0$, $x = 1$, and $x = -1$.

d. The factor x^2 implies the function has a zero of multiplicity 2 at the origin so the curve just touches at the origin without crossing through the origin as in (ii).

2. **a.** The only curve with zeros at $x = 1$, $x = -2$ and $x = 2$ is (iii).

b. The factor x^2 implies the function has a zero of multiplicity 2 at the origin so the curve just touches at the origin without crossing through the origin as in (iv).

c. The equation implies the curve has a zero of multiplicity 2 at $x = 2$ so the curve just touches at $x = 2$ without crossing the x-axis at $x = 2$ as in (i).

d. Since $y = x(x + 1)(2 - x) = -x(x + 1)(x - 2)$, the end behavior is similar to that of $-x^3$ as in (ii).

3. The graph shows a zero of at least multiplicity 2 at the origin and another positive zero so the lowest possible degree for the polynomial is 3.

4. The graphs shows 5 distinct zeros, so the lowest possible degree for the polynomial is 5.

5. The graph can be shifted downward so that the shifted graph has 4 distinct zeros so the lowest possible degree for the polynomial is 4.

6. The graph can be shifted downward so that the shifted graph has a zero of multiplicity at least 3 (the flattened area on the left) and 2 other distinct zeros so the lowest possible degree for the polynomial is 5.

7. Vertically stretch the graph of $y = x^3$ by a factor of 3, and then shift the resulting graph downward 2 units.

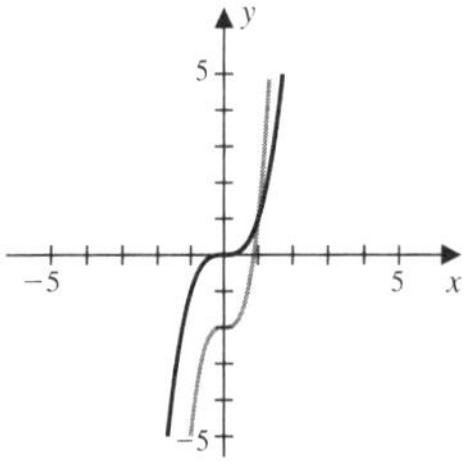

8. Vertically shrink the graph of $y = x^3$ by a factor of 2, and then shift the resulting graph upward 2 units.

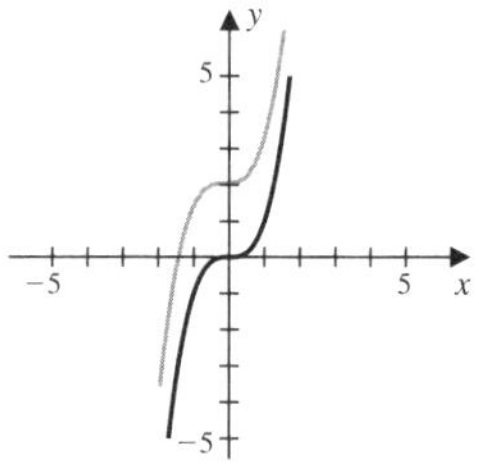

9. Shift the graph of $y = x^4$ to the right 2 units, and then shift the result upward 1 unit.

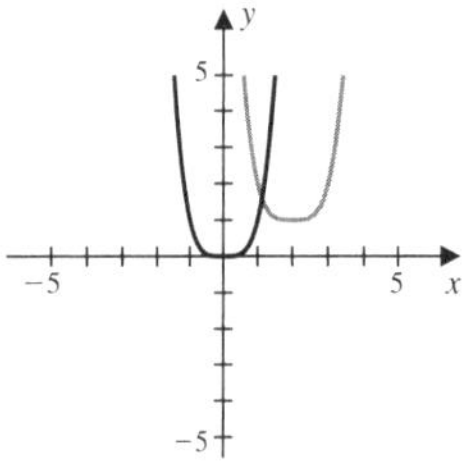

10. Shift the graph of $y = x^3$ to the left 1 unit, and then shift the result downward 2 units.

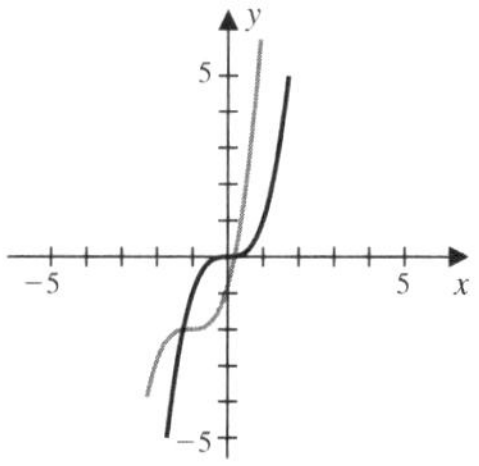

11. Vertically shrink the graph of $y = x^3$ by a factor of 2, reflect about the x-axis, shift to the right 3 units, and then shift downward 3 units.

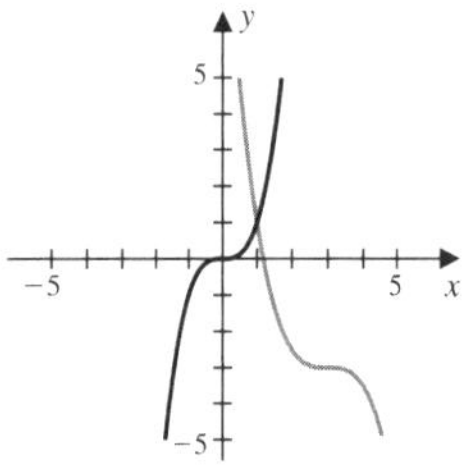

12. Vertically stretch the graph of $y = x^4$ by a factor of 2, reflect about the x-axis, shift to the left 2 units, and then shift upward 3 units.

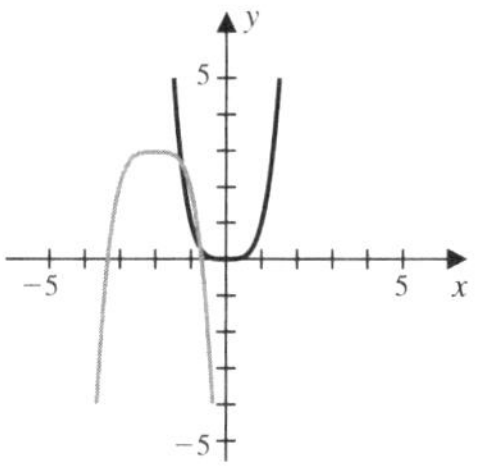

13. $f(x) = (x - 1)(x + 1)(x - 2)$

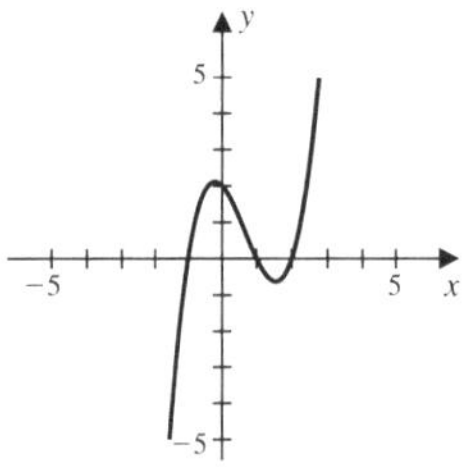

14. $f(x) = (x + 3)(x - 2)(x - 1)$

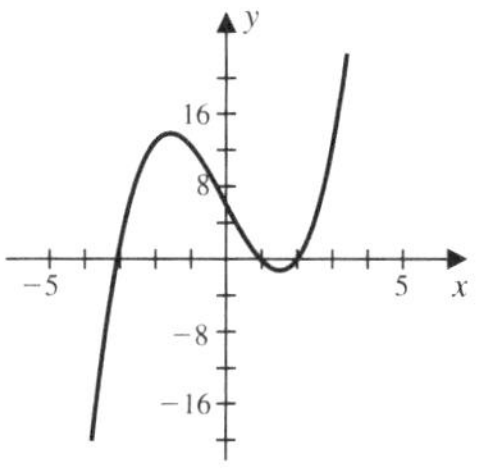

15. $f(x) = x(x+1)(x-1)(x-2)$

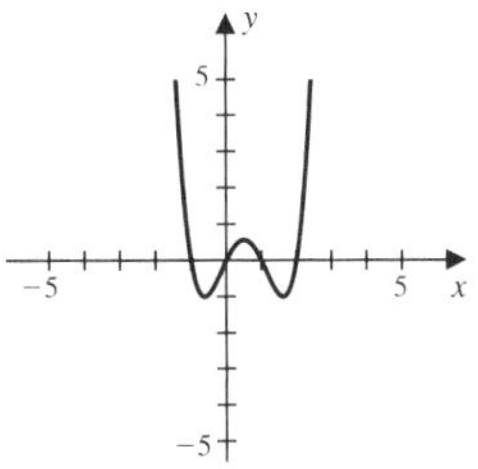

16. $f(x) = -x(x+1)(x+2)(x-3)$

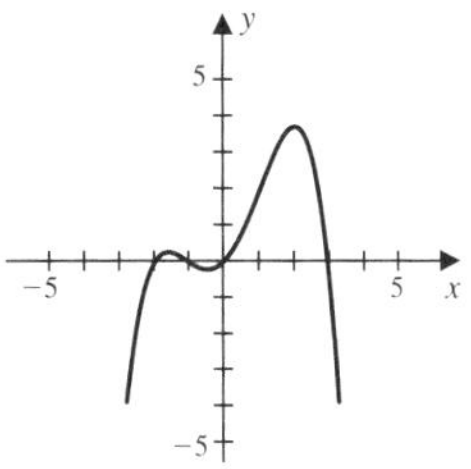

17. $f(x) = -x^3(x-1)$

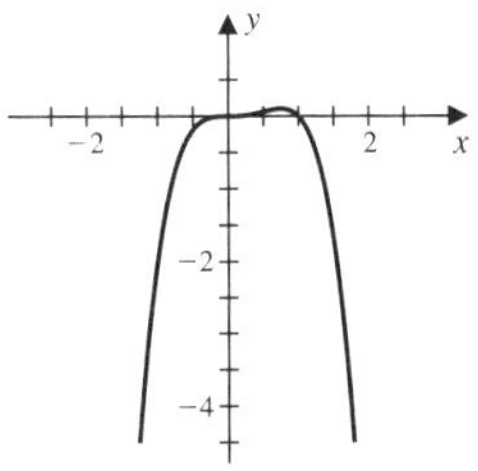

18. $f(x) = x^2(x+2)(x-1)$

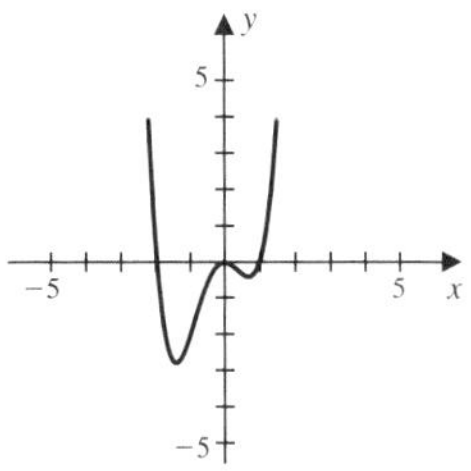

19. $f(x) = -x^2(x+1)^2$

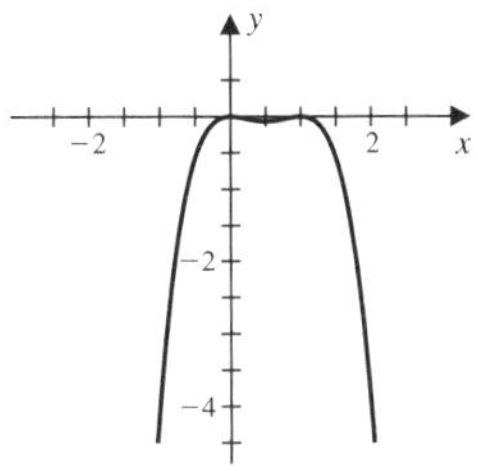

20. $f(x) = (x-1)^3(x+1)^3$

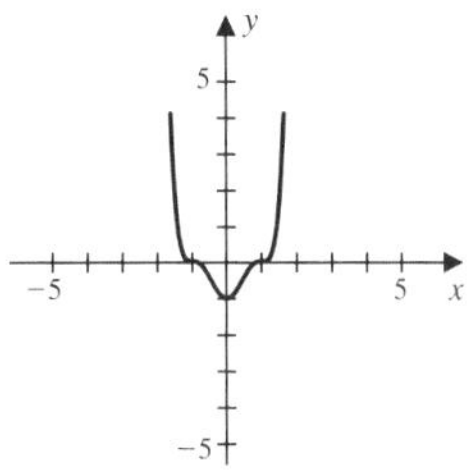

21.

$$\begin{aligned} f(x) =& x^3 - x^2 - 6x \\ =& x(x^2 - x - 6) \\ =& x(x-3)(x+2) \end{aligned}$$

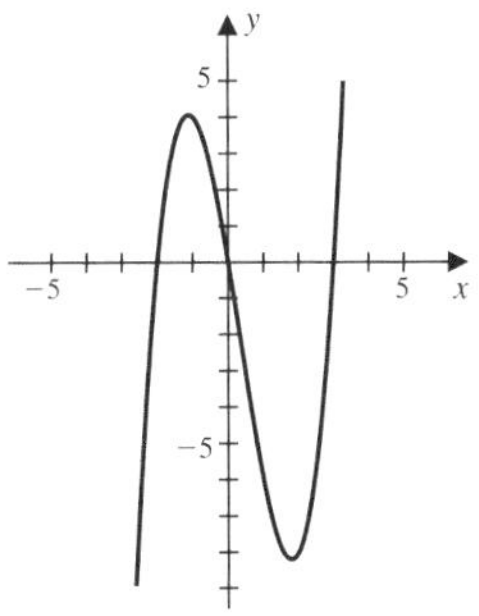

22.

$$\begin{aligned} f(x) =& x^4 - x^3 - 2x^2 \\ =& x^2(x^2 - x - 2) \\ =& x^2(x-2)(x+1) \end{aligned}$$

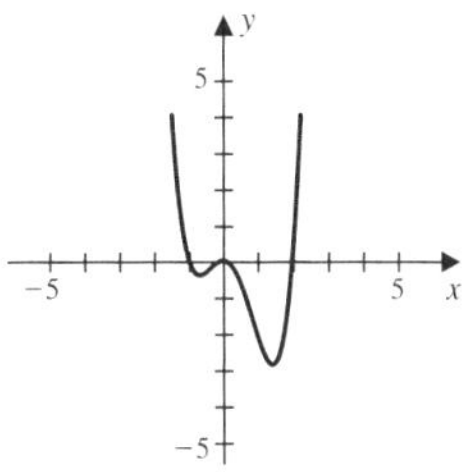

23. $y = f(x - 2)$

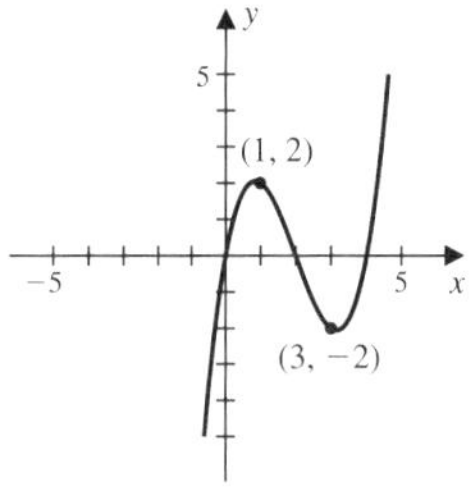

24. $y = f(x + 1)$

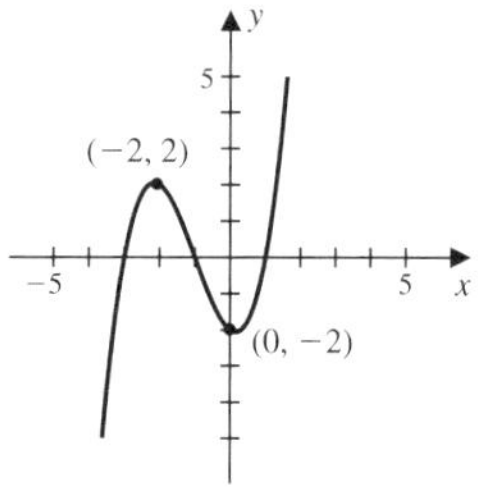

25. $y = f(x - 1) + 2$

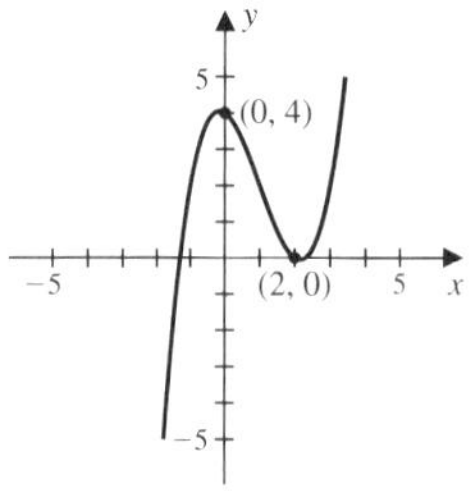

26. $y = -f(x)$

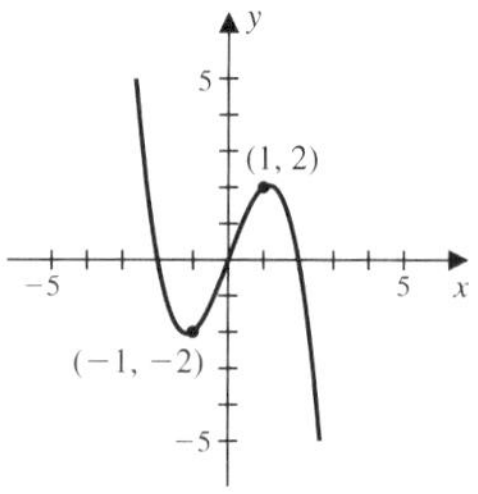

27. $y = f(-x)$

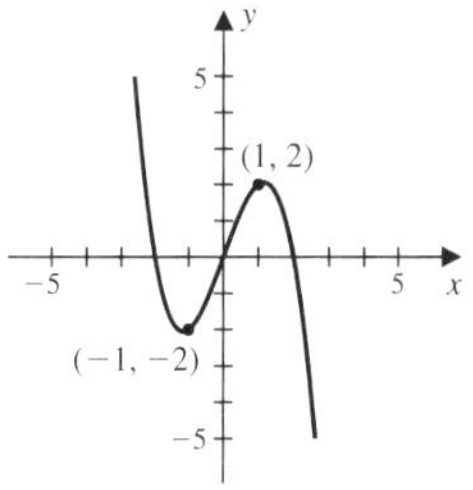

28. $y = -f(-x - 2) = -f(-(x + 2))$

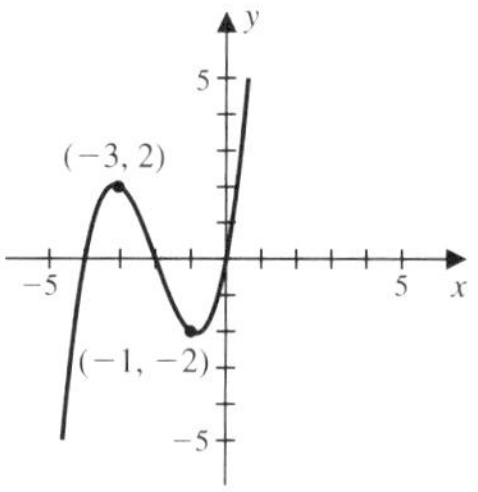

29. $y = |f(x)|$

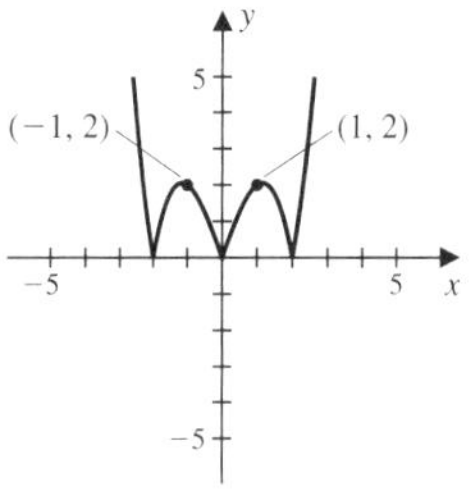

30. $y = f(|x|)$

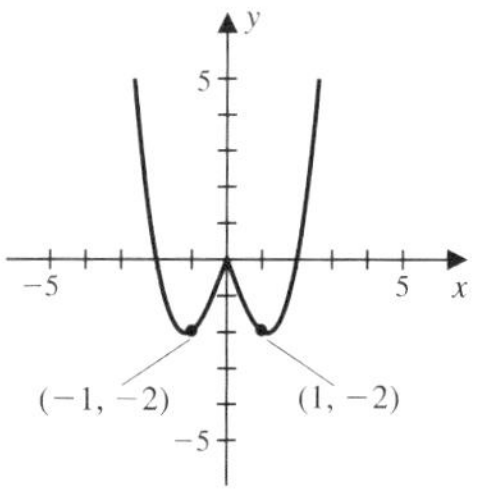

31. **a.** The graph of the polynomial is shown.

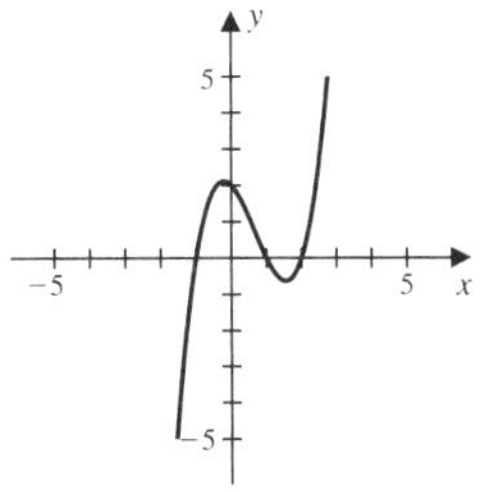

b. The least possible degree of the polynomial is 3.

32. **a.** The graph of the polynomial is shown.

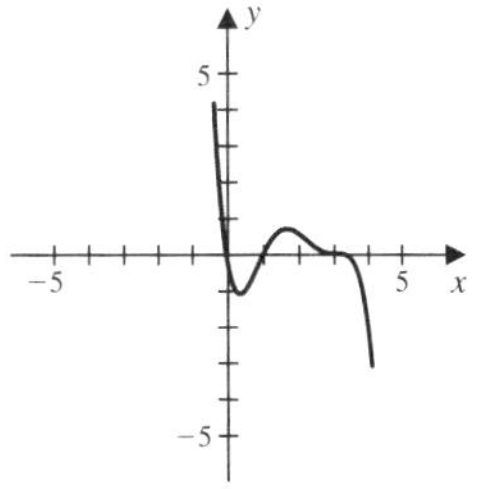

b. The least possible degree of the polynomial is 5.

33. The polynomial has the form

$$P(x) = a(x-1)^2(x-2),$$

and

$$4 = P(-1) = a(-2)^2(-3) = -12a$$

implies $a = -\frac{1}{3}$.
So

$$\begin{aligned} P(x) &= -\frac{1}{3}(x-1)^2(x-2) \\ &= -\frac{1}{3}x^3 + \frac{4}{3}x^2 - \frac{5}{3}x + \frac{2}{3}. \end{aligned}$$

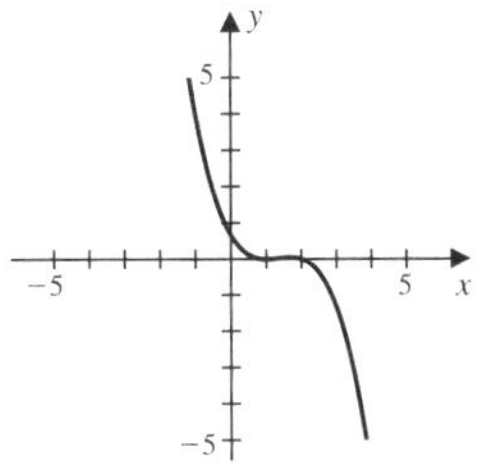

34. First, we must have for some constant a,

$$P(x) = a(x-1)(x-2)(x+2).$$

Since the y- intercept is 2, the graph passes through the point $(0, 2)$. So

$$2 = P(0) = a(-1)(-2)(2)$$

implies $2 = 4a$, so $a = \frac{1}{2}$
and

$$\begin{aligned} P(x) &= \frac{1}{2}(x-1)(x-2)(x+2) \\ &= \frac{1}{2}x^3 - \frac{1}{2}x^2 - 2x + 2. \end{aligned}$$

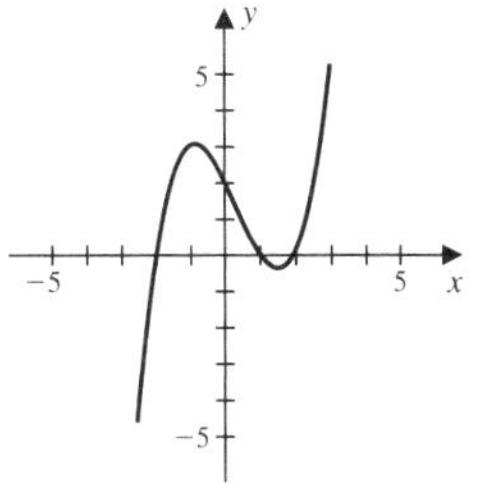

35. First, we must have for some constant a,

$$P(x) = a(x+2)x^2(x-2) = a(x^4 - 4x^2),$$

and since $(1, 2)$ is on the graph,

$$2 = P(1) = a(1-4) = -3a \text{ implies } a = -\frac{2}{3}.$$

The polynomial is

$$P(x) = -\frac{2}{3}x^4 + \frac{8}{3}x^2.$$

36. First, we must have for some constant a,

$$P(x) = a(x+1)x(x-2)^3,$$

and since $(1, -1)$ is on the graph,

$$-1 = P(1) = a(2)(1)(-1)^3 = -2a$$

implies $a = \frac{1}{2}$.
The polynomial is

$$\begin{aligned} P(x) &= \frac{1}{2}(x+1)x(x-2)^3 \\ &= \frac{1}{2}x^5 - \frac{5}{2}x^4 + 3x^3 + 2x^2 - 4x. \end{aligned}$$

37. For $f_n(x) = x^n$, if $0 < x < 1$, $f_n(x) = x^n > x^{n+1} = f_{n+1}(x)$ and if $x > 1$, $f_n(x) = x^n < x^{n+1} = f_{n+1}(x)$.

38. The graph shows plots of $y = P(x) = x^4 + x^3 - x^2 - x$ and $y = Q(x) = 4 - x^2$. The points where $P(x) < Q(x)$ correspond to the portions of the graph of $y = P(x)$ that lie below the graph of $y = Q(x)$. The points of intersection of the two graphs are approximately $x = -1.6$ and $x = 1.3$, so $P(x) < Q(x)$ for $-1.6 < x < 1.3$.

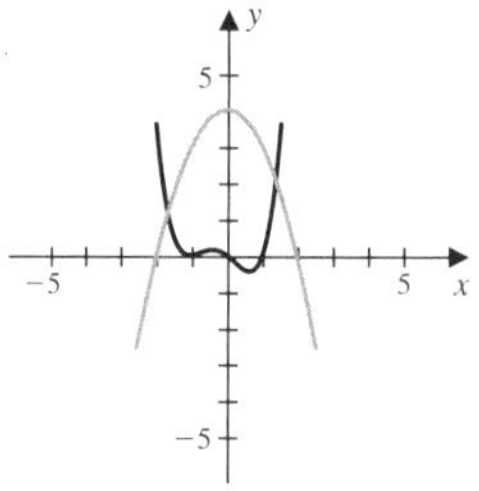

39. a. $f(x) = x^4 + x^3 - 2x^2$ is increasing on $(-1.4, 0) \cup (0.7, \infty)$ and is decreasing on $(-\infty, -1.4) \cup (0, 0.7)$.
b. Local minima: $(-1.4, -2.8)$, $(0.7, -0.4)$; local maximum: $(0, 0)$

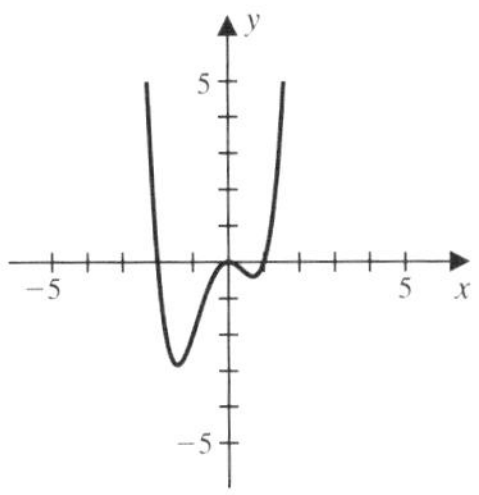

40. a. $f(x) = x^4 - 2x^3$ is increasing on $(1.5, \infty)$ and is decreasing on $(-\infty, 1.5)$.
b. Local minimum: $(1.5, -1.7)$; local maximum: none

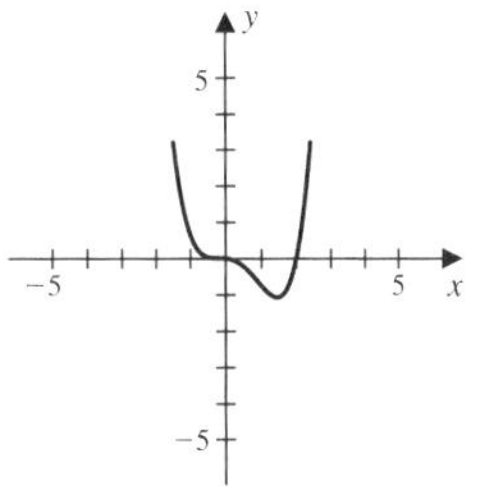

41. a. $f(x) = \frac{1}{5}x^5 - \frac{5}{4}x^4 + \frac{5}{3}x^3 + \frac{5}{2}x^2 - 6x + 1$ is increasing on $(-\infty, -1) \cup (1, 2) \cup (3, \infty)$ and is decreasing on $(-1, 1) \cup (2, 3)$.
b. Local minima: $(1, -1.9)$, $(3, -2.2)$; local maxima: $(-1, 6.4)$, $(2, -1.3)$

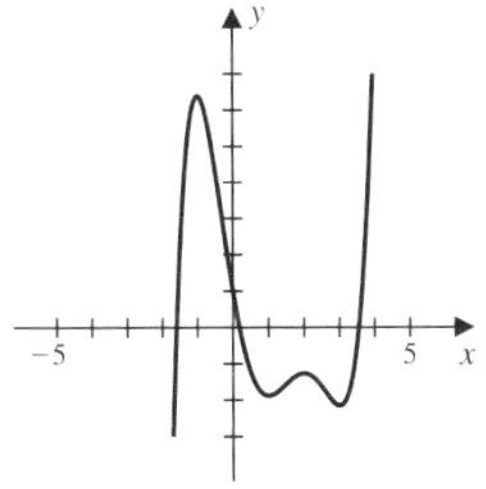

42. a. $f(x) = x^5 - 2x^4 + 2x^2 - x$ is increasing on $(-\infty, -0.7) \cup (0.3, \infty)$ and is decreasing on $(-0.7, 0.3)$.
b. Local minimum: $(0.3, -0.2)$; local maximum: $(-0.7, 1)$

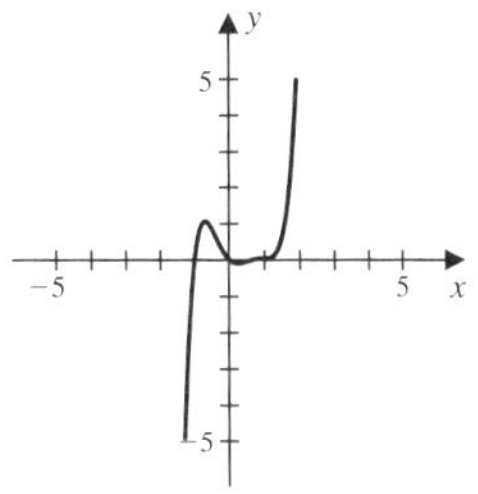

43. For $f(x) = x^3 - 2x^2 + x + 2$ with $a = -3$ and $b = 2$.
a. $y = f(x)$ **b.** $y = -f(x)$ **c.** $y = f(-x)$ **d.** $y = |f(x)|$ **e.** $y = f(x-3) + 2$

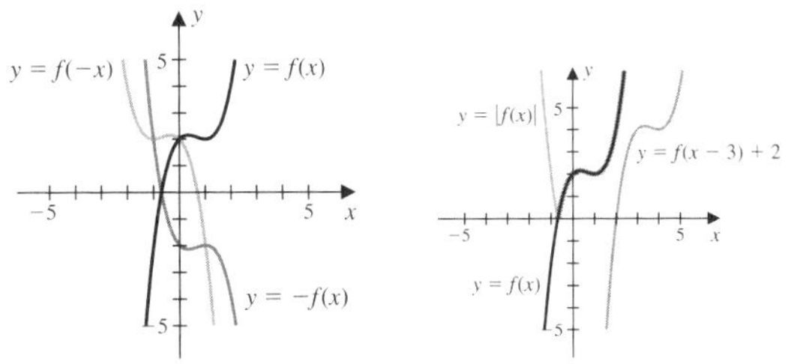

44. For $f(x) = -x^3 + 7x^2 - 36$ with $a = 7$ and $b = -30$.

a. $y = f(x)$ **b.** $y = -f(x)$ **c.** $y = f(-x)$ **d.** $y = |f(x)|$ **e.** $y = f(x + 7) - 30$

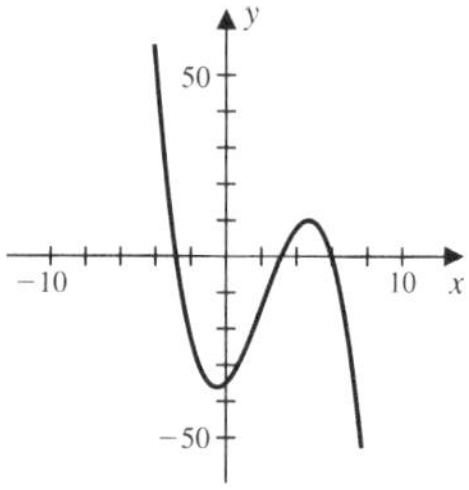

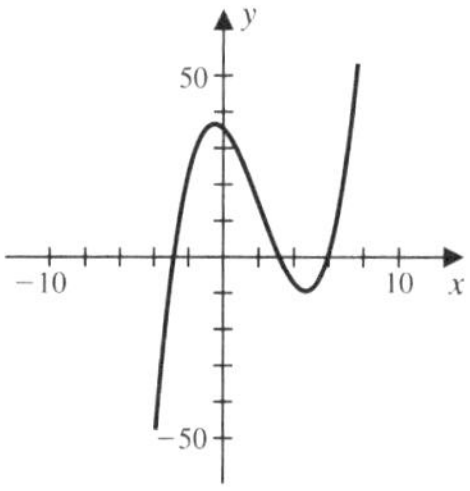

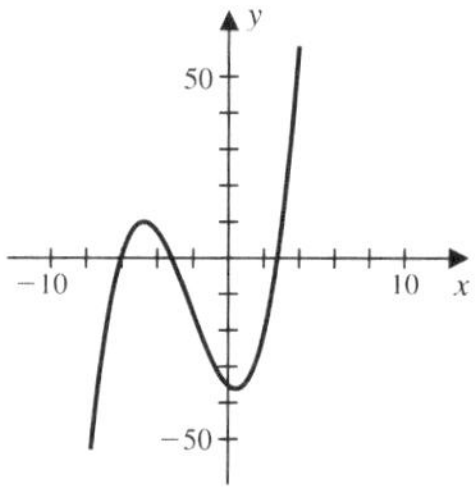

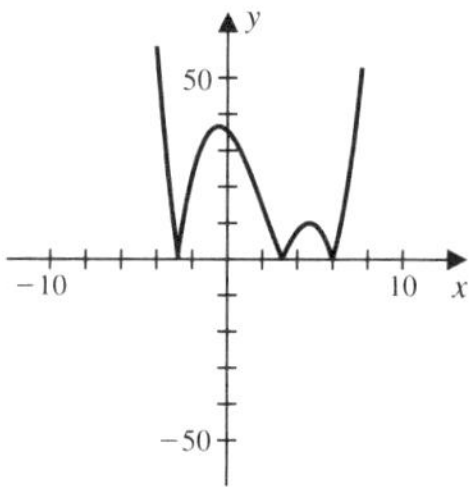

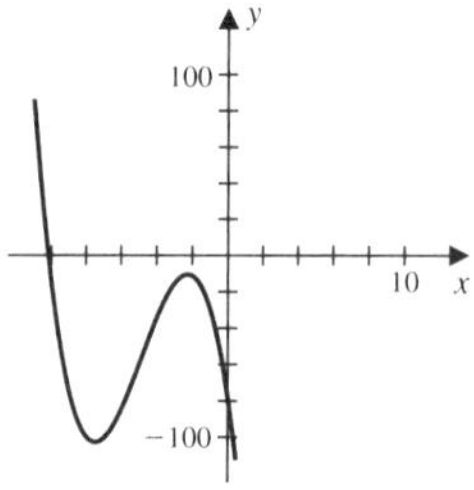

45. For $f(x) = x^4 - 400x^2$ with $a = -10$ and $b = 30000$.

a. $y = f(x)$ **b.** $y = -f(x)$ **c.** $y = f(-x)$ **d.** $y = |f(x)|$ **e.** $y = f(x - 10) + 30000$

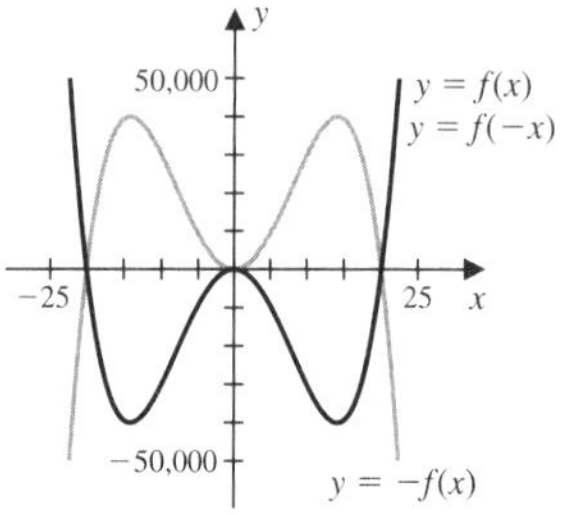

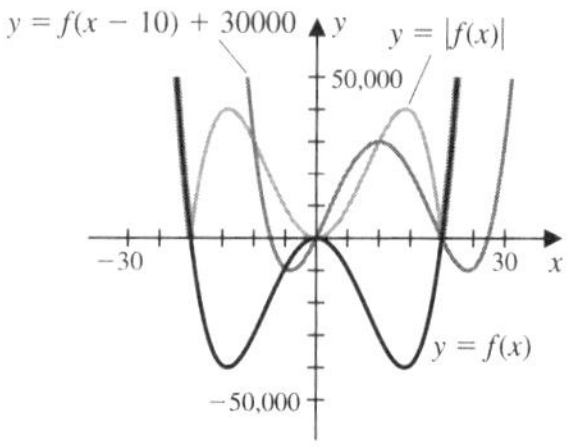

46. For $f(x) = x^5 + x^4 - 13x^3 - x^2 + 48x - 36$ with $a = 5$ and $b = -10$.

a. $y = f(x)$ **b.** $y = -f(x)$

c. $y = f(-x)$ **d.** $y = |f(x)|$

e. $y = f(x+5) - 10$

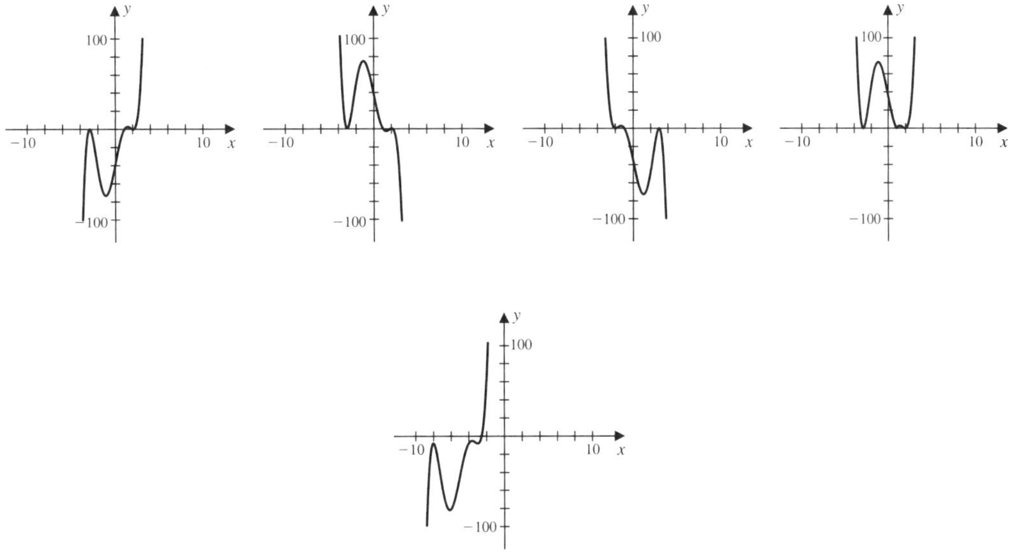

47. **a.** $V(x) = x(20 - 2x)(20 - 2x) = 400x - 80x^2 + 4x^3$

b. The graph shows $y = -500 + 400x - 80x^2 + 4x^3$. The graph crosses the x-axis at $x = 5$ and $x \approx 1.9$, which are values of x that produce a volume of 500 in^3. Note the graph also crosses the x-axis at $x \approx 13$. However this value is ignored since $0 \le x \le 10$.

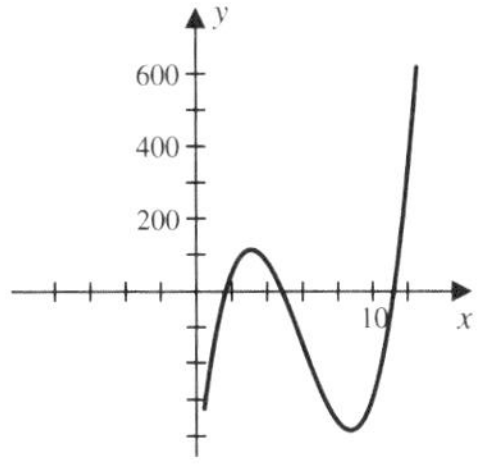

48. The volume of the cylindrical section is $\pi r^2 h = 10\pi r^2$ and the two hemispherical parts together form a complete sphere with volume $\frac{4}{3}\pi r^3$. So the volume of the tank is $V(r) = \frac{4}{3}\pi r^3 + 10\pi r^2$ and from the graph of $y = \frac{4}{3}\pi r^3 + 10\pi r^2 - 50$ we have $V(r) = 50$ when $r \approx 1.2$ meters.

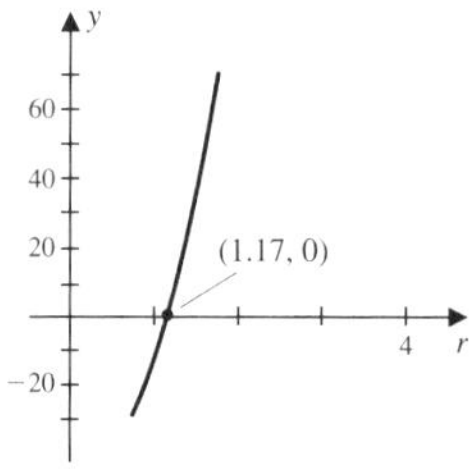

Exercise Set 3.3 (Page 144)

1. The quotient is $Q(x) = 3x+1$ and remainder is $R(x) = 3$.

$$
\begin{array}{r|rrrrr}
 & 3x & + & 1 & & \\
\cline{2-6}
x-1 & 3x^2 & - & 2x & + & 2 \\
 & 3x^2 & - & 3x & & \\
\cline{2-4}
 & & & x & + & 2 \\
 & & & x & - & 1 \\
\cline{4-6}
 & & & & & 3
\end{array}
$$

2. The quotient is $Q(x) = 2x-7$ and remainder is $R(x) = 18$.

$$
\begin{array}{r|rrrrr}
 & 2x & - & 7 & & \\
\cline{2-6}
x+2 & 2x^2 & - & 3x & + & 4 \\
 & 2x^2 & + & 4x & & \\
\cline{2-4}
 & & & -7x & + & 4 \\
 & & & -7x & - & 14 \\
\cline{4-6}
 & & & & & 18
\end{array}
$$

3. The quotient is $Q(x) = x^2 + 2x + 2$ and remainder is $R(x) = 0$.

$$
\begin{array}{r|rrrrrrr}
 & x^2 & + & 2x & + & 2 & & \\
\cline{2-8}
x-1 & x^3 & + & x^2 & + & 0x & - & 2 \\
 & x^3 & - & x^2 & & & & \\
\cline{2-4}
 & & & 2x^2 & + & 0x & & \\
 & & & 2x^2 & - & 2x & & \\
\cline{4-6}
 & & & & & 2x & - & 2 \\
 & & & & & 2x & - & 2 \\
\cline{6-8}
 & & & & & & & 0
\end{array}
$$

4. The quotient is $Q(x) = 3x^2 - 4x + 11$ and remainder is $R(x) = -27x + 13$.

$$
\begin{array}{r|rrrrrrrrr}
 & 3x^2 & - & 4x & + & 11 & & & & \\ \cline{2-10}
x^2+2x-1 & 3x^4 & + & 2x^3 & + & 0x^2 & - & x & + & 2 \\
 & 3x^4 & + & 6x^3 & - & 3x^2 & & & & \\ \cline{2-6}
 & & - & 4x^3 & + & 3x^2 & - & x & & \\
 & & - & 4x^3 & - & 8x^2 & + & 4x & & \\ \cline{3-8}
 & & & & & 11x^2 & - & 5x & + & 2 \\
 & & & & & 11x^2 & + & 22x & - & 11 \\ \cline{6-10}
 & & & & & & - & 27x & + & 13
\end{array}
$$

5. The quotient is $Q(x) = 2x^2 + 6$ and remainder is $R(x) = 7x + 4$.

$$
\begin{array}{r|rrrrrrrrr}
 & 2x^2 & + & 6 & & & & & & \\ \cline{2-10}
x^2-x-1 & 2x^4 & - & 2x^3 & + & 4x^2 & + & x & - & 2 \\
 & 2x^4 & - & 2x^3 & - & 2x^2 & & & & \\ \cline{2-6}
 & & & & & 6x^2 & + & x & - & 2 \\
 & & & & & 6x^2 & - & 6x & - & 6 \\ \cline{6-10}
 & & & & & & & 7x & + & 4
\end{array}
$$

6. The quotient is $Q(x) = 3x + 4$ and remainder is $R(x) = 9x^3 + 2x^2 - 8x - 5$.

$$
\begin{array}{r|rrrrrrrrrrr}
 & 3x & + & 4 & & & & & & & & \\ \cline{2-12}
x^4-2x^3+x+1 & 3x^5 & - & 2x^4 & + & x^3 & + & 5x^2 & - & x & - & 1 \\
 & 3x^5 & - & 6x^4 & & & + & 3x^2 & + & 3x & & \\ \cline{2-10}
 & & & 4x^4 & + & x^3 & + & 2x^2 & - & 4x & - & 1 \\
 & & & 4x^4 & - & 8x^3 & & & + & 4x & + & 4 \\ \cline{3-12}
 & & & & & 9x^3 & + & 2x^2 & - & 8x & - & 5
\end{array}
$$

7. We have $P(x) = x^3 - 5x^2 + 8x - 4$ and $P(1) = 1^3 - 5(1)^2 + 8(1) - 4 = 0$, so $x - 1$ is a factor. Dividing gives $P(x) = (x-1)(x^2 - 4x + 4) = (x-1)(x-2)(x-2) = (x-1)(x-2)^2$.

$$
\begin{array}{r|rrrrrrr}
 & x^2 & - & 4x & + & 4 & & \\ \cline{2-8}
x-1 & x^3 & - & 5x^2 & + & 8x & - & 4 \\
 & x^3 & - & x^2 & & & & \\ \cline{2-4}
 & & - & 4x^2 & + & 8x & & \\
 & & - & 4x^2 & + & 4x & & \\ \cline{3-6}
 & & & & & 4x & - & 4 \\
 & & & & & 4x & - & 4 \\ \cline{6-8}
 & & & & & & & 0
\end{array}
$$

8. We have $P(x) = 3x^4 + 5x^3 - 5x^2 - 5x + 2$ and

$$P(1) = 3(1)^4 + 5(1)^3 - 5(1)^2 - 5(1) + 2 = 0,$$

so $x - 1$ is a factor. Dividing gives $P(x) = (x - 1)(3x^3 + 8x^2 + 3x - 2)$.

$$
\begin{array}{r|rrrrrrrrr}
 & 3x^3 & + & 8x^2 & + & 3x & - & 2 & & \\ \cline{2-10}
x-1 & 3x^4 & + & 5x^3 & - & 5x^2 & - & 5x & + & 2 \\
 & 3x^4 & - & 3x^3 & & & & & & \\ \cline{2-4}
 & & & 8x^3 & - & 5x^2 & & & & \\
 & & & 8x^3 & - & 8x^2 & & & & \\ \cline{4-6}
 & & & & & 3x^2 & - & 5x & & \\
 & & & & & 3x^2 & - & 3x & & \\ \cline{6-8}
 & & & & & & - & 2x & + & 2 \\
 & & & & & & - & 2x & + & 2 \\ \cline{7-10}
 & & & & & & & & & 0
\end{array}
$$

If $x = -2$ is a zero of $P(x)$, then it will also be a zero of $Q(x) = 3x^3 + 8x^2 + 3x - 2$. Since

$$Q(-2) = 3(-2)^3 + 8(-2)^2 + 3(-2) - 2 = 0,$$

$x + 2$ is a factor of $Q(x)$ and also of $P(x)$. Then
$P(x) = (x - 1)(x + 2)(3x^2 + 2x - 1) = (x - 1)(x + 2)(3x - 1)(x + 1)$.

$$
\begin{array}{r|rrrrrrr}
 & 3x^2 & + & 2x & - & 1 & & \\ \cline{2-8}
x+2 & 3x^3 & + & 8x^2 & + & 3x & - & 2 \\
 & 3x^3 & + & 6x^2 & & & & \\ \cline{2-4}
 & & & 2x^2 & + & 3x & & \\
 & & & 2x^2 & + & 4x & & \\ \cline{3-6}
 & & & & - & x & - & 2 \\
 & & & & - & x & - & 2 \\ \cline{5-8}
 & & & & & & & 0
\end{array}
$$

9. We have $P(x) = 2x^4 + 3x^3 - 12x^2 - 7x + 6$ and $P(-1) = 2 - 3 - 12 + 7 + 6 = 0$,
so $x + 1$ is a factor. Dividing gives $P(x) = (x + 1)(2x^3 + x^2 - 13x + 6)$.

$$
\begin{array}{r|rrrrrrrrr}
 & 2x^3 & + & x^2 & - & 13x & + & 6 & & \\ \cline{2-10}
x+1 & 2x^4 & + & 3x^3 & - & 12x^2 & - & 7x & + & 6 \\
 & 2x^4 & + & 2x^3 & & & & & & \\ \cline{2-4}
 & & & x^3 & - & 12x^2 & & & & \\
 & & & x^3 & + & x^2 & & & & \\ \cline{4-6}
 & & & & - & 13x^2 & - & 7x & & \\
 & & & & - & 13x^2 & - & 13x & & \\ \cline{5-8}
 & & & & & & & 6x & + & 6 \\
 & & & & & & & 6x & + & 6 \\ \cline{8-10}
 & & & & & & & & & 0
\end{array}
$$

If $x = -3$ is a zero of $P(x)$, then it will also be a zero of $Q(x) = 2x^3 + x^2 - 13x + 6$. Since

$$Q(-3) = 2(-3)^3 + (-3)^2 - 13(-3) + 6 = 0,$$

$x + 3$ is a factor of $Q(x)$ and also of $P(x)$. Then
$P(x) = (x + 1)(x + 3)(2x^2 - 5x + 2) = (x + 1)(x + 3)(2x - 1)(x - 2)$.

$$\begin{array}{r|rrrr}
 & 2x^2 & -\ 5x & +\ 2 & \\ \cline{2-5}
x+3 & 2x^3 & +\ x^2 & -\ 13x & +\ 6 \\
 & 2x^3 & +\ 6x^2 & & \\ \cline{2-3}
 & & -\ 5x^2 & -\ 13x & \\
 & & -\ 5x^2 & -\ 15x & \\ \cline{3-4}
 & & & 2x & +\ 6 \\
 & & & 2x & +\ 6 \\ \cline{4-5}
 & & & & 0
\end{array}$$

10. We have $P(x) = 6x^4 - 13x^3 + 13x - 6$ and

$$P(-1) = 6 + 13 - 13 - 6 = 0 \text{ and } P(1) = 6 - 13 + 13 - 6 = 0,$$

so $x + 1$ and $x - 1$ are factors. Then $(x + 1)(x - 1) = x^2 - 1$ is a factor and dividing gives
$P(x) = (x + 1)(x - 1)(6x^2 - 13x + 6) = (x + 1)(x - 1)(3x - 2)(2x - 3)$.

$$\begin{array}{r|rrrrr}
 & 6x^2 & -\ 13x & +\ 6 & & \\ \cline{2-6}
x^2-1 & 6x^4 & -\ 13x^3 & +\ 0x^2 & +\ 13x & -\ 6 \\
 & 6x^4 & & -\ 6x^2 & & \\ \cline{2-4}
 & & -\ 13x^3 & +\ 6x^2 & +\ 13x & \\
 & & -\ 13x^3 & & +\ 13x & \\ \cline{3-5}
 & & & 6x^2 & & -\ 6 \\
 & & & 6x^2 & & -\ 6 \\ \cline{4-6}
 & & & & & 0
\end{array}$$

11. We have $P(x) = 3x^3 + x^2 - 8x + 4$ and

$$P\left(\frac{2}{3}\right) = 3\left(\frac{8}{27}\right) + \frac{4}{9} - \frac{16}{3} + 4 = \frac{8}{9} + \frac{4}{9} - \frac{48}{9} + \frac{36}{9} = 0,$$

so $x - \frac{2}{3}$ is a factor. Dividing gives

$$\begin{aligned}
P(x) = &\left(x - \frac{2}{3}\right)(3x^2 + 3x - 6) = \left(x - \frac{2}{3}\right)3(x^2 + x - 2) \\
=&3\left(x - \frac{2}{3}\right)(x + 2)(x - 1) = 3\frac{1}{3}(3x - 2)(x + 2)(x - 1) \\
=&(3x - 2)(x + 2)(x - 1).
\end{aligned}$$

$$
\begin{array}{r|rrrrrrr}
\multicolumn{1}{r}{} & 3x^2 & + & 3x & - & 6 & & \\
\cline{2-8}
x-\frac{2}{3} & 3x^3 & + & x^2 & - & 8x & + & 4 \\
 & 3x^3 & - & 2x^2 & & & & \\
\cline{2-4}
 & & & 3x^2 & - & 8x & & \\
 & & & 3x^2 & - & 2x & & \\
\cline{4-6}
 & & & & - & 6x & + & 4 \\
 & & & & - & 6x & + & 4 \\
\cline{6-8}
 & & & & & & & 0
\end{array}
$$

12. We have $P(x) = 2x^3 - 5x^2 - 4x + 3$ and

$$P\left(\frac{1}{2}\right) = \frac{2}{8} - \frac{5}{4} - 2 + 3 = \frac{1}{4} - \frac{5}{4} + 1 = 0,$$

so $x - \frac{1}{2}$ is a factor. Dividing gives

$$
\begin{aligned}
P(x) &= \left(x - \frac{1}{2}\right)(2x^2 - 4x - 6) = \left(x - \frac{1}{2}\right) 2(x^2 - 2x - 3) \\
&= 2\left(x - \frac{1}{2}\right)(x-3)(x+1) = 2\frac{1}{2}(2x-1)(x-3)(x+1) \\
&= (2x-1)(x-3)(x+1).
\end{aligned}
$$

$$
\begin{array}{r|rrrrrrr}
\multicolumn{1}{r}{} & 2x^2 & - & 4x & - & 6 & & \\
\cline{2-8}
x-\frac{1}{2} & 2x^3 & - & 5x^2 & - & 4x & + & 3 \\
 & 2x^3 & - & x^2 & & & & \\
\cline{2-4}
 & & - & 4x^2 & - & 4x & & \\
 & & - & 4x^2 & + & 2x & & \\
\cline{4-6}
 & & & & - & 6x & + & 3 \\
 & & & & - & 6x & + & 3 \\
\cline{6-8}
 & & & & & & & 0
\end{array}
$$

13. The Rational Zero Test gives the possible rational zeros

$$\frac{\text{factors of } 6}{\text{factors of } 1} = \frac{\pm 1, \pm 2, \pm 3, \pm 6}{\pm 1} = \pm 1, \pm 2, \pm 3, \pm 6.$$

14. The Rational Zero Test gives gives the possible rational zeros

$$\frac{\text{factors of } 16}{\text{factors of } 1} = \frac{\pm 1, \pm 2, \pm 4, \pm 8, \pm 16}{\pm 1} = \pm 1, \pm 2, \pm 4, \pm 8, \pm 16.$$

15. Since the rational zeros will be the same if we divide the equation by 2, consider the equation $5x^5 - 7x^3 + 9x^2 + 3x - 2 = 0$. The possible rational zeros are

$$\frac{\text{factors of } 2}{\text{factors of } 5} = \frac{\pm 1, \pm 2}{\pm 1, \pm 5} = \pm 1, \pm 2, \pm\frac{1}{5}, \pm\frac{2}{5}.$$

16. Dividing the equation by 2, the possible rational zeros are

$$\frac{\text{factors of } 2}{\text{factors of } 3} = \frac{\pm 1, \pm 2}{\pm 1, \pm 3} = \pm 1, \pm 2, \pm\frac{1}{3}, \pm\frac{2}{3}.$$

17. Since the rational zeros will be the same if we multiply the equation by 3, consider the equation $2x^3 - 19x^2 + 54x - 45 = 0$. The possible rational zeros are

$$\frac{\text{factors of } 45}{\text{factors of } 2} = \frac{\pm 1, \pm 3, \pm 5, \pm 9, \pm 15, \pm 45}{\pm 1, \pm 2}$$
$$= \pm\, 1, \pm 3, \pm 5, \pm 9, \pm 15, \pm 45, \pm\frac{1}{2}, \pm\frac{3}{2}, \pm\frac{5}{2}, \pm\frac{9}{2}, \pm\frac{15}{2}, \pm\frac{45}{2}.$$

18. Since the rational zeros will be the same if we multiply the equation by 8, consider the equation $6x^3 + x^2 - 18x + 8 = 0$.The possible rational zeros are

$$\frac{\text{factors of } 8}{\text{factors of } 6} = \frac{\pm 1, \pm 2, \pm 4, \pm 8}{\pm 1, \pm 2, \pm 3, \pm 6} = \pm 1, \pm 2, \pm 4, \pm 8, \pm\frac{1}{2}, \pm\frac{1}{3}, \pm\frac{1}{6}, \pm\frac{2}{3}, \pm\frac{4}{3}, \pm\frac{8}{3}.$$

19. For the polynomial $P(x) = x^3 - 3x^2 + 4$, the possible rational zeros are $\pm 1, \pm 2, \pm 4$. By substitution, the polynomial has zeros at $x = -1$, and 2 with the zero at $x = 2$ of multiplicity 2. Dividing $(x + 1)$ into $P(x)$ gives

$$P(x) = (x + 1)(x^2 - 4x + 4) = (x + 1)(x - 2)^2.$$

20. For the polynomial $P(x) = 3x^3 - 8x^2 - 5x + 6$, the possible rational zeros are $\pm 1, \pm 2, \pm 3, \pm 6, \pm\frac{1}{3}, \pm\frac{2}{3}$. By substitution, the polynomial has zeros at $x = -1$, $\frac{2}{3}$, and 3. Dividing $(x + 1)$ into $P(x)$ gives

$$P(x) = (x + 1)(3x^2 - 11x + 6) = (x + 1)(3x - 2)(x - 3).$$

21. Since

$$P(x) = 2x^4 - 3x^3 - 3x^2 + 2x = x(2x^3 - 3x^2 - 3x + 2),$$

$P(x)$ has a zero at $x = 0$. The possible rational zeros of $Q(x) = 2x^3 - 3x^2 - 3x + 2$ are

$$\frac{\pm 1, \pm 2}{\pm 1, \pm 2} = \pm 1, \pm 2, \pm\frac{1}{2}.$$

By substitution, the polynomial $Q(x)$ has zeros at $x = -1, \frac{1}{2}, 2$. Dividing $(x + 1)$ into $Q(x)$ gives

$$P(x) = x(x + 1)(2x^2 - 5x + 2) = x(x + 1)(2x - 1)(x - 2).$$

22. Since

$$P(x) = 3x^4 - 13x^3 + 13x^2 - 3x = x(3x^3 - 13x^2 + 13x - 3),$$

$P(x)$ has a zero at $x = 0$. The possible rational zeros of $Q(x) = 3x^3 - 13x^2 + 13x - 3$ are

$$\frac{\pm 1, \pm 3}{\pm 1, \pm 3} = \pm 1, \pm 3, \pm\frac{1}{3}.$$

By substitution, the polynomial $Q(x)$ has zeros at $x = 1, \frac{1}{3}, 3$. Dividing $(x - 1)$ into $Q(x)$ gives

$$P(x) = x(x - 1)(3x^2 - 10x + 3) = x(x - 1)(3x - 1)(x - 3).$$

23. For the polynomial

$$P(x) = 2x^5 + x^4 - 12x^3 + 10x^2 + 2x - 3,$$

the possible rational zeros are

$$\frac{\pm 1, \pm 2, \pm 3}{\pm 1, \pm 2} = \pm 1, \pm 2, \pm 3, \pm\frac{1}{2}, \pm\frac{3}{2}.$$

By substitution, the polynomial has zeros at $x = 1$, of multiplicity 3, and zeros at $x = -3, -\frac{1}{2}$. So

$$P(x) = (x - 1)^3(2x + 1)(x + 3).$$

24. For the polynomial

$$P(x) = x^4 - 4x^3 + 3x^2 + 4x - 4,$$

the possible rational zeros are $\pm 1, \pm 2, \pm 4$. By substitution, the polynomial has zeros at $x = -1, 1, 2$. The zero at $x = 2$ is of multiplicity 2, so

$$P(x) = (x + 1)(x - 1)(x - 2)^2.$$

25. For the polynomial $P(x) = x^3 + x^2 - 3x + 1$, the possible rational zeros are ± 1. By substitution, the polynomial has a zero at $x = 1$. Dividing $(x - 1)$ into $P(x)$ gives

$$P(x) = (x - 1)(x^2 + 2x - 1).$$

To factor the quadratic requires the quadratic formula to find the zeros of $x^2 + 2x - 1 = 0$. That is,

$$x = \frac{-2 \pm \sqrt{4 - 4(1)(-1)}}{2} = \frac{-2 \pm \sqrt{8}}{2} = \frac{-2 \pm 2\sqrt{2}}{2} = -1 \pm \sqrt{2},$$

and

$$P(x) = (x - 1)\left(x - \left(-1 + \sqrt{2}\right)\right)\left(x - \left(-1 - \sqrt{2}\right)\right).$$

26. Since

$$P(x) = x^4 - x^3 - 5x^2 + 6x = x(x^3 - x^2 - 5x + 6),$$

$P(x)$ has a zero at $x = 0$. The possible rational zeros of $Q(x) = x^3 - x^2 - 5x + 6$ are $\pm 1, \pm 2, \pm 3, \pm 6$. By substitution, the polynomial $Q(x)$ has a zero at $x = 2$. Dividing $(x - 2)$ into $Q(x)$ gives

$$P(x) = x(x - 2)(x^2 + x - 3).$$

To factor the quadratic requires the quadratic formula to find the zeros of $x^2 + x - 3 = 0$. That is,

$$x = \frac{-1 \pm \sqrt{1 - 4(1)(-3)}}{2} = \frac{-1 \pm \sqrt{13}}{2},$$

and

$$P(x) = x(x - 2)\left(x - \left(\frac{-1 + \sqrt{13}}{2}\right)\right)\left(x - \left(\frac{-1 - \sqrt{13}}{2}\right)\right).$$

27. For the polynomial $P(x) = 4x^3 - 12x^2 + 9x - 1$, the possible rational zeros are $\pm 1, \pm\frac{1}{2}, \pm\frac{1}{4}$. By substitution, the polynomial has a zero at $x = 1$. Dividing $(x - 1)$ into $P(x)$ gives

$$P(x) = (x - 1)(4x^2 - 8x + 1).$$

To factor the quadratic requires the quadratic formula to find the zeros of $4x^2 - 8x + 1 = 0$. That is,

$$x = \frac{8 \pm \sqrt{64 - 4(4)(1)}}{8} = \frac{8 \pm \sqrt{48}}{8} = \frac{8 \pm 4\sqrt{3}}{8} = 1 \pm \frac{\sqrt{3}}{2},$$

and

$$P(x) = 4(x - 1)\left(x - \left(1 + \frac{\sqrt{3}}{2}\right)\right)\left(x - \left(1 - \frac{\sqrt{3}}{2}\right)\right).$$

28. For the polynomial $P(x) = 9x^3 - 3x^2 - 7x + 1$, the possible rational zeros are $\pm 1, \pm\frac{1}{3}, \pm\frac{1}{9}$. By substitution, the polynomial has a zero at $x = 1$. Dividing $(x - 1)$ into $P(x)$ gives

$$P(x) = (x - 1)(9x^2 + 6x - 1).$$

To factor the quadratic requires the quadratic formula to find the zeros of $9x^2 + 6x - 1 = 0$. That is,

$$x = \frac{-6 \pm \sqrt{36 - 4(9)(-1)}}{18} = \frac{-6 \pm \sqrt{72}}{18} = \frac{-6 \pm 6\sqrt{2}}{18} = \frac{-1 \pm \sqrt{2}}{3}$$

and

$$P(x) = (x - 1)\left(x - \left(\frac{-1 + \sqrt{2}}{3}\right)\right)\left(x - \left(\frac{-1 - \sqrt{2}}{3}\right)\right).$$

29. The curves $y = f(x) = x^3$ and $y = g(x) = 2x^2 + x - 2$ intersect if and only if $f(x) = g(x)$ which implies $x^3 = 2x^2 + x - 2$ so $x^3 - 2x^2 - x + 2 = 0$. For the polynomial $P(x) = x^3 - 2x^2 - x + 2$, the possible rational zeros are $\pm 1, \pm 2$. By substitution, the polynomial has a zero at $x = 1$. Dividing $(x - 1)$ into $P(x)$ gives

$$P(x) = (x - 1)(x^2 - x - 2) = (x - 1)(x + 1)(x - 2).$$

The points of intersection are $(-1, -1)$, $(1, 1)$, and $(2, 8)$. To sketch the parabola, the quadratic in standard form is

$$2x^2 + x - 2 = 2\left(x^2 + \frac{1}{2}x\right) - 2 = 2\left(x^2 + \frac{1}{2}x + \frac{1}{16} - \frac{1}{16}\right) - 2 = 2\left(x + \frac{1}{4}\right)^2 - \frac{17}{8}.$$

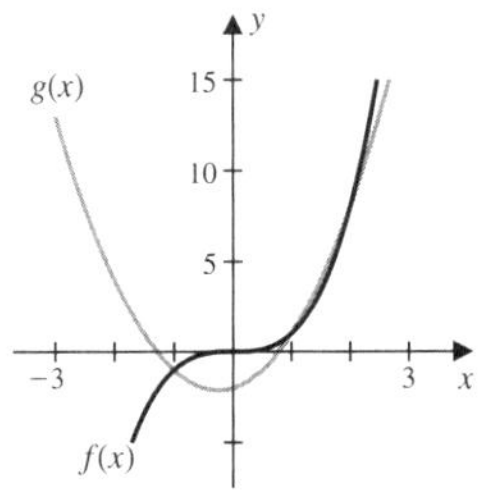

30. The curves $y = f(x) = x^3 + 3$ and $y = g(x) = 2x^2 + 5x - 3$ intersect if and only if $f(x) = g(x)$ which implies $x^3 + 3 = 2x^2 + 5x - 3$ so $x^3 - 2x^2 - 5x + 6 = 0$. For the polynomial $P(x) = x^3 - 2x^2 - 5x + 6$, the possible rational zeros are $\pm 1, \pm 2, \pm 3, \pm 6$. By substitution, the polynomial has a zero at $x = 1$. Dividing $(x - 1)$ into $P(x)$ gives

$$P(x) = (x - 1)(x^2 - x - 6) = (x - 1)(x - 3)(x + 2).$$

The points of intersection are $(-2, -5)$, $(1, 4)$, and $(3, 30)$. To sketch the parabola, the quadratic in standard form is

$$2x^2 + 5x - 3 = 2\left(x^2 + \frac{5}{2}x\right) - 3 = 2\left(x^2 + \frac{5}{2}x + \frac{25}{16} - \frac{25}{16}\right) - 3 = 2\left(x + \frac{5}{4}\right)^2 - \frac{49}{8}.$$

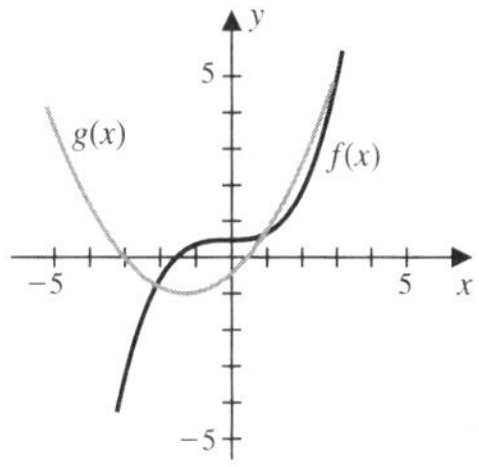

31. The curves $y = f(x) = x^3 - 1$ and $y = g(x) = 6x + 3$ intersect if and only if $f(x) = g(x)$ which implies $x^3 - 1 = 6x + 3$ so $x^3 - 6x - 4 = 0$. For the polynomial $P(x) = x^3 - 6x - 4$, the possible rational zeros are $\pm 1, \pm 2, \pm 4$. By substitution, the polynomial has a zero at $x = -2$. Dividing $(x + 2)$ into $P(x)$ gives

$$P(x) = (x + 2)(x^2 - 2x - 2).$$

To factor the quadratic requires the quadratic formula to find the zeros of $x^2 - 2x - 2 = 0$. That is,

$$x = \frac{2 \pm \sqrt{4 - 4(1)(-2)}}{2} = \frac{2 \pm \sqrt{12}}{2} = \frac{2 \pm 2\sqrt{3}}{2} = 1 \pm \sqrt{3}.$$

and

$$P(x) = (x + 2)\left(x - \left(1 + \sqrt{3}\right)\right)\left(x - \left(1 - \sqrt{3}\right)\right).$$

The points of intersection are $(-2, -9)$, $\left(1 + \sqrt{3}, \left(1 + \sqrt{3}\right)^3 - 1\right)$, and $\left(1 - \sqrt{3}, \left(1 - \sqrt{3}\right)^3 - 1\right)$.

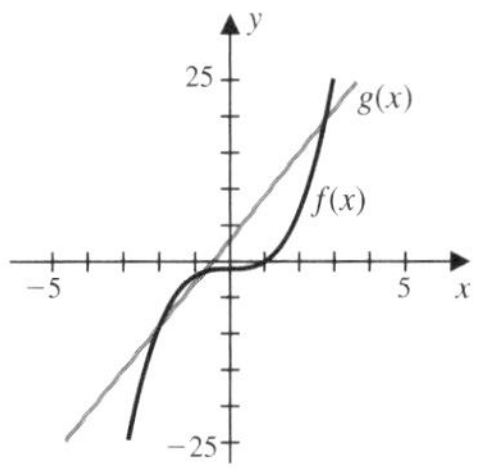

32. The curves $y = f(x) = x^3 - x^2$ and $y = g(x) = 3x - 2$ intersect if and only if $f(x) = g(x)$ which implies $x^3 - x^2 = 3x - 2$ so $x^3 - x^2 - 3x + 2 = 0$. For the polynomial $P(x) = x^3 - x^2 - 3x + 2$, the possible rational zeros are ± 1, ± 2. By substitution, the polynomial has a zero at $x = 2$. Dividing $(x - 2)$ into $P(x)$ gives

$$P(x) = (x - 2)(x^2 + x - 1).$$

To factor the quadratic requires the quadratic formula to find the zeros of $x^2 + x - 1 = 0$. That is,

$$x = \frac{-1 \pm \sqrt{1 - 4(1)(-1)}}{2} = \frac{-1 \pm \sqrt{5}}{2},$$

and

$$P(x) = (x - 2)\left(x - \left(\frac{-1 + \sqrt{5}}{2}\right)\right)\left(x - \left(\frac{-1 - \sqrt{5}}{2}\right)\right).$$

The points of intersection are $(2, 4)$, $\left(-\frac{1}{2} + \frac{1}{2}\sqrt{5}, -\frac{7}{2} - \frac{3}{2}\sqrt{5}\right)$, and $\left(-\frac{1}{2} - \frac{1}{2}\sqrt{5}, -\frac{7}{2} + \frac{3}{2}\sqrt{5}\right)$.

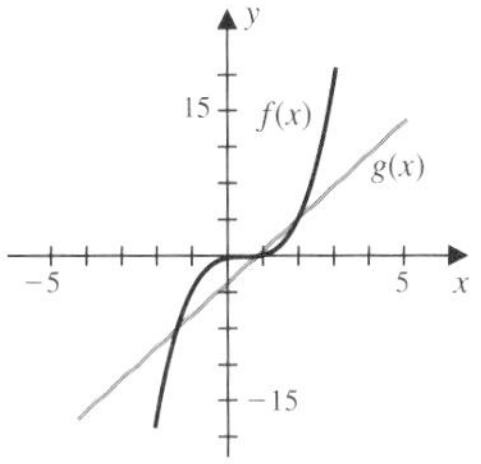

33. Since the polynomial has zeros of multiplicity 1 at $x = -3$, $x = -1$, $x = 1$, and $x = 4$, we have

$$P(x) = a(x + 3)(x + 1)(x - 1)(x - 4).$$

Since the graph passes through $(2, -3)$, we have

$$-3 = P(2) = a(5)(3)(1)(-2) = -30a \text{ so } a = \frac{1}{10}.$$

Thus

$$P(x) = \frac{1}{10}(x + 3)(x + 1)(x - 1)(x - 4) = \frac{1}{10}x^4 - \frac{1}{10}x^3 - \frac{13}{10}x^2 + \frac{1}{10}x + \frac{6}{5}.$$

34. A third-degree polynomial with zeros $x = -1, 1, 2$ and hence, the factors $(x + 1)$, $(x - 1)$, $(x - 2)$, is

$$Q(x) = (x + 1)(x - 1)(x - 2) = x^3 - 2x^2 - x + 2.$$

To have the coefficient of the x term be 3, multiply the polynomial by -3. So

$$P(x) = -3(x^3 - 2x^2 - x + 2) = -3x^3 + 6x^2 + 3x - 6.$$

35. A fourth-degree polynomial with zeros $x = -2, -1, 0, 1$ and hence, the factors $(x + 2), (x + 1), x, (x - 1)$, is

$$Q(x) = (x + 1)x(x - 1)(x + 2) = x^4 + 2x^3 - x^2 - 2x.$$

To have the coefficient of the x^2 term be 5, multiply the polynomial by -5. So

$$P(x) = -5(x^4 + 2x^3 - x^2 - 2x) = -5x^4 - 10x^3 + 5x^2 + 10x.$$

36. We have $P(x) = (x - 1)^2(x - 4)(x^2 + x + 1) = x^5 - 5x^4 + 4x^3 - x^2 + 5x - 4$.

37. We have $P(x) = (x - 2)^3x^2(x - 1) = x^6 - 7x^5 + 18x^4 - 20x^3 + 8x^2$.

38. **a.** We have $P(x) = x^4 - 4x^3 + 4x^2 - 1 = (x - 1)^2 \left(x - \left(1 + \sqrt{2}\right)\right)\left(x - \left(1 - \sqrt{2}\right)\right)$.

b. The graph of $y = Q(x) = P(x) + c$ is a vertical shift of the graph of $y = P(x)$. If $c < 0$, the graph is shifted downward and $Q(x)$ will then have only 2 zeros. If $0 < c < 1$, the graph is shifted upward an amount so the local minima will still remain below the x-axis, and $Q(x)$ will have 4 zeros. If $c = 1$, the graph will be shifted upward so the two minima sit on the x-axis, and $Q(x)$ will have 2 zeros. If $c > 1$, the graph will be shifted completely above the x-axis, and $Q(x)$ will have no zeros.

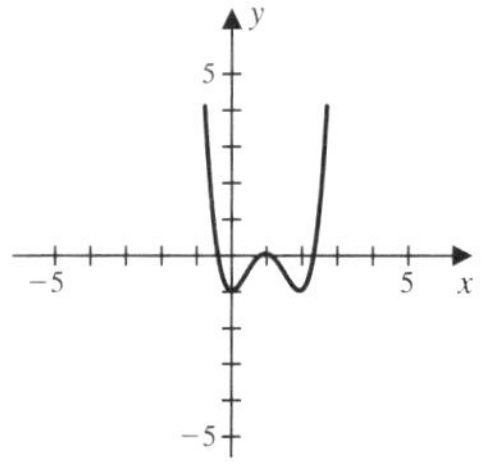

39. If n is a positive integer with $n \geq 1$, then $(1)^n - 1 = 1 - 1 = 0$, and hence $x - 1$ is a factor of $x^n - 1$.

40. If n is an odd positive integer with $n \geq 1$, then $(-1)^n + 1 = -1 + 1 = 0$, and hence $x + 1$ is a factor of $x^n + 1$. If n is an even positive integer, then $(-1)^n + 1 = 2$ and $x + 1$ is not a factor.

Exercise Set 3.4 (Page 157)

1. **a.** $\text{Domain}(f) = \{x \mid x \neq 1\} = (-\infty, 1) \cup (1, \infty)$

b. $\text{Range}(f) = \{x \mid x \neq 0\} = (-\infty, 0) \cup (0, \infty)$

c. Vertical asymptote: $x = 1$; horizontal asymptote: $y = 0$

2. **a.** $\text{Domain}(f) = \{x \mid x \neq -1, x \neq 1\} = (-\infty, -1) \cup (-1, 1) \cup (1, \infty)$

b. $\text{Range}(f) = (-\infty, 0] \cup (1, \infty)$

c. Vertical asymptotes: $x = -1, x = 1$; horizontal asymptote: $y = 1$

3. **a.** $\text{Domain}(f) = \{x \mid x \neq -1, x \neq 1\} = (-\infty, -1) \cup (-1, 1) \cup (1, \infty)$

b. $\text{Range}(f) = (-\infty, 1) \cup [2, \infty)$

c. Vertical asymptotes: $x = -1, x = 1$; horizontal asymptote: $y = 1$

4. **a.** $\text{Domain}(f) = \{x \mid x \neq -3, x \neq 3\} = (-\infty, -3) \cup (-3, 3) \cup (3, \infty)$

b. $\text{Range}(f) = (-\infty, \infty)$

c. Vertical asymptotes: $x = -3, x = 3$; horizontal asymptote: $y = 0$

5. **a.** $\text{Domain}(f) = (-\infty, 1)$

b. $\text{Range}(f) = (-\infty, \infty)$

c. Vertical asymptote: $x = 1$; horizontal asymptote: none

6. **a.** $\text{Domain}(f) = (-\infty, \infty)$

b. $\text{Range}(f) = [0, 2)$

c. Vertical asymptote: none; horizontal asymptote: $y = 2$

7. For $f(x) = \frac{x-3}{x+1}$, we have:

a. $\text{Domain}(f) = \{x \mid x \neq -1\} = (-\infty, -1) \cup (-1, \infty)$

b. Solve

$$\frac{x-3}{x+1} = 0 \text{ which implies } x - 3 = 0, \text{ so } x = 3.$$

x-intercepts: $(3, 0)$

Set $x = 0$ to get

$$y = \frac{0-3}{0+1} = -3.$$

y-intercept: $(0, -3)$

c. Since for x large in magnitude

$$\frac{x-1}{x+1} \approx \frac{x}{x} = 1.$$

Vertical asymptote: $x = -1$; horizontal asymptote: $y = 1$.

8. For $f(x) = \frac{(x-1)(2x+2)}{(x+3)(x-4)}$, we have:

a. Domain$(f) = \{x \mid x \neq -3, x \neq 4\} = (-\infty, -3) \cup (-3, 4) \cup (4, \infty)$

b. Solve

$$\frac{(x-1)(2x+2)}{(x+3)(x-4)} = 0 \text{ which implies } (x-1)(2x+2) = 0, \text{ so } x = 1, x = -1.$$

x-intercepts: $(-1, 0)$, $(1, 0)$

Set $x = 0$ to get

$$y = \frac{(0-1)(0+2)}{(0+3)(0-4)} = \frac{-2}{-12} = \frac{1}{6}.$$

y-intercept: $\left(0, \frac{1}{6}\right)$

c. Since for x large in magnitude

$$\frac{(x-1)(2x+2)}{(x+3)(x-4)} \approx \frac{2x^2}{x^2} = 2.$$

Vertical asymptotes: $x = -3$, $x = 4$; horizontal asymptote: $y = 2$.

9. For

$$f(x) = \frac{2x^2 - 5x + 3}{x^2 - 4} = \frac{(2x-3)(x-1)}{(x-2)(x+2)},$$

we have:

a. Domain$(f) = \{x \mid x \neq -2, x \neq 2\} = (-\infty, -2) \cup (-2, 2) \cup (2, \infty)$

b. Solve

$$\frac{(2x-3)(x-1)}{(x-2)(x+2)} = 0 \text{ which implies } 2x - 3 = 0, x - 1 = 0, \text{ so } x = \frac{3}{2}, x = 1.$$

x-intercepts: $\left(\frac{3}{2}, 0\right)$, $(1, 0)$

Set $x = 0$ to get

$$y = \frac{(-3)(-1)}{(-2)(2)} = -\frac{3}{4}.$$

y-intercept: $\left(0, -\frac{3}{4}\right)$;

c. Since for x large in magnitude

$$\frac{(2x-3)(x-1)}{(x-2)(x+2)} \approx \frac{2x^2}{x^2} = 2.$$

Vertical asymptotes: $x = -2$, $x = 2$; horizontal asymptote: $y = 2$.

10. For

$$f(x) = \frac{x^2+3x-10}{x^2-4x+4} = \frac{(x+5)(x-2)}{(x-2)^2} = \frac{x+5}{x-2},$$

we have:

a. Domain$(f) = \{x \mid x \neq 2\} = (-\infty, 2) \cup (2, \infty)$

b. Solve

$$\frac{x+5}{x-2} = 0 \text{ which implies } x+5=0, \text{ so } x=-5.$$

x-intercepts: $(-5, 0)$
Set $x = 0$ to get

$$y = \frac{(0+5)}{(0-2)} = -\frac{5}{2}.$$

y-intercept: $\left(0, -\frac{5}{2}\right)$

c. Since for x large in magnitude

$$\frac{x+5}{x-2} \approx \frac{x}{x} = 1.$$

Vertical asymptotes: $x = 2$; horizontal asymptote: $y = 1$.

11. For

$$f(x) = \frac{x^3-2x^2-x+2}{x^2} = \frac{(x+1)(x-1)(x-2)}{x^2},$$

we have:

a. Domain$(f) = \{x \mid x \neq 0\} = (-\infty, 0) \cup (0, \infty)$

b. Solve

$$\frac{(x+1)(x-1)(x-2)}{x^2} = 0 \text{ which implies } (x+1)(x-1)(x-2)=0, \text{ so } x=-1, 1, 2.$$

x-intercepts: $(-1, 0)$, $(1, 0)$, and $(2, 0)$

y-intercept: None, since $x = 0$ is not in the domain of the function.

c. For x large in magnitude

$$\frac{(x+1)(x-1)(x-2)}{x^2} \approx \frac{x^3}{x^2} = x.$$

Vertical asymptotes: $x = 0$; horizontal asymptote: None

12. For

$$f(x) = \frac{x^5 - 2x^4 - x + 2}{x^3 - 1} = \frac{(x+1)(x-1)(x-2)\left(x^2+1\right)}{(x-1)\left(x^2+x+1\right)} = \frac{(x+1)(x-2)\left(x^2+1\right)}{x^2+x+1},$$

we have:

a. $\text{Domain}(f) = \{x \mid x \neq 1\} = (-\infty, 1) \cup (1, \infty)$

b. Solve

$$\frac{(x+1)(x-2)\left(x^2+1\right)}{x^2+x+1} = 0 \text{ which implies } (x+1)(x-2)\left(x^2+1\right) = 0, \text{ so } x = -1, 2.$$

x-intercepts: $(-1, 0)$, $(2, 0)$

y-intercept: Set $x = 0$ to get $y = -2$.

c. Vertical asymptotes: None; horizontal asymptote: None.

13. For $f(x) = \frac{2}{x-1}$. Vertical asymptotes: $x = 1$; horizontal asymptote: $y = 0$; x-intercepts: none; y-intercept: $(0, -2)$

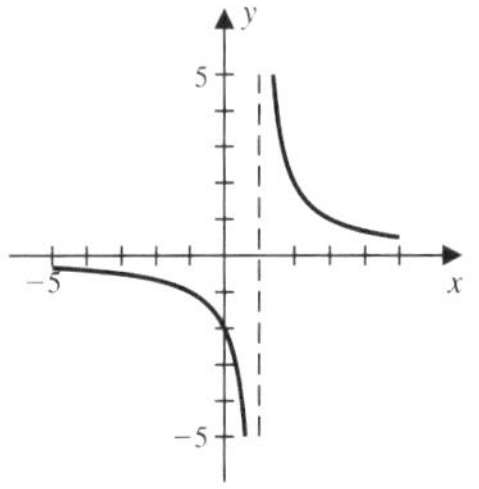

14. For $f(x) = \frac{4}{x+3}$. Vertical asymptotes: $x = -3$; horizontal asymptote: $y = 0$; x-intercepts: none; y-intercept: $(0, 4/3)$

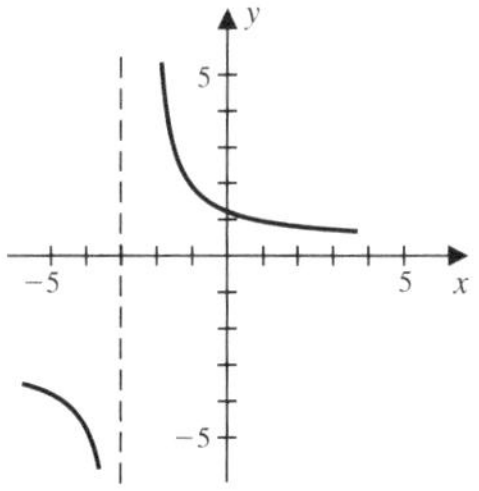

15. For $f(x) = \frac{x}{x-3}$. Vertical asymptotes: $x = 3$; horizontal asymptote: $y = 1$; x-intercepts: $(0, 0)$; y-intercept: $(0, 0)$

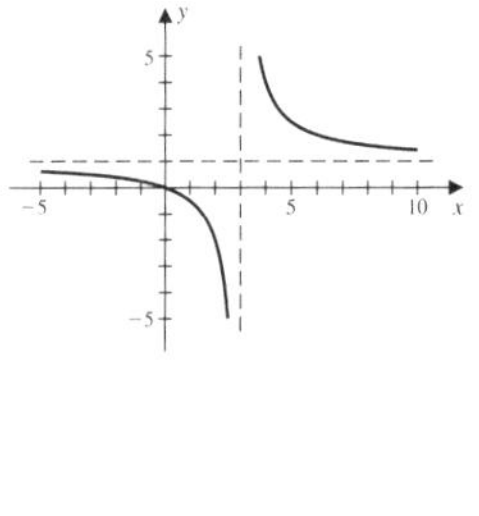

16. For $f(x) = \frac{2x}{3x-1}$. Vertical asymptotes: $x = \frac{1}{3}$; horizontal asymptote: $y = \frac{2}{3}$; x-intercepts: $(0, 0)$; y-intercept: $(0, 0)$

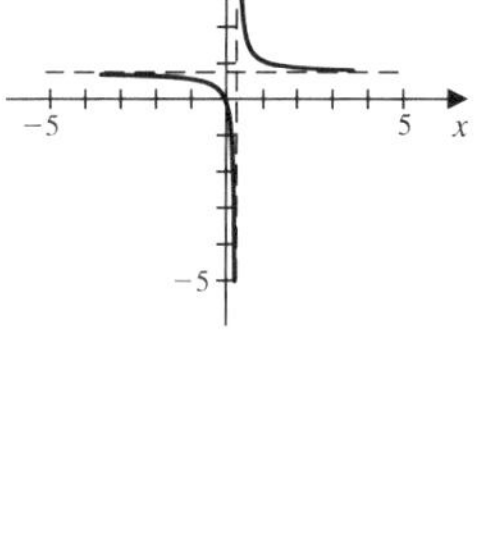

17. For $f(x) = \frac{(2x+1)(x-2)}{(3x-1)(x+3)}$. Vertical asymptotes: $x = -3$, $x = \frac{1}{3}$; horizontal asymptote: $y = \frac{2}{3}$; x-intercepts: $\left(-\frac{1}{2}, 0\right)$, $(2, 0)$; y-intercept: $\left(0, \frac{2}{3}\right)$

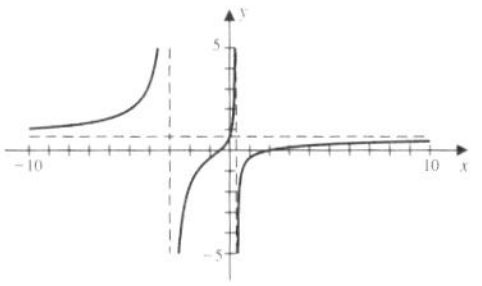

18. For $f(x) = \frac{x(x+4)}{(x-1)(x+2)}$. Vertical asymptotes: $x = -2$, $x = 1$; horizontal asymptote: $y = 1$; x-intercepts: $(-4, 0)$, $(0, 0)$; y-intercept: $(0, 0)$

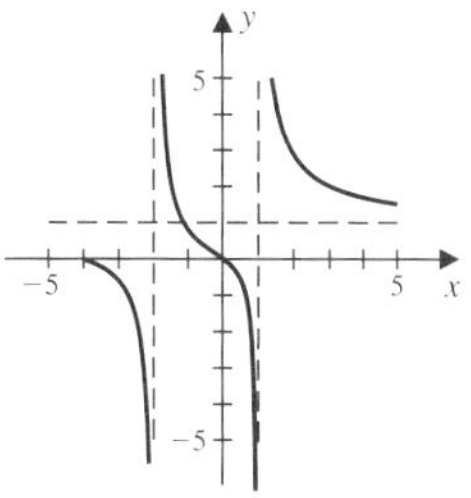

19. For

$$f(x) = \frac{x^2 - 9}{x^2 - 16} = \frac{(x-3)(x+3)}{(x-4)(x+4)}.$$

Vertical asymptotes: $x = 4$, $x = -4$; horizontal asymptote: $y = 1$; x-intercepts: $(-3, 0)$, $(3, 0)$; y-intercept: $(0, 9/16)$

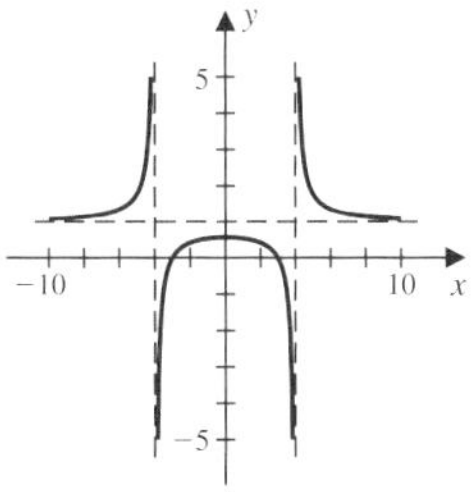

20. For

$$f(x) = \frac{x^2 - 16}{x^2 - 9} = \frac{(x-4)(x+4)}{(x-3)(x+3)}.$$

Vertical asymptotes: $x = -3$, $x = 3$; horizontal asymptote: $y = 1$; x-intercepts: $(-4, 0)$, $(4, 0)$; y-intercept: $(0, 16/9)$

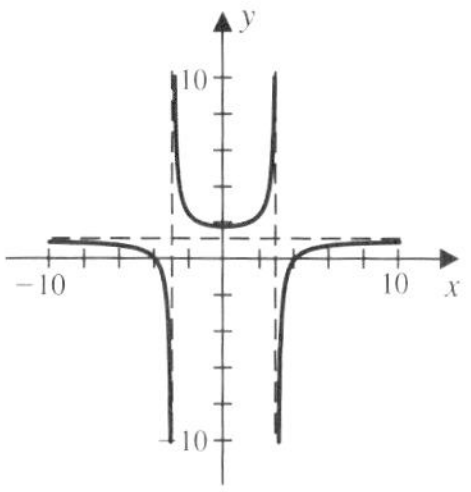

21. For

$$f(x) = \frac{x^2 + 2x - 3}{x - 1} = \frac{(x+3)(x-1)}{x-1}$$
$$= x + 3, \text{ for } x \neq 1.$$

Vertical asymptotes: None; horizontal asymptote: None; x-intercept: $(-3, 0)$; y-intercept: $(0, 3)$

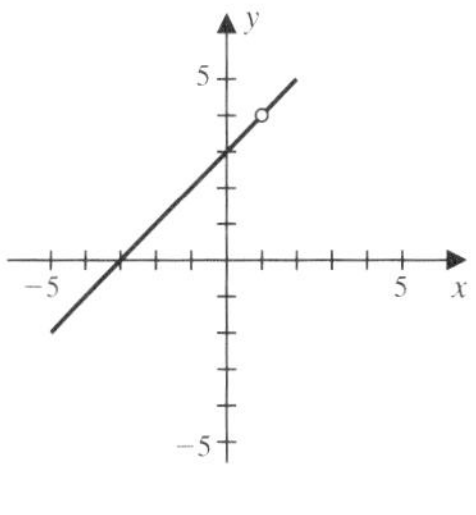

22. For

$$f(x) = \frac{x^2 + 5x + 6}{x + 2} = \frac{(x+2)(x+3)}{x+2}$$
$$= x + 3, \text{ for } x \neq -2.$$

Vertical asymptotes: None; horizontal asymptote: None; x-intercept: $(-3, 0)$; y-intercept: $(0, 3)$

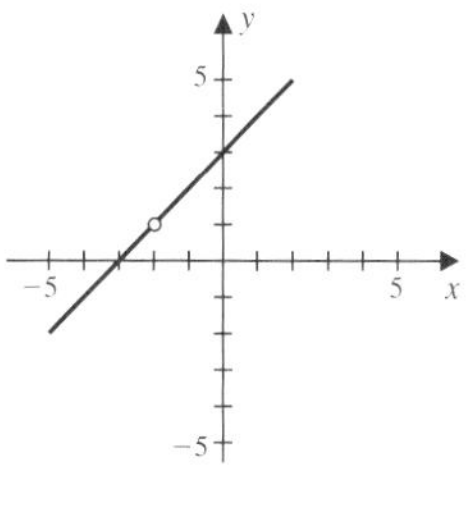

23. For

$$f(x) = \frac{x^2 - x - 2}{x^2 - 2x - 3} = \frac{(x-2)(x+1)}{(x-3)(x+1)}$$
$$= \frac{x-2}{x-3}, \text{ for } x \neq -1.$$

Vertical asymptote: $x = 3$; horizontal asymptote: $y = 1$; x-intercepts: $(2, 0)$; y-intercept: $(0, 2/3)$

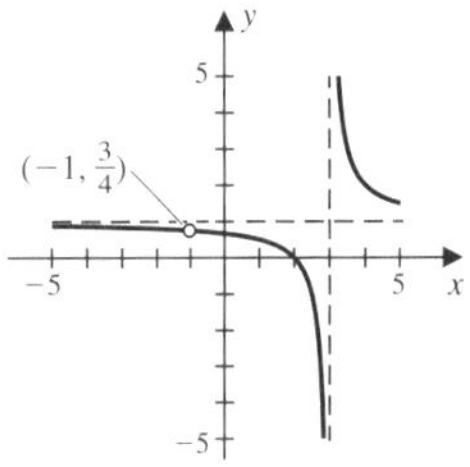

24. For

$$f(x) = \frac{x^2 - 1}{x^2 - 3x + 2} = \frac{(x-1)(x+1)}{(x-2)(x-1)}$$
$$= \frac{x+1}{x-2}, \text{ for } x \neq 1.$$

Vertical asymptote: $x = 2$; horizontal asymptote: $y = 1$; x-intercepts: $(-1, 0)$; y-intercept: $(0, -1/2)$

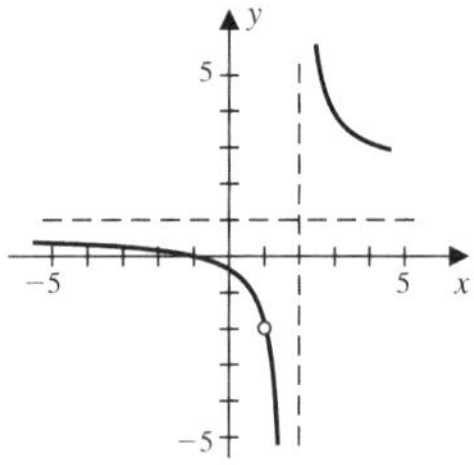

25. For

$$f(x) = \frac{2x-3}{x^2 - x - 6} = \frac{2x-3}{(x-3)(x+2)}.$$

Vertical asymptotes: $x = -2$, $x = 3$; horizontal asymptote: $y = 0$; x-intercepts: $(3/2, 0)$; y-intercept: $(0, 1/2)$

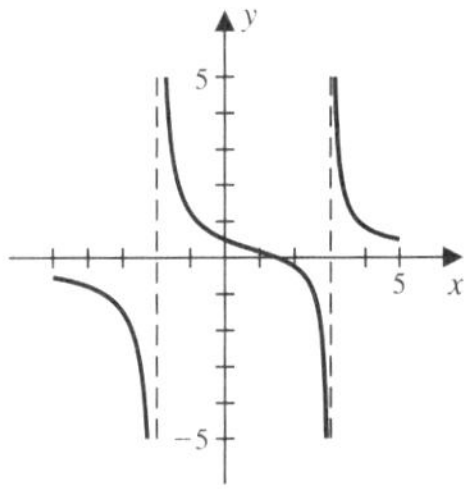

26. For

$$f(x) = \frac{-2x}{x^2 - 9} = \frac{-2x}{(x-3)(x+3)}.$$

Vertical asymptotes: $x = -3$, $x = 3$; horizontal asymptote: $y = 0$; x-intercepts: $(0, 0)$; y-intercept: $(0, 0)$

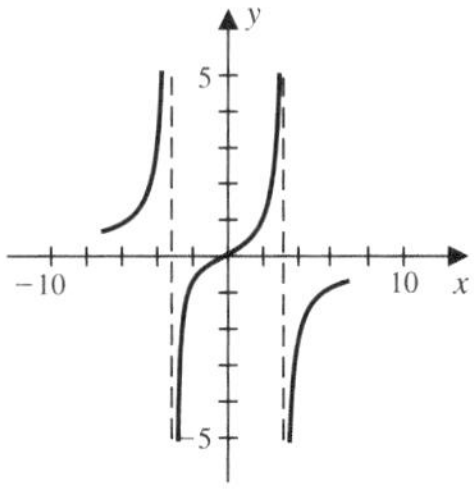

27. One example is $f(x) = \dfrac{1}{x-1}$.

28. One example is $f(x) = \dfrac{4}{x+2}$.

29. One example is

$$f(x) = \frac{(x-3)(x+4)}{(x-2)(x+3)}.$$

30. One example is

$$f(x) = \frac{(x-3)(x-4)}{(x-2)(x+3)}.$$

31. One example is

$$f(x) = \frac{3(x-2)}{x} = \frac{3x-6}{x}.$$

32. One example is

$$f(x) = \frac{2(x-4)(x+4)}{(x-3)(x+3)} = \frac{2x^2 - 32}{x^2 - 9}.$$

33. $f(x) = \frac{x^2}{x-1} = x + 1 + \dfrac{1}{x-1}$

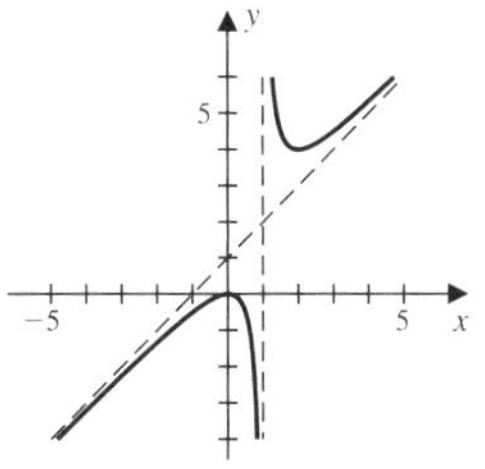

34. $f(x) = \dfrac{x^2 + x}{x+2} = x - 1 + \dfrac{2}{x+2}$

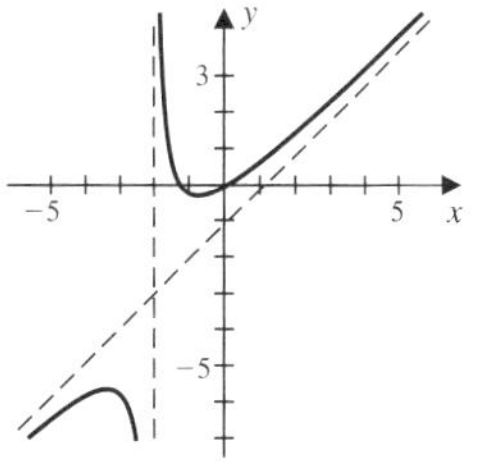

35. $f(x) = \dfrac{x^2 + x - 2}{x+1} = x - \dfrac{2}{x+1}$

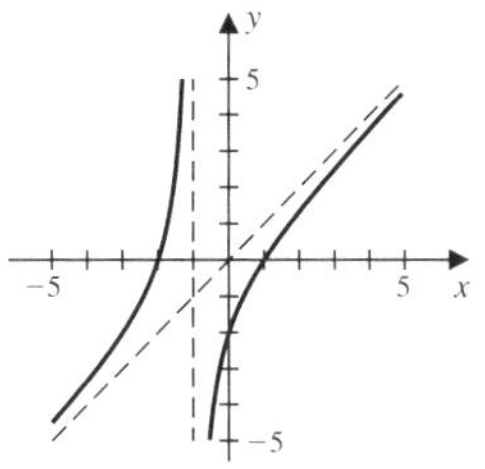

36. $f(x) = \dfrac{x^2 - 2x - 3}{x-1} = x - 1 - \dfrac{4}{x-1}$

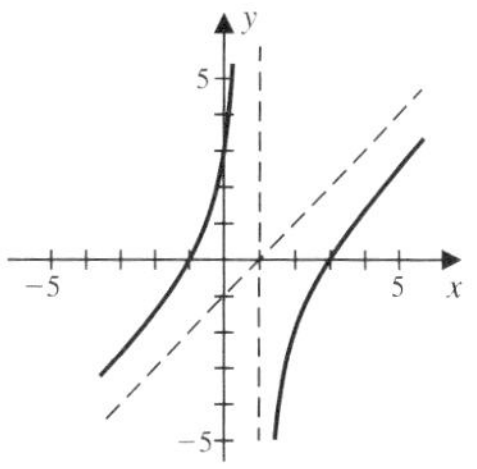

37. $f(x) = \dfrac{x}{x^2 - 3x + 2} = \dfrac{x}{(x-1)(x-2)}$

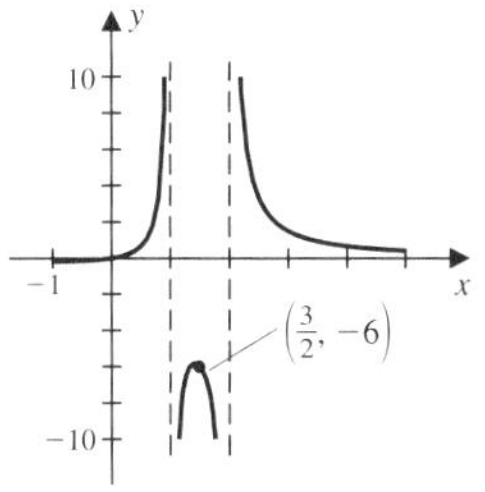

38. $f(x) = \dfrac{x^2 - 6x + 8}{x^2 - 4x + 3} = \dfrac{(x-2)(x-4)}{(x-1)(x-3)}$

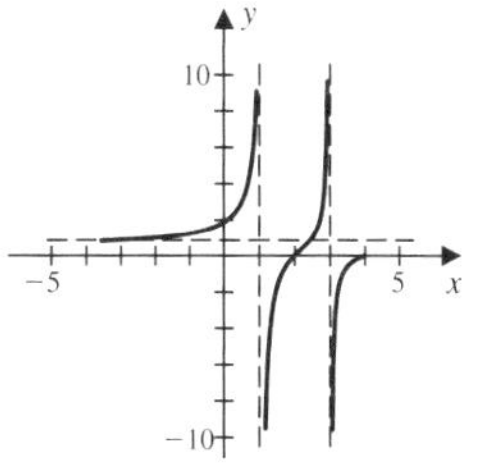

39. $f(x) = \dfrac{x^2}{x^2 - 5x + 6} = \dfrac{x^2}{(x-2)(x-3)}$

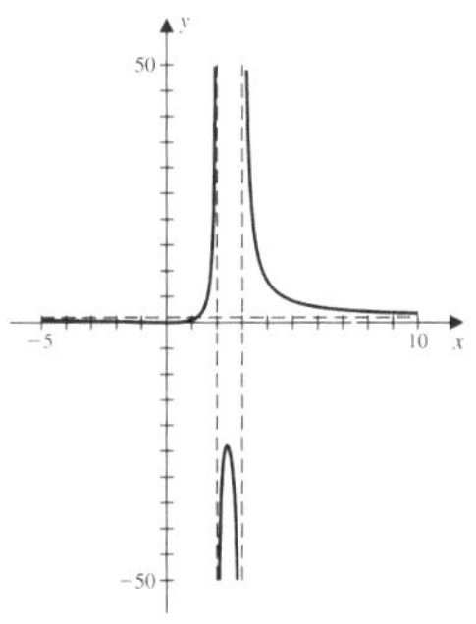

40.

$$f(x) = \frac{3 - x^2}{2x^2 + x - 3} = \frac{(\sqrt{3} - x)(\sqrt{3} + x)}{(2x+3)(x-1)}$$

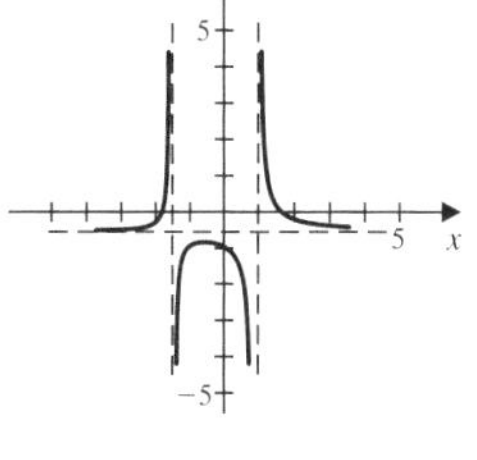

41. We have $A = \pi r^2 + 2\pi r h$ and $318 = \pi r^2 h$ which implies $h = \frac{318}{\pi r^2}$, so

$$A(r) = \pi r^2 + 2\pi r\left(\frac{318}{\pi r^2}\right) = \pi r^2 + \frac{636}{r}.$$

From the graph of $y = A(r)$, the minimum point can be estimated to be

$$r \approx 4.66, \text{ so } h = \frac{318}{\pi r^2} \approx 4.66.$$

42. To determine the cost of the box the surface area is required, which is given by

$$A = (\text{area of top and bottom}) + (\text{area of the 4 sides}) = 2x^2 + 4xh.$$

To eliminate one of the variables the volume of 20 feet^3 can be used. That is,

$$20 = V = x^2 h, \text{ so } h = \frac{20}{x^2} \quad \text{and} \quad x = \sqrt{\frac{20}{h}}$$

a. $C = 2x^2(0.20) + 4xh(0.08) = 0.4x^2 + 0.32xh$

b. $C(x) = 0.4x^2 + 0.32x\left(\frac{20}{x^2}\right) = 0.4x^2 + \frac{6.4}{x}$

c.

$$C(h) = 0.4\left(\sqrt{\frac{20}{h}}\right)^2 + 0.32\left(\sqrt{\frac{20}{h}}\right)h = \frac{8}{h} + 0.32\sqrt{20}\sqrt{h} = \frac{8}{h} + 0.64\sqrt{5}\sqrt{h}.$$

d. From the computer generated graph of $y = 0.4x^2 + \frac{6.4}{x}$, the minimum cost occurs at

$$x \approx 2, \text{ so } h = \frac{20}{x^2} \approx 5.$$

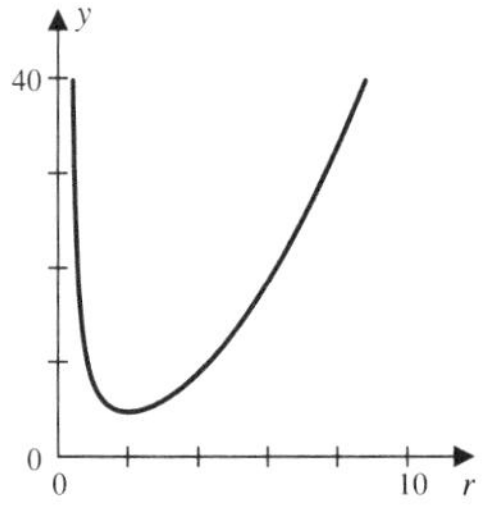

43. The number of bacteria at time t is $n = 10{,}000\left(\frac{3t^2 + 1}{t^2 + 1}\right)$.

a. For t large,

$$n = 10{,}000\left(\frac{3t^2+1}{t^2+1}\right) \approx 10{,}000\left(\frac{3t^2}{t^2}\right) = 30{,}000$$

so the bacteria colony stabilizes.

b. The bacteria colony stabilizes to the level 30,000.

c. To find when the number of bacteria exceeds 22,000, solve

$$10{,}000\left(\frac{3t^2+1}{t^2+1}\right) > 22{,}000 \text{ or } 3t^2+1 > 2.2t^2+2.2$$

This implies that

$$0.8t^2 > 1.2, \text{ so } t^2 - 1.5 > 0 \text{ and } t > \sqrt{1.5} = \frac{\sqrt{6}}{2} \approx 1.2.$$

44. a. For $c = \dfrac{at}{t^2+b}$,
the left figure shows $b = 1$, $a = 1, 2, 3, -1, -2$,
the middle figure shows $a = 1$, $b = 1, 2, 3, \frac{1}{2}, \frac{1}{3}$, and
the right figure shows $a = 1$, $b = -1, -2, -3$.

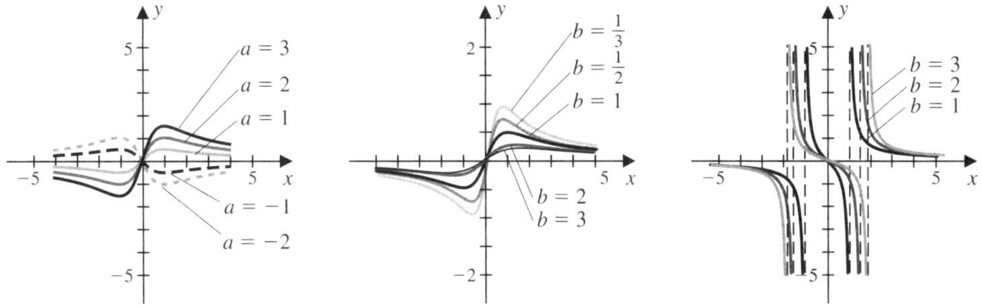

b. The figure shows the highest point on the curve $c = \dfrac{3t}{t^2+1}$ at approximately $t = 1$.

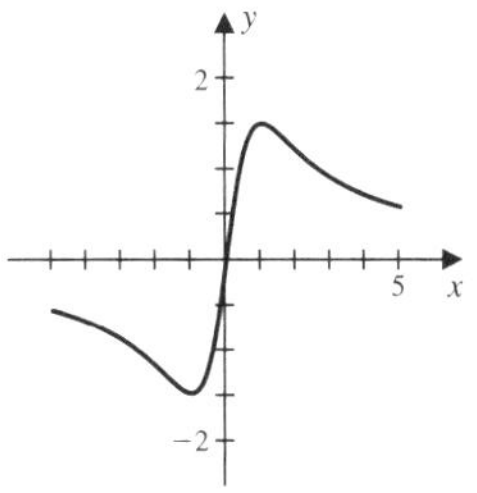

c. As time increases the concentration of the drug in the bloodstream decreases towards 0.

d. The drug concentration will reach a level of 0.2 when

$$\frac{3t}{t^2+1} = 0.2, \text{ so } t \approx 15.$$

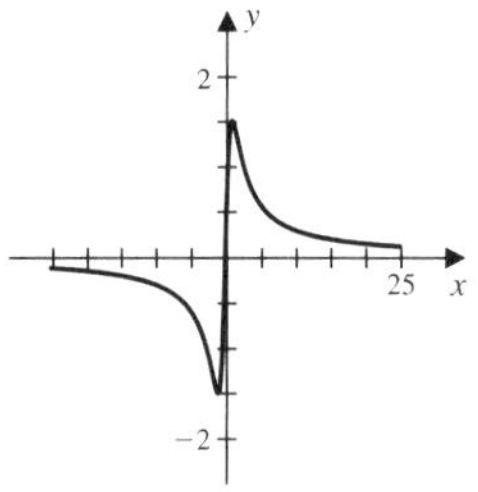

45. **a.** For $P(t) = \dfrac{at^2}{t^2+1}$, the parameter a determines the height of the curve.
The left figure shows $a = 1, 2, 3, \frac{1}{2}, \frac{1}{3}$, and the right figure shows $a = -1, -2, -3, -\frac{1}{2}, -\frac{1}{3}$.

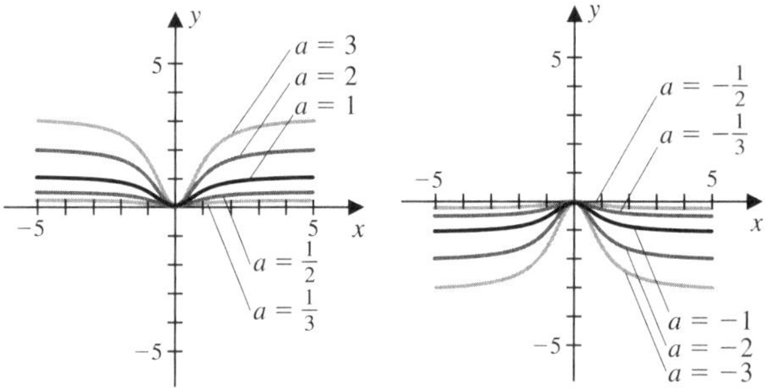

b. As t increases the population stabilizes to the value a.

Exercise Set 3.5 (Page 163)

1. The domain of $f(x) = \sqrt{x^2+4x-12}$ consists of all x satisfying $x^2+4x-12 \geq 0$ which implies $(x+6)(x-2) \geq 0$, so $x \leq -6$ or $x \geq 2$, that is, $(-\infty, -6] \cup [2, \infty)$.

2. The domain of $f(x) = \sqrt{-6+5x-x^2}$ consists of all x satisfying $-6+5x-x^2 \geq 0$

or $x^2 - 5x + 6 \le 0$ which implies $(x-3)(x-2) \le 0$, so $2 \le x \le 3$, that is, $[2, 3]$.

3. The domain of $f(x) = \sqrt{(x-1)^2(x+2)}$ consists of all x satisfying $(x-1)^2(x+2) \ge 0$ which implies $(x+2) \ge 0$, so $x \ge -2$, that is, $[-2, \infty)$.

4. The domain of $f(x) = \sqrt{x^2(x-3)}$ consists of all x satisfying $x^2(x-3) \ge 0$ which implies $(x-3) \ge 0$, so $x \ge 3$, that is, $[3, \infty)$.

5. The domain of $f(x) = \sqrt{(x-1)(x-4)(x+3)}$ consists of all x satisfying $(x-1)(x-4)(x+3) \ge 0$.
The solution to this inequality can be found from the following sign chart.
The domain is where the last inequality is positive or 0, that is, $[-3, 1] \cup [4, \infty)$.

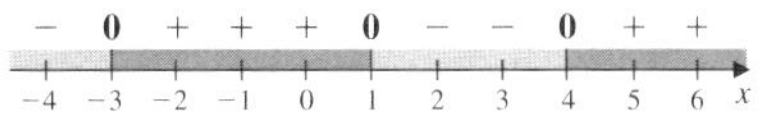

6. The domain of $f(x) = \sqrt{(x+1)(x-2)(x-3)}$ consists of all x satisfying $(x+1)(x-2)(x-3) \ge 0$.
The solution to this inequality can be found from the following sign chart.
The domain is where the last inequality is positive or 0, that is, $[-1, 2] \cup [3, \infty)$.

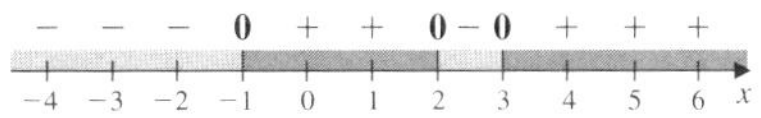

7. The domain of

$$f(x) = \sqrt{\frac{1-x}{x+3}}$$

consists of all x satisfying $\dfrac{1-x}{x+3} \ge 0$ which implies $\dfrac{x-1}{x+3} \le 0$ so $-3 < x \le 1$, that is, $(-3, 1]$.

8. The domain of

$$f(x) = \sqrt{\frac{7-2x}{x-5}}$$

consists of all x satisfying $\dfrac{7-2x}{x-5} \ge 0$ which implies $\dfrac{2x-7}{x-5} \le 0$ so $\dfrac{7}{2} \le x < 5$, that is, $\left[\frac{7}{2}, 5\right)$.

9. The domain of

$$f(x) = \sqrt{\frac{x+1}{x^2+2x-3}} = \sqrt{\frac{x+1}{(x+3)(x-1)}}$$

consists of all x satisfying $\dfrac{x+1}{(x+3)(x-1)} \geq 0$. The solution to this inequality can be found from the following sign chart. The domain is the where the last inequality is positive or 0, that is, $(-3, -1] \cup (1, \infty)$.

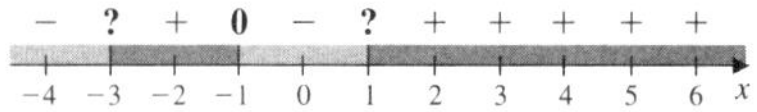

10. The domain of

$$f(x) = \sqrt{\frac{x^2-3x-28}{x^2-4}} = \sqrt{\frac{(x-7)(x+4)}{(x-2)(x+2)}}$$

consists of all x satisfying $\dfrac{(x-7)(x+4)}{(x-2)(x+2)} \geq 0$. The solution to this inequality can be found from the following sign chart. The domain is where the last inequality is 0 or positive, that is $(-\infty, -4] \cup (-2, 2) \cup [7, \infty)$.

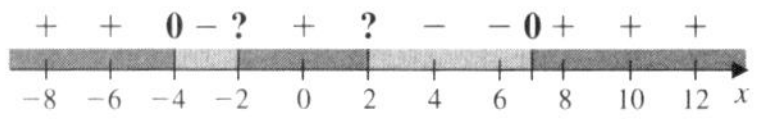

11. **a.** iii **b.** vi **c.** i **d.** v **e.** iv **f.** ii

12. **a.** iv **b.** v **c.** i **d.** vi **e.** ii **f.** iii

13. $f(x) = x^{3/2} - 1$

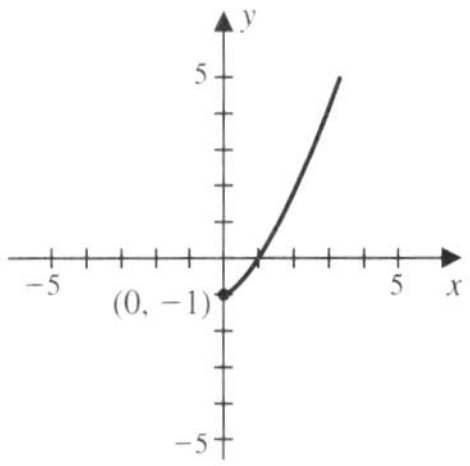

14. $f(x) = x^{2/3} + 1$

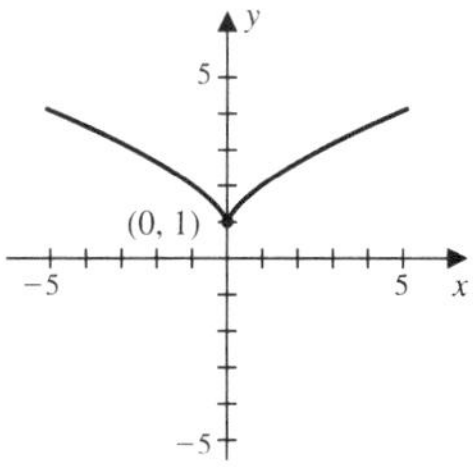

15. $f(x) = (x-1)^{1/3} + 1$

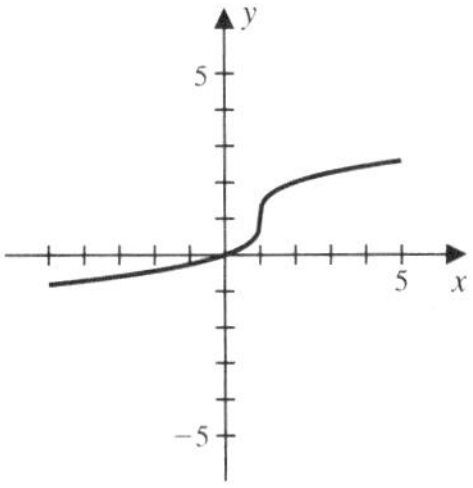

16. $f(x) = (x+1)^{1/4} - 1$

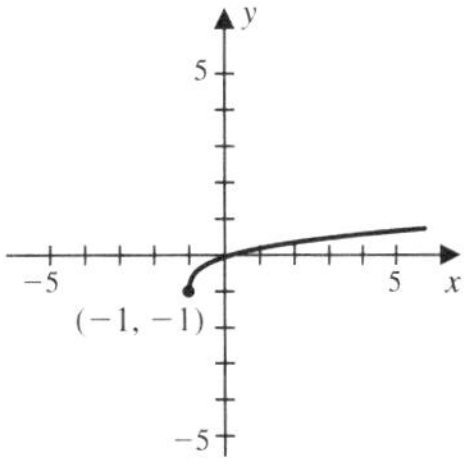

17. $f(x) = -2(x+1)^{2/3} - 2$

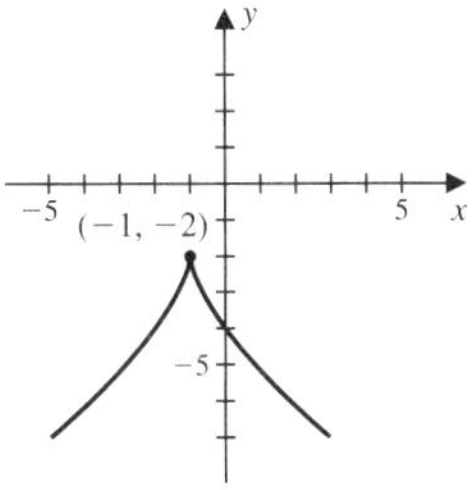

18. $f(x) = -3(x-2)^{3/4} + 1$

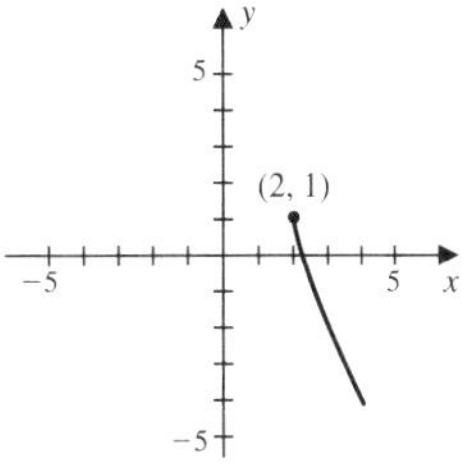

19.

$$\begin{aligned} f(x) &= (1-x)^{4/3} - 1 \\ &= (-(x-1))^{4/3} - 1 \\ &= (x-1)^{4/3} - 1 \end{aligned}$$

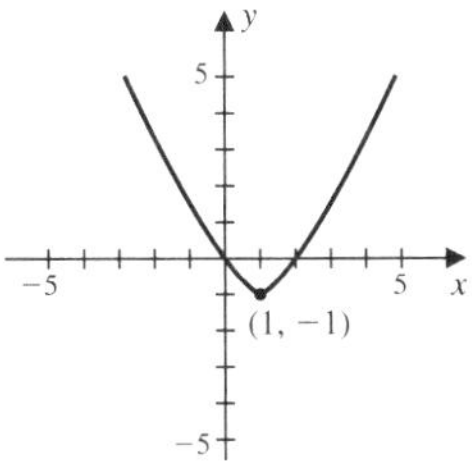

20. $f(x) = (4-x)^{3/2} + 1$

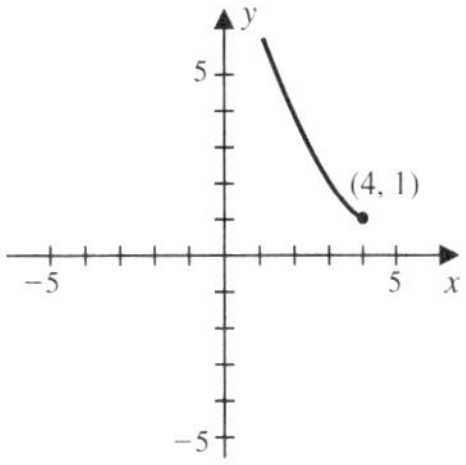

21. $f(x) = x^{-1/2} - 2$

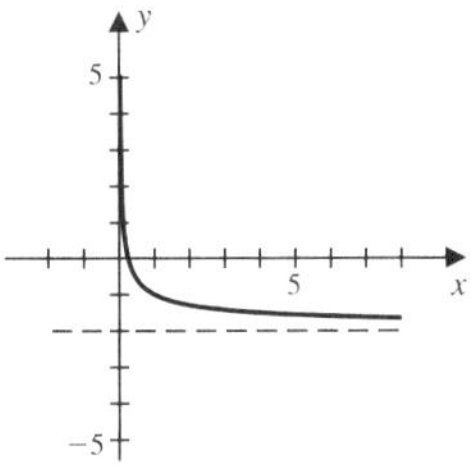

22. $f(x) = x^{-1/3} + 1$

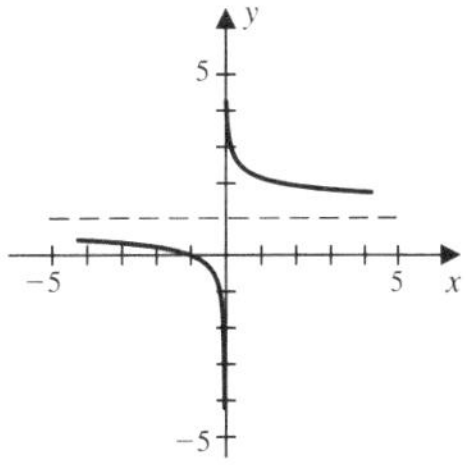

23. $f(x) = \sqrt{\dfrac{x+2}{x-1}}$

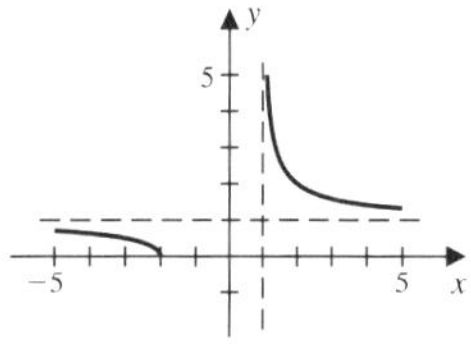

24. $f(x) = \sqrt{\dfrac{x-1}{x+2}}$

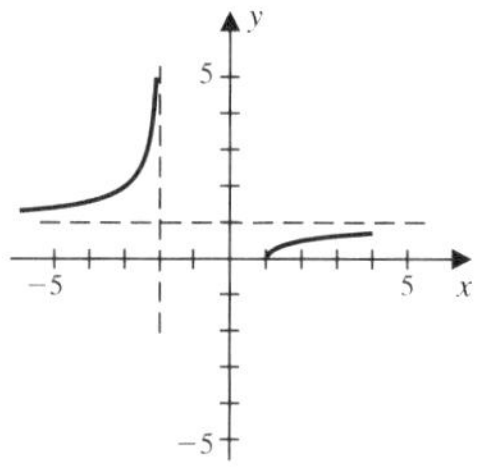

25. $f(x) = \sqrt{\dfrac{2-x}{x+2}}$

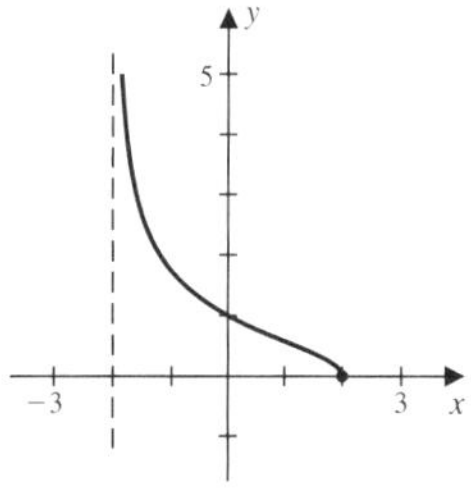

26. $f(x) = \sqrt{\dfrac{1-x}{x+3}}$

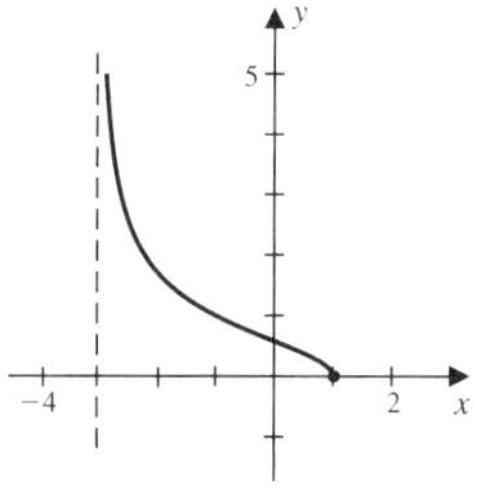

27. $f(x) = \dfrac{x}{\sqrt{x^2+1}}$

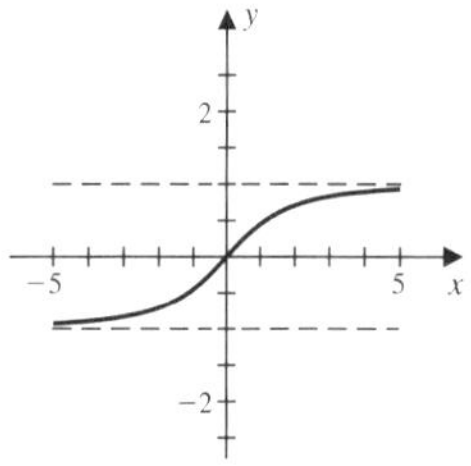

28. $f(x) = \dfrac{3x}{\sqrt{4x^2+9}}$

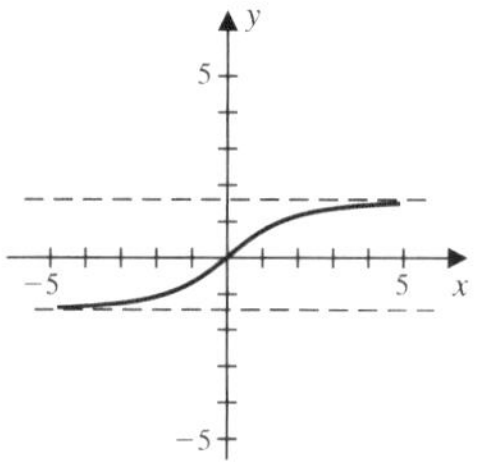

29. $f(x) = \dfrac{x-1}{\sqrt{(x+1)(x-2)}}$

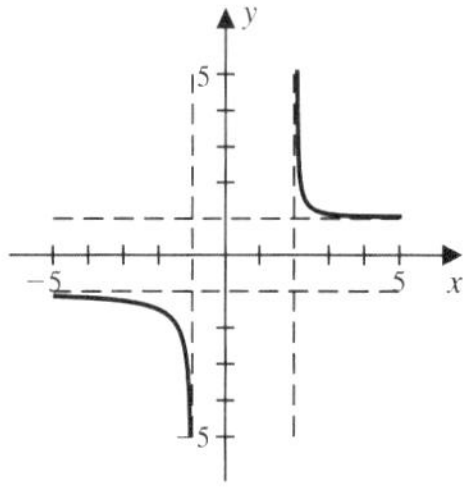

30.

$$f(x) = \frac{x}{\sqrt{x^2+3x+2}} = \frac{x}{\sqrt{(x+2)(x+1)}}$$

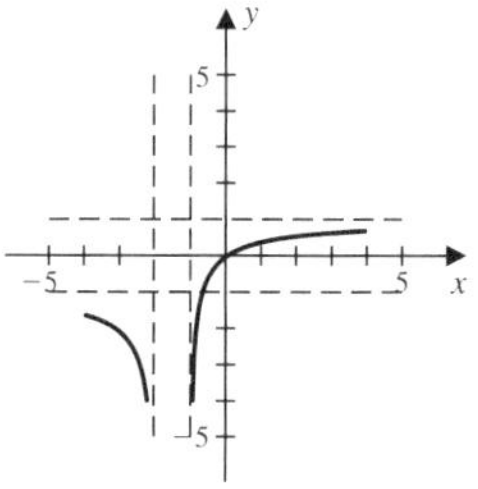

31. $f(x) = \dfrac{x}{\sqrt{4-x^2}} = \dfrac{x}{\sqrt{(2-x)(2+x)}}$

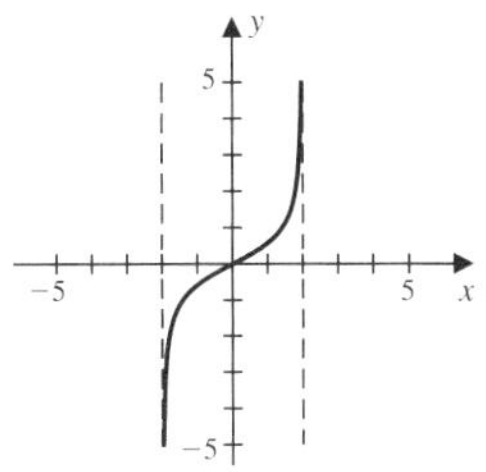

32. $f(x) = -\dfrac{x}{\sqrt{9-x^2}} = -\dfrac{x}{\sqrt{(3-x)(3+x)}}$

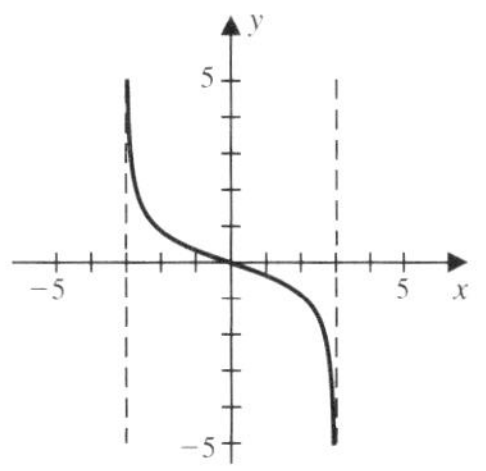

Exercise Set 3.6 (Page 172)

1. $(2+i)+(-3-2i) = (2-3)+(1-2)i$
$= -1-i$

2. $(3-i)+(2+4i) = (3+2)+(-1+4)i$
$= 5+3i$

3. $(-3+5i)-(2-3i) =$
$(-3-2)+(5+3)i = -5+8i$

4. $(5-7i)-(2-4i) = (5-2)+(-7+4)i$
$= 3-3i$

5. $2i \cdot (3+i) = 6i + 2i^2 = -2+6i$

6. $i \cdot (-2-i) = -2i - i^2 = 1-2i$

7. $(2-i)\cdot(3+i) = 6+2i-3i-i^2 = 7-i$

8. $(5-2i)\cdot(4-i) = 20-5i-8i+2i^2$
$= 18-13i$

9. $(6+5i)\cdot(-3-2i) =$
$-18-12i-15i-10i^2 = -8-27i$

10. $(7-6i)\cdot(2-3i) =$
$14-21i-12i+18i^2 = -4-33i$

11. $(2-3i)\cdot(2+3i) = 4+6i-6i-9i^2 = 13$

12. $(-3+4i)\cdot(-3-4i) =$
$9+12i-12i-16i^2 = 25$

13. $(3-8i)\cdot\overline{(2+i)} = (3-8i)(2-i) =$
$6-3i-16i+8i^2 = -2-19i$

14. $(4-i)\cdot\overline{(4-i)} = (4-i)(4+i) =$
$16+4i-4i-i^2 = 17$

15. $i^5 = i^4 \cdot i = i^2 \cdot i^2 \cdot i = (-1)(-1)i = i$

16. $i^6 = i^4 \cdot i^2 = 1\cdot(-1) = -1$

17. $i^{100} = (i^4)^{25} = 1^{25} = 1$

18. $i^{101} = i^{100}\cdot i = 1\cdot i = i$

19. $\sqrt{-9} = 3\sqrt{-1} = 3i$

20. $\sqrt{-25} = 5\sqrt{-1} = 5i$

21. $\left(2+\sqrt{-5}\right)\left(1+\sqrt{-1}\right)=\left(2+\sqrt{5}i\right)(1+i)$ which reduces to $2+2i+\sqrt{5}i+\sqrt{5}i^2=\left(2-\sqrt{5}\right)+\left(2+\sqrt{5}\right)i.$

22. $\left(1-\sqrt{-3}\right)\left(1+\sqrt{-3}\right)=\left(1-\sqrt{3}i\right)\left(1+\sqrt{3}i\right)$ which reduces to $1+\sqrt{3}i-\sqrt{3}i-3i^2=4.$

23. $\sqrt{-\frac{16}{9}}=\frac{4}{3}\sqrt{-1}=\frac{4}{3}i$

24. $\sqrt{-\frac{25}{4}}=\frac{5}{2}\sqrt{-1}=\frac{5}{2}i$

25. $\frac{1}{1-i}=\frac{1}{1-i}\cdot\frac{1+i}{1+i}=\frac{1+i}{2}=\frac{1}{2}+\frac{1}{2}i$

26. $\frac{1}{i}=\frac{1}{i}\cdot\frac{i}{i}=\frac{i}{-1}=-i$

27. $\frac{1}{2+3i}=\frac{1}{2+3i}\cdot\frac{2-3i}{2-3i}=\frac{2-3i}{13}=\frac{2}{13}-\frac{3}{13}i$

28. $\frac{1}{3-4i}=\frac{1}{3-4i}\cdot\frac{3+4i}{3+4i}=\frac{3+4i}{25}=\frac{3}{25}+\frac{4}{25}i$

29. $\frac{2-3i}{2+3i}=\frac{2-3i}{2+3i}\cdot\frac{2-3i}{2-3i}=\frac{4-12i+9i^2}{13}=\frac{-5-12i}{13}=-\frac{5}{13}-\frac{12}{13}i$

30. $\frac{5+6i}{5-6i}=\frac{5+6i}{5-6i}\cdot\frac{5+6i}{5+6i}=\frac{25+60i+36i^2}{61}=\frac{-11+60i}{61}=-\frac{11}{61}+\frac{60}{61}i$

31. **a.** Solving for the zeros we have

$$x^2+4=0 \text{ implies } x^2=-4 \text{ so } x=\pm\sqrt{-4}=\pm 2i.$$

b. $f(x)=x^2+4=(x-2i)(x+2i)$

32. **a.** Solving for the zeros we have

$$2x^2+18=0 \text{ implies } 2x^2=-18 \text{ so } x^2=-9 \text{ and } x=\pm\sqrt{-9}=\pm 3i.$$

b. $f(x)=(x-3i)(x+3i)$

33. **a.** Solving for the zeros we have

$$x^2-2x+2=0 \text{ implies } x=\frac{2\pm\sqrt{(-2)^2-4(1)(2)}}{2}=\frac{2\pm\sqrt{-4}}{2}=\frac{2\pm 2i}{2}=1\pm i.$$

b. $f(x)=(x-(1-i))\,(x-(1+i))$

34. **a.** Solving for the zeros we have $x^2-x+1=0$ which implies that

$$x=\frac{1\pm\sqrt{(-1)^2-4(1)(1)}}{2}=\frac{1\pm\sqrt{-3}}{2}=\frac{1\pm\sqrt{3}i}{2}.$$

b.

$$f(x)=\left(x-\left(\frac{1-\sqrt{3}i}{2}\right)\right)\left(x-\left(\frac{1+\sqrt{3}i}{2}\right)\right)$$

35. a. Solving for the zeros we have

$$2x^2 - x + 2 = 0 \text{ implies } x = \frac{1 \pm \sqrt{(-1)^2 - 4(2)(2)}}{2(2)} = \frac{1 \pm \sqrt{-15}}{4} = \frac{1 \pm \sqrt{15}i}{4}.$$

b.

$$f(x) = \left(x - \left(\frac{1 - \sqrt{15}i}{4}\right)\right)\left(x - \left(\frac{1 + \sqrt{15}i}{4}\right)\right)$$

36. a. Solving for the zeros we have $3x^2 + 2x + 1 = 0$ which implies that

$$x = \frac{-2 \pm \sqrt{(2)^2 - 4(3)(1)}}{2(3)} = \frac{-2 \pm \sqrt{-8}}{6} = \frac{-2 \pm 2\sqrt{2}i}{6} = \frac{-1 \pm \sqrt{2}i}{3}.$$

b.

$$f(x) = \left(x - \left(\frac{-1 - \sqrt{2}i}{3}\right)\right)\left(x - \left(\frac{-1 + \sqrt{2}i}{3}\right)\right)$$

37. We have

$$x^4 - 1 = (x^2 - 1)(x^2 + 1) = (x - 1)(x + 1)(x - i)(x + i) = 0,$$

and the zeros are $-1, 1, -i, i$.

38. We have

$$x^4 - 9 = (x^2 - 3)(x^2 + 3) = \left(x - \sqrt{3}\right)\left(x + \sqrt{3}\right)\left(x - \sqrt{3}i\right)\left(x + \sqrt{3}i\right) = 0,$$

and the zeros are $-\sqrt{3}, \sqrt{3}, -\sqrt{3}i, \sqrt{3}i$.

39. We have

$$x^4 - x^2 - 6 = (x^2 - 3)(x^2 + 2) = \left(x - \sqrt{3}\right)\left(x + \sqrt{3}\right)\left(x - \sqrt{2}i\right)\left(x + \sqrt{2}i\right) = 0,$$

and the zeros are $-\sqrt{3}, \sqrt{3}, -\sqrt{2}i, \sqrt{2}i$.

40. We have

$$x^4 + 2x^2 - 8 = (x^2 - 2)(x^2 + 4) = \left(x - \sqrt{2}\right)\left(x + \sqrt{2}\right)(x - 2i)(x + 2i) = 0,$$

and the zeros are $-\sqrt{2}, \sqrt{2}, -2i, 2i$.

41. By inspection, $x = -2$ is a solution to $x^3 + 8 = 0$, so $x + 2$ is a factor. Dividing $x^3 + 1$ by $x + 2$ gives $x^3 + 1 = (x + 2)(x^2 - 2x + 4)$. Then solving for the zeros of $x^2 - 2x + 4 = 0$ gives

$$x = \frac{2 \pm \sqrt{4 - 4(1)(4)}}{2} = \frac{2 \pm \sqrt{-12}}{2} = \frac{2 \pm 2\sqrt{3}i}{2} = 1 \pm \sqrt{3}i.$$

The zeros are $-2, 1 + \sqrt{3}i, 1 - \sqrt{3}i$.

42. By inspection, $x = 3$ is a solution to $x^3 - 27 = 0$, so $x - 3$ is a factor. Dividing $x^3 - 27$ by $x - 3$ gives $x^3 - 27 = (x - 3)(x^2 + 3x + 9)$. Then solving for the zeros of $x^2 + 3x + 9 = 0$ gives

$$x = \frac{-3 \pm \sqrt{9 - 4(1)(9)}}{2} = \frac{-3 \pm \sqrt{-27}}{2} = \frac{-3 \pm 3\sqrt{3}i}{2}.$$

The zeros are $3, \frac{-3+3\sqrt{3}i}{2}, \frac{-3-3\sqrt{3}i}{2}$.

43. For $f(x) = x^3 - 2x^2 + 9x - 18$, we have

$$f(3i) = (3i)^3 - 2(3i)^2 + 9(3i) - 18 = -27i + 18 + 27i - 18 = 0.$$

Since $3i$ is a zero, the conjugate $-3i$ is also a zero, so $(x - 3i)(x + 3i) = x^2 + 9$ is a factor. Dividing $x^2 + 9$ into $f(x)$ gives

$$f(x) = (x^2 + 9)(x - 2).$$

The third solution is $x = 2$.

44. For $f(x) = x^3 + 3x^2 + 16x - 20$, we have

$$\begin{aligned} f(-2 + 4i) =& (-2 + 4i)^3 + 3(-2 + 4i)^2 + 16(-2 + 4i) - 20 \\ =& (-2 + 4i)[(-2 + 4i)^2 + 3(-2 + 4i) + 16] - 20 \\ =& (-2 + 4i)[4 - 16i - 16 - 6 + 12i + 16] - 20 \\ =& (-2 + 4i)(-2 - 4i) - 20 = 20 - 20 = 0. \end{aligned}$$

Since $-2 + 4i$ is a zero, the conjugate $-2 - 4i$ is also a zero, so

$$(x - (-2 + 4i))(x - (-2 - 4i)) = x^2 + 4x + 20$$

is a factor. Dividing $x^2 + 4x + 20$ into $f(x)$ gives

$$f(x) = (x^2 + 4x + 20)(x - 1).$$

The third solution is $x = 1$.

45. For $f(x) = x^4 - 2x^3 - 2x^2 - 2x - 3$, we have

$$f(i) = (i)^4 - 2(i)^3 - 2(i)^2 - 2(i) - 3 = 1 + 2i + 2 - 2i - 3 = 0.$$

Since i is a zero, the conjugate $-i$ is also a zero, so

$$(x - i)(x + i) = x^2 + 1$$

is a factor. Dividing $x^2 + 1$ into $f(x)$ gives

$$f(x) = (x^2 + 1)(x^2 - 2x - 3) = (x^2 + 1)(x - 3)(x + 1).$$

The third and fourth solutions are $x = -1, x = 3$.

46. For

$$f(x) = x^5 - 2x^4 + x^3 + 2x^2 - 2x = x(x^4 - 2x^3 + x^2 + 2x - 2),$$

$x = 0$ is a solution. To find the other solutions it is sufficient to consider the factors of

$$g(x) = x^4 - 2x^3 + x^2 + 2x - 2.$$

Note that

$$\begin{aligned} g(1 - i) =& (1 - i)^4 - 2(1 - i)^3 + (1 - i)^2 + 2(1 - i) - 2 \\ & (1 - i)^2[(1 - i)^2 - 2(1 - i) + 1] + 2 - 2i - 2 \\ =& (1 - i)^2[1 - 2i - 1 - 2 + 2i + 1] - 2i \\ =& (1 - i)^2(-1) - 2i \\ =& 2i - 2i = 0. \end{aligned}$$

Since $1 - i$ is a zero, the conjugate $1 + i$ is also a zero, so

$$(x - (1 - i))(x - (1 + i)) = x^2 - 2x + 2$$

is a factor. Dividing $x^2 - 2x + 2$ into $g(x)$ gives

$$g(x) = (x^2 - 2x + 2)(x^2 - 1) = (x^2 - 2x + 2)(x - 1)(x + 1),$$

so

$$f(x) = x(x^2 - 2x + 2)(x - 1)(x + 1).$$

The fourth and fifth solutions are $x = -1$ and $x = 1$.

47. The possible rational zeros of $f(x) = x^3 - 3x^2 + 9x - 27$ are

$$\pm 1, \pm 3, \pm 9, \pm 27,$$

and by substitution, $x = 3$ is a zero. Then

$$f(x) = (x - 3)(x^2 + 9) = (x - 3)(x - 3i)(x + 3i).$$

48. The possible rational zeros of $f(x) = x^3 + 2x^2 + 2x + 1$ are $x = \pm 1$,and by substitution, $x = -1$ is a zero. Then

$$f(x) = (x+1)(x^2 + x + 1).$$

Solving $x^2 + x + 1 = 0$ we have

$$x = \frac{-1 \pm \sqrt{1 - 4(1)(1)}}{2} = \frac{-1 \pm \sqrt{-3}}{2} = \frac{-1 \pm \sqrt{3}i}{2}$$

and

$$f(x) = (x+1)\left(x - \left(\frac{-1+\sqrt{3}i}{2}\right)\right)\left(x - \left(\frac{-1-\sqrt{3}i}{2}\right)\right).$$

49. Since $f(x) = x^4 - x^2 - 2x + 2$ has a zero of multiplicity 2 at $x = 1$, we have

$$f(x) = (x-1)^2(x^2 + 2x + 2) = (x-1)^2(x - (-1 - i))(x - (-1 + i)).$$

50. Since $f(x) = x^4 + 3x^3 + 4x^2 + 3x + 1$ has a zero of multiplicity 2 at $x = -1$, we have

$$\begin{aligned} f(x) =& (x+1)^2(x^2 + x + 1) \\ =& (x+1)^2\left(x - \left(\frac{-1+\sqrt{3}i}{2}\right)\right)\left(x - \left(\frac{-1-\sqrt{3}i}{2}\right)\right). \end{aligned}$$

51. Since $2i$ is a zero, the conjugate $-2i$ is also a zero and

$$f(x) = (x-2)(x-2i)(x+2i) = (x-2)(x^2+4) = x^3 - 2x^2 + 4x - 8.$$

52. Since $1 - i$ is a zero, the conjugate $1 + i$ is also a zero and

$$f(x) = (x-1)(x - (1+i))(x - (1-i)) = (x-1)(x^2 - 2x + 2) = x^3 - 3x^2 + 4x - 2.$$

53. Since $\sqrt{3}i$ and $3i$ are zeros, the conjugates $-\sqrt{3}i$ and $-3i$ are also zeros, so a polynomial with the specified zeros is

$$f(x) = (x - \sqrt{3}i)(x + \sqrt{3}i)(x - 3i)(x + 3i) = (x^2 + 3)(x^2 + 9) = x^4 + 12x^2 + 27.$$

54. Since $x = 1$ is a zero of multiplicity 2, $(x-1)^2$ is a factor, and since $2+i$ is a zero, the conjugate $2-i$ is also a zero. So

$$\begin{aligned} f(x) &= (x-1)^2(x-(2+i))(x-(2-i)) \\ &= (x^2-2x+1)(x^2-4x+5) = x^4-6x^3+14x^2-14x+5. \end{aligned}$$

55. Since $x = -2$ is a zero of multiplicity 2, $(x+2)^2$ is a factor, and since $1+2i$ is a zero, the conjugate $1-2i$ is also a zero and since $x = 0$ is a zero x is a factor. So

$$\begin{aligned} f(x) &= x(x+2)^2(x-(1+2i))(x-(1-2i)) \\ &= x(x^2+4x+4)(x^2-2x+5) = x^5+2x^4+x^3+12x^2+20x. \end{aligned}$$

56. Since i and $3-i$ are zeros, the conjugates $-i$ and $3+i$ are also zeros. Since the graph passes through the origin, $x = 0$ is another zero and x is a factor, and

$$f(x) = x(x-i)(x+i)(x-(3-i))(x-(3+i)) = x^5-6x^4+11x^3-6x^2+10x.$$

Review Exercises for Chapter 3 (Page 173)

1. **a.** (ii) The graph crosses the x-axis at $x = 1$ and $x = 2$. Since the zero at $x = -1$ is of multiplicity 2, the curve just touches and turns without crossing the x-axis at $x = -1$.

b. (iii) The graph crosses the x-axis at $x = 1$ and $x = -1$. Since the zero at $x = 2$ is of multiplicity 2, the curve just touches and turns without crossing the x-axis at $x = 2$.

c. (iv) The graph crosses the x-axis at $x = 2$ and $x = -1$. Since the zero at $x = 1$ is of multiplicity 2, the curve just touches and turns without crossing the x-axis at $x = 1$.

d. (i) The graph touches without crossing the x-axis at $x = 1$ and $x = -1$ since both zeros are of multiplicity 2.

2. **a.** (ii) The graph has vertical asymptotes at $x = 1$ and $x = -1$, and horizontal asymptote $y = 0$.

b. (iv) The graph has vertical asymptotes at $x = 1$ and $x = -1$, and horizontal asymptote $y = 1$.

c. (i) The graph has vertical asymptote $x = -2$, and horizontal asymptote $y = 2$.

d. (iii) The graph has vertical asymptote $x = -2$, and horizontal asymptote $y = -1$.

3. $f(x) = -2(x-1)^2 + 2$

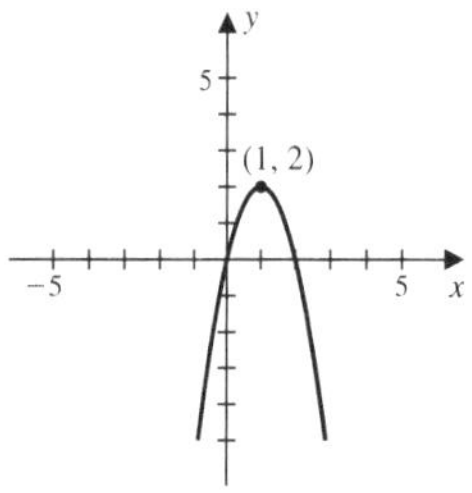

4. $f(x) = -2(x-2)^3 - 1$

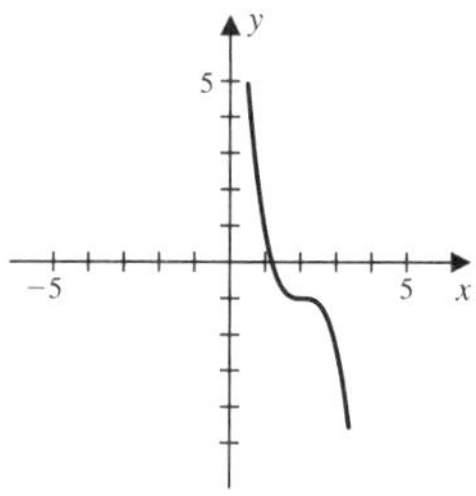

5. $f(x) = -x^4 - 3$

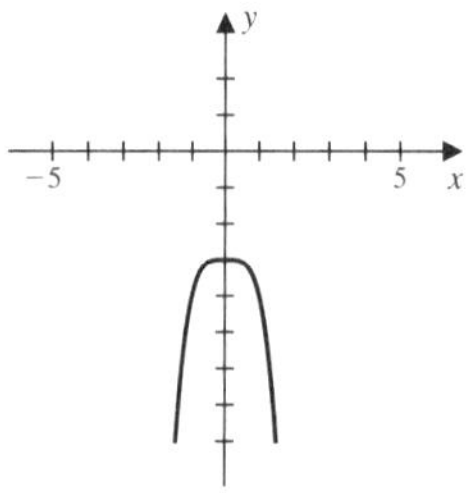

6. $f(x) = (x+1)^4 + 2$

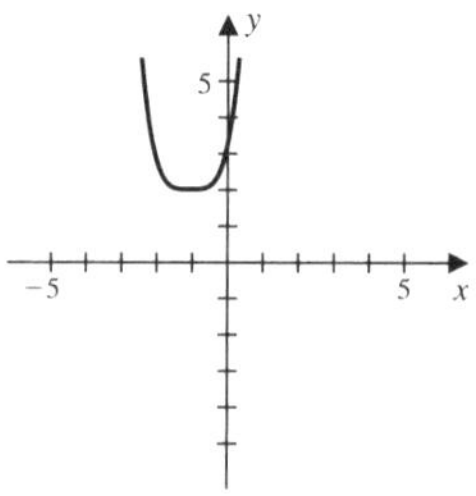

7. $f(x) = (x+1)(x+2)(x-3)$

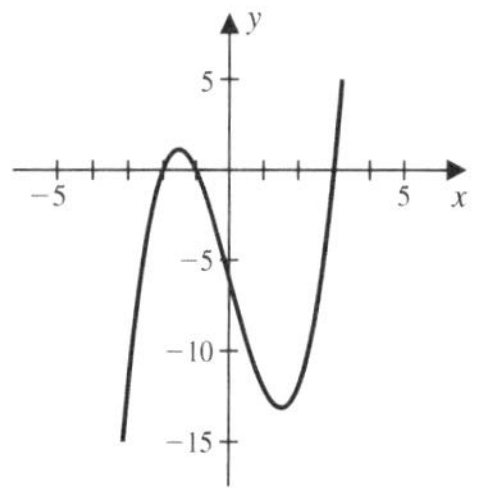

8. $f(x) = x^2(x-1)$

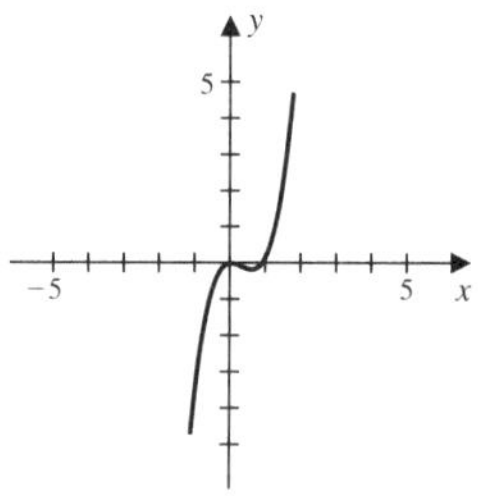

9. $f(x) = \frac{1}{2}(x-1)^3(x+2)$

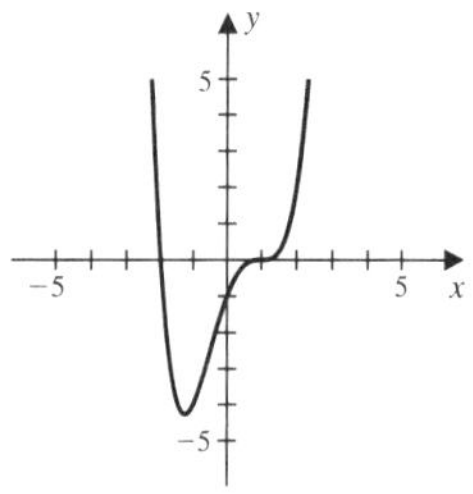

10. $f(x) = -\frac{1}{16}(x-2)^3(x+1)(x+2)$

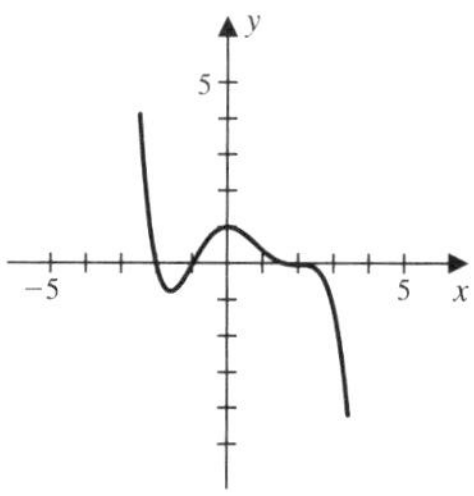

11.
$$\begin{aligned} f(x) &= x^3 - \frac{1}{2}x^2 - \frac{1}{2}x \\ &= \frac{1}{2}x(2x^2 - x - 1) \\ &= \frac{1}{2}x(2x+1)(x-1) \end{aligned}$$

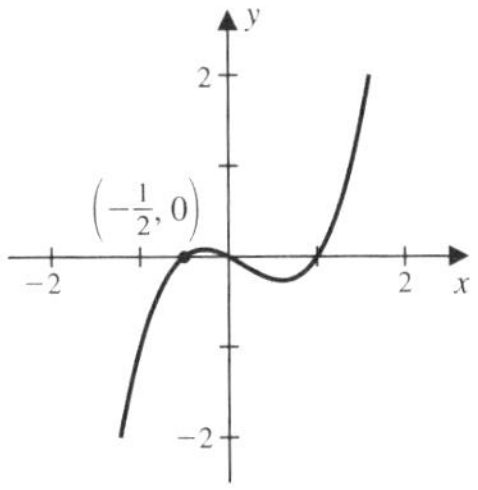

12.
$$\begin{aligned} f(x) &= x^5 + 2x^4 + 4x + 8 \\ &= (x+2)(x^2 - 2x + 2) \\ &\quad (x^2 + 2x + 2) \end{aligned}$$

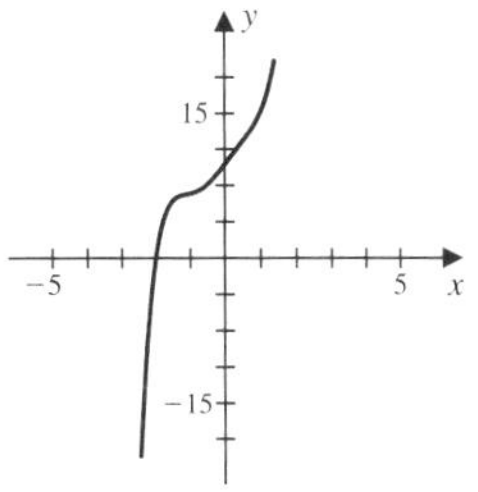

13. Degree 3, leading coefficient positive. The graph shows a zero of at least multiplicity 2, and one other zero of multiplicity 1, so the degree is at least 3. The end behavior is the same as that for $y = x^3$, so the leading coefficient is positive.

14. Degree 5, leading coefficient positive. The graph shows five zeros of multiplicity 1, so the degree is at least 5, and the end behavior is the same as that for $y = x^5$ so the leading coefficient is positive.

15. Degree 4, leading coefficient negative. Adding or subtracting a positive constant from a polynomial shifts the curve upward or downward but does not change the degree. The curve can be shifted downward so as to appear to have 4 zeros of multiplicity 1, so the degree of the original polynomial is at least 4. Since the end behavior is the opposite of that of $y = x^4$ the leading coefficient is negative.

16. Degree 6, leading coefficient negative. Adding or subtracting a positive constant from a polynomial shifts the curve upward or downward but does not change the degree. The curve can be shifted upward so as to appear to have 3 zeros of multiplicity 1 and a zero of multiplicity at least 3, so the degree of the original polynomial is at least 6. Since the end behavior is the opposite of that of $y = x^6$ the leading coefficient is negative.

17. $Q(x) = 4x + 10$ and $R(x) = 19$

18. $Q(x) = -\frac{3}{2}x + \frac{5}{4}$ and $R(x) = \frac{23}{4}$

19. $Q(x) = 3x^2 - 4x + 6$ and $R(x) = -11$

20. $Q(x) = x + 3$ and
$R(x) = 6x^3 - 6x^2 - 6x - 6$

21. $P(x) = 3x^4 - 9x^3 - 2x^2 + 5x + 3$

a.
$$\frac{\pm 1, \pm 3}{\pm 1, \pm 3} = \pm 1, \pm\frac{1}{3}, \pm 3$$

b. Since

$$P(3) = 3(3)^4 - 9(3)^3 - 2(3)^2 + 5(3) + 3 = 3^5 - 3^5 - 18 + 15 + 3 = 0,$$

$x - 3$ is a factor.

c. $P(x) = (x-3)(3x^3 - 2x - 1) = (x-3)(x-1)(3x^2 + 3x + 1)$

22. $P(x) = x^4 + 4x^3 + 6x^2 + 5x + 2$

a.
$$\frac{\pm 1, \pm 2}{\pm 1} = \pm 1, \pm 2$$

b. $P(-2) = (-2)^4 + 4(-2)^3 + 6(-2)^2 + 5(-2) + 2 = 0$, so $x + 2$ is a factor.

c. $P(x) = (x+2)(x^3 + 2x^2 + 2x + 1) = (x+2)(x+1)(x^2 + x + 1)$

23. $P(x) = x^5 - 3x^4 - 5x^3 + 27x^2 - 32x + 12$

a.
$$\frac{\pm 1, \pm 2, \pm 3, \pm 4, \pm 6, \pm 12}{\pm 1} = \pm 1, \pm 2, \pm 3, \pm 4, \pm 6, \pm 12$$

b. $P(-3) = (-3)^5 - 3(-3)^4 - 5(-3)^3 + 27(-3)^2 - 32(-3) + 12 = 0$, so $x + 3$ is a factor.

c. $P(x) = (x+3)(x^4 - 6x^3 + 13x^2 - 12x + 4) = (x+3)(x-1)^2(x-2)^2$

24. $P(x) = x^6 - 5x^5 + 5x^4 + 9x^3 - 14x^2 - 4x + 8$

a.
$$\frac{\pm 1, \pm 2, \pm 4, \pm 8}{\pm 1} = \pm 1, \pm 2, \pm 4, \pm 8$$

b. $P(1) = 1^6 - 5(1)^5 + 5(1)^4 + 9(1)^3 - 14(1)^2 - 4(1) + 8 = 0$, so $x - 1$ is a factor.

c. $P(x) = x^6 - 5x^5 + 5x^4 + 9x^3 - 14x^2 - 4x + 8 = (x-1)(x+1)^2(x-2)^3$

25. $P(x) = x^4 - 5x^3 + 7x^2 + 3x - 10$

Since $2 + i$ is a zero, so is $2 - i$ and

$$(x - (2+i))(x - (2-i)) = x^2 - 4x + 5.$$

Then

$$P(x) = (x^2 - 4x + 5)(x^2 - x - 2) = (x - (2+i))(x - (2-i))(x-2)(x+1).$$

26. $P(x) = x^4 - 3x^3 + 13x^2 - 12x + 36$

Since $2i$ is a zero, so is $-2i$ and

$$(x - 2i)(x + 2i) = x^2 + 4.$$

Then

$$P(x) = (x^2 + 4)(x^2 - 3x + 9) = (x - 2i)(x + 2i)\left(x - \left(\frac{3 + 3\sqrt{3}i}{2}\right)\right)\left(x - \left(\frac{3 - 3\sqrt{3}i}{2}\right)\right).$$

27. $P(x) = x^5 - x^4 + 10x^3 - 10x^2 + 9x - 9$

Since i is a zero, so is $-i$ and

$$(x - i)(x + i) = x^2 + 1.$$

Then

$$\begin{aligned} P(x) &= (x^2 + 1)(x^3 - x^2 + 9x - 9) = (x^2 + 1)(x - 1)(x^2 + 9) \\ &= (x - i)(x + i)(x - 1)(x - 3i)(x + 3i). \end{aligned}$$

28. $P(x) = x^5 - x^4 + 4x - 4$

Since $1 + i$ is a zero, so is $1 - i$ and

$$(x - (1 + i))(x - (1 - i)) = x^2 - 2x + 2.$$

Then

$$\begin{aligned} P(x) &= (x^2 - 2x + 2)(x^3 + x^2 - 2) \\ &= (x^2 - 2x + 2)(x - 1)(x^2 + 2x + 2) \\ &= (x - (1 + i))(x - (1 - i))(x - 1)(x + 1)^2. \end{aligned}$$

29. For $f(x) = \dfrac{x - 4}{x - 1}$.

a. $\text{Domain}(f) = \{x \mid x \neq 1\} = (-\infty, 1) \cup (1, \infty)$

b. x-intercepts: Solve

$$\frac{x - 4}{x - 1} = 0 \text{ which implies } x - 4 = 0, \text{ so } x = 4;$$

y-intercept: Set $x = 0$ to get $y = 4$.

c. Vertical asymptotes: Solve $x - 1 = 0$ implies $x = 1$; horizontal asymptotes: For x large in magnitude

$$\frac{x-4}{x-1} \approx \frac{x}{x} = 1, \quad \text{so} \quad y = 1.$$

30. For $f(x) = \dfrac{(x-2)(x+2)}{(x-3)(x+1)}$.

a. $\text{Domain}(f) = \{x \mid x \neq -1, x \neq 3\} = (-\infty, -1) \cup (-1, 3) \cup (3, \infty)$

b. x-intercepts: Solve

$$\frac{(x-2)(x+2)}{(x-3)(x+1)} = 0 \text{ which implies } x - 2 = 0,\ x + 2 = 0, \text{ so } x = 2,\ x = -2;$$

y-intercept: Set $x = 0$ to give $y = \frac{4}{3}$.

c. Vertical asymptotes: Solve $(x-3)(x+1) = 0$ implies $x = 3$, $x = -1$; horizontal asymptotes: For x large in magnitude

$$\frac{(x-2)(x+2)}{(x-3)(x+1)} \approx \frac{x^2}{x^2} = 1, \quad \text{so} \quad y = 1.$$

31. For $f(x) = \dfrac{x^2 - 2x + 1}{2x^2 - 18} = \dfrac{(x-1)^2}{2(x-3)(x+3)}$.

a. $\text{Domain}(f) = \{x \mid x \neq -3, x \neq 3\} = (-\infty, -3) \cup (-3, 3) \cup (3, \infty)$

b. x-intercepts: Solve

$$\frac{(x-1)^2}{2(x-3)(x+3)} = 0 \text{ which implies } (x-1)^2 = 0, \text{ so } x = 1;$$

y-intercept: Set $x = 0$ to give $y = -\frac{1}{18}$.

c. Vertical asymptotes: Solve $(x-3)(x+3) = 0$ implies $x = -3$, $x = 3$; horizontal asymptotes: For x large in magnitude,

$$\frac{(x-1)^2}{2(x-3)(x+3)} \approx \frac{x^2}{2x^2} = \frac{1}{2}, \quad \text{so} \quad y = \frac{1}{2}.$$

32. For $f(x) = \dfrac{3x^2 + 7x - 6}{x^2 - x - 6} = \dfrac{(3x-2)(x+3)}{(x-3)(x+2)}$.

a. $\text{Domain}(f) = \{x \mid x \neq -2, x \neq 3\} = (-\infty, -2) \cup (-2, 3) \cup (3, \infty)$

b. x-intercepts: Solve

$$\frac{(3x-2)(x+3)}{(x-3)(x+2)} = 0 \text{ which implies } 3x-2=0,\ x+3=0,\ \text{ so } x=\frac{2}{3},\ x=-3;$$

y-intercept: Set $x=0$ to give $y=1$.

c. Vertical asymptotes: Solve $(x-3)(x+2)=0$ implies $x=3$, $x=-2$; horizontal asymptotes: For x large in magnitude

$$\frac{(3x-2)(x+3)}{(x-3)(x+2)} \approx \frac{3x^2}{x^2} = 3, \quad \text{so} \quad y=3.$$

33. For $f(x) = \dfrac{x^3+2x^2-x-2}{x^3} = \dfrac{(x+2)(x+1)(x-1)}{x^3}$.

a. $\text{Domain}(f) = \{x \mid x \neq 0\} = (-\infty, 0) \cup (0, \infty)$

b. x-intercepts: Solve

$$\frac{(x+2)(x+1)(x-1)}{x^3} = 0$$

implies

$$x+2=0,\ x+1=0,\ x-1=0,\ \text{ so } x=-2,\ x=-1,\ x=1;$$

y-intercept: None, since 0 is not in the domain.

c. Vertical asymptotes: Solve $x^3=0$ to get $x=0$;

horizontal asymptotes: For x large in magnitude

$$\frac{(x+2)(x+1)(x-1)}{x^3} \approx \frac{x^3}{x^3} = 1, \quad \text{so} \quad y=1.$$

34. For

$$\begin{aligned} f(x) &= \frac{x^4-2x^3+x^2}{x^3-1} = \frac{x^2(x^2-2x+1)}{x^3-1} = \frac{x^2(x-1)^2}{x^3-1} \\ &= \frac{x^2(x-1)^2}{(x-1)(x^2+x+1)} = \frac{x^2(x-1)}{x^2+x+1}. \end{aligned}$$

a. $\text{Domain}(f) = \{x \mid \neq 1\} = (-\infty, 1) \cup (1, \infty)$

b. x-intercepts: Solve

$$\frac{x^2(x-1)}{x^2+x+1} = 0 \text{ implies } x^2=0,\ (x-1)=0,\ \text{ so } x=0,\ x=1;$$

y-intercept: Set $x = 0$ to get $y = 0$.

c. Vertical asymptotes: None, since $x^2 + x + 1 > 0$;

horizontal asymptotes: For x large in magnitude

$$\frac{x^2(x-1)}{x^2+x+1} \approx \frac{x^3}{x^2} = x,$$

so there is no horizontal asymptote. However,

$$\frac{x(x^2-1)}{x^2+x+1} = x - 2 - \frac{x+2}{x^2+x+1}$$

and

$$\frac{x+2}{x^2+x+1} \to 0 \quad \text{as} \quad x \to \infty,$$

so $y = x - 2$ is a slant asymptote.

35. For $f(x) = \dfrac{3}{x-2}$
Horizontal asymptotes: $y = 0$; vertical asymptotes: $x = 2$; x-intercepts: none; y-intercept: $(0, -3/2)$

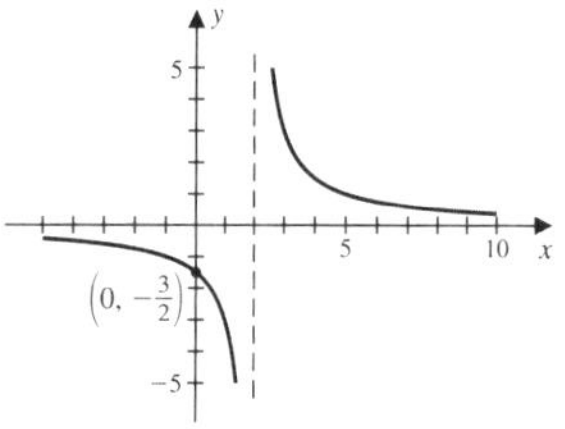

36. For $f(x) = -\dfrac{4}{x+2}$.
Horizontal asymptotes: $y = 0$; vertical asymptotes: $x = -2$; x-intercepts: none; y-intercept: $(0, -2)$

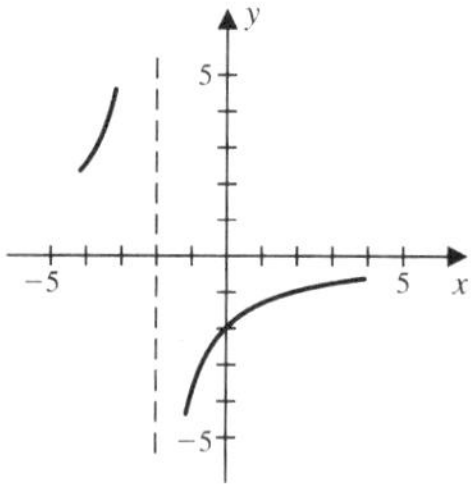

37. For $f(x) = -\dfrac{2}{(x-1)(x-2)}$.
Horizontal asymptotes: $y = 0$; vertical asymptotes: $x = 1, x = 2$; x-intercepts: none; y-intercept: $(0, -1)$

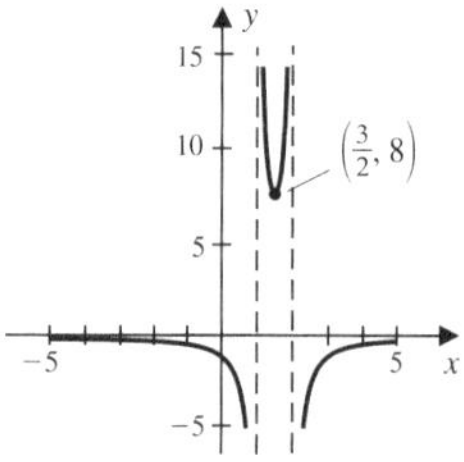

38. For $f(x) = \dfrac{3}{(x+3)(x-4)}$.
Horizontal asymptotes: $y = 0$; vertical asymptotes: $x = -3, x = 4$; x-intercepts: none; y-intercept: $(0, -1/4)$

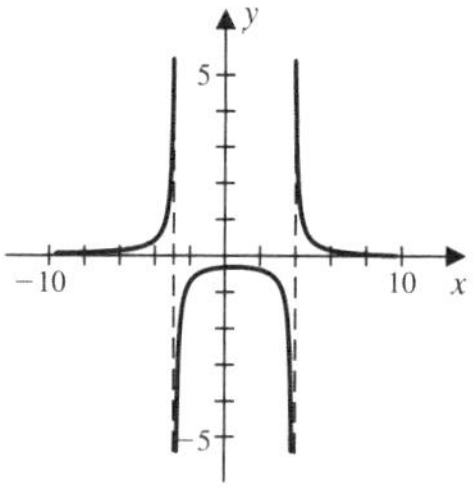

39. For

$$f(x) = \frac{4}{x^2 - 4} = \frac{4}{(x-2)(x+2)}.$$

Horizontal asymptotes: $y = 0$; vertical asymptotes: $x = -2$, $x = 2$; x-intercepts: none; y-intercept: $(0, -1)$

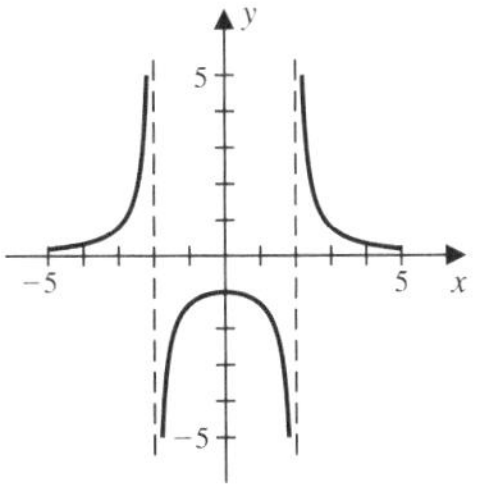

40. For

$$f(x) = \frac{x-5}{x^2 - 2x - 3} = \frac{x-5}{(x-3)(x+1)}.$$

Horizontal asymptotes: $y = 0$; vertical asymptotes: $x = -1$, $x = 3$; x-intercepts: $(5, 0)$; y-intercept: $(0, 5/3)$

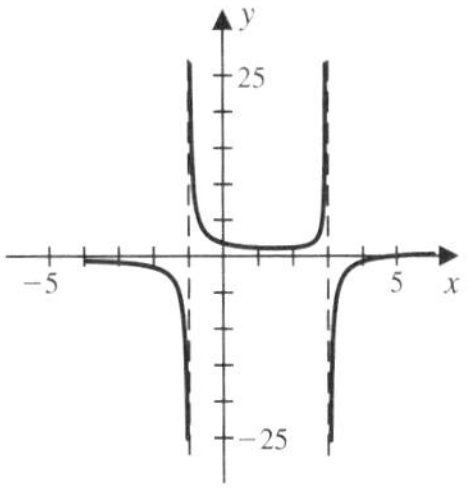

41. For

$$f(x) = \frac{x^2 - 1}{x^2 + 2x} = \frac{(x-1)(x+1)}{x(x+2)}.$$

Horizontal asymptotes: $y = 1$; vertical asymptotes: $x = -2$, $x = 0$; x-intercepts: $(1, 0)$, $(-1, 0)$; y-intercept: none

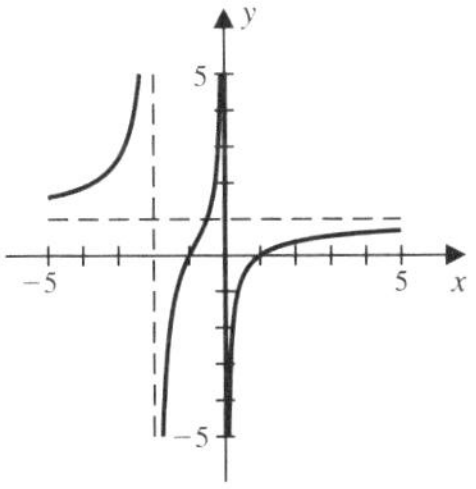

42. For

$$f(x) = \frac{4 - x^2}{x^2 - 9} = \frac{(2-x)(2+x)}{(x-3)(x+3)}.$$

Horizontal asymptotes: $y = -1$; vertical asymptotes: $x = -3$, $x = 3$; x-intercepts: $(2, 0)$, $(-2, 0)$; y-intercept: $\left(0, -\frac{4}{9}\right)$

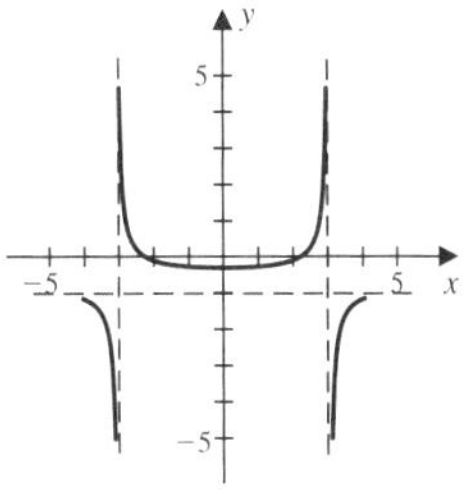

43. For

$$f(x) = \frac{x^2 - 2x + 1}{x + 1} = x - 3 + \frac{4}{x + 1},$$

we have the following graph.

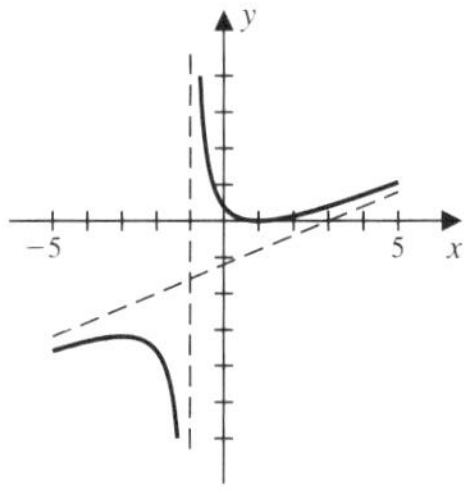

44. For

$$f(x) = \frac{x^3 - 2x^2 + 4x - 3}{x^2 - 3x + 2}$$
$$= x + 1 + \frac{5x - 5}{x^2 - 3x + 2},$$

we have the following graph.

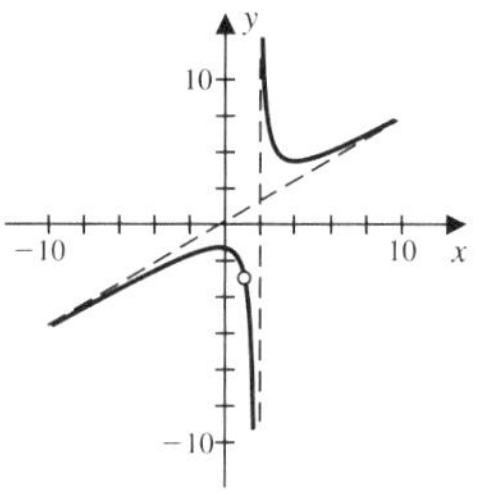

45. For $f(x) = \sqrt{\dfrac{x - 2}{x + 1}}$.

a. Domain$(f) = (-\infty, -1) \cup [2, \infty)$

b. The graph of the function is shown.

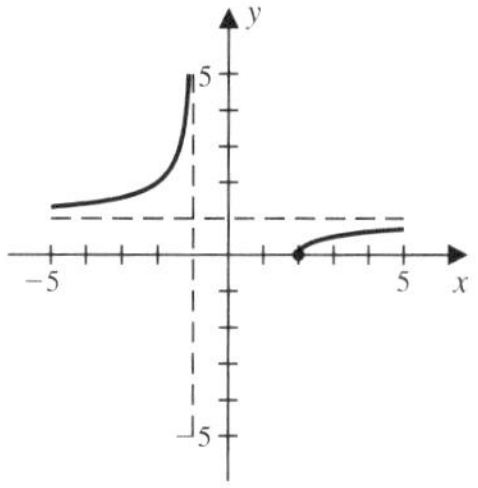

c. Horizontal asymptotes: $y = 1$; vertical asymptotes: $x = -1$; x-intercepts: $(2, 0)$; y-intercept: none

46. For $f(x) = \sqrt{\dfrac{x - 3}{x + 3}}$.

a. Domain$(f) = (-\infty, -3) \cup [3, \infty)$

b. The graph of the function is shown.

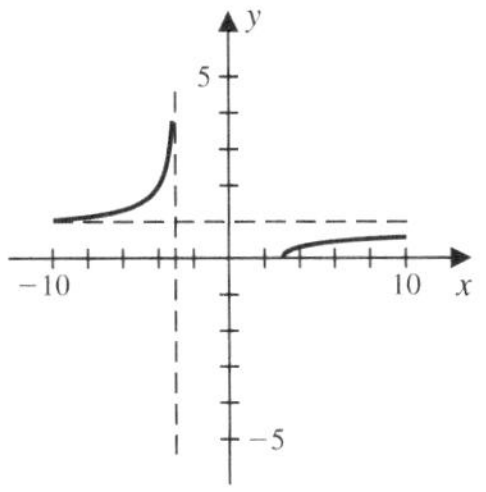

c. Horizontal asymptotes: $y = 1$; vertical asymptotes: $x = -3$; x-intercepts: $(3, 0)$; y-intercept: none

47. For $f(x) = \sqrt{\dfrac{4-x}{x+4}}$.

a. $\text{Domain}(f) = (-4, 4]$

b. The graph of the function is shown.

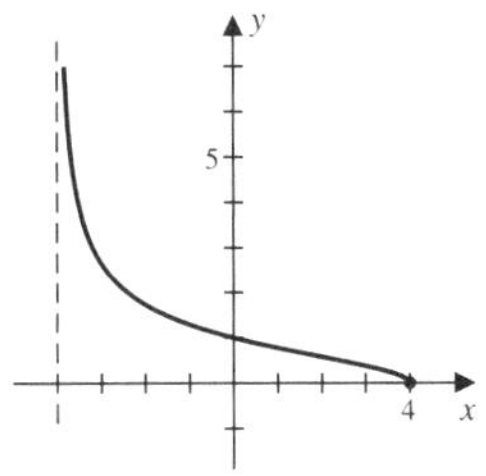

c. Horizontal asymptotes: none; vertical asymptotes: $x = -4$; x-intercepts: $(4, 0)$; y-intercept: $(0, 1)$

48. For $f(x) = \dfrac{x-3}{\sqrt{(x-1)(x+2)}}$.

a. $\text{Domain}(f) = (-\infty, -2) \cup (1, \infty)$

b. The graph of the function is shown.

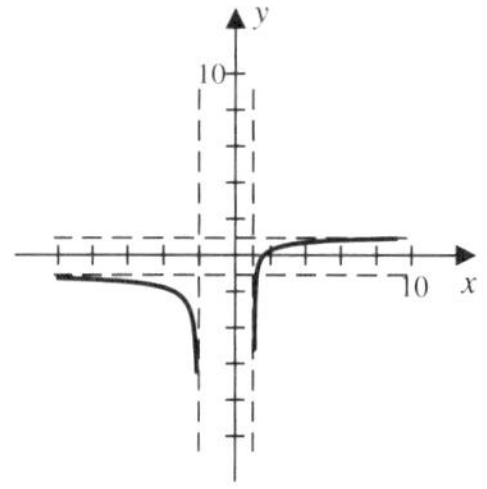

c. Horizontal asymptotes: $y = 1$, $y = -1$

Since for x large in magnitude

$$\frac{x-3}{\sqrt{(x-1)(x+2)}} \approx \frac{x}{\sqrt{x^2}} = \frac{x}{|x|};$$

vertical asymptotes: $x = 1$, $x = -2$; x-intercepts: $(3, 0)$; y-intercept: none

49. For $f(x) = \dfrac{x^2}{\sqrt{4-x^2}} = \dfrac{x^2}{\sqrt{(2-x)(2+x)}}$.

a. $\text{Domain}(f) = (-2, 2)$

b. The graph of the function is shown.

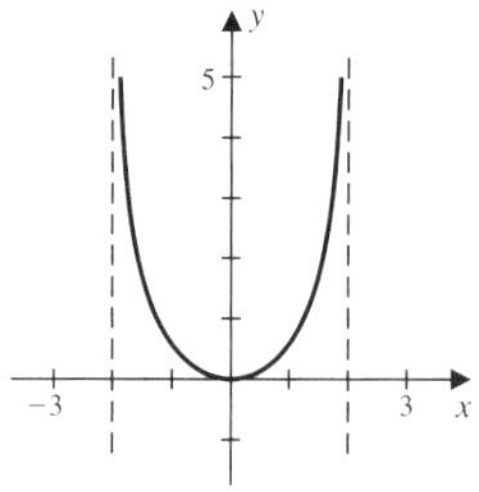

c. Horizontal asymptotes: none; vertical asymptotes: $x = 2$, $x = -2$; x-intercept: $(0, 0)$; y-intercept: $(0, 0)$

50. For $f(x) = \dfrac{\sqrt{x^2-9}}{x-2} = \dfrac{\sqrt{(x-3)(x+3)}}{x-2}$.

a. $\text{Domain}(f) = (-\infty, -3] \cup [3, \infty)$

b. The graph of the function is shown.

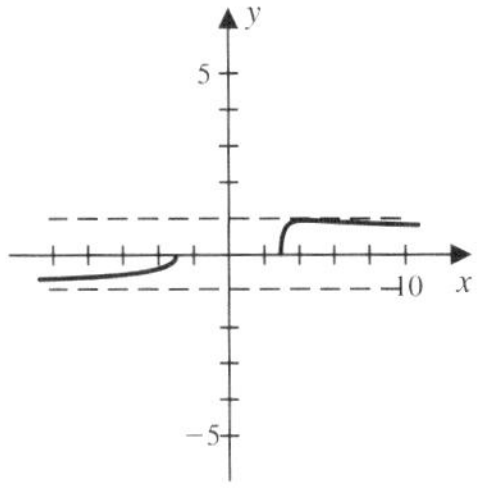

c. Horizontal asymptotes: $y = 1$, $y = -1$

Since for x large in magnitude

$$\frac{\sqrt{x^2 - 9}}{x - 2} \approx \frac{\sqrt{x^2}}{x} = \frac{|x|}{x};$$

vertical asymptotes: None, since the domain of the function does not include 2; x-intercepts: $(-3, 0)$, $(3, 0)$; y-intercept: none

51. $f(x) = x^3 - 2x^2 - x + 2$
a. increasing: $(-\infty, -0.2) \cup (1.5, \infty)$; decreasing: $(-0.2, 1.5)$
b. local maximum: $(-0.2, 2.1)$; local minimum: $(1.5, -0.6)$

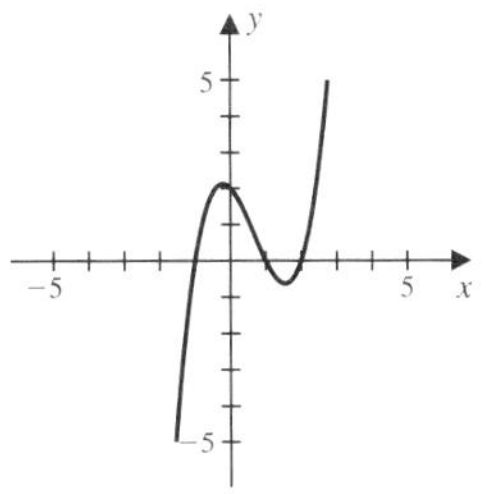

52. $f(x) = x^4 - 2x^3$
a. increasing: $(1.5, \infty)$; decreasing: $(-\infty, 1.5)$
b. local maximum: none; local minimum: $(1.5, -1.7)$

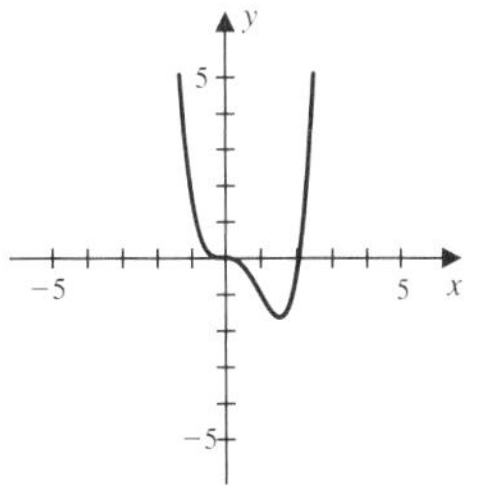

53. $f(x) = \frac{2}{3}x^3 + \frac{7}{2}x^2 - 12x$
a. increasing: $(-\infty, -4.7) \cup (1.3, \infty)$; decreasing: $(-4.7, 1.3)$
b. local maximum: $(-4.7, 64.5)$; local minimum: $(1.3, -8.2)$

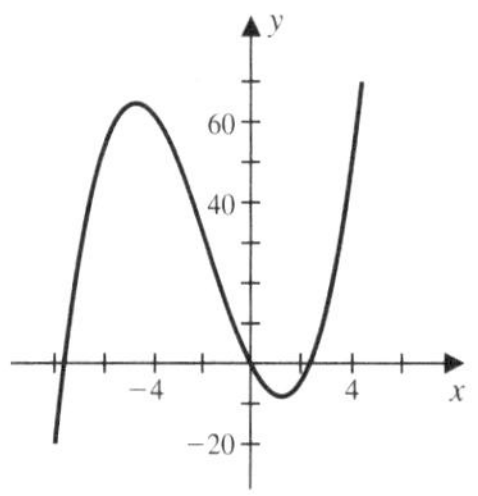

54. $f(x) = x^5 - 2x^4 + 2x^2 - x$
a. increasing: $(-\infty, -0.7) \cup (0.3, \infty)$; decreasing: $(-0.7, 0.3)$
b. local maximum: $(-0.7, 1)$; local minimum: $(0.3, -0.1)$

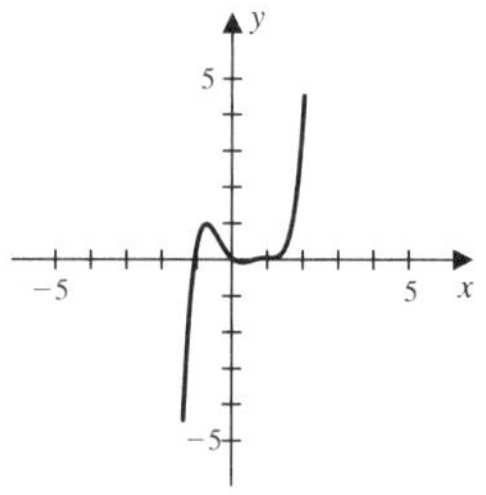

55. $\frac{2 + i\sqrt{2}}{4} = \frac{1}{2} + \frac{\sqrt{2}}{4}i$

56. $\frac{-5 - \sqrt{-4}}{6} = \frac{-5 - 2i}{6} = -\frac{5}{6} - \frac{1}{3}i$

57. $(3 - i) - (2 - 3i) = (3 - 2) + (-1 + 3)i = 1 + 2i$

58. $(-2 + 6i) + (-3 + i) = (-2 - 3) + (6 + 1)i = -5 + 7i$

59. $(2 - i) \cdot \overline{(2 + i)} = (2 - i) \cdot (2 - i) = 4 - 4i + i^2 = 3 - 4i$

60. $(4 - 6i) \cdot \overline{(3 - 2i)} = (4 - 6i) \cdot (3 + 2i) = 12 + 8i - 18i - 12i^2 = 24 - 10i.$

61. $i^{20} = (i^4)^5 = 1^5 = 1$

62. $i^{21} = i^{20} \cdot i = 1 \cdot i = i$

63. $$\frac{2 + 3i}{4 - 7i} = \frac{2 + 3i}{4 - 7i} \cdot \frac{4 + 7i}{4 + 7i} = \frac{8 + 14i + 12i + 21i^2}{16 + 49} = \frac{-13 + 26i}{65} = -\frac{1}{5} + \frac{2}{5}i$$

64. $$\frac{-5 + 3i}{2 - 3i} = \frac{-5 + 3i}{2 - 3i} \cdot \frac{2 + 3i}{2 + 3i} = \frac{-10 - 15i + 6i + 9i^2}{4 + 9} = \frac{-19 - 9i}{13} = -\frac{19}{13} - \frac{9}{13}i$$

65. a. $y = f(x-1)$ **b.** $y = f(x-1) - 1$ **c.** $y = -f(x+1) + 1$ **d.** $y = |f(x)|$

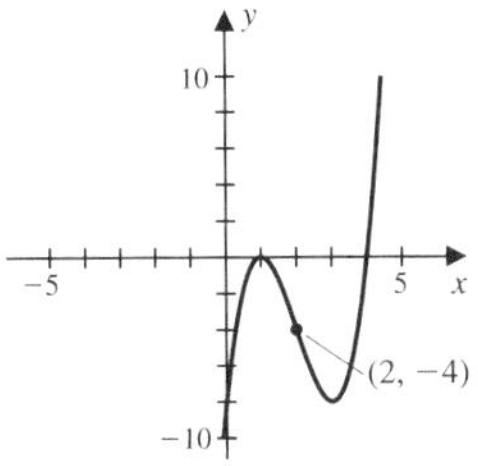

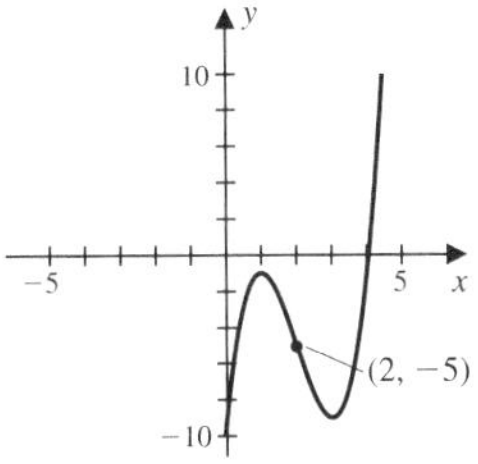

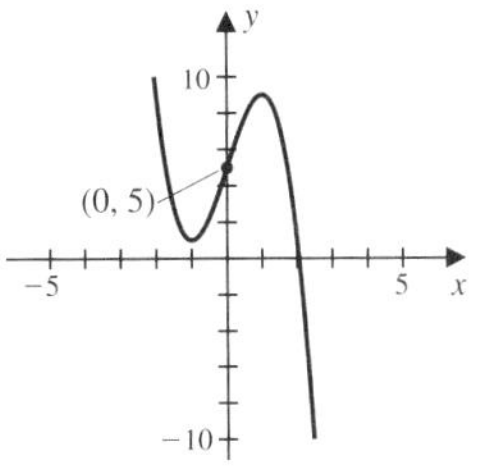

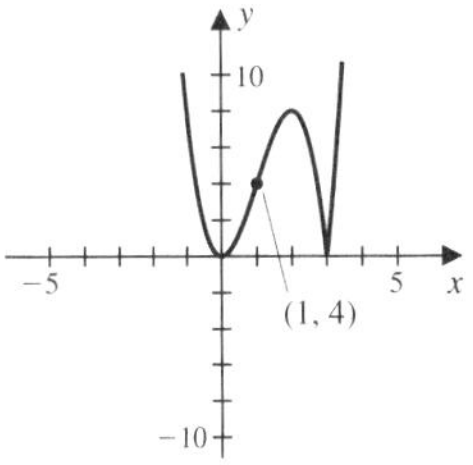

66. a. $y = f(x-1) + 1$ **b.** $y = -f(x) + 1$ **c.** $y = f(-x) - 2$ **d.** $y = |f(x)|$

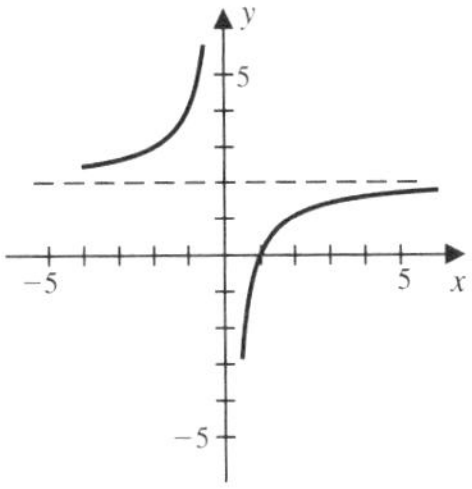

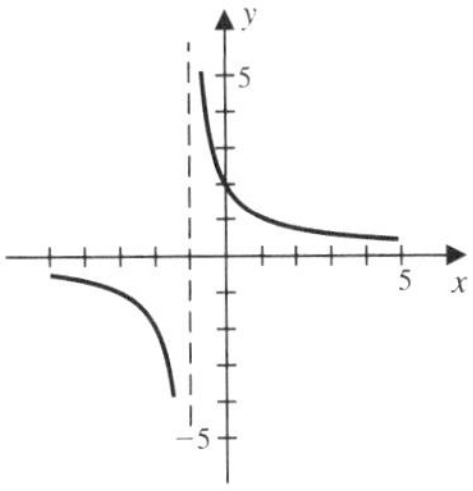

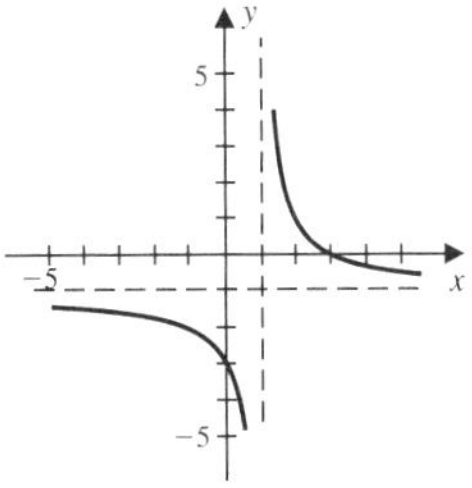

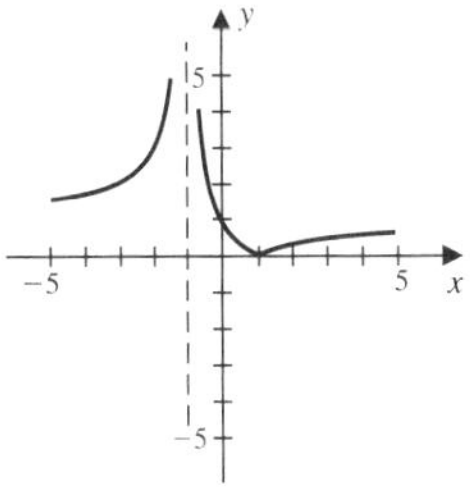

67. A possible graph is shown.

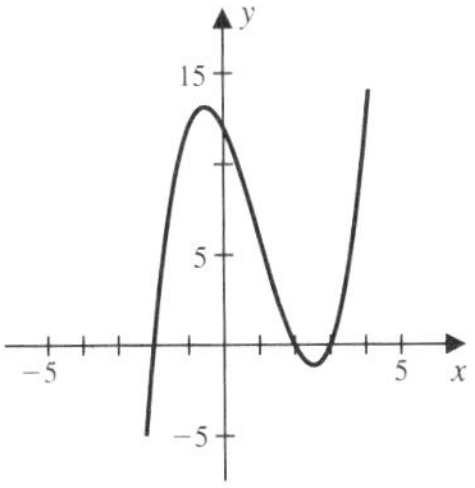

68. A possible graph is shown.

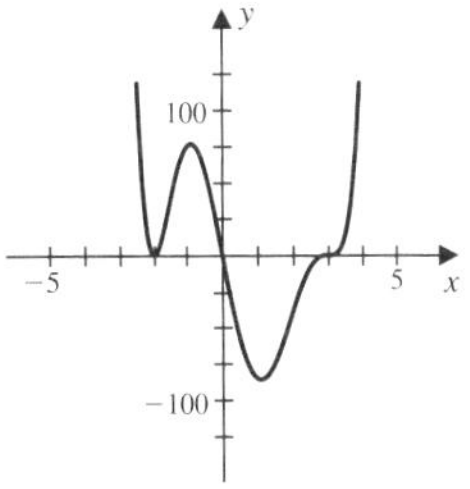

69. First we have

$$P(x) = a(x-1)(x-3)(x+1) = a(x^3 - 3x^2 - x + 3).$$

If $P(0) = 1$, then

$$1 = a((-1)(-3)(1)) \text{ implies } 3a = 1 \text{ so } a = \frac{1}{3}.$$

And

$$P(x) = \frac{1}{3}x^3 - x^2 - \frac{1}{3}x + 1.$$

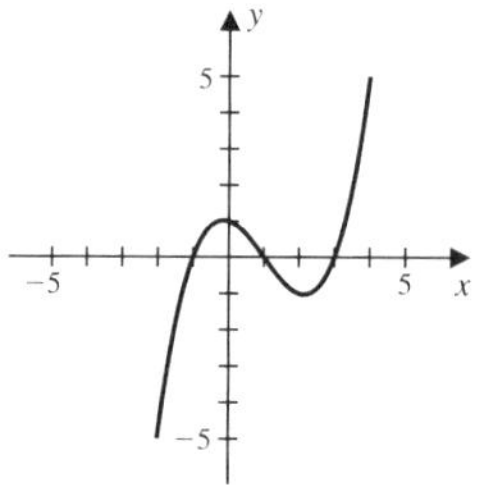

70. First we have

$$P(x) = a(x-1)(x+1)(x-2)^2 = a(x^4 - 4x^3 + 3x^2 + 4x - 4).$$

If $P(-2) = 2$, then

$$2 = a((-2-1)(-2+1)(-2-2)^2) = a(-3)(-1)(16) = 48a \text{ so}$$
$$a = \frac{2}{48} = \frac{1}{24}.$$

And

$$P(x) = \frac{1}{24}x^4 - \frac{1}{6}x^3 + \frac{1}{8}x^2 + \frac{1}{6}x - \frac{1}{6}.$$

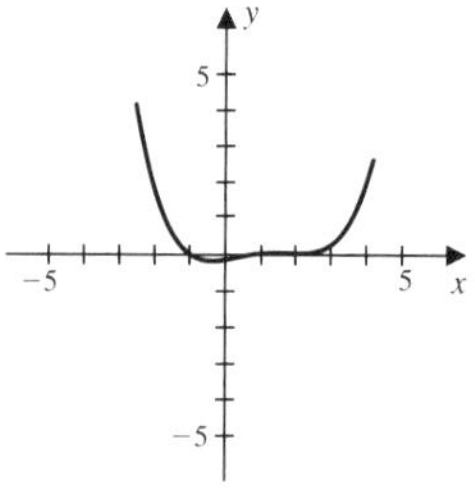

71. **a.** The graph of the function is shown.

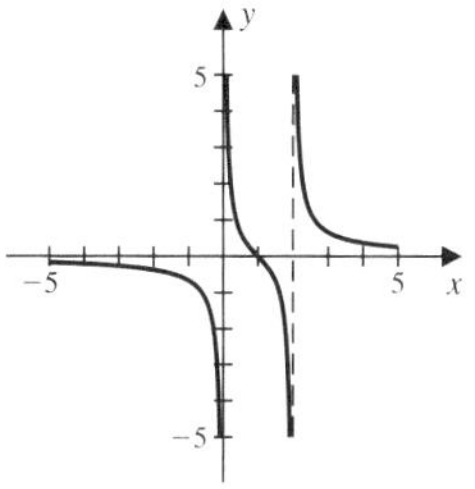

b. The graph will have vertical asymptotes $x = 0$ and $x = 2$ if the denominator of the function has factors x and $x - 2$. To have a horizontal asymptote $y = 0$, the degree of the numerator must be less than the degree of the denominator. If the numerator is $x - 1$, then $f(1) = 0$, and a possible definition for the function is

$$f(x) = \frac{x-1}{x(x-2)}.$$

72. **a.** The graph of the function is shown.

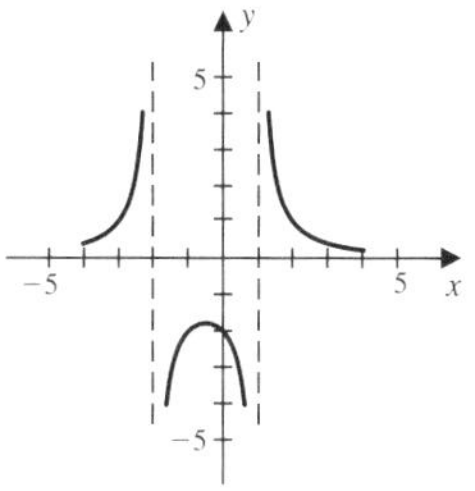

b. The function

$$f(x) = \frac{a}{(x-1)(x+2)}$$

will have vertical asymptotes $x = 1$ and $x = -2$. Since

$$-2 = f(-1) = \frac{a}{-2} \text{ we have } a = 4.$$

A possible definition for the function is $f(x) = \dfrac{4}{(x-1)(x+2)}$.

73. To find the points of intersection of the curves

$$f(x) = x^3 \quad \text{and} \quad g(x) = 2x^2 + x - 2,$$

solve $x^3 = 2x^2 + x - 2$ which implies

$$x^3 - 2x^2 - x + 2 = (x-1)(x-2)(x+1) = 0,$$

so $x = -1, x = 1,$ and $x = 2$. The points of intersection are $(-1, -1)$, $(1, 1)$, $(2, 8)$.

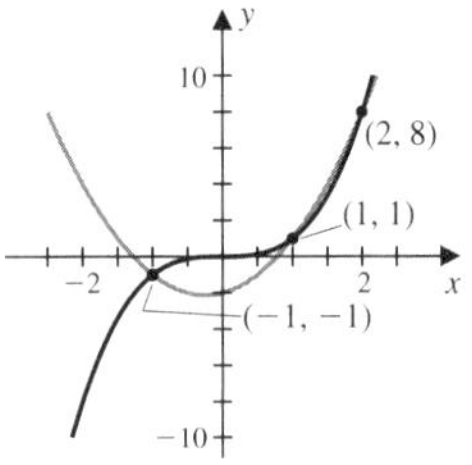

74. To find the points of intersection of the curves

$$f(x) = x^3 - x \quad \text{and} \quad g(x) = -x^2 + 1,$$

solve $x^3 - x = -x^2 + 1$ which implies

$$x^3 + x^2 - x - 1 = (x-1)(x+1)^2 = 0, \text{ so } x = -1 \text{ and } x = 1.$$

The points of intersection are $(-1, 0)$, $(1, 0)$.

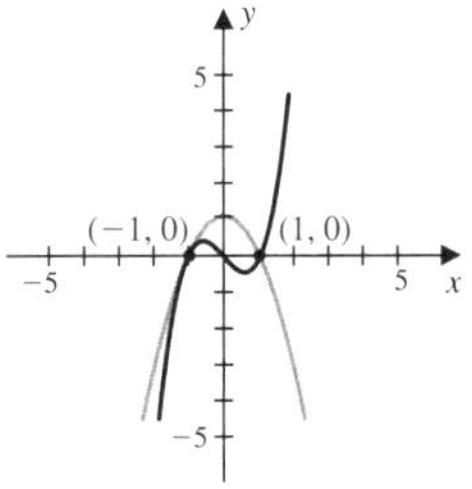

75. If the polynomial has integer coefficients and a zero $-i$, then the conjugate i is also a zero. The polynomial is of degree three with zeros, $1, i, -i$, so

$$P(x) = (x-1)(x-i)(x+i) = (x-1)(x^2+1) = x^3 - x^2 + x - 1.$$

76. If the polynomial has integer coefficients and a zero $3 - i$, then the conjugate $3 + i$ is also a zero. The polynomial is of degree four with zeros, $3 - i$, $3 + i$ and -2, of multiplicity 2, so

$$P(x) = (x-(3-i))(x-(3+i))(x+2)^2 = (x^2-6x+10)(x^2+4x+4) = x^4-2x^3-10x^2+16x+40.$$

77. If the polynomial has integer coefficients and zeros at $\sqrt{2}i$ and $2i$, then the conjugates $-\sqrt{2}i$ and $-2i$ are also zeros. The polynomial has the form

$$P(x) = a\left(x - \sqrt{2}i\right)\left(x + \sqrt{2}i\right)(x - 2i)(x + 2i) = a(x^2 + 2)(x^2 + 4) = a(x^4 + 6x^2 + 8),$$

and if the constant term is to be 8, then $a = 1$, so $P(x) = x^4 + 6x^2 + 8$.

78. If the polynomial has integer coefficients and zeros at i and $2 + i$, then the conjugates $-i$ and $2 - i$ are also zeros. Since the polynomial is to have degree 5, choose another zero for the polynomial, other than 0, since then the graph would pass through $(0, 0)$ and could not pass through $(0, 5)$. Let the fifth zero be $x = 1$, so

$$\begin{aligned} P(x) =& a(x - 1)(x - i)(x + i)(x - (2 - i))(x - (2 + i)) \\ =& a(x - 1)(x^2 + 1)(x^2 - 4x + 5) = a(x^5 - 5x^4 + 10x^3 - 10x^2 + 9x - 5). \end{aligned}$$

If $P(0) = 5$, then $-5a = 5$ implies $a = -1$, so

$$P(x) = -x^5 + 5x^4 - 10x^3 + 10x^2 - 9x + 5.$$

79. If the length and width of the rectangle are l and w, respectively, and the perimeter is 20, then

$$2l + 2w = 20 \text{ implies } l + w = 10, \text{ so } w = 10 - l.$$

The area is

$$A = lw = l(10 - l) = 10l - l^2.$$

To find l so the area is a maximum find the vertex of the parabola. Completing the square gives

$$-l^2 + 10l = -(l^2 - 10l + 25 - 25) = -(l - 5)^2 + 25$$

and the vertex is $(5, 25)$. So $l = 5$ implies $w = 10 - 5 = 5$, and the rectangle with maximum area is a square of side 5ft and area $25 ft^2$.

80. **a.** $V(x) = x(2 - 2x)(4 - 2x) = 4x^3 - 12x^2 + 8x$

b. The domain of V is the interval $[0, 1]$, since the largest cutout can not be more than half the shortest side.

c. From the figure, the maximum volume occurs when $x \approx 0.42$.

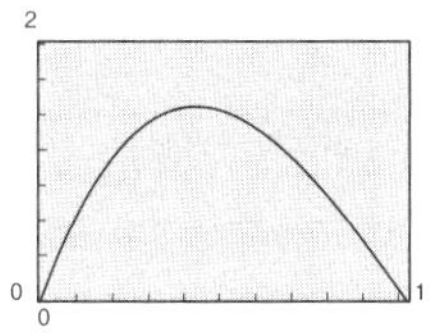

Chapter 3 Exercises for Calculus (Page 176)

1. a. First we have $P(x) = 2x^3 - 3x^2 = x^2(2x - 3)$.

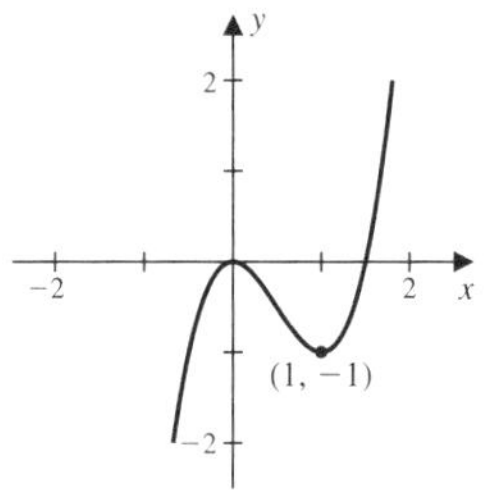

$P(x) = 2x^3 - 3x^2$

b. Adding a constant, C, to $P(x)$ shifts the graph upward C units, if $C > 0$, and downward $|C|$ units, if $C < 0$. If $C < 0$, $Q(x) = P(x) + C$ has 1 zero. If $0 < C < 1$, $Q(x) = P(x) + C$ has 3 zeros. If $C = 1$, $Q(x) = P(x) + C$ has 2 zeros. If $C > 1$, $Q(x) = P(x) + C$ has 1 zero. If $C = 0$, $Q(x) = P(x) + C$ has two real zeros.

2. a. A possible graph of the polynomial is shown.

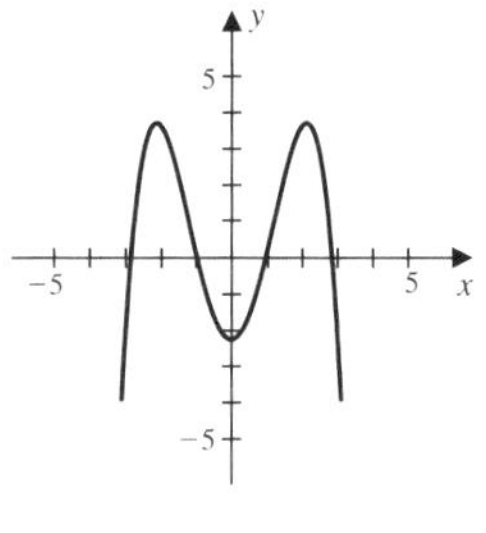

b. The minimum degree of the polynomial is 4.

c. The minimum number of real zeros the polynomial can have is 0.

d. The maximum number of real zeros is 4, as shown in the figure in part (a).

3. a. Let $P(x) = mx + b$. Then

$$(P \circ P)(x) = m(mx + b) + b = m^2x + b(m + 1),$$

which is a linear polynomial with positive slope since the slope is $m^2 > 0$.

b. Let $P(x) = ax^2 + bx + c.$, Then

$$(P \circ P)(x) = a(ax^2 + bx + c)^2 + b(ax^2 + bx + c) + c$$

$$=a^3x^4 + 2a^2bx^3 + (2a^2c + ab^2 + ab)x^2 + (2abc + b^2)x + (ac^2 + bc + c)$$

which is a polynomial of degree 4.

4. **a.** The domain is $\{x \mid x \neq \pm 1\} = (-\infty, -1) \cup (-1, 1) \cup (1, \infty)$.

b. The range is $(-\infty, 1) \cup [4, \infty)$.

c. Increasing on $(0, 1)$ and $(1, \infty)$.

d. Decreasing on $(-\infty, -1)$ and $(-1, 0)$.

e. A horizontal asymptote $y = 1$.

f. Vertical asymptotes $x = -1$ and $x = 1$.

g. No, the function does not have an inverse on the interval $(-1, 1)$, since every horizontal line that crosses the curve intersects it in two points.

h. Yes, the function does have an inverse on the interval $(1, \infty)$.

5. The possible rational zeros of $f(x) = x^3 + x^2 + kx - 3$ are ± 1 and ± 3. Then

$$f(1) = 0 \text{ implies } k = 1, \qquad f(-1) = 0 \text{ implies } k = -3,$$
$$f(3) = 0 \text{ implies } k = -11, \quad \text{and} \quad f(-3) = 0 \text{ implies } k = -7.$$

For

$$k = 1: \; f(x) = (x-1)(x^2 + 2x + 3);$$

$$k = -3: \; f(x) = (x+1)(x^2 - 3) = (x+1)\left(x - \sqrt{3}\right)\left(x + \sqrt{3}\right);$$

$$k = -11: \; f(x) = (x-3)(x^2 + 4x + 1) = (x-3)\left(x - \left(-2 + \sqrt{3}\right)\right)\left(x - \left(-2 - \sqrt{3}\right)\right);$$

$$k = -7: \; f(x) = (x+3)(x^2 - 2x - 1) = (x+3)\left(x - \left(1 + \sqrt{2}\right)\right)\left(x - \left(1 - \sqrt{2}\right)\right).$$

So $f(x)$ has at least one rational zero for $k = 1, -3, -11,$ and -7.

6. If $P(x) = x^3 + bx^2 + cx + d$, and r_1, r_2, and r_3 are zeroes, then

$$P(x) = (x - r_1)(x - r_2)(x - r_3) = x^3 - (r_1 + r_2 + r_3)x^2 + (r_1r_2 + r_1r_3 + r_2r_3)x - r_1r_2r_3.$$

So

$$b = -(r_1 + r_2 + r_3), \quad c = r_1r_2 + r_1r_3 + r_2r_3, \quad \text{and} \quad d = -r_1r_2r_3.$$

7. **a.** Since

$$P(x_1) = \frac{x_1 - x_2}{x_1 - x_2}y_1 + \frac{x_1 - x_1}{x_2 - x_1}y_2 = y_1,$$

the point (x_1, y_1) is on the line. Similarly,

$$P(x_2) = \frac{x_2 - x_2}{x_1 - x_2} y_1 + \frac{x_2 - x_1}{x_2 - x_1} y_2 = y_2,$$

so (x_2, y_2) is on the line.

b. We have

$$y = \frac{x-1}{-1-1}(6) + \frac{x-(-1)}{1-(-1)}(-2) = -3(x-1) - (x+1) = -4x + 2$$

8. a. The form of the polynomial is

$$P(x) = \frac{(x-x_2)(x-x_3)}{(x_1-x_2)(x_1-x_3)} y_1 + \frac{(x-x_1)(x-x_3)}{(x_2-x_1)(x_2-x_3)} y_2 + \frac{(x-x_1)(x-x_2)}{(x_3-x_1)(x_3-x_2)} y_3.$$

b. We have

$$\begin{aligned} P(x) &= \frac{(x-1)(x-2)}{(-2)(-3)}(6) + \frac{(x+1)(x-2)}{(2)(-1)}(-2) + \frac{(x+1)(x-1)}{(3)(1)}(3) \\ &= (x-1)(x-2) + (x+1)(x-2) + (x+1)(x-1) = 3x^2 - 4x - 1. \end{aligned}$$

9. The height of an object above the ground is given by $s(t) = v_0 t - \frac{g}{2}t^2$.

a. The initial height of the object occurs when $t = 0$, and equals $s(0) = 0$.

b. To determine how long the object is in the air we need to find the time when the object strikes the ground. Solving

$$0 = s(t) = v_0 t - \frac{g}{2}t^2 = t\left(v_0 - \frac{g}{2}t\right)$$

gives $t = 0$ and $t = 2v_0/g$. Hence $t = 2v_0/g$ is the amount of time the object is in the air.

c. To determine the maximum height reached by the object find the vertex of the parabola. Completing the square we have

$$-\frac{g}{2}t^2 + v_0 t = -\frac{g}{2}\left(t^2 - \frac{2v_0}{g}t\right) = -\frac{g}{2}\left(t^2 - \frac{2v_0}{g}t + \frac{v_0^2}{g^2} - \frac{v_0^2}{g^2}\right) = -\frac{g}{2}\left(t - \frac{v_0}{g}\right)^2 + \frac{v_0^2}{2g}.$$

So the vertex is $\left(v_0/g, v_0^2/2g\right)$ and the maximum height reached is $v_0^2/2g$.

d. The object reaches the maximum height after $t = v_0/g$.

10. The cost to construct the rectangular box depends on the surface area. If x is the side of the square base and h is the height, then since the box does not have a top, the surface area is

$$S = x^2 + 4xh.$$

To eliminate the variable h, we can use the fact that the volume of the box is 10 cubic feet. Then

$$10 = V = x^2 h \Rightarrow h = \frac{10}{x^2}$$

and

$$S(x) = x^2 + 4x\left(\frac{10}{x^2}\right) = x^2 + \frac{40}{x}.$$

Since the cost of the material for the base is \$15 and for the sides is \$6, the cost to construct the box is

$$C(x) = 15x^2 + 6\frac{40}{x} = 15x^2 + \frac{240}{x}.$$

Using a graphing device, we can show that the minimum cost occurs when $x \approx 2$ which gives a value of $h \approx \frac{10}{4} = 2.5$.

11. The dimensions on the figure imply that $V(x) = x(20 - 2x)\left(\frac{50-3x}{2}\right) = x(10 - x)(50 - 3x)$.

12. **a.** Since $y = 50 - x$ and $A = xy$, we have $A(x) = x(50 - x)$.

b. The graph of

$$y = A(x) = 50x - x^2$$

is a parabola that opens downward and hence has a maximum at the vertex. To find the vertex place the parabola in standard form by completing the square.

$$-x^2 + 50x = -(x^2 - 50x + (25)^2 - (25)^2) = -(x - 25)^2 + 625$$

So the vertex is (25, 625), and the maximum area of the inscribed rectangle is 625 when

$$x = 25, \text{ so } y = 50 - x = 50 - 25 = 25,$$

and the inscribed rectangle is a square of side 25.

Chapter 3 Chapter Test (Page 178)

1. True.

2. False. If the number c is a zero of the polynomial $P(x)$, then $x - c$ is a factor of the polynomial.

3. False. The polynomial

$$P(x) = x^3 - x^2 + x - 1$$

has a factor $x - 1$.

4. True.

5. False. The polynomial

$$P(x) = x^2 + 1,$$

has no real numbers with $P(x) = 0$.

6. False. The polynomial

$$\begin{aligned} P(x) =& x^4 - 3x^3 + 2x^2 \\ =& x^2(x^2 - 3x + 2) \\ =& x^2(x - 1)(x - 2) \end{aligned}$$

has 3 distinct real zeros, at $x = 0$, $x = 1$, and $x = 2$.

7. True.

8. False. The graph of the polynomial

$$P(x) = x(x - 1)^3(x + 1)^2$$

crosses the x-axis twice. It touches the axis at $x = -1$, but does not cross.

9. True.

10. False. The end-behavior of the polynomial

$$P(x) = -x^6 - x^5 + 3x^2 - 2x + 7$$

satisfies

$$P(x) \to -\infty, \text{ as } x \to -\infty$$

and

$$P(x) \to -\infty, \text{ as } x \to \infty$$

11. True.

12. False. The polynomials

$$P(x) = x^4 - 2x^3 + 6x - 9$$

and

$$Q(x) = x^4 - 2x^3 + 6x - 5$$

do not have the same zeros, since one is not a multiple of the other.

13. False. The possible rational zeros of the polynomial

$$P(x) = 2x^5 - 2x^3 + x^2 - 3x + 5$$

are

$$\pm 1, \pm\frac{1}{2}, \pm 5, \pm\frac{5}{2}.$$

14. True.

15. False. The polynomial shown in the figure has a zero of even multiplicity at $x = 0$.

16. False. The polynomial shown in the figure has a zero of odd multiplicity at $x = 2$.

17. True.

18. True.

19. True.

20. False. The end-behavior of the polynomial shown in the figure is similar to the end-behavior of $-x^3$, but the given graph approaches the end-behavior more rapidly because its degree is higher.

21. True.

22. False. A polynomial with the same zeros but with y-intercept $(0, 1)$ is

$$P(x) = -\frac{1}{4}(x+1)^2(x-1)^3(x-2)^2.$$

23. True.

24. False. One zero of the polynomial

$$P(x) = x^5 - 5x^4 + 6x^3 + 4x^2 - 8x$$

is $x = 0$, and another is $x = -1$. Factoring $P(x)$ using these observations gives

$$P(x) = x(x+1)(x^3 - 6x^2 + 12x - 8)$$

and the right-most term has a zero at $x = 2$. Continuing we have

$$x^3 - 6x^2 + 12x - 8 = (x-2)^3$$

and

$$P(x) = x(x+1)(x-2)^3.$$

So, $P(x)$ has a zero of multiplicity 3 at $x = 2$.

25. False. The rational function

$$f(x) = \frac{1-x^2}{x^2+1}$$

has no vertical asymptotes, but it has a horizontal asymptote at $y = -1$.

26. False. Since

$$f(x) = \frac{x^2 - 3x + 2}{x - 2} = \frac{(x-2)(x-1)}{x-2} = x - 1,$$

for $x \neq 2$, the function does not have a vertical asymptote. The graph is given by the line with equation, $y = x - 1$, except that there is a hole in the graph at $(2, 1)$.

27. True.

28. False. The rational function

$$f(x) = \frac{x^2 + 5x + 4}{x^2 - 2x - 3} = \frac{(x+4)(x+1)}{(x-3)(x+1)} = \frac{x+4}{x-3}$$

has a vertical asymptote at $x = 3$.

29. True.

30. True.

31. True.

32. False. The rational function

$$f(x) = \frac{x-1}{x^2-4}$$

has vertical asymptotes $x = 2$ and $x = -2$, as shown in the figure, but it has a horizontal asymptote $y = 0$ and y-intercept $(0, \frac{1}{4})$, which does not agree with the figure. Since the figure has a horizontal asymptote $y = 1$ and y-intercept $(0, \frac{5}{4})$, a possible equation is

$$f(x) = \frac{x-1}{x^2-4} + 1.$$

33. True.

34. False. The graph of the rational function has a horizontal asymptote $y = 2$.

35. False. The domain of the rational function is $(-\infty, -1) \cup (-1, 1) \cup (1, \infty)$.

36. False. The range of the rational function is $(-\infty, 1] \cup (2, \infty)$.

37. False. If the curve is given by $y = f(x)$, then

$$f(x) \to 2, \text{ as } x \to \infty$$
$$f(x) \to 2, \text{ as } x \to -\infty$$
$$f(x) \to \infty, \text{ as } x \to 1^+$$
$$f(x) \to -\infty, \text{ as } x \to 1^-.$$

38. False. A possible definition for the function is

$$f(x) = \frac{2x^2 - 1}{x^2 - 1},$$

since there is a horizontal asymptote at $y = 2$.

39. False. The function

$$f(x) = \frac{x^2 + 1}{x^2 - 3x + 2}$$

satisfies

$$f(x) \to -\infty, \text{ as } x \to 1^+$$
$$f(x) \to \infty, \text{ as } x \to 1^-.$$

40. True.

41. False. The function

$$f(x) = \frac{3x^4 - 2x^3 - x - 1}{x^4 + 2x + 1}$$

has a horizontal asymptote $y = 3$.

42. True.

43. False. For large values of x,

$$\frac{100x^2 - 2x + 3}{x^3 + x + 1} < \frac{x^3 - 2x + 5}{100x^3 + 3x + 4},$$

since

$$\frac{100x^2 - 2x + 3}{x^3 + x + 1} \to 0, \text{ as } x \to \infty$$

and

$$\frac{x^3 - 2x + 5}{100x^3 + 3x + 4} \to \frac{1}{100}, \text{ as } x \to \infty.$$

44. True.

Exercise Set 4.2 (Page 187)

1. $60° = 60\frac{\pi}{180}$ radians $= \frac{\pi}{3}$ radians

2. $135° = 135\frac{\pi}{180}$ radians $= \frac{3\pi}{4}$ radians

3. $150° = 150\frac{\pi}{180}$ radians $= \frac{5\pi}{6}$ radians

4. $40° = 40\frac{\pi}{180}$ radians $= \frac{2\pi}{9}$ radians

5. $225° = 225\frac{\pi}{180}$ radians $= \frac{5\pi}{4}$ radians

6. $300° = 300\frac{\pi}{180}$ radians $= \frac{5\pi}{3}$ radians

7. $-72° = -72\frac{\pi}{180}$ radians $= -\frac{2\pi}{5}$ radians

8. $-270° = -270\frac{\pi}{180}$ radians $= -\frac{3\pi}{2}$ radians

9. $\frac{3\pi}{4}$ radians $= \frac{3\pi}{4}\frac{180}{\pi}$ degrees $= 135°$

10. $-\frac{\pi}{6}$ radians $= -\frac{\pi}{6}\frac{180}{\pi}$ degrees $= -30°$

11. $-\frac{11\pi}{6}$ radians $= -\frac{11\pi}{6}\frac{180}{\pi}$ degrees $= -330°$

12. $\frac{2\pi}{3}$ radians $= \frac{2\pi}{3}\frac{180}{\pi}$ degrees $= 120°$

13. $\frac{9\pi}{2}$ radians $= \frac{9\pi}{2}\frac{180}{\pi}$ degrees $= 810°$

14. 7π radians $= 7\pi\frac{180}{\pi}$ degrees $= 1260°$

15. The angle 405° is 45° beyond 360° so the angle coincides with 45°.

16. The angle 570° is 30° beyond $540° = 360° + 180°$, which coincides with 180°. But $-120°$ is 60° from 180° so the angles do not coincide.

17. Since $2\pi = \frac{12\pi}{6}$, the angles $\frac{\pi}{6}$ and $\frac{13\pi}{6}$ coincide.

18. The angle $\frac{3\pi}{4}$ is in quadrant II and $-\frac{7\pi}{4}$ is in quadrant I, so the angles do not coincide.

19. The angle 150° is 30° from 180°, but $-240°$ is 60° from 180°, so the angles do not coincide.

20. The angles coincide.

21. **a.** $P\left(\frac{\pi}{2}\right)$ **b.** $P(\pi)$ **c.** $P(2\pi)$ **d.** $P\left(\frac{3\pi}{2}\right)$

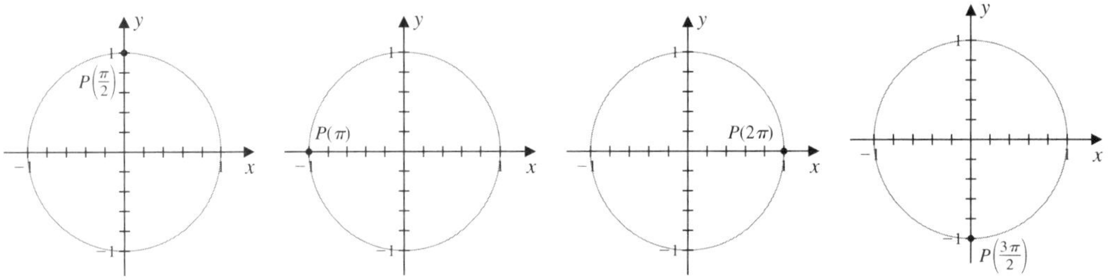

22. **a.** $P\left(\frac{5\pi}{6}\right)$ **b.** $P\left(\frac{3\pi}{4}\right)$ **c.** $P\left(\frac{4\pi}{3}\right)$ **d.** $P\left(\frac{7\pi}{6}\right)$

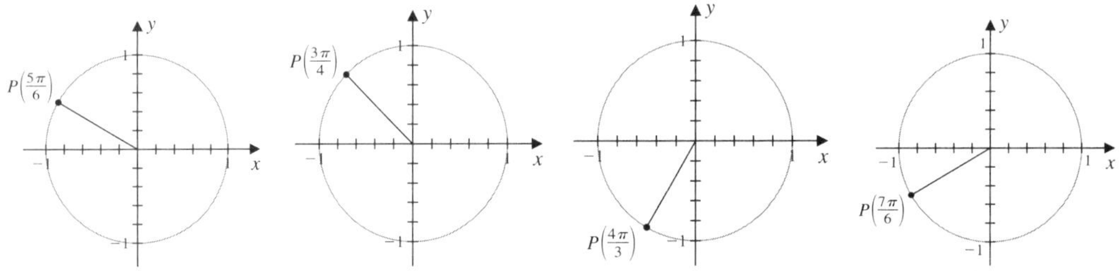

23. **a.** $P\left(-\frac{\pi}{4}\right)$ **b.** $P\left(-\frac{4\pi}{3}\right)$ **c.** $P\left(-\frac{37\pi}{6}\right)$ **d.** $P\left(-\frac{7\pi}{4}\right)$

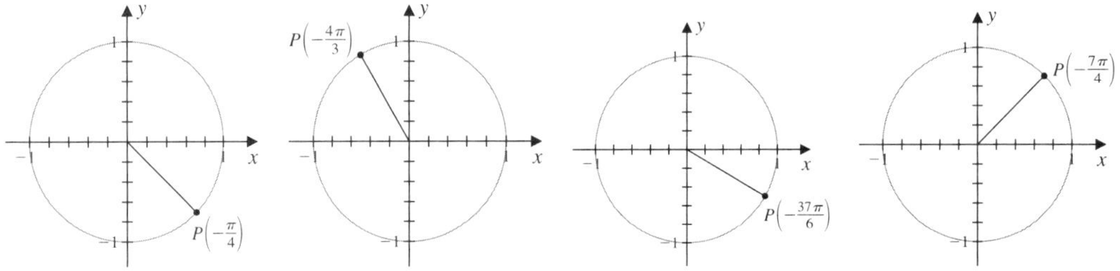

24. **a.** $P\left(\frac{21\pi}{2}\right)$ **b.** $P\left(\frac{317\pi}{4}\right)$ **c.** $P\left(-\frac{33\pi}{2}\right)$ **d.** $P\left(-\frac{19\pi}{6}\right)$

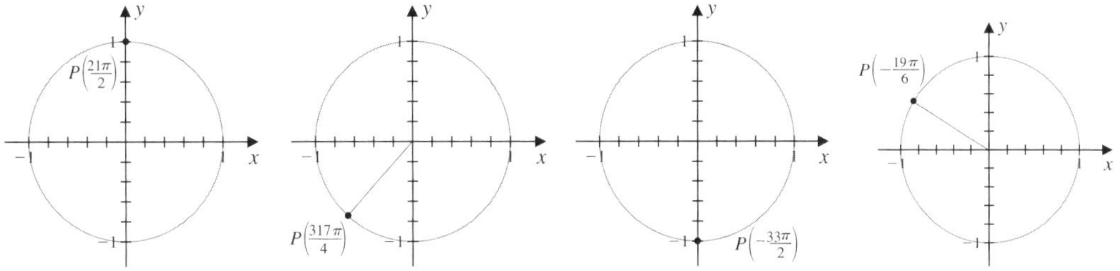

25. Let $P(t) = \left(\frac{3}{5}, \frac{4}{5}\right)$.

a. Since $P(t)$ is in quadrant I, adding π to t moves the point to quadrant III, so $P(t+\pi) = \left(-\frac{3}{5}, -\frac{4}{5}\right)$.

b. Since $P(t)$ is in quadrant I, $P(-t)$ is in quadrant IV, so $P(-t) = \left(\frac{3}{5}, -\frac{4}{5}\right)$.

c. Since $P(t)$ is in quadrant I, subtracting π from t moves the point to quadrant III, so $P(t-\pi) = \left(-\frac{3}{5}, -\frac{4}{5}\right)$.

d. Since $P(t)$ is in quadrant I, $P(-t)$ is in quadrant IV so subtracting π from $-t$ moves the point to quadrant II, so $P(-t-\pi) = \left(-\frac{3}{5}, \frac{4}{5}\right)$.

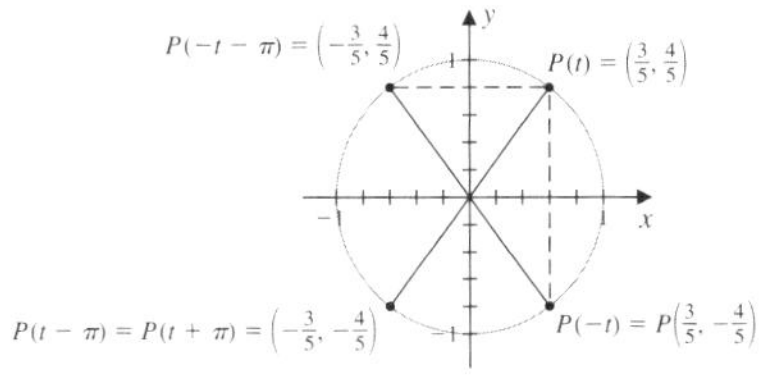

26. **a.** $P(t+2\pi) = P(t) = \left(\frac{3}{4}, -\frac{\sqrt{7}}{4}\right)$

b. Since $P(t)$ is in quadrant IV, $P(-t)$ is in quadrant I, so $P(-t) = \left(\frac{3}{4}, \frac{\sqrt{7}}{4}\right)$.

c. Since $P(t-\pi)$ corresponds to the reflection of $P(t)$ through the origin $P(t-\pi) = \left(-\frac{3}{4}, \frac{\sqrt{7}}{4}\right)$.

d. $P(t-3\pi) = P(t-\pi) = \left(-\frac{3}{4}, \frac{\sqrt{7}}{4}\right)$

27. **a.** $P(t+\pi) = \left(\frac{\sqrt{5}}{3}, -\frac{2}{3}\right)$

b. $P(-t) = -\left(\frac{\sqrt{5}}{3}, -\frac{2}{3}\right)$

c. $P(t-\pi) = P(t+\pi) = \left(\frac{\sqrt{5}}{3}, -\frac{2}{3}\right)$

d. $P(-t-\pi) = \left(\frac{\sqrt{5}}{3}, \frac{2}{3}\right)$

28. a. $P(t+2\pi) = P(t) = \left(-\frac{2\sqrt{13}}{13}, -\frac{3\sqrt{13}}{13}\right)$

b. Since $P(t)$ is in quadrant III, $P(-t)$ is in quadrant II, so $P(-t) = P(t) = \left(-\frac{2\sqrt{13}}{13}, \frac{3\sqrt{13}}{13}\right)$.

c. Since $P(t-\pi)$ is in quadrant I, $P(t-\pi) = P(t) = \left(\frac{2\sqrt{13}}{13}, \frac{3\sqrt{13}}{13}\right)$.

d. $P(t-3\pi) = P(t-\pi) = \left(\frac{2\sqrt{13}}{13}, \frac{3\sqrt{13}}{13}\right)$.

29. Since $45° = \frac{\pi}{4}$ radians, the length of the arc is $s = r\theta = (8)\frac{\pi}{4} = 2\pi$.

30. Since $30° = \frac{\pi}{6}$ radians, the length of the arc is $s = r\theta = (2)\frac{\pi}{6} = \frac{\pi}{3}$.

31. $\theta = \frac{s}{r} = \frac{6}{3} = 2$ radians $= 2\frac{180}{\pi}^{\circ} = \frac{360}{\pi}^{\circ}$

32. $\theta = \frac{s}{r} = \frac{3}{5}$ radians $= \frac{3}{5}\frac{180}{\pi}^{\circ} = \frac{108}{\pi}^{\circ}$

33. Since $110° = 110\frac{\pi}{180}$ radians $= \frac{11\pi}{18}$ radians, the length of the arc is $s = r\theta = (4)\frac{11\pi}{18} = \frac{22\pi}{9}$.

34. $\theta = \frac{s}{r} = \frac{9}{4}$

35. $A = \frac{1}{2}r^2\theta = \frac{1}{2}(10)^2\frac{1}{2} = 25$

36. Since $30° = \frac{\pi}{6}$ radians, $A = \frac{1}{2}r^2\theta = \frac{1}{2}(4)^2\frac{\pi}{6} = \frac{4\pi}{3}$.

37. Since $45° = \frac{\pi}{4}$ radians, $r = \sqrt{\frac{2A}{\theta}} = \sqrt{\frac{(2)(2)}{\frac{\pi}{4}}} = \frac{4}{\sqrt{\pi}}$.

38. $\theta = \frac{2A}{r^2} = \frac{2(10)}{9} = \frac{20}{9}$ radians

39. Since $65° = 65\frac{\pi}{180}$ radians $= \frac{13\pi}{36}$ radians, $A = \frac{1}{2}r^2\theta = \frac{1}{2}(16)\frac{13\pi}{36} = \frac{26\pi}{9}$.

40. $A = \frac{1}{2}r^2\theta = \frac{1}{2}6^2(1.5) = 18(1.5) = 27$

41. Since the angle between the two cities measured north along the common meridian is $43 - 36.5 = 6.5° = 6.5\frac{\pi}{180}$ radians, the distance between the two cities is approximately $3960\left(\frac{6.5\pi}{180}\right) \approx 449$ miles.

42. $3960\left(\frac{35.9\pi}{180}\right) \approx 2481$ miles.

Exercise Set 4.3 (Page 200)

1. **a.** $\frac{\pi}{4}$ **b.** $\frac{\pi}{3}$ **c.** $\pi - \frac{2\pi}{3} = \frac{\pi}{3}$ **d.** $\pi - \frac{3\pi}{4} = \frac{\pi}{4}$

2. **a.** $\frac{7\pi}{6} - \pi = \frac{\pi}{6}$ **b.** $\frac{4\pi}{3} - \pi = \frac{\pi}{3}$ **c.** $2\pi - \frac{7\pi}{4} = \frac{\pi}{4}$ **d.** $2\pi - \frac{11\pi}{6} = \frac{\pi}{6}$

3. **a.** $\frac{\pi}{3}$ **b.** $\frac{7\pi}{6} - \pi = \frac{\pi}{6}$ **c.** $\pi - \frac{2\pi}{3} = \frac{\pi}{3}$ **d.** $\pi - \frac{3\pi}{4} = \frac{\pi}{4}$

4. **a.** $\frac{\pi}{6}$ **b.** $\frac{\pi}{3}$ **c.** $\frac{\pi}{3}$ **d.** $\frac{\pi}{4}$

5. **a.** $45°$ **b.** $180° - 150° = 30°$ **c.** $240° - 180° = 60°$ **d.** $360° - 330° = 30°$

6. **a.** $60°$ **b.** $180° - 135° = 45°$ **c.** $210° - 180° = 30°$ **d.** $360° - 315° = 45°$

7. $\sin\frac{\pi}{6} = \frac{1}{2}$; $\cos\frac{\pi}{6} = \frac{\sqrt{3}}{2}$

8. $\sin\frac{\pi}{4} = \frac{\sqrt{2}}{2}$; $\cos\frac{\pi}{4} = \frac{\sqrt{2}}{2}$

9. $\sin\frac{3\pi}{4} = \sin\frac{\pi}{4} = \frac{\sqrt{2}}{2}$; $\cos\frac{3\pi}{4} = -\cos\frac{\pi}{4} = -\frac{\sqrt{2}}{2}$

10. $\sin\frac{2\pi}{3} = \sin\frac{\pi}{3} = \frac{\sqrt{3}}{2}$; $\cos\frac{2\pi}{3} = -\cos\frac{\pi}{3} = -\frac{1}{2}$

11. $\sin\frac{4\pi}{3} = -\sin\frac{\pi}{3} = -\frac{\sqrt{3}}{2}$; $\cos\frac{4\pi}{3} = -\cos\frac{\pi}{3} = -\frac{1}{2}$

12. $\sin\frac{7\pi}{6} = -\sin\frac{\pi}{6} = -\frac{1}{2}$; $\cos\frac{7\pi}{6} = -\cos\frac{\pi}{6} = -\frac{\sqrt{3}}{2}$

13. $\sin\frac{11\pi}{6} = -\sin\frac{\pi}{6} = -\frac{1}{2}$; $\cos\frac{11\pi}{6} = \cos\frac{\pi}{6} = \frac{\sqrt{3}}{2}$

14. $\sin\frac{7\pi}{4} = -\sin\frac{\pi}{4} = -\frac{\sqrt{2}}{2}$; $\cos\frac{7\pi}{4} = \cos\frac{\pi}{4} = \frac{\sqrt{2}}{2}$

15. $\sin\left(-\frac{\pi}{3}\right) = -\sin\frac{\pi}{3} = -\frac{\sqrt{3}}{2}$; $\cos\left(-\frac{\pi}{3}\right) = \cos\frac{\pi}{3} = \frac{1}{2}$

16. $\sin\left(-\frac{\pi}{6}\right) = -\sin\frac{\pi}{6} = -\frac{1}{2}$; $\cos\left(-\frac{\pi}{6}\right) = \cos\frac{\pi}{6} = \frac{\sqrt{3}}{2}$

17. $\sin\left(-\frac{5\pi}{6}\right) = -\sin\frac{\pi}{6} = -\frac{1}{2}$; $\cos\left(-\frac{5\pi}{6}\right) = -\cos\frac{\pi}{6} = -\frac{\sqrt{3}}{2}$

18. $\sin\left(-\frac{3\pi}{4}\right) = -\sin\frac{\pi}{4} = -\frac{\sqrt{2}}{2}$; $\cos\left(-\frac{3\pi}{4}\right) = -\cos\frac{\pi}{4} = -\frac{\sqrt{2}}{2}$

19. $\sin\left(-\frac{5\pi}{4}\right) = \sin\frac{\pi}{4} = \frac{\sqrt{2}}{2}$; $\cos\left(-\frac{5\pi}{4}\right) = -\cos\frac{\pi}{4} = -\frac{\sqrt{2}}{2}$

20. $\sin\left(-\frac{5\pi}{3}\right) = \sin\frac{\pi}{3} = \frac{\sqrt{3}}{2}$; $\cos\left(-\frac{5\pi}{3}\right) = \cos\frac{\pi}{3} = \frac{1}{2}$

21. $\sin\left(-\frac{5\pi}{2}\right) = \sin\frac{3\pi}{2} = -1$; $\cos\left(-\frac{5\pi}{2}\right) = \cos\frac{3\pi}{2} = 0$

22. $\sin\left(-\frac{7\pi}{2}\right) = \sin\frac{\pi}{2} = 1$; $\cos\left(-\frac{7\pi}{2}\right) = \cos\frac{\pi}{2} = 0$

23. $\sin(-7\pi) = \sin(\pi) = 0$; $\cos(-7\pi) = \cos(\pi) = -1$

24. $\sin(8\pi) = \sin(0) = 0$; $\cos(8\pi) = \cos(0) = 1$

25. $\sin 45^\circ = \frac{\sqrt{2}}{2}$, $\cos 45^\circ = \frac{\sqrt{2}}{2}$

26. $\sin 60^\circ = \frac{\sqrt{3}}{2}$, $\cos 60^\circ = \frac{1}{2}$

27. $\sin 150^\circ = \sin 30^\circ = \frac{1}{2}$, $\cos 150^\circ = -\cos 30^\circ = -\frac{\sqrt{3}}{2}$

28. $\sin 135^\circ = \sin 45^\circ = \frac{\sqrt{2}}{2}$, $\cos 135^\circ = -\cos 45^\circ = -\frac{\sqrt{2}}{2}$

29. $\sin 240° = -\sin 60° = -\frac{\sqrt{3}}{2}$, $\cos 240° = -\cos 60° = -\frac{1}{2}$

30. $\sin 210° = -\sin 30° = -\frac{1}{2}$, $\cos 210° = -\cos 30° = -\frac{\sqrt{3}}{2}$

31. $\sin 330° = -\sin 30° = -\frac{1}{2}$, $\cos 330° = \cos 30° = \frac{\sqrt{3}}{2}$

32. $\sin 315° = -\sin 45° = -\frac{\sqrt{2}}{2}$, $\cos 315° = \cos 45° = \frac{\sqrt{2}}{2}$

33. $\sin(-45°) = -\sin 45° = -\frac{\sqrt{2}}{2}$, $\cos(-45°) = \cos 45° = \frac{\sqrt{2}}{2}$

34. $\sin(-150°) = -\sin 30° = -\frac{1}{2}$, $\cos(-150°) = -\cos 30° = -\frac{\sqrt{3}}{2}$

35. $\sin(-240°) = \sin 60° = \frac{\sqrt{3}}{2}$, $\cos(-240°) = -\cos 60° = -\frac{1}{2}$

36. $\sin(-300°) = \sin 60° = \frac{\sqrt{3}}{2}$, $\cos(-300°) = \cos 60° = \frac{1}{2}$

37. $\sin 540° = \sin 180° = 0$, $\cos 540° = \cos 180° = -1$

38. $\sin(-720°) = \sin 0° = 0$, $\cos(720°) = \cos 0° = 1$

39. $\cos t = \frac{\sqrt{2}}{2}$ implies $t = \frac{\pi}{4}, \frac{7\pi}{4}$

40. $\sin t = \frac{\sqrt{3}}{2}$ implies $t = \frac{\pi}{3}, \frac{2\pi}{3}$

41. $\sin t = -\frac{1}{2}$ implies $t = \frac{7\pi}{6}, \frac{11\pi}{6}$

42. $\cos t = -\frac{\sqrt{3}}{2}$ implies $t = \frac{5\pi}{6}, \frac{7\pi}{6}$

43. $\cos t = 1$ implies $t = 0, 2\pi$

44. $\sin t = -1$ implies $t = \frac{3\pi}{2}$

45. If $0 \le t \le 2\pi$, then $0 \le \frac{t}{2} \le \pi$ and $\cos\frac{t}{2} = \frac{1}{2}$ implies $\frac{t}{2} = \frac{\pi}{3}$ so $t = \frac{2\pi}{3}$.

46. If $0 \le t \le 2\pi$, then $0 \le 3t \le 6\pi$ and $\sin 3t = -\frac{\sqrt{2}}{2}$. Thus
$3t = \frac{5\pi}{4}, \frac{7\pi}{4}, \frac{13\pi}{4}, \frac{15\pi}{4}, \frac{21\pi}{4}, \frac{23\pi}{4}$ implies $t = \frac{5\pi}{12}, \frac{7\pi}{12}, \frac{13\pi}{12}, \frac{15\pi}{12}, \frac{21\pi}{12}, \frac{23\pi}{12}$.

47. The function $f(x) = (\cos x)^2$ is even since
$f(-x) = (\cos(-x))^2 = (\cos x)^2 = f(x)$.

48. The function $f(x) = x^3 \sin x$ is even since
$f(-x) = (-x)^3 \sin(-x) = -x^3(-\sin x) = x^3 \sin x = f(x)$.

49. The function $f(x) = |x| \sin x$ is odd since
$f(-x) = |-x| \sin(-x) = |x|(-\sin x) = -|x| \sin x = -f(x)$.

50. The function $f(x) = \cos(\sin x)$ is even since
$f(-x) = \cos(\sin(-x)) = \cos(-\sin x) = \cos(\sin x) = f(x)$.

51. Since $\sin t = -\frac{2\sqrt{2}}{3}$ and $\cos t = \frac{1}{3}$, the terminal point $P(t) = \left(\frac{1}{3}, -\frac{2\sqrt{2}}{3}\right)$ and lies in quadrant IV.

a. $P(t+\pi)$ lies in quadrant II, and $\sin(t+\pi) = \frac{2\sqrt{2}}{3}$, $\cos(t+\pi) = -\frac{1}{3}$.

b. $P(-t)$ lies in quadrant I, and $\sin(-t) = \frac{2\sqrt{2}}{3}$, $\cos(-t) = \frac{1}{3}$.

c. $P\left(t+\frac{\pi}{2}\right)$ lies in quadrant I and the coordinates of the point are interchanged, so
$\sin\left(t+\frac{\pi}{2}\right) = \frac{1}{3}$, $\cos\left(t+\frac{\pi}{2}\right) = \frac{2\sqrt{2}}{3}$.

d. $P\left(-t+\frac{\pi}{2}\right)$ lies in quadrant II and the coordinates of the point are interchanged, so
$\sin\left(-t+\frac{\pi}{2}\right) = \frac{1}{3}$, $\cos\left(-t+\frac{\pi}{2}\right) = -\frac{2\sqrt{2}}{3}$.

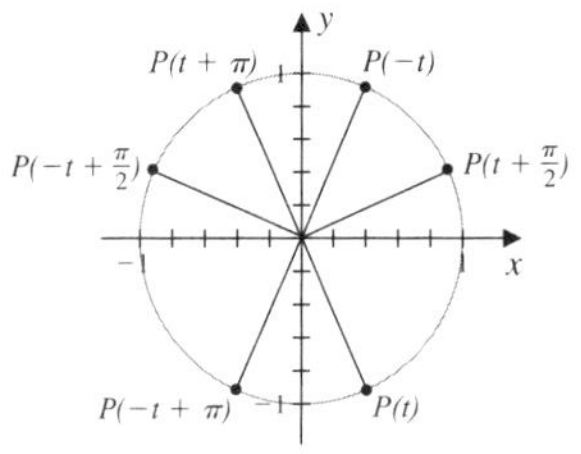

52. We have

$$(\cos t)^2 + \cos t - 2 = (\cos t + 2)(\cos t - 1) = 0,$$

which holds when $\cos t + 2 = 0$ or $\cos t - 1 = 0$. Hence $\cos t = -2$, which never holds, or $\cos t = 1$ which implies $t = 0, 2\pi$.

53. We have

$$2(\sin t)^2 + \sin t - 1 = (2\sin t - 1)(\sin t + 1) = 0,$$

which holds when $2\sin t - 1 = 0$ or $\sin t + 1 = 0$. Hence $\sin t = \frac{1}{2}$ or $\sin t = -1$ which implies $t = \frac{\pi}{6}, \frac{5\pi}{6}, \frac{3\pi}{2}$.

54. We have

$$0 = \sin t \cos t - \sin t - \cos t + 1 = \sin t(\cos t - 1) - (\cos t - 1)$$
$$= (\sin t - 1)(\cos t - 1).$$

So $\sin t - 1 = 0$ or $\cos t - 1 = 0$, which implies that $\sin t = 1$ or $\cos t = 1$. Hence $t = 0, \frac{\pi}{2}, 2\pi$.

55. Squaring both sides of the equation $\sin t + \cos t = 1$ gives

$$1 = (\sin t + \cos t)^2 = (\sin t)^2 + 2\sin t \cos t + (\cos t)^2 = 1 + 2\sin t \cos t,$$

So $\sin t \cos t = 0$.

Hence, the equation $\sin t + \cos t = 1$ is satisfied precisely when one of $\cos t$ or $\sin t$ is 0 and the other is 1. So $t = 0, \frac{\pi}{2}, 2\pi$.

56. a. The graph of the function is shown.

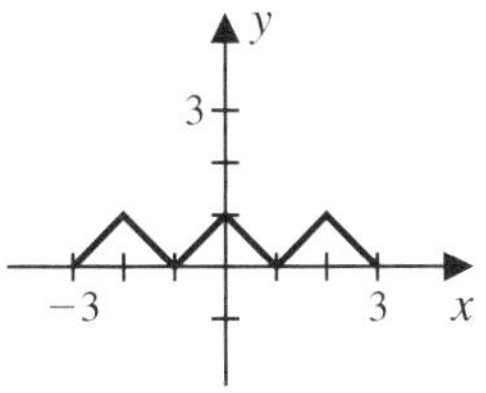

b. We have $f(x) = 0$, for $x = \pm 1, \pm 3, \pm 5, \ldots$ and $f(x) = 1$, for $x = 0, \pm 2, \pm 4, \pm 6, \ldots$.
The range of the function is the interval $[0, 1]$.

Exercise Set 4.4 (Page 210)

1. a. For $y = 2\cos x$, amplitude $= 2$, period $= 2\pi$

b. For $y = -3\cos x$, amplitude $= 3$, period $= 2\pi$

c. For $y = \frac{1}{2}\cos 2x$, amplitude $= \frac{1}{2}$, period$= \frac{2\pi}{2} = \pi$

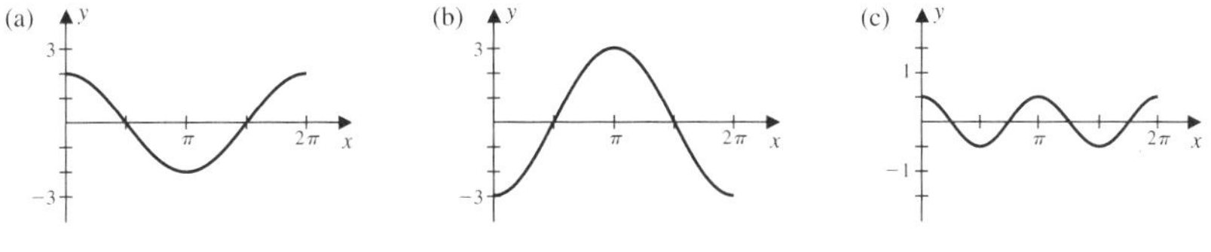

2. **a.** For $y = \sin 2x$, amplitude $= 1$, period $= \frac{2\pi}{2} = \pi$

b. For $y = \sin \frac{1}{2}x$, amplitude $= 1$, period $= \frac{2\pi}{1/2} = 4\pi$

c. For $y = -2\sin 3x$, amplitude $= 2$, period $= \frac{2\pi}{3}$

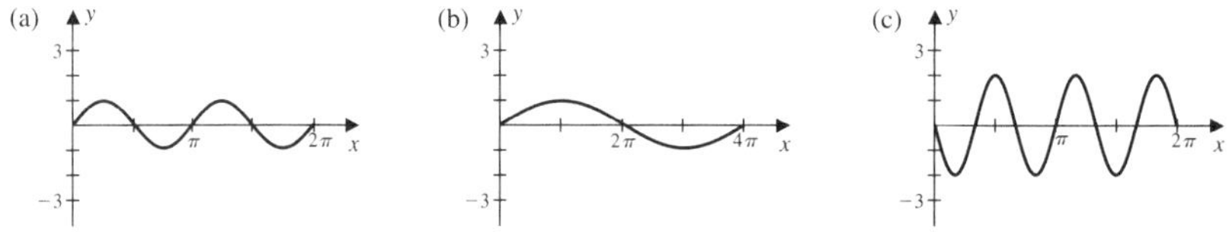

3. **a.** For $y = 2\cos \pi x$, amplitude $= 2$, period $= \frac{2\pi}{\pi} = 2$

b. For $y = \cos 2\pi x$, amplitude $= 1$, period $= \frac{2\pi}{2\pi} = 1$

c. For $y = -2\cos \frac{\pi}{2}x$, amplitude $= 2$, period $= \frac{2\pi}{\pi/2} = 4$

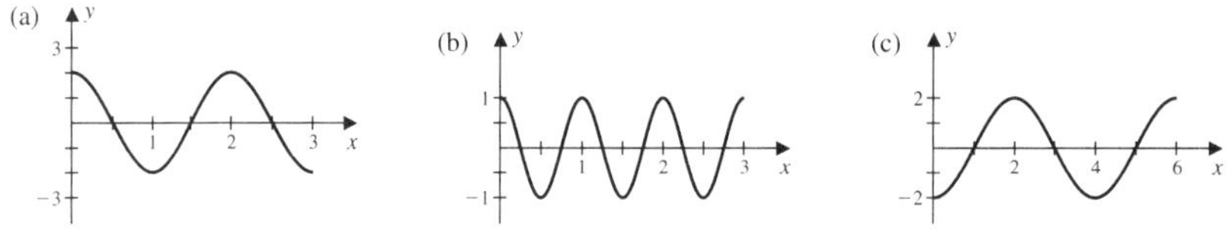

4. **a.** For $y = -2\sin \pi x$, amplitude $= 2$, period $= \frac{2\pi}{\pi} = 2$

b. For $y = \sin \frac{\pi}{2}x$,amplitude $= 1$, period $= \frac{2\pi}{\pi/2} = 4$

c. For $y = -3\sin 2\pi x$, amplitude $= 3$, period $= \frac{2\pi}{2\pi} = 1$

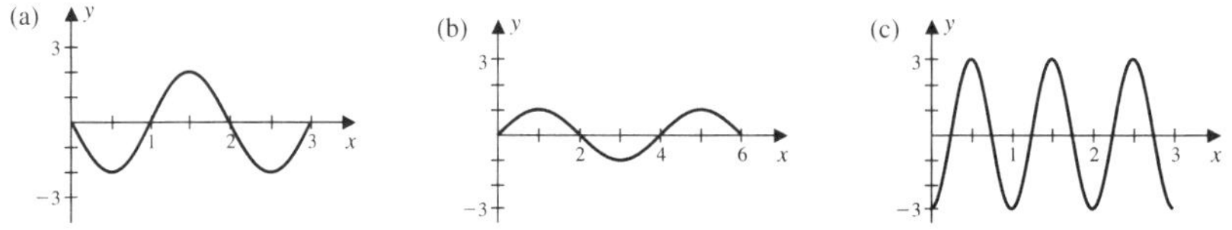

5. $y = 1 + \cos x$

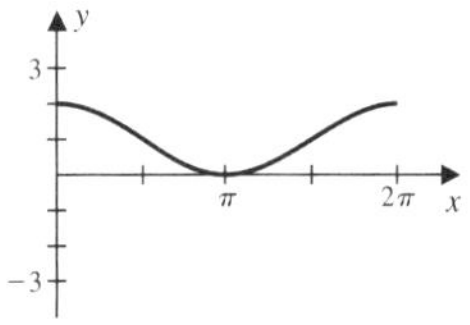

6. $y = -2 + \sin x$

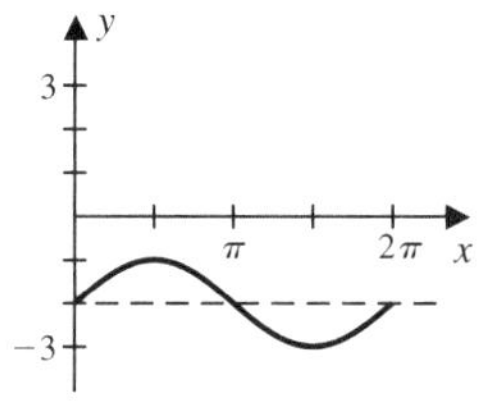

7. $y = \cos\left(x - \frac{\pi}{2}\right)$

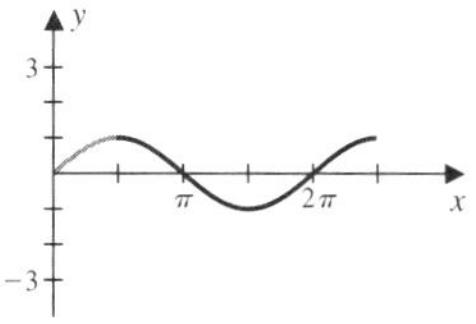

8. $y = \sin(x - \pi)$

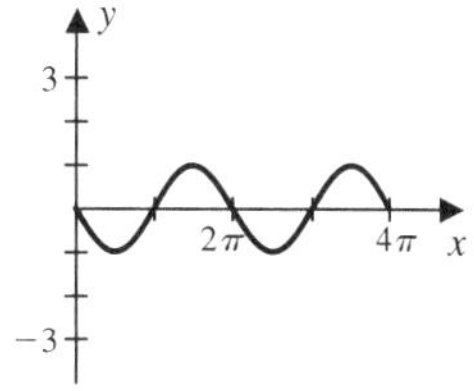

9. $y = -2 + \cos \pi x$

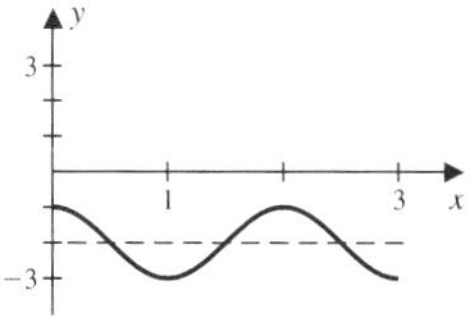

10. $y = 2 + \sin \pi x$

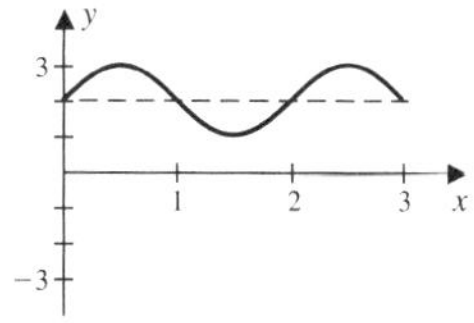

11. $y = 1 + \cos(2x + \pi) = 1 + \cos 2\left(x + \frac{\pi}{2}\right)$

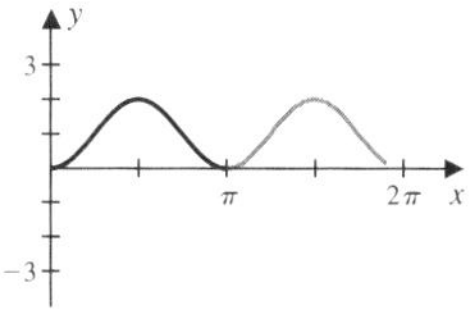

12.
$y = -1 + \sin(3x + \pi) = -1 + \sin 3\left(x + \frac{\pi}{3}\right)$

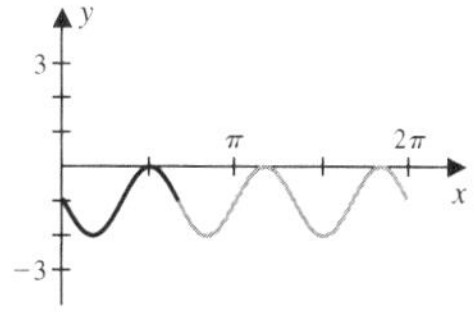

13. $y = -2 + 3\sin\left(3x - \frac{\pi}{2}\right) =$
$-2 + 3\sin 3\left(x - \frac{\pi}{6}\right)$

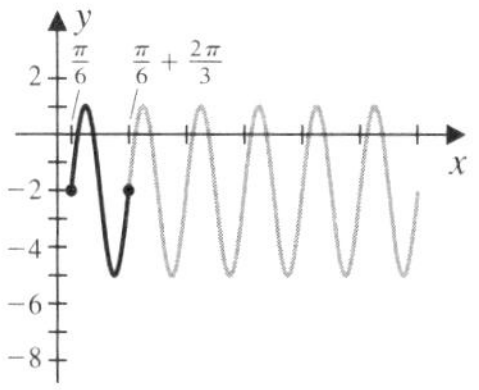

14. $y = \frac{1}{2}\cos\left(\frac{\pi}{2} - 2x\right) = \frac{1}{2}\cos\left(2x - \frac{\pi}{2}\right) =$
$\frac{1}{2}\cos 2\left(x - \frac{\pi}{4}\right)$

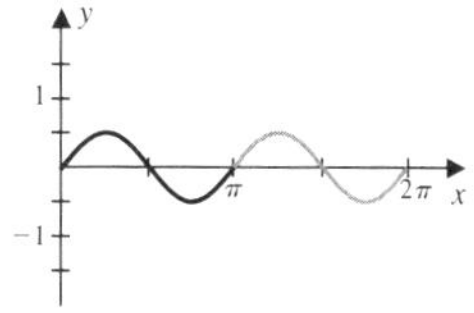

15. $y = -2\sin(x - 1) + 3$

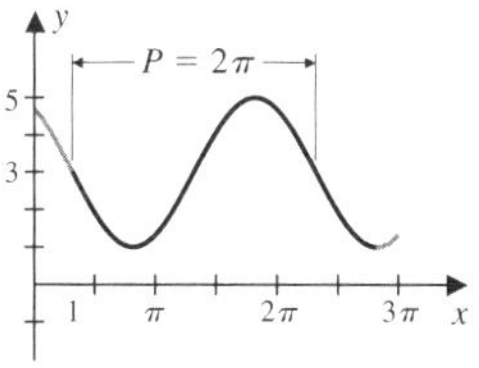

16. $y = 2 - \cos(x - 1)$

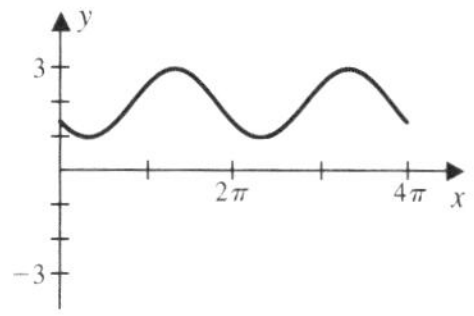

17. $y = |\cos x|$

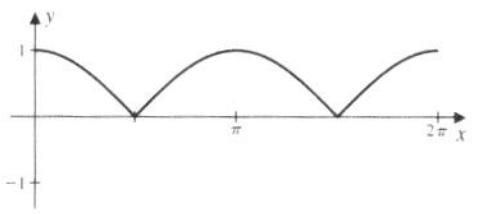

18. $y = |\sin x|$

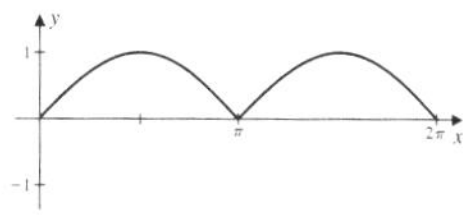

19. **a.** The equation of the graph has the form $y = A\cos(Bx)$, where the amplitude is 3 and the period is π. So $A = 3$, $\frac{2\pi}{B} = \pi$ which implies $B = 2$, and $y = 3\cos 2x$.

b. The equation can also be given as a sine curve shifted to the left by $\frac{\pi}{4}$ units. With amplitude 3, period π, and phase shift $-\frac{\pi}{4}$, the sine wave has equation $y = 3\sin 2\left(x + \frac{\pi}{4}\right)$.

20. **a.** The equation of the graph has the form $y = A\cos(Bx) + D$, where the amplitude is 2, the period is 4, and the curve has been shifted downward 1 unit. So $A = 2$, $\frac{2\pi}{B} = 4$ which implies $B = \frac{\pi}{2}$, $D = -1$, and $y = 2\cos\left(\frac{\pi}{2}x\right) - 1$.

b. The equation can also be given as a sine curve shifted to the right by 1 unit. That is, $y = -1 + 2\sin\frac{\pi}{2}(x + 1)$.

21. **a.** The equation of the graph has the form

$$y = A\cos(Bx + C) + D = A\cos B\left(x + \frac{C}{B}\right) + D,$$

where the amplitude is 1 the period is 6, the curve has been shifted horizontally to the right $\frac{5}{2}$ units, and shifted upward 2 units. So $A = 1$, $\frac{2\pi}{B} = 6$, which implies $B = \frac{\pi}{3}$, $\frac{C}{B} = -\frac{5}{2}$, $D = 2$, and $y = \cos\frac{\pi}{3}\left(x - \frac{5}{2}\right) + 2$.

b. The equation of the graph can also have the form

$$y = A\sin(Bx + C) + D = A\sin B\left(x + \frac{C}{B}\right) + D,$$

where the amplitude is 1, the period is 6, the curve has been shifted horizontally to the right 1 unit, and the curve has been shifted upward 2 units. So $A = 1$ and $\frac{2\pi}{B} = 6$, which implies $B = \frac{\pi}{3}$, $\frac{C}{B} = -1$, $D = 2$, and $y = \sin\frac{\pi}{3}(x - 1) + 2$.

22. **a.** The equation of the graph has the form $y = A\cos Bx$, and the curve has been reflected through the x-axis. The amplitude is $\frac{1}{2}$, and the period is $\frac{4\pi}{3}$. So, $A = -\frac{1}{2}$, $\frac{2\pi}{B} = \frac{4\pi}{3}$ which implies $B = \frac{3}{2}$, and $y = -\frac{1}{2}\cos\frac{3}{2}x$.

b. As a sine wave the equation of the graph has the form

$$y = A\sin(Bx + C) = A\sin B\left(x + \frac{C}{B}\right),$$

where the amplitude is $\frac{1}{2}$, the period is $\frac{4\pi}{3}$, the curve has been shifted horizontally to the left $\frac{\pi}{3}$ units, and the curve has been reflected through the x-axis. So $A = -\frac{1}{2}$, $\frac{2\pi}{B} = \frac{4\pi}{3}$, which implies $B = \frac{3}{2}$, $\frac{C}{B} = \frac{\pi}{3}$, and $y = -\frac{1}{2}\sin\frac{3}{2}\left(x + \frac{\pi}{3}\right)$.

23. The amplitude of the waves in (a) and (c) is 2, which could only match (iii) and (iv), and the amplitude in (b) and (d) is $\frac{1}{2}$, which could only match (i) and (ii). Setting $x = 0$ in each of the equations, the y-intercepts are $2\sin\left(0 - \frac{\pi}{2}\right) = -2$, $2\cos\left(0 - \frac{\pi}{2}\right) = 0$, $\frac{1}{2}\sin\left(0 + \frac{\pi}{2}\right) = \frac{1}{2}$, and $\frac{1}{2}\cos\left(0 + \frac{\pi}{2}\right) = 0$. The matching is then (a) iv, (b) i, (c) iii, and (d) ii.

24. **a.** ii **b.** i **c.** iv **d.** iii

25. $y = -1 + \frac{1}{2}\sin(x - 2)$

26. $y = 3 + \cos 2(x + 1)$

27. The period of $f(x) = \cos(100x)$ is $\frac{2\pi}{100} = \frac{\pi}{50} \approx 0.06$, and the amplitude is 1, so a reasonable viewing rectangle is $[-\pi/50, \pi/50] \times [-1, 1]$.

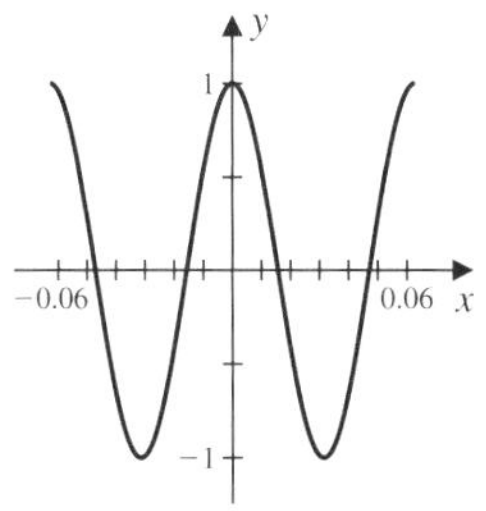

28. The period of $f(x) = -5\sin(20x)$ is $\frac{2\pi}{20} = \frac{\pi}{10} \approx 0.314$, and the amplitude is 5, so a reasonable viewing rectangle is $[-\pi/10, \pi/10] \times [-5, 5]$.

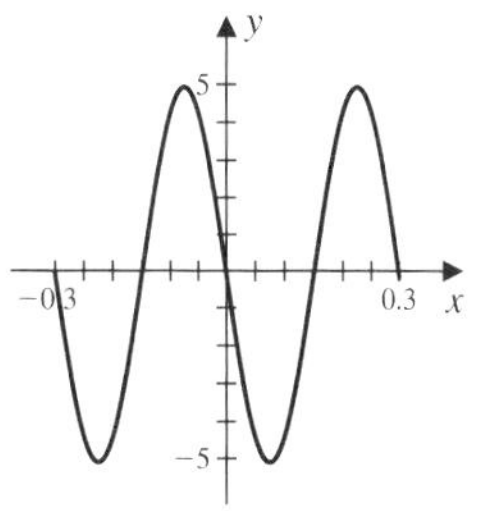

29. The period of $f(x) = \sin\left(\frac{x}{50}\right)$ is $\frac{2\pi}{1/50} = 100\pi$, and the amplitude is 1, so a reasonable viewing rectangle is $[-100\pi, 100\pi] \times [-1, 1]$.

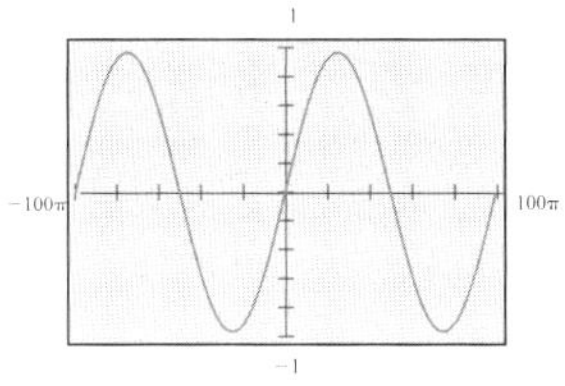

30. The period of $y = 5\sin(10x)$ is $\frac{2\pi}{10} = \frac{\pi}{5} \approx 0.63$, and the amplitude is 5, so a reasonable viewing rectangle for the sine function is $[-\pi/5, \pi/5] \times [-5, 5]$. To see the effect of adding x^2, expand the viewing rectangle to $-5 \le x \le 5$, and since $x^2 \ge 0$, $-5 \le x^2 + 5\sin(10x) \le 30$. A reasonable viewing rectangle is $[-5, 5] \times [-6, 30]$.

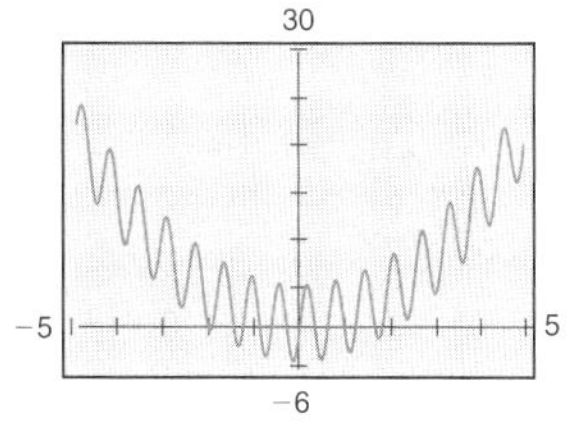

31. $\cos x = x$ so $x \approx 0.7$

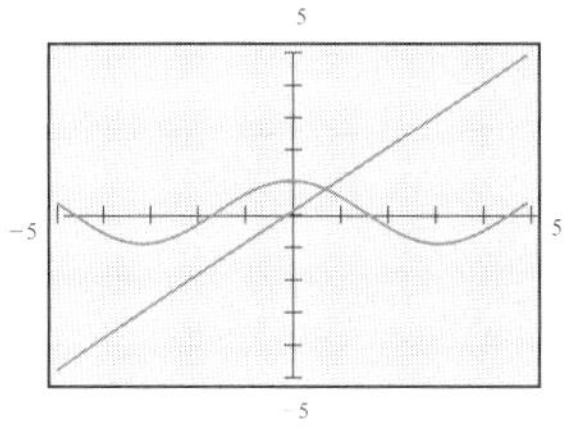

32. $\cos x = x^2$ so $x \approx 0.8$, $x \approx -0.8$

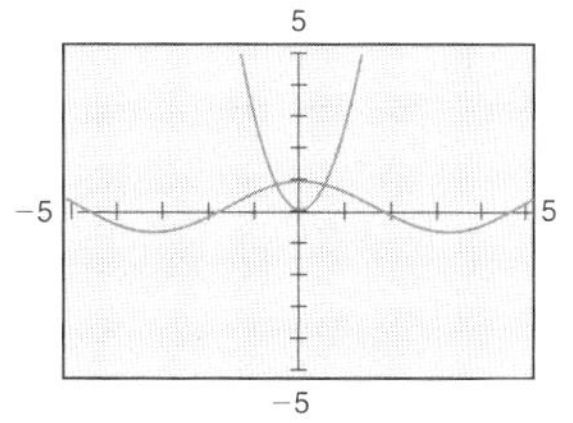

33. $\sin x + \cos x = x$ so $x \approx 1.3$

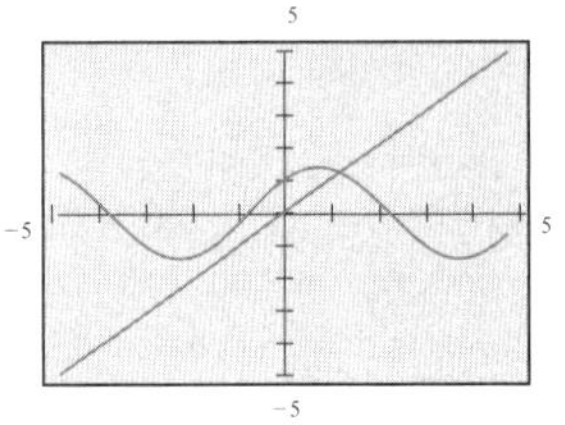

34. $\sin x + x = x^3$ so $x \approx 1.3$, $x \approx -1.3$

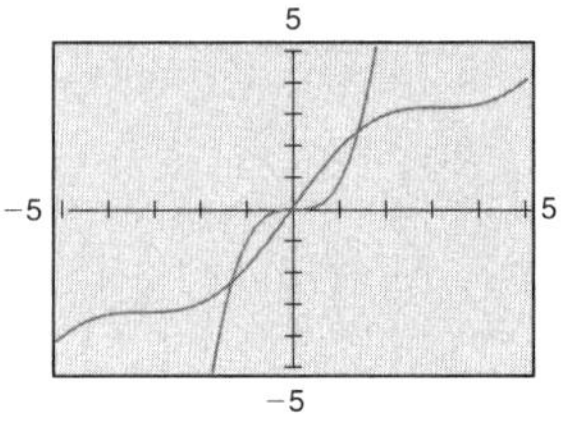

35. **a.** The smallest positive value of t for which

$$0 = f(t) = 2\sin(3t - \frac{\pi}{4})$$

occurs when

$$3t - \frac{\pi}{4} = 0, \text{ so } t = \frac{\pi}{12}.$$

b. A maximum value first occurs when

$$\sin(3t - \frac{\pi}{4}) = 1,$$

so

$$3t - \frac{\pi}{4} = \frac{\pi}{2} \text{ and } t = \frac{\pi}{4}.$$

c. A minimum value first occurs when

$$\sin(3t - \frac{\pi}{4}) = -1,$$

so

$$3t - \frac{\pi}{4} = \frac{3\pi}{2} \text{ and } t = \frac{7\pi}{12}.$$

36. **a.** The smallest positive value of t for which

$$0 = f(t) = -3\cos(2t + \frac{\pi}{6})$$

occurs when

$$2t + \frac{\pi}{6} = \frac{\pi}{2}, \text{ so } t = \frac{\pi}{6}.$$

b. A maximum value first occurs when

$$\cos(2t + \frac{\pi}{6}) = -1,$$

so

$$2t + \frac{\pi}{6} = \pi \text{ and } t = \frac{5\pi}{12}.$$

c. A minimum value first occurs when

$$\cos(2t + \frac{\pi}{6}) = 1,$$

so

$$2t + \frac{\pi}{6} = 2\pi \text{ and } t = \frac{11\pi}{12}.$$

Note: A minimum also occurs when $2t + \frac{\pi}{6} = 0$, but in this case $t = -\frac{\pi}{12}$ is negative.

37. The amplitude is 2 and the line on which the graph is centered is $y = 1$. The period is 4, so the points on the graph are given by

$$y = 1 + 2\sin\frac{\pi}{2}x.$$

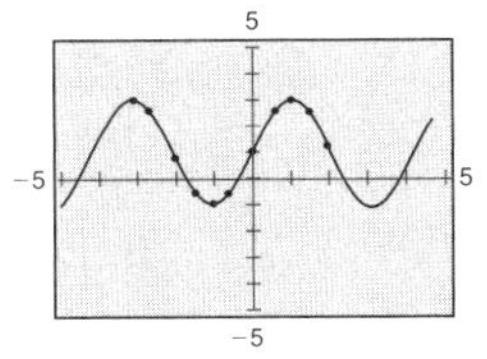

38. The amplitude is 1 and the line on which the graph is centered is $y = 2$. The period is 6 and the graph is shifted left 2 units, so the points on the graph are given by

$$y = 2 + 1\cos\frac{\pi}{3}(x + 2).$$

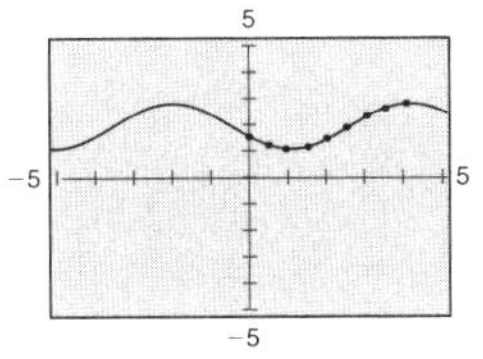

39. The graphs are shown for each part.

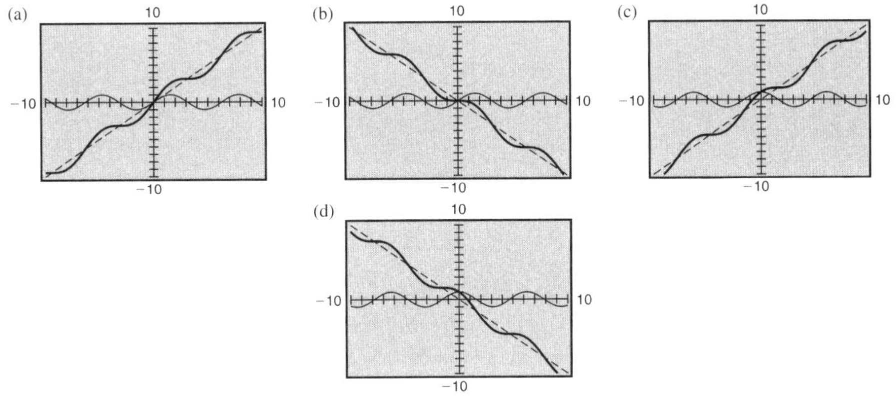

40. The graphs are shown for each part.

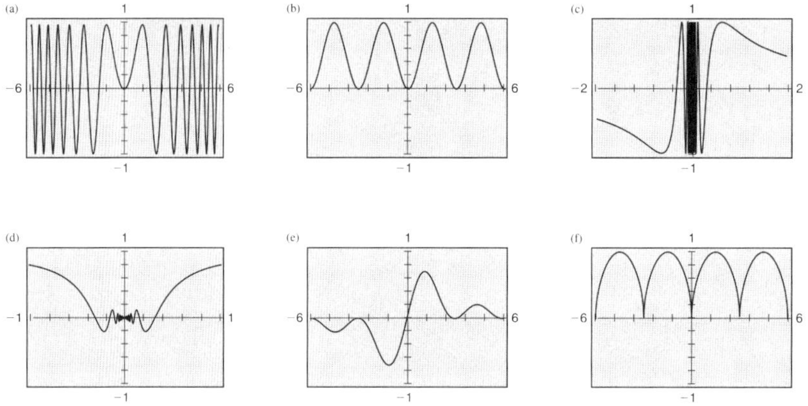

41. **a.** It vertically stretches the curve by a factor of 2, so the amplitude is doubled. **b.** It horizontally compresses the curve by a factor of 2, so the period is halved. **c.** The horizontal shift is doubled.

42. **a.** Since the difference between the largest data value and the smallest data value is about $77 - 39 = 38$ the amplitude is half this amount or 19. One wave appears to occur on an interval of length 12, so $\frac{2\pi}{B} = 12$ and $B = \frac{\pi}{6}$. The wave is also shifted to the right about 4.5 units and upward 57 units. A possible equation for the curve is $f(x) = 19 \sin\left(\frac{\pi}{6}(x - 4.5)\right) + 57$.

b.

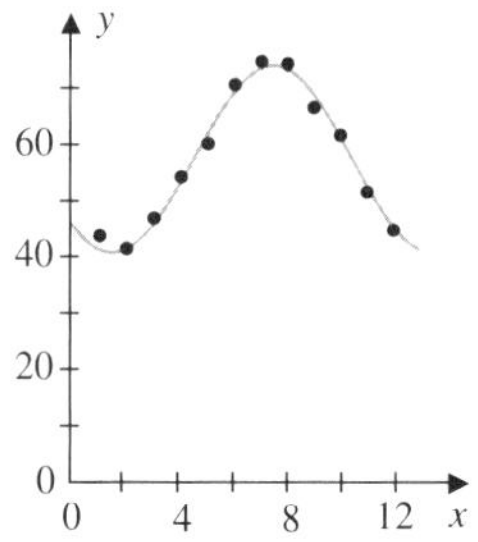

43. **a.** Since the difference between the largest data value and the smallest data value is about $70 - 26 = 44$ the amplitude is half this amount or 22. One wave appears to occur on an interval of length 12, so $\frac{2\pi}{B} = 12$ and $B = \frac{\pi}{6}$. The wave is also shifted to the right about 4.5 units and upward 48 units. A possible equation for the curve is

$f(x) = 22 \sin\left(\frac{\pi}{6}(x - 4.5)\right) + 48.$

b.

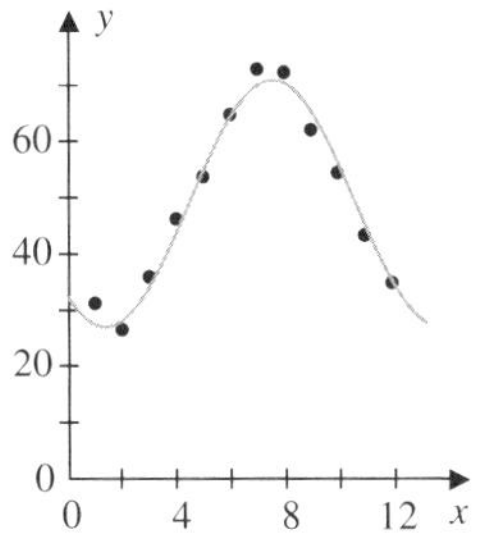

Exercise Set 4.5 (Page 219)

1. $\sin t < 0$ and $\cos t > 0$ so $P(t)$ is in quadrant IV.

2. $\tan t < 0$ and $\cos t < 0$ implies $\frac{\sin t}{\cos t} < 0$ and $\cos t < 0$ so $\sin t > 0$ and $\cos t < 0$ and $P(t)$ is in quadrant II.

3. $\sin t < 0$ and $\cot t > 0$ implies $\sin t < 0$ and $\frac{\cos t}{\sin t} > 0$ so $\sin t < 0$ and $\cos t < 0$ and $P(t)$ is in quadrant III.

4. $\sec t > 0$ and $\tan t < 0$ implies $\frac{1}{\cos t} > 0$ and $\frac{\sin t}{\cos t} < 0$ so $\cos t > 0$ and $\sin t < 0$ and $P(t)$ is in quadrant IV.

5. $\sin t > 0$ and $\tan t > 0$ implies $\sin t > 0$ and $\frac{\sin t}{\cos t} > 0$ so $\sin t > 0$ and $\cos t > 0$ and $P(t)$ is in quadrant I.

6. $\sin t < 0$ and $\sec t < 0$ implies $\sin t < 0$ and $\frac{1}{\cos t} < 0$ so $\sin t < 0$ and $\cos t < 0$ and $P(t)$ is in quadrant III.

7. A dash in the table indicates the trigonometric function is undefined at the given value.

	$\frac{7\pi}{6}$	$\frac{5\pi}{4}$	$\frac{4\pi}{3}$	$\frac{3\pi}{2}$	$\frac{5\pi}{3}$	$\frac{7\pi}{4}$	$\frac{11\pi}{6}$	2π
$\sin t$	$-\frac{1}{2}$	$-\frac{\sqrt{2}}{2}$	$-\frac{\sqrt{3}}{2}$	-1	$-\frac{\sqrt{3}}{2}$	$-\frac{\sqrt{2}}{2}$	$-\frac{1}{2}$	0
$\cos t$	$-\frac{\sqrt{3}}{2}$	$-\frac{\sqrt{2}}{2}$	$-\frac{1}{2}$	0	$\frac{1}{2}$	$\frac{\sqrt{2}}{2}$	$\frac{\sqrt{3}}{2}$	1
$\tan t$	$\frac{\sqrt{3}}{3}$	1	$\sqrt{3}$	–	$-\sqrt{3}$	-1	$-\frac{\sqrt{3}}{3}$	0
$\cot t$	$\sqrt{3}$	1	$\frac{\sqrt{3}}{3}$	0	$-\frac{\sqrt{3}}{3}$	-1	$-\sqrt{3}$	–
$\sec t$	$-\frac{2\sqrt{3}}{3}$	$-\sqrt{2}$	-2	–	2	$\sqrt{2}$	$\frac{2\sqrt{3}}{3}$	1
$\csc t$	-2	$-\sqrt{2}$	$-\frac{2\sqrt{3}}{3}$	-1	$-\frac{2\sqrt{3}}{3}$	$-\sqrt{2}$	-2	–

8. A dash in the table indicates the trigonometric function is undefined at the given value.

	$-\frac{\pi}{6}$	$-\frac{\pi}{4}$	$-\frac{\pi}{3}$	$-\frac{\pi}{2}$	$-\frac{2\pi}{3}$	$-\frac{3\pi}{4}$	$-\frac{5\pi}{6}$	$-\pi$
$\sin t$	$-\frac{1}{2}$	$-\frac{\sqrt{2}}{2}$	$-\frac{\sqrt{3}}{2}$	-1	$-\frac{\sqrt{3}}{2}$	$-\frac{\sqrt{2}}{2}$	$-\frac{1}{2}$	0
$\cos t$	$\frac{\sqrt{3}}{2}$	$\frac{\sqrt{2}}{2}$	$\frac{1}{2}$	0	$-\frac{1}{2}$	$-\frac{\sqrt{2}}{2}$	$-\frac{\sqrt{3}}{2}$	-1
$\tan t$	$-\frac{\sqrt{3}}{3}$	-1	$-\sqrt{3}$	–	$\sqrt{3}$	1	$\frac{\sqrt{3}}{3}$	0
$\cot t$	$-\sqrt{3}$	-1	$-\frac{\sqrt{3}}{3}$	0	$\frac{\sqrt{3}}{3}$	1	$\sqrt{3}$	–
$\sec t$	$\frac{2\sqrt{3}}{3}$	$\sqrt{2}$	2	–	-2	$-\sqrt{2}$	$-\frac{2\sqrt{3}}{3}$	-1
$\csc t$	-2	$-\sqrt{2}$	$-\frac{2\sqrt{3}}{3}$	-1	$-\frac{2\sqrt{3}}{3}$	$-\sqrt{2}$	-2	–

9. Since $\sin t = \frac{4}{5}$ and $(\cos t)^2 + (\sin t)^2 = 1$, we have

$$(\cos t)^2 = 1 - \left(\frac{4}{5}\right)^2 = \frac{9}{25}, \text{ which implies } \cos t = \pm\sqrt{\frac{9}{25}} = \pm\frac{3}{5},$$

and since t is in quadrant I, $\cos t > 0$, so $\cos t = \frac{3}{5}$. Then

$$\tan t = \frac{4/5}{3/5} = \frac{4}{3},\ \cot t = \frac{3/5}{4/5} = \frac{3}{4},\ \sec t = \frac{1}{3/5} = \frac{5}{3},\ \csc t = \frac{1}{4/5} = \frac{5}{4}.$$

10. Since $\cos t = -\frac{1}{2}$, with $\pi \le t \le 2\pi$ implies t is in quadrant III, so $\sin t = -\frac{\sqrt{3}}{2}$. Then

$$\tan t = \frac{-\sqrt{3}/2}{-1/2} = \sqrt{3},\ \cot t = \frac{1}{\sqrt{3}} = \frac{\sqrt{3}}{3},$$

and

$$\sec t = \frac{1}{-1/2} = -2,\ \csc t = \frac{1}{-\sqrt{3}/2} = -\frac{2}{\sqrt{3}} = -\frac{2\sqrt{3}}{3}.$$

11. Since $\cos t = -\frac{4}{5}$, with $\pi \le t \le \frac{3\pi}{2}$ implies t is in quadrant III, so $\sin t < 0$. Then since $(\cos t)^2 + (\sin t)^2 = 1$, we have

$$(\sin t)^2 = 1 - \left(-\frac{4}{5}\right)^2 = 1 - \frac{16}{25} = \frac{9}{25}, \text{ which implies } \sin t = \pm\sqrt{\frac{9}{25}} = \pm\frac{3}{5},$$

and since $\sin t < 0$, $\sin t = -\frac{3}{5}$. Then

$$\tan t = \frac{-3/5}{-4/5} = \frac{3}{4},\ \cot t = \frac{1}{\tan t} = \frac{4}{3},\ \sec t = \frac{1}{-4/5} = -\frac{5}{4},\ \csc t = \frac{1}{-3/5} = -\frac{5}{3}.$$

12. Since $\sin t = \frac{1}{3}$ and $(\cos t)^2 + (\sin t)^2 = 1$ we have

$$(\cos t)^2 = 1 - \left(\frac{1}{3}\right)^2 = \frac{8}{9}, \text{ which implies } \cos t = \pm\sqrt{\frac{8}{9}} = \pm\frac{2\sqrt{2}}{3},$$

and since t is in quadrant II, $\cos t < 0$, so $\cos t = -\frac{2\sqrt{2}}{3}$. Then

$$\tan t = \frac{1/3}{-2\sqrt{2}/3} = -\frac{1}{2\sqrt{2}} = -\frac{\sqrt{2}}{4}, \quad \cot t = -\frac{4}{\sqrt{2}} = -2\sqrt{2},$$

$$\sec t = \frac{1}{-2\sqrt{2}/3} = -\frac{3}{2\sqrt{2}} = -\frac{3\sqrt{2}}{4}, \quad \csc t = \frac{1}{1/3} = 3.$$

13. Since $\tan t = \frac{\sin t}{\cos t} = 2$ implies $\sin t = 2\cos t$ or $\sqrt{1 - (\cos t)^2} = 2\cos t$,

so

$$1 - (\cos t)^2 = 4(\cos t)^2 \text{ implies } 5(\cos t)^2 = 1 \text{ so } \cos t = \pm\sqrt{\frac{1}{5}} = \pm\frac{\sqrt{5}}{5},$$

and since $0 < t < \frac{\pi}{2}$, $\cos t > 0$, we have $\cos t = \frac{\sqrt{5}}{5}$ and $\sin t = 2\cos t = \frac{2\sqrt{5}}{5}$. Then

$$\cot t = \frac{1}{\tan t} = \frac{1}{2},\ \sec t = \frac{1}{\sqrt{5}/5} = \frac{5}{\sqrt{5}} = \sqrt{5},\ \csc t = \frac{1}{2\sqrt{5}/5} = \frac{5}{2\sqrt{5}} = \frac{\sqrt{5}}{2}.$$

14. Since $\csc t = -2$ implies $\sin t = -\frac{1}{2}$, and since $\frac{3\pi}{2} \le t \le 2\pi$ implies t is in quadrant IV, and we have $\cos t = \frac{\sqrt{3}}{2}$. Then

$$\tan t = \frac{-1/2}{\sqrt{3}/2} = -\frac{1}{\sqrt{3}} = -\frac{\sqrt{3}}{3},\ \cot t = -\frac{3}{\sqrt{3}} = -\sqrt{3},\ \sec t = \frac{1}{\sqrt{3}/2} = \frac{2}{\sqrt{3}} = \frac{2\sqrt{3}}{3}.$$

15. Since $\sec t = 3$ implies $\cos t = \frac{1}{3}$, and since t is in quadrant IV, $\sin t < 0$. Since $(\cos t)^2 + (\sin t)^2 = 1$, we have

$$(\sin t)^2 = 1 - \left(\frac{1}{3}\right)^2 = 1 - \frac{1}{9} = \frac{8}{9} \text{ implies } \sin t = \pm\sqrt{\frac{8}{9}} = \pm\frac{2\sqrt{2}}{3},$$

and since $\sin t < 0$, $\sin t = -\frac{2\sqrt{2}}{3}$. Then

$$\tan t = \frac{-2\sqrt{2}/3}{1/3} = -2\sqrt{2},\ \cot t = -\frac{1}{2\sqrt{2}} = -\frac{\sqrt{2}}{4},\ \csc t = -\frac{1}{2\sqrt{2}/3} = -\frac{3\sqrt{2}}{4}.$$

16. Since $\cot t = \frac{\cos t}{\sin t} = 3$ implies $\cos t = 3\sin t$ or $\sqrt{1 - (\sin t)^2} = 3\sin t$.
So

$$1 - (\sin t)^2 = 9(\sin t)^2 \text{ implies } 10(\sin t)^2 = 1 \text{ so } \sin t = \pm\sqrt{\frac{1}{10}} = \pm\frac{\sqrt{10}}{10},$$

and since t is in quadrant III, $\sin t = -\frac{\sqrt{10}}{10}$ and $\cos t = 3\sin t = -\frac{3\sqrt{10}}{10}$. Then

$$\tan t = \frac{1}{\cot t} = \frac{1}{3}, \quad \sec t = \frac{10}{-3\sqrt{10}} = -\frac{\sqrt{10}}{3}, \quad \csc t = \frac{10}{-\sqrt{10}} = -\sqrt{10}.$$

17. $y = 2\tan x$

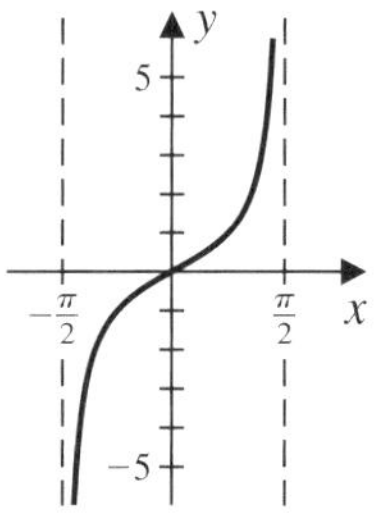

18. $y = \frac{1}{2}\cot x$

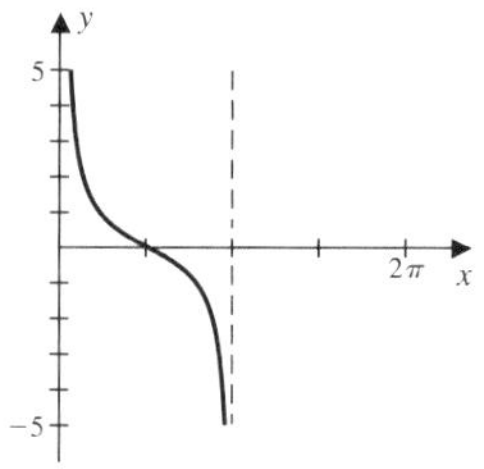

19. $y = -2\sec x$

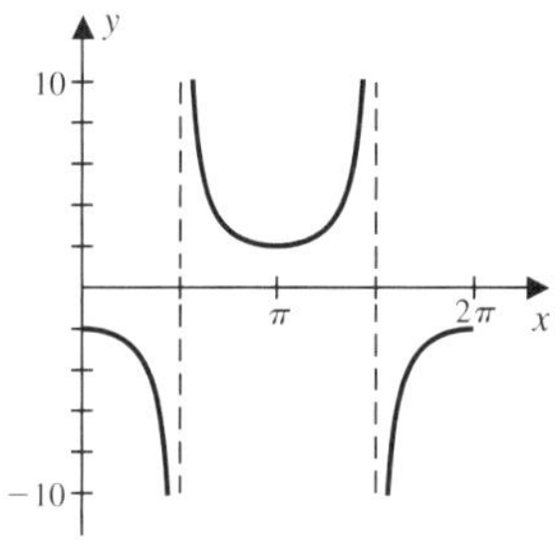

20. $y = -\frac{1}{2}\csc x$

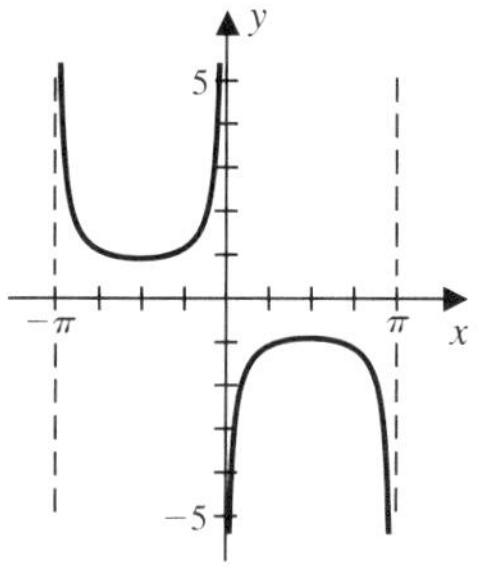

21. $y = \tan(x + \pi/4)$

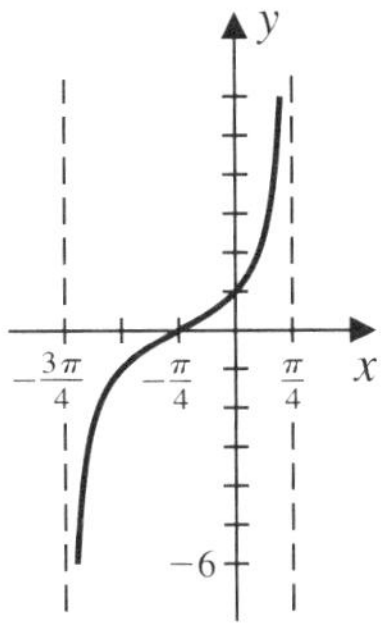

22. $y = \tan(x - \pi/2)$

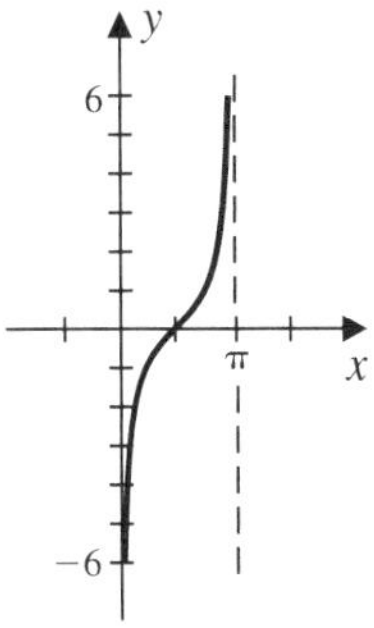

23. $y = \frac{1}{2}\sec(x + \pi/4)$

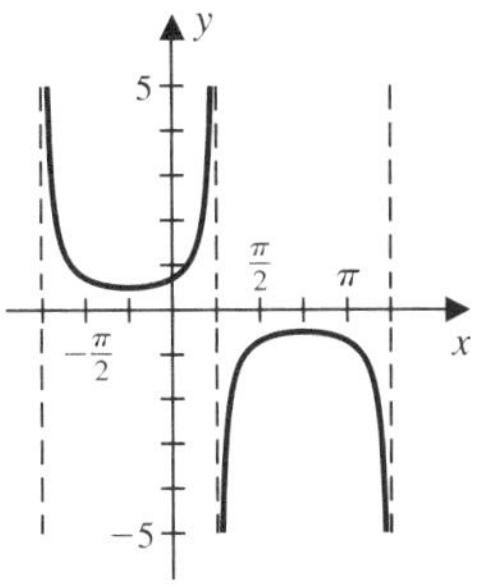

24. $y = 2\csc(x - \pi/2)$

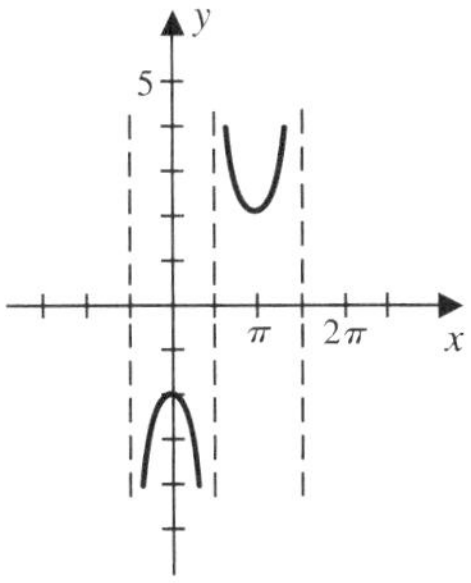

25. For $y = \tan \pi x$, the period is $\frac{\pi}{\pi} = 1$.

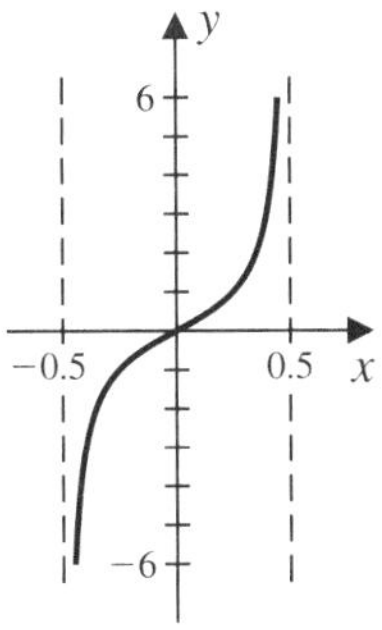

26. For $y = \cot(\pi x/2)$, the period is $\frac{\pi}{\pi/2} = 2$.

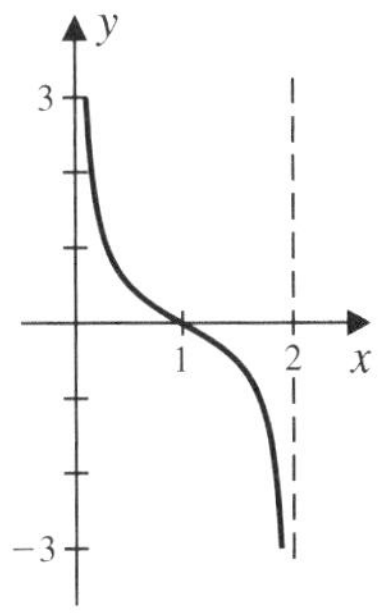

27. For $y = \tan\left(2x - \frac{\pi}{2}\right) = \tan 2\left(x - \frac{\pi}{4}\right)$, the period is $\frac{\pi}{2}$.

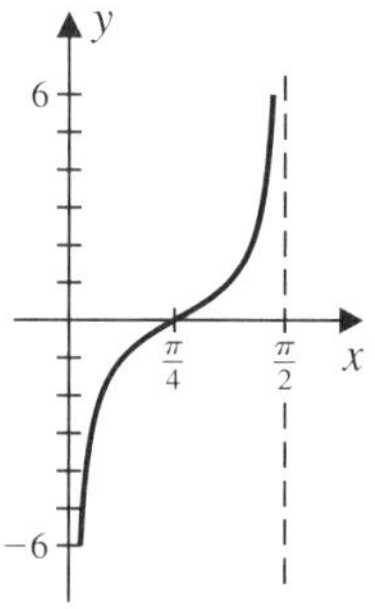

28. For $y = \sec(2x + \pi) = \sec 2\left(x + \frac{\pi}{2}\right)$, the period is $\frac{2\pi}{2} = \pi$.

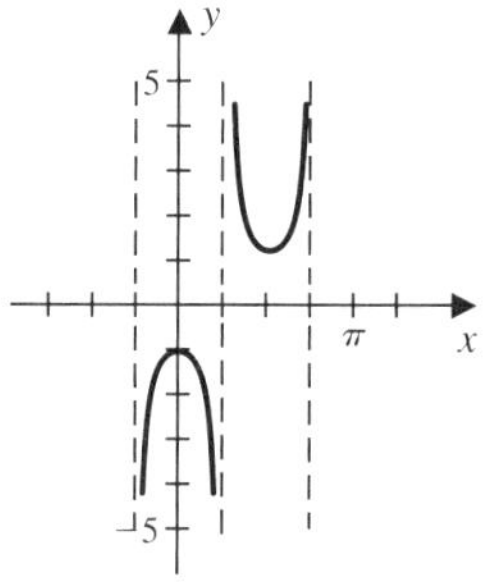

29. $\tan t + 1 = 0$ implies $\tan t = -1$ so $t = \frac{3\pi}{4}, \frac{7\pi}{4}$

30. $\cot t + \sqrt{3} = 0$ implies $\cot t = -\sqrt{3}$ and $\tan t = -\frac{1}{\sqrt{3}}$. So $t = \frac{5\pi}{6}, \frac{11\pi}{6}$

31. $(\tan t)^2 = \frac{1}{3}$ implies $\tan t = \pm\frac{\sqrt{3}}{3}$ so $t = \frac{\pi}{6}, \frac{5\pi}{6}, \frac{7\pi}{6}, \frac{11\pi}{6}$

32. Since

$$(\cot t)^2 = \frac{(\cos t)^2}{(\sin t)^2} = 3 \text{ implies } (\cos t)^2 = 3(\sin t)^2 \text{ so } (\cos t)^2 = 3(1 - (\cos t)^2),$$

we have

$$4(\cos t)^2 = 3 \text{ implies } (\cos t)^2 = \frac{3}{4} \text{ so } \cos t = \pm\frac{\sqrt{3}}{2} \text{ and } t = \frac{\pi}{6}, \frac{5\pi}{6}, \frac{7\pi}{6}, \frac{11\pi}{6}.$$

33. $|\tan t| = 1$ implies $\tan t = 1$ or $\tan t = -1$ so $t = \frac{\pi}{4}, \frac{3\pi}{4}, \frac{5\pi}{4}, \frac{7\pi}{4}$

34. $|\sec t| = 1$ implies $\sec t = 1$ or $\sec t = -1$ so $\cos t = 1$ or $\cos t = -1$ and $t = 0, \pi, 2\pi$

35. Since

$$2\sin 2t - \sqrt{2}\tan 2t = 2\sin 2t - \sqrt{2}\frac{\sin 2t}{\cos 2t} = 0 \text{ implies } \frac{2\sin 2t\cos 2t - \sqrt{2}\sin 2t}{\cos 2t} = 0,$$

we have

$$2\sin 2t\cos 2t - \sqrt{2}\sin 2t = \sin 2t(2\cos 2t - \sqrt{2}) = 0 \text{ so } \sin 2t = 0 \text{ or } \cos 2t = \frac{\sqrt{2}}{2}.$$

If $0 \le t \le 2\pi$, then $0 \le 2t \le 4\pi$, so $2t = 0, \pi, 2\pi, 3\pi, 4\pi$ or $2t = \frac{\pi}{4}, \frac{7\pi}{4}, \frac{9\pi}{4}, \frac{15\pi}{4}$, which implies that $t = 0, \frac{\pi}{2}, \pi, \frac{3\pi}{2}, 2\pi, \frac{\pi}{8}, \frac{7\pi}{8}, \frac{9\pi}{8}, \frac{15\pi}{8}$.

36. Since

$$\tan t - 3\cot t = \tan t - 3\frac{1}{\tan t} = \frac{(\tan t)^2 - 3}{\tan t} = 0 \text{ implies } (\tan t)^2 - 3 = 0,$$

$\tan t = \pm\sqrt{3}$, and we have $t = \frac{\pi}{3}, \frac{2\pi}{3}, \frac{4\pi}{3}, \frac{5\pi}{3}$.

37. The period of $f(x) = \tan(5x)$ is $\frac{\pi}{5}$, so a reasonable viewing rectangle is $\left[-\frac{\pi}{10}, \frac{\pi}{10}\right] \times [-5, 5]$.

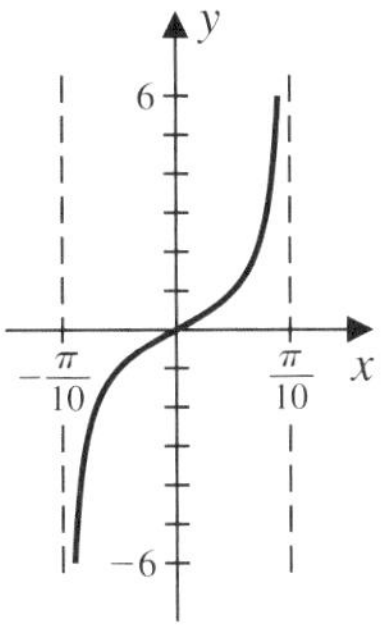

38. The period of $f(x) = \tan(8x - 10) = \tan 8\left(x - \frac{5}{4}\right)$ is $\frac{\pi}{8}$, and a reasonable viewing rectangle of $y = \tan(8x)$ is $\left[-\frac{\pi}{16}, \frac{\pi}{16}\right] \times [-5, 5]$. Since the graph is also shifted right $\frac{5}{4}$ units a reasonable viewing rectangle is $\left[-\frac{\pi}{8}, \frac{\pi}{8}\right] \times [-5, 5]$.

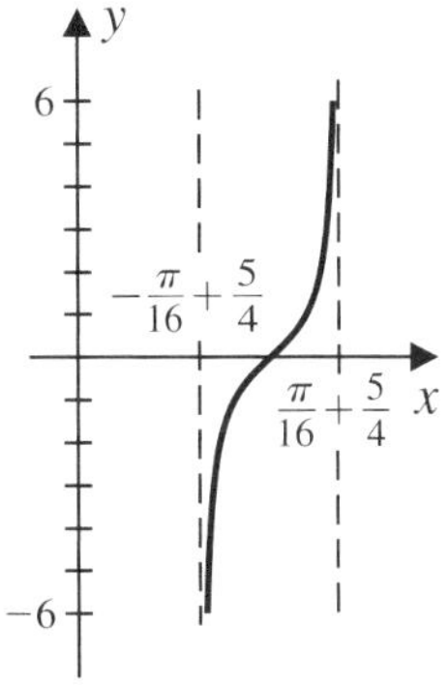

39. The period of $f(x) = \csc 100x$ is $\frac{2\pi}{100} = \frac{\pi}{50}$, so a reasonable viewing rectangle is $\left[-\frac{\pi}{50}, \frac{\pi}{50}\right] \times [-5, 5]$.

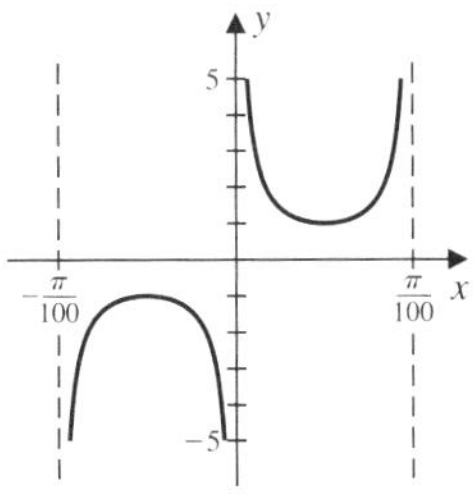

40. The period of $f(x) = \sec\left(\frac{x}{50}\right)$ is $\frac{2\pi}{1/50} = 100\pi$, so a reasonable viewing rectangle is $[-100\pi, 100\pi] \times [-5, 5]$.

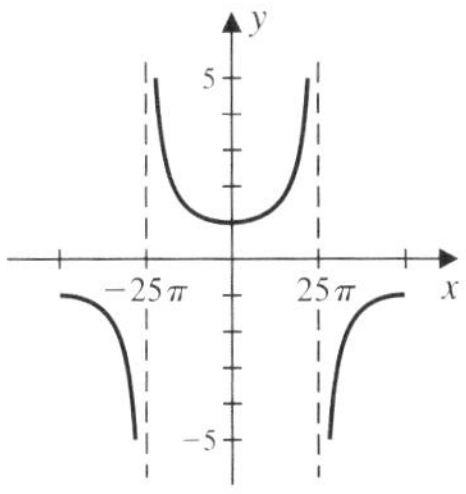

41. The period of $f(x) = \tan\left(\frac{x}{100}\right)$ is $\frac{\pi}{1/100} = 100\pi$, so a reasonable viewing rectangle is $[-50\pi, 50\pi] \times [-5, 5]$.

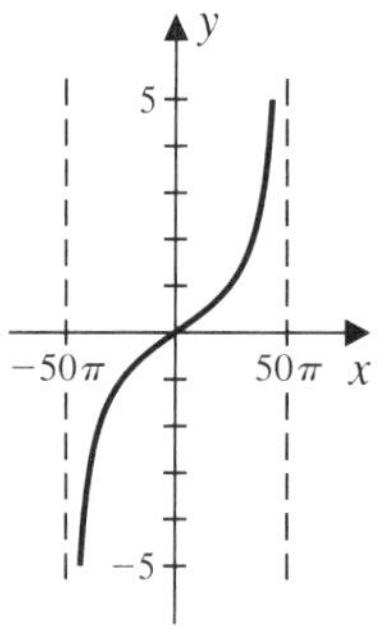

42. Since the period of the tangent is π and the period of the cosecant is 2π, the period of the function $f(x) = \tan(25x) - \csc(25x)$ is $\frac{2\pi}{25}$, so a reasonable viewing rectangle is $\left[-\frac{2\pi}{25}, \frac{2\pi}{25}\right] \times [-5, 5]$.

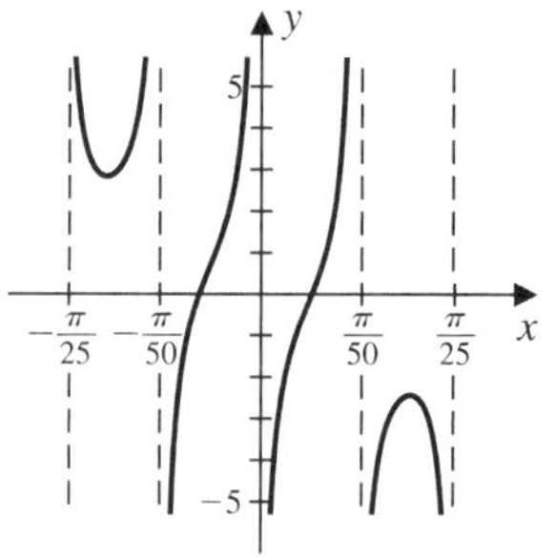

43. Let $P(t) = (x, y)$. Since $P(t)$ lies on the unit circle $x^2 + y^2 = 1$, and since the point lies on the line $y = -2x$, we have

$$x^2 + (-2x)^2 = 1 \text{ implies } 5x^2 = 1 \text{ so } x = \pm\sqrt{\frac{1}{5}} = \pm\frac{\sqrt{5}}{5}.$$

Since $P(t)$ lies in quadrant IV, $x > 0$ and $y < 0$, so we have $x = \frac{\sqrt{5}}{5}$, $y = -\frac{2\sqrt{5}}{5}$. Then

$$\cos t = \frac{\sqrt{5}}{5}, \quad \sin t = -\frac{2\sqrt{5}}{5}, \quad \tan t = \frac{-2\sqrt{5}/5}{\sqrt{5}/5} = -2,$$

$$\cot t = -\frac{1}{2}, \quad \sec t = \frac{5}{\sqrt{5}} = \sqrt{5}, \quad \csc t = -\frac{5}{2\sqrt{5}} = -\frac{\sqrt{5}}{2}.$$

44. To show the two triangles are congruent, show all three corresponding angles are equal. Angles ODC and OAB are both right angles and so are equal. Let α be the angle from the positive y-axis to the ray OC and β the angle from the ray OC to the negative x-axis. Then

$$t+\frac{\pi}{2}+\beta = \pi \text{ implies } \beta = \frac{\pi}{2}-t, \quad \alpha+\beta = \frac{\pi}{2} \text{ implies } \beta = \frac{\pi}{2}-\alpha, \text{ and } \frac{\pi}{2}-t = \frac{\pi}{2}-\alpha \text{ implies } t = \alpha,$$

so angles OCD and AOB are equal, and DOC and AOB are congruent triangles. Hence, $\overline{AB} = \overline{OD}$ and $\overline{DC} = \overline{OA}$, and since the coordinates of point B are $(\cos t, \sin t)$, $\overline{AB} = \sin t = \overline{OD}$, $\overline{DC} = \cos t = \overline{OA}$. This gives that the coordinates of point C are $(-\sin t, \cos t)$, so

a. $\sin\left(t + \frac{\pi}{2}\right) = \cos t$

b. $\cos\left(t + \frac{\pi}{2}\right) = -\sin t$

c. $\tan\left(t+\frac{\pi}{2}\right)=\frac{\sin(t+\frac{\pi}{2})}{\cos(t+\frac{\pi}{2})}=\frac{\cos t}{-\sin t}=-\cot t.$

Exercise Set 4.6 (Page 231)

1. $\sin\left(\frac{\pi}{3}-\frac{5\pi}{4}\right)=\sin\frac{\pi}{3}\cos\frac{5\pi}{4}-\cos\frac{\pi}{3}\sin\frac{5\pi}{4}=\frac{\sqrt{3}}{2}\left(-\frac{\sqrt{2}}{2}\right)-\frac{1}{2}\left(-\frac{\sqrt{2}}{2}\right)=-\frac{\sqrt{2}}{4}\left(\sqrt{3}-1\right)=\frac{\sqrt{2}}{4}\left(1-\sqrt{3}\right)$

2. $\cos\left(\frac{5\pi}{6}+\frac{\pi}{4}\right)=\cos\frac{5\pi}{6}\cos\frac{\pi}{4}-\sin\frac{5\pi}{6}\sin\frac{\pi}{4}=-\frac{\sqrt{3}}{2}\left(\frac{\sqrt{2}}{2}\right)-\frac{1}{2}\left(\frac{\sqrt{2}}{2}\right)=-\frac{\sqrt{2}}{4}\left(\sqrt{3}+1\right)$

3. $\cos\left(\frac{7\pi}{12}\right)=\cos\left(\frac{\pi}{3}+\frac{\pi}{4}\right)=\cos\frac{\pi}{3}\cos\frac{\pi}{4}-\sin\frac{\pi}{3}\sin\frac{\pi}{4}=\frac{1}{2}\left(\frac{\sqrt{2}}{2}\right)-\frac{\sqrt{3}}{2}\left(\frac{\sqrt{2}}{2}\right)=\frac{\sqrt{2}}{4}\left(1-\sqrt{3}\right)$

4. $\sin\left(-\frac{\pi}{12}\right)=\sin\left(\frac{\pi}{4}-\frac{\pi}{3}\right)=\sin\frac{\pi}{4}\cos\frac{\pi}{3}-\cos\frac{\pi}{4}\sin\frac{\pi}{3}=\frac{\sqrt{2}}{2}\left(\frac{1}{2}\right)-\frac{\sqrt{2}}{2}\left(\frac{\sqrt{3}}{2}\right)=\frac{\sqrt{2}}{4}\left(1-\sqrt{3}\right)$

5. $\tan\left(\frac{\pi}{12}\right)=\tan\left(\frac{\pi}{3}-\frac{\pi}{4}\right)=\frac{\tan(\frac{\pi}{3})-\tan(\frac{\pi}{4})}{1+\tan(\frac{\pi}{3})\tan(\frac{\pi}{4})}=\frac{\sqrt{3}-1}{1+\sqrt{3}(1)}=\frac{\sqrt{3}-1}{\sqrt{3}+1}$

6. $\cot\left(-\frac{\pi}{12}\right)=\cot\left(\frac{\pi}{4}-\frac{\pi}{3}\right)=\frac{1}{\tan(\frac{\pi}{4}-\frac{\pi}{3})}=\frac{1+\tan(\frac{\pi}{4})\tan(\frac{\pi}{3})}{\tan(\frac{\pi}{4})-\tan(\frac{\pi}{3})}=\frac{1+(1)\sqrt{3}}{1-\sqrt{3}}=\frac{\sqrt{3}+1}{1-\sqrt{3}}$

7. Since $\left(\sin\frac{7\pi}{12}\right)^2=\frac{1}{2}\left(1-\cos\frac{7\pi}{6}\right)=\frac{1}{2}\left(1-\left(-\frac{\sqrt{3}}{2}\right)\right)=\frac{1}{2}\left(1+\frac{\sqrt{3}}{2}\right)$ and since $\frac{\pi}{2}<\frac{7\pi}{12}<\pi$, the sine is positive, so

$$\sin\frac{7\pi}{12}=\sqrt{\frac{1}{2}\left(1+\frac{\sqrt{3}}{2}\right)}=\sqrt{\frac{2+\sqrt{3}}{4}}=\frac{\sqrt{2+\sqrt{3}}}{2}.$$

8. Since $\left(\cos\frac{\pi}{8}\right)^2=\frac{1}{2}\left(1+\cos\frac{\pi}{4}\right)=\frac{1}{2}\left(1+\frac{\sqrt{2}}{2}\right)$ and since $0<\frac{\pi}{8}<\frac{\pi}{2}$, the cosine is positive, so

$$\cos\frac{\pi}{8}=\sqrt{\frac{1}{2}\left(1+\frac{\sqrt{2}}{2}\right)}=\sqrt{\frac{2+\sqrt{2}}{4}}=\frac{\sqrt{2+\sqrt{2}}}{2}.$$

9. Since $\left(\cos\frac{3\pi}{8}\right)^2 = \frac{1}{2}\left(1+\cos\frac{3\pi}{4}\right) = \frac{1}{2}\left(1-\frac{\sqrt{2}}{2}\right)$ and since $0 < \frac{3\pi}{8} < \frac{\pi}{2}$, the cosine is positive, so

$$\cos\frac{3\pi}{8} = \sqrt{\frac{1}{2}\left(1-\frac{\sqrt{2}}{2}\right)} = \sqrt{\frac{2-\sqrt{2}}{4}} = \frac{\sqrt{2-\sqrt{2}}}{2}.$$

10. Since $\left(\sin\frac{5\pi}{12}\right)^2 = \frac{1}{2}\left(1-\cos\frac{5\pi}{6}\right) = \frac{1}{2}\left(1-\left(-\frac{\sqrt{3}}{2}\right)\right) = \frac{1}{2}\left(1+\frac{\sqrt{3}}{2}\right)$ and since $0 < \frac{5\pi}{12} < \frac{\pi}{2}$, the sine is positive, so

$$\sin\frac{5\pi}{12} = \sqrt{\frac{1}{2}\left(1+\frac{\sqrt{3}}{2}\right)} = \sqrt{\frac{2+\sqrt{3}}{4}} = \frac{\sqrt{2+\sqrt{3}}}{2}.$$

11. Since $\left(\sin\frac{13\pi}{12}\right)^2 = \frac{1}{2}\left(1-\cos\frac{13\pi}{6}\right) = \frac{1}{2}\left(1-\frac{\sqrt{3}}{2}\right)$ and since $\pi < \frac{13\pi}{12} < \frac{3\pi}{2}$, the sine is negative, so

$$\sin\frac{13\pi}{12} = -\sqrt{\frac{1}{2}\left(1-\frac{\sqrt{3}}{2}\right)} = -\sqrt{\frac{2-\sqrt{3}}{4}} = -\frac{\sqrt{2-\sqrt{3}}}{2}.$$

12. Since $\left(\cos\frac{11\pi}{8}\right)^2 = \frac{1}{2}\left(1+\cos\frac{11\pi}{4}\right) = \frac{1}{2}\left(1+\left(-\frac{\sqrt{2}}{2}\right)\right) = \frac{1}{2}\left(1-\frac{\sqrt{2}}{2}\right)$ and since $\pi < \frac{11\pi}{8} < \frac{3\pi}{2}$, the cosine is negative, so

$$\cos\frac{11\pi}{8} = -\sqrt{\frac{1}{2}\left(1-\frac{\sqrt{2}}{2}\right)} = -\sqrt{\frac{2-\sqrt{2}}{4}} = -\frac{\sqrt{2-\sqrt{2}}}{2}.$$

13. We are given $\cos t = \frac{3}{5}$, $0 < t < \frac{\pi}{2}$. Since $0 < t < \frac{\pi}{2}$, $\sin t > 0$.

a. $\cos 2t = 2(\cos t)^2 - 1 = 2\left(\frac{3}{5}\right)^2 - 1 = 2\left(\frac{9}{25}\right) - 1 = -\frac{7}{25}$

b. To use the formula $\sin 2t = 2\sin t\cos t$, first find $\sin t$. That is,

$$(\cos t)^2 + (\sin t)^2 = 1 \text{ implies } (\sin t)^2 = 1 - \left(\frac{3}{5}\right)^2 = \frac{16}{25} \text{ so } \sin t = \frac{4}{5}.$$

Then $\sin 2t = 2\sin t\cos t = 2\left(\frac{4}{5}\right)\left(\frac{3}{5}\right) = \frac{24}{25}$.

c. $\left(\cos \frac{t}{2}\right)^2 = \frac{1}{2}(1 + \cos t) = \frac{1}{2}\left(1 + \frac{3}{5}\right) = \frac{4}{5}$ so $\cos \frac{t}{2} = \pm\sqrt{\frac{4}{5}}$. Since $0 < t < \frac{\pi}{2}$ implies $0 < \frac{t}{2} < \frac{\pi}{4}$, we have $\cos \frac{t}{2} > 0$ so $\cos \frac{t}{2} = \sqrt{\frac{4}{5}} = \frac{2}{\sqrt{5}} = \frac{2\sqrt{5}}{5}$.

d. $\left(\sin \frac{t}{2}\right)^2 = \frac{1}{2}(1 - \cos t) = \frac{1}{2}\left(1 - \frac{3}{5}\right) = \frac{1}{5}$ so $\sin \frac{t}{2} = \pm\sqrt{\frac{1}{5}}$. Since $0 < t < \frac{\pi}{2}$ implies $0 < \frac{t}{2} < \frac{\pi}{4}$, we have $\sin \frac{t}{2} > 0$ so $\sin \frac{t}{2} = \sqrt{\frac{1}{5}} = \frac{1}{\sqrt{5}} = \frac{\sqrt{5}}{5}$.

14. We are given $\sin t = -\frac{3}{5}, \pi < t < \frac{3\pi}{2}$. Since $\pi < t < \frac{3\pi}{2}$, $\cos t < 0$.

a. $\cos 2t = 1 - 2(\sin t)^2 = 1 - 2\left(-\frac{3}{5}\right)^2 = 1 - 2\left(\frac{9}{25}\right) = \frac{7}{25}$

b. To use the formula $\sin 2t = 2 \sin t \cos t$, first find $\cos t$. That is,

$$(\cos t)^2 + (\sin t)^2 = 1 \text{ implies } (\cos t)^2 = 1 - \left(-\frac{3}{5}\right)^2 = \frac{16}{25} \text{ so } \cos t = -\frac{4}{5}.$$

Then $\sin 2t = 2 \sin t \cos t = 2\left(-\frac{3}{5}\right)\left(-\frac{4}{5}\right) = \frac{24}{25}$.

c. $\left(\cos \frac{t}{2}\right)^2 = \frac{1}{2}(1 + \cos t) = \frac{1}{2}\left(1 - \frac{4}{5}\right) = \frac{1}{10}$ so $\cos \frac{t}{2} = \pm\sqrt{\frac{1}{10}}$. Since $\pi < t < \frac{3\pi}{2}$ implies $\frac{\pi}{2} < \frac{t}{2} < \frac{3\pi}{4}$, we have $\cos \frac{t}{2} < 0$ so $\cos \frac{t}{2} = -\sqrt{\frac{1}{10}} = -\frac{1}{\sqrt{10}} = -\frac{\sqrt{10}}{10}$.

d. $\left(\sin \frac{t}{2}\right)^2 = \frac{1}{2}(1 - \cos t) = \frac{1}{2}\left(1 + \frac{4}{5}\right) = \frac{9}{10}$ so $\sin \frac{t}{2} = \pm\sqrt{\frac{9}{10}}$. Since $\pi < t < \frac{3\pi}{2}$ implies $\frac{\pi}{2} < \frac{t}{2} < \frac{3\pi}{4}$, we have $\sin \frac{t}{2} > 0$ so $\sin \frac{t}{2} = \sqrt{\frac{9}{10}} = \frac{3}{\sqrt{10}} = \frac{3\sqrt{10}}{10}$.

15. We are given $\tan t = \frac{5}{12}$, $\sin t < 0$. Since $\tan t > 0$ and $\sin t < 0$ this implies $\cos t < 0$. First determine $\sin t$ and $\cos t$. From the identity,

$$(\tan t)^2 + 1 = (\sec t)^2 = \frac{1}{(\cos t)^2} \text{ implies } \left(\frac{5}{12}\right)^2 + 1 = \frac{1}{(\cos t)^2},$$

and $(\cos t)^2 = \frac{144}{169}$ so $\cos t = -\frac{12}{13}$. Then

$$(\sin t)^2 = 1 - (\cos t)^2 = 1 - \frac{144}{169} = \frac{25}{169} \text{ so } \sin t = -\sqrt{\frac{25}{169}} = -\frac{5}{13}.$$

a.

$$\cos 2t = 2(\cos t)^2 - 1 = 2\left(\frac{144}{169}\right) - 1 = \frac{119}{169}$$

b.

$$\sin 2t = 2 \sin t \cos t = 2\left(-\frac{5}{13}\right)\left(-\frac{12}{13}\right) = \frac{120}{169}$$

c.

$$\left(\cos\frac{t}{2}\right)^2 = \frac{1}{2}(1+\cos t) = \frac{1}{2}\left(1-\frac{12}{13}\right) = \frac{1}{26} \text{ so } \cos\frac{t}{2} = \pm\sqrt{\frac{1}{26}}$$

Since $\cos t < 0$ and $\sin t < 0$, t is in quadrant III, so $\pi < t < \frac{3\pi}{2}$ implies $\frac{\pi}{2} < \frac{t}{2} < \frac{3\pi}{4}$, and we have

$$\cos\frac{t}{2} < 0 \text{ so } \cos\frac{t}{2} = -\frac{1}{\sqrt{26}} = -\frac{\sqrt{26}}{26}.$$

d.

$$\left(\sin\frac{t}{2}\right)^2 = \frac{1}{2}(1-\cos t) = \frac{1}{2}\left(1+\frac{12}{13}\right) = \frac{25}{26} \text{ so } \sin\frac{t}{2} = \pm\sqrt{\frac{25}{26}}$$

Since $\pi < t < \frac{3\pi}{2}$ implies $\frac{\pi}{2} < \frac{t}{2} < \frac{3\pi}{4}$, and we have

$$\sin\frac{t}{2} > 0 \text{ so } \sin\frac{t}{2} = \sqrt{\frac{25}{26}} = \frac{5\sqrt{26}}{26}.$$

16. We are given $\cot t = -\frac{24}{7}$, $\cos t > 0$. Since $\cot t < 0$ and $\cos t > 0$ this implies $\sin t < 0$. First determine $\sin t$ and $\cos t$. From the identity

$$(\cot t)^2 + 1 = (\csc t)^2 = \frac{1}{(\sin t)^2} \text{ implies } \left(-\frac{24}{7}\right)^2 + 1 = \frac{1}{(\sin t)^2},$$

and $(\sin t)^2 = \frac{49}{625}$ so $\sin t = -\frac{7}{25}$. Then

$$(\cos t)^2 = 1-(\sin t)^2 = 1-\frac{49}{625} = \frac{576}{625} \text{ so } \cos t = \sqrt{\frac{576}{625}} = \frac{24}{25}.$$

a.

$$\cos 2t = 2(\cos t)^2 - 1 = 2\left(\frac{576}{625}\right) - 1 = \frac{527}{625}$$

b.

$$\sin 2t = 2\sin t\cos t = 2\left(-\frac{7}{25}\right)\left(\frac{24}{25}\right) = -\frac{336}{625}$$

c.

$$\left(\cos\frac{t}{2}\right)^2 = \frac{1}{2}(1+\cos t) = \frac{1}{2}\left(1+\frac{24}{25}\right) = \frac{49}{50} \text{ so } \cos\frac{t}{2} = \pm\sqrt{\frac{49}{50}}$$

Since $\cos t > 0$ and $\sin t < 0$, t is in quadrant IV, so $\frac{3\pi}{2} < t < 2\pi$ implies $\frac{3\pi}{4} < \frac{t}{2} < \pi$, and we have

$$\cos\frac{t}{2} < 0 \text{ so } \cos\frac{t}{2} = -\frac{7}{5\sqrt{2}} = -\frac{7\sqrt{2}}{10}.$$

d.

$$\left(\sin\frac{t}{2}\right)^2 = \frac{1}{2}(1-\cos t) = \frac{1}{2}\left(1-\frac{24}{25}\right) = \frac{1}{50} \text{ so } \sin\frac{t}{2} = \pm\sqrt{\frac{1}{50}}$$

Since $\frac{3\pi}{2} < t < 2\pi$ implies $\frac{3\pi}{4} < \frac{t}{2} < \pi$, and we have

$$\sin\frac{t}{2} > 0 \text{ so } \sin\frac{t}{2} = \sqrt{\frac{1}{50}} = \frac{1}{5\sqrt{2}} = \frac{\sqrt{2}}{10}.$$

17. $\sin(t+\pi) = \sin t\cos\pi + \cos t\sin\pi = (\sin t)(-1) + (\cos t)(0) = -\sin t$

18. $\cos(t+\pi) = \cos t\cos\pi - \sin t\sin\pi = (\cos t)(-1) - (\sin t)(0) = -\cos t$

19. $\sin\left(t+\frac{\pi}{2}\right) = \sin t\cos\frac{\pi}{2} + \cos t\sin\frac{\pi}{2} = (\sin t)(0) + (\cos t)(1) = \cos t$

20. $\cos\left(t+\frac{\pi}{2}\right) = \cos t\cos\frac{\pi}{2} - \sin t\sin\frac{\pi}{2} = (\cos t)(0) - (\sin t)(1) = -\sin t$

21. $\sin(\pi-t) = \sin\pi\cos t - \cos\pi\sin t = (0)(\cos t) - (-1)(\sin t) = \sin t$

22. $\cos(\pi-t) = \cos\pi\cos t + \sin\pi\sin t = (-1)(\cos t) + (0)(\sin t) = -\cos t$

23. $(\cos 2x)^2 = \frac{1+\cos 4x}{2} = \frac{1}{2} + \frac{1}{2}\cos 4x$

24. $(\sin 3x)^2 = \frac{1-\cos 6x}{2} = \frac{1}{2} - \frac{1}{2}\cos 6x$

25.

$$\begin{aligned}(\sin x)^4 &= ((\sin x)^2)^2 = \left(\frac{1-\cos 2x}{2}\right)^2 = \frac{1}{4}\left(1-2\cos 2x+(\cos 2x)^2\right)\\ &= \frac{1}{4}\left(1-2\cos 2x+\frac{1+\cos 4x}{2}\right) = \frac{3}{8}-\frac{1}{2}\cos 2x+\frac{1}{8}\cos 4x\end{aligned}$$

26.
$$\begin{aligned}(\sin 3x)^2(\cos 3x)^2 &= \left(\frac{1-\cos 6x}{2}\right)\left(\frac{1+\cos 6x}{2}\right) = \frac{1}{4}(1-(\cos 6x)^2)\\ &= \frac{1}{4}\left(1-\frac{1+\cos 12x}{2}\right) = \frac{1}{8}-\frac{1}{8}\cos 12x\end{aligned}$$

27. $\sin 6t \cos 5t = \frac{1}{2}(\sin(6t+5t)+\sin(6t-5t)) = \frac{1}{2}(\sin 11t + \sin t)$

28.
$$\begin{aligned}\cos 5t \sin 8t &= \frac{1}{2}(\sin(5t+8t)-\sin(5t-8t)) = \frac{1}{2}(\sin 13t - \sin(-3t))\\ &= \frac{1}{2}(\sin 13t + \sin 3t)\end{aligned}$$

29. $\cos 2t \cos 3t = \frac{1}{2}(\cos(2t+3t)+\cos(2t-3t)) = \frac{1}{2}(\cos 5t + \cos(-t)) = \frac{1}{2}(\cos 5t + \cos t)$

30.
$$\begin{aligned}\sin 5t \sin 7t &= \frac{1}{2}(\cos(5t-7t)-\cos(5t+7t)) = \frac{1}{2}(\cos(-2t)-\cos 12t)\\ &= \frac{1}{2}(\cos 2t - \cos 12t)\end{aligned}$$

31. $(1-(\cos x)^2)(\sec x)^2 = (\sin x)^2\frac{1}{(\cos x)^2} = (\tan x)^2$

32. $\cot x + \tan x = \frac{\cos x}{\sin x} + \frac{\sin x}{\cos x} = \frac{(\cos x)^2+(\sin x)^2}{\sin x \cos x} = \frac{1}{\sin x \cos x} = \frac{1}{\sin x}\frac{1}{\cos x} = \sec x \csc x$

33. $\cot x - \tan x = \frac{\cos x}{\sin x} - \frac{\sin x}{\cos x} = \frac{(\cos x)^2-(\sin x)^2}{\sin x \cos x} = \frac{\cos 2x}{\sin x \cos x} = \frac{2\cos 2x}{\sin 2x} = 2\cot 2x$

34. $(\sin x+\cos x)^2 = (\sin x)^2+2\sin x\cos x+(\cos x)^2 = (\sin x)^2+(\cos x)^2+2\sin x \cos x = 1+\sin 2x$

35. $\sin x \sin 2x + \cos x \cos 2x = \cos(2x-x) = \cos x$

36. $(\tan x)^2 - (\sin x)^2 = \frac{(\sin x)^2}{(\cos x)^2} - (\sin x)^2 = \frac{(\sin x)^2-(\sin x)^2(\cos x)^2}{(\cos x)^2} = \frac{(\sin x)^2(1-(\cos x)^2)}{(\cos x)^2} =$ $\frac{(\sin x)^2}{(\cos x)^2}(\sin x)^2 = (\tan x)^2(\sin x)^2$

37. $\sec x - \cos x = \frac{1}{\cos x} - \cos x = \frac{1-(\cos x)^2}{\cos x} = \frac{(\sin x)^2}{\cos x} = \sin x\frac{\sin x}{\cos x} = \sin x \tan x$

38. $\cos x(\cot x + \tan x) = \cos x\left(\frac{\cos x}{\sin x} + \frac{\sin x}{\cos x}\right) = \cos x\left(\frac{(\cos x)^2+(\sin x)^2}{\sin x \cos x}\right) = \cos x\left(\frac{1}{\sin x \cos x}\right) = \frac{1}{\sin x} = \csc x$

39. $\sin 2x = \sin x$ so $\sin 2x - \sin x = 2\sin x \cos x - \sin x = \sin x(2\cos x - 1) = 0$ implies $\sin x = 0$ or $\cos x = \frac{1}{2}$ so $x = 0, \pi, 2\pi, \frac{\pi}{3}, \frac{5\pi}{3}$

40.

$\sin 2x = \cos x$ so $\sin 2x - \cos x = 2\sin x \cos x - \cos x = \cos x(2\sin x - 1) = 0$ implies $\cos x = 0$ or $\sin x = \frac{1}{2}$ so $x = \frac{\pi}{2}, \frac{3\pi}{2}, \frac{\pi}{6}, \frac{5\pi}{6}$

41. $2(\sin x)^2 + \cos x - 1 = 2(1 - (\cos x)^2) + \cos x - 1 = -2(\cos x)^2 + \cos x + 1 =$ $2(\cos x)^2 - \cos x - 1 = (2\cos x + 1)(\cos x - 1) = 0$ implies $\cos x = -\frac{1}{2}$ or $\cos x = 1$ so

$x = \frac{2\pi}{3}, \frac{4\pi}{3}, 0, 2\pi$

42. We have $(\cos x)^2 - 3\sin x - 3 = (1 - (\sin x)^2) - 3\sin x - 3 = -(\sin x)^2 - 3\sin x - 2 =$ $(\sin x)^2 + 3\sin x + 2 = (\sin x + 1)(\sin x + 2) = 0$ implies $\sin x = -1$ or $\sin x = -2$.

Since $\sin x = -2$ does not exist, the only correct solution is $\sin x = -1$ so $x = \frac{3\pi}{2}$.

43. We have $\tan x + \cot x = \frac{2}{\sin 2x}$ implies $\frac{\sin x}{\cos x} + \frac{\cos x}{\sin x} = \frac{(\cos x)^2+(\sin x)^2}{\cos x \sin x} = \frac{2}{\sin 2x}$ so $\sin 2x = 2\sin x \cos x$. So the equation is an identity and hence holds for all applicable x.

44. $2(\cot x)^2 + (\csc x)^2 - 2 = 2((\csc x)^2 - 1) + (\csc x)^2 - 2 = 3(\csc x)^2 - 4 = 0$ implies $(\csc x)^2 = \frac{4}{3}$ so $(\sin x)^2 = \frac{3}{4}$ and $\sin x = \pm\frac{\sqrt{3}}{2}$ so $x = \frac{\pi}{3}, \frac{2\pi}{3}, \frac{4\pi}{3}, \frac{5\pi}{3}$

45. $2\sin\frac{x+y}{2}\cos\frac{x-y}{2} = 2\cdot\frac{1}{2}\left(\sin\left(\frac{x+y}{2} - \frac{x-y}{2}\right) + \sin\left(\frac{x+y}{2} + \frac{x-y}{2}\right)\right) = \sin y + \sin x$

46. $2\cos\frac{x+y}{2}\sin\frac{x-y}{2} = 2\cdot\frac{1}{2}\left(\sin\left(\frac{x+y}{2} + \frac{x-y}{2}\right) - \sin\left(\frac{x+y}{2} - \frac{x-y}{2}\right)\right) = \sin x - \sin y$

47. $2\cos\frac{x+y}{2}\cos\frac{x-y}{2} = 2\cdot\frac{1}{2}\left(\cos\left(\frac{x+y}{2} + \frac{x-y}{2}\right) + \cos\left(\frac{x+y}{2} - \frac{x-y}{2}\right)\right) = \cos x + \cos y$

48. $-2\sin\frac{x+y}{2}\sin\frac{x-y}{2} = -2\cdot\frac{1}{2}\left(\cos\left(\frac{x+y}{2} - \frac{x-y}{2}\right) - \cos\left(\frac{x+y}{2} + \frac{x-y}{2}\right)\right) = -(\cos y - \cos x) = \cos x - \cos y$

49. It is an identity since the graphs of the two functions coincide.

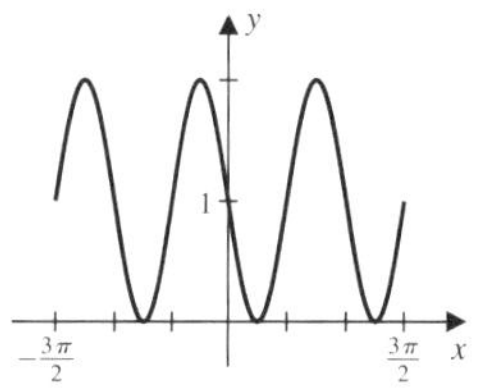

50. It is an identity.

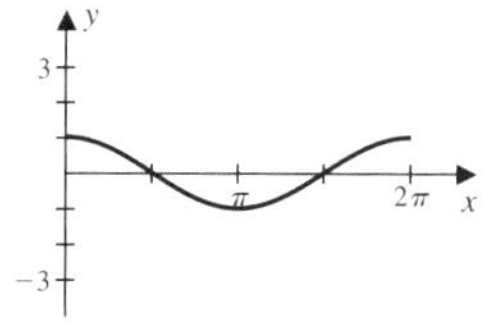

51. It is an identity.

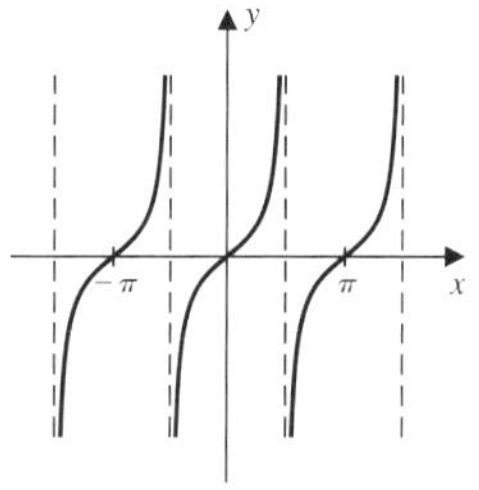

52. It is an identity.

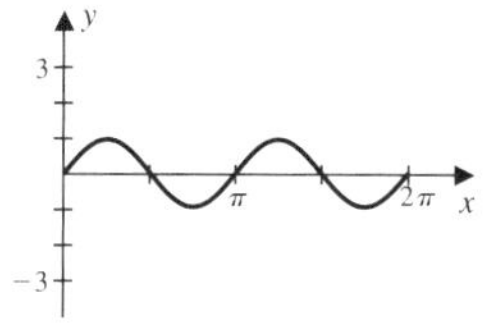

53. It is not an identity since the graphs of the two functions do not coincide.

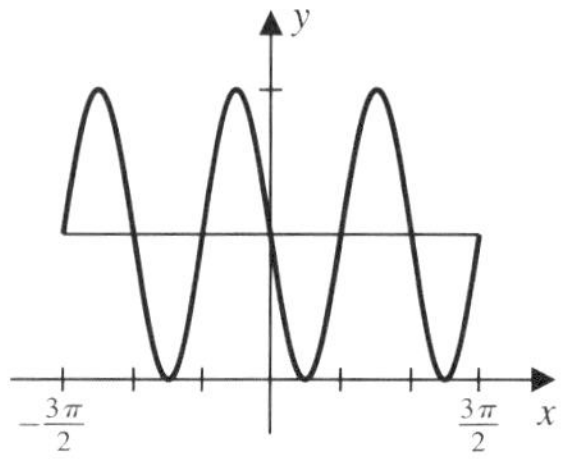

54. It is not an identity since the graphs of the two functions do not coincide.

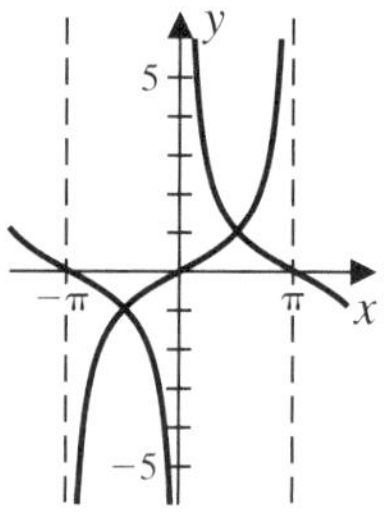

Exercise Set 4.7 (Page 237)

1. The hypotenuse is $\sqrt{4^2+3^2} = \sqrt{25} = 5$, so

$$\cos\theta = \frac{4}{5},\ \sin\theta = \frac{3}{5},\ \tan\theta = \frac{3}{4}, \cot\theta = \frac{4}{3},\ \sec\theta = \frac{5}{4},\ \csc\theta = \frac{5}{3}.$$

2. Let x be the missing side. Then $x^2 + 6 = 9$ so $x = \sqrt{3}$, and

$$\cos\theta = \frac{\sqrt{6}}{3}, \quad \sin\theta = \frac{\sqrt{3}}{3}, \quad \tan\theta = \frac{\sqrt{3}}{\sqrt{6}} = \frac{\sqrt{2}}{2},$$

$$\cot\theta = \frac{2}{\sqrt{2}} = \sqrt{2}, \quad \sec\theta = \frac{3}{\sqrt{6}} = \frac{\sqrt{6}}{2}, \quad \csc\theta = \frac{3}{\sqrt{3}} = \sqrt{3}.$$

3. The hypotenuse is $\sqrt{3+1} = 2$, so

$$\cos\theta = \frac{\sqrt{3}}{2}, \ \sin\theta = \frac{1}{2}, \ \tan\theta = \frac{\sqrt{3}}{3},$$

$$\cot\theta = \frac{3}{\sqrt{3}} = \sqrt{3}, \ \sec\theta = \frac{2}{\sqrt{3}} = \frac{2\sqrt{3}}{3}, \ \csc\theta = 2.$$

4. Let x be the missing side. Then $x^2 + 1 = 2$ so $x = 1$, and

$$\cos\theta = \frac{1}{\sqrt{2}} = \frac{\sqrt{2}}{2}, \quad \sin\theta = \frac{1}{\sqrt{2}} = \frac{\sqrt{2}}{2}, \quad \tan\theta = 1,$$

$$\cot\theta = 1, \quad \sec\theta = \sqrt{2}, \quad \csc\theta = \sqrt{2}.$$

5. Let x be the missing side. Then $x^2 + 25 = 169$ so $x = \sqrt{144} = 12$, and

$$\cos\theta = \frac{12}{13}, \quad \sin\theta = \frac{5}{13}, \quad \tan\theta = \frac{5}{12},$$

$$\cot\theta = \frac{12}{5}, \quad \sec\theta = \frac{13}{12}, \quad \csc\theta = \frac{13}{5}.$$

6. Let x be the missing side. Then $x^2 + 16 = 25$ so $x = \sqrt{9} = 3$, and

$$\cos\theta = \frac{3}{5}, \quad \sin\theta = \frac{4}{5}, \quad \tan\theta = \frac{4}{3},$$

$$\cot\theta = \frac{3}{4}, \quad \sec\theta = \frac{5}{3}, \quad \csc\theta = \frac{5}{4}.$$

7. Since $\cos 30° = \frac{x}{16}$, it implies that $x = 16\cos 30° = 16\frac{\sqrt{3}}{2} = 8\sqrt{3}$.

8. Since $\sin 45° = \frac{x}{25}$, it implies that $x = 25\sin 45° = 25\frac{\sqrt{2}}{2} = \frac{25\sqrt{2}}{2}$.

9. Since $\sin 45° = \frac{5}{x}$, it implies that $x = \frac{5}{\sin 45°} = \frac{5}{\sqrt{2}/2} = \frac{10}{\sqrt{2}} = 5\sqrt{2}$.

10. Since $\sin 60° = \frac{x}{8}$, it implies that $x = 8\sin 60° = 8\frac{\sqrt{3}}{2} = 4\sqrt{3}$.

11. Since $\tan 60° = \frac{6}{x}$, it implies that $x = \frac{6}{\tan 60°} = \frac{6}{\sqrt{3}} = \frac{6\sqrt{3}}{3} = 2\sqrt{3}$.

12. Since $\tan 30° = \frac{x}{7}$, it implies that $x = 7\tan 30° = 7\frac{\sqrt{3}}{3}$.

13. The angle $\beta = 60°$, and

$$\tan 30° = \frac{\overline{BC}}{\overline{AB}} = \frac{4}{\overline{AB}} \text{ so } \overline{AB} = \frac{4}{\sqrt{3}/3} = 4\sqrt{3}.$$

Also,

$$\sin 30° = \frac{\overline{BC}}{\overline{AC}} = \frac{4}{\overline{AC}} \text{ so } \overline{AC} = \frac{4}{1/2} = 8.$$

Note that once $\overline{AB}$ is computed $\overline{AC}$ can be found using the Pythagorean Theorem as well. That is, $\overline{AC} = \sqrt{16+48} = \sqrt{64} = 8$.

14. The angle $\beta = 30°$, and

$$\tan 60° = \frac{\overline{BC}}{\overline{AB}} = \frac{\overline{BC}}{12} \text{ so } \overline{BC} = 12\sqrt{3}.$$

Also,

$$\cos 60° = \frac{12}{\overline{AC}} \text{ so } \overline{AC} = \frac{12}{\cos 60°} = \frac{12}{1/2} = 24.$$

15. The angle $\alpha = 45°$, and

$$\cos 45° = \frac{\overline{AB}}{\overline{AC}} = \frac{\overline{AB}}{5} \text{ so } \overline{AB} = \frac{5\sqrt{2}}{2}.$$

Also,

$$\sin 45° = \frac{\overline{BC}}{5} \text{ so } \overline{BC} = \frac{5\sqrt{2}}{2}.$$

16. The angle $\alpha = 40^\circ$, and

$$\tan 50^\circ = \frac{\overline{AB}}{\overline{BC}} = \frac{\overline{AB}}{10} \text{ so } \overline{AB} = 10 \tan 50^\circ \approx 11.9.$$

Also,

$$\cos 50^\circ = \frac{\overline{BC}}{\overline{AC}} = \frac{10}{\overline{AC}} \text{ so } \overline{AC} = \frac{10}{\cos 50^\circ} \approx 15.6.$$

17. The angle $\beta = 90^\circ - 27.6^\circ = 62.4^\circ$, and

$$\tan 27.6^\circ = \frac{\overline{BC}}{\overline{AB}} = \frac{15.3}{\overline{AB}} \text{ so } \overline{AB} = \frac{15.3}{\tan 27.6^\circ} \approx 29.3.$$

Also,

$$\sin 27.6^\circ = \frac{\overline{BC}}{\overline{AC}} = \frac{15.3}{\overline{AC}} \text{ so } \overline{AC} = \frac{15.3}{\sin 27.6^\circ} \approx 33.$$

18. The angle $\alpha = 90^\circ - 52.6^\circ = 37.4^\circ$, and

$$\sin 52.6^\circ = \frac{\overline{AB}}{\overline{AC}} = \frac{\overline{AB}}{12.3} \text{ so } \overline{AB} = 12.3 \sin 52.6^\circ \approx 9.8.$$

Also,

$$\cos 52.6^\circ = \frac{\overline{BC}}{\overline{AC}} = \frac{\overline{BC}}{12.3} \text{ so } \overline{BC} = 12.3 \cos 52.6^\circ \approx 7.5.$$

19. a. The missing side is $\sqrt{9-4} = \sqrt{5}$, and

$$\cos\theta = \frac{\sqrt{5}}{3}, \tan\theta = \frac{2}{\sqrt{5}} = \frac{2\sqrt{5}}{5}, \cot\theta = \frac{\sqrt{5}}{2},$$

$$\sec\theta = \frac{3}{\sqrt{5}} = \frac{3\sqrt{5}}{5}, \csc\theta = \frac{3}{2}.$$

b. The missing side is $\sqrt{25-1} = \sqrt{24} = 2\sqrt{6}$, and

$$\sin\theta = \frac{2\sqrt{6}}{5}, \tan\theta = 2\sqrt{6}, \cot\theta = \frac{1}{2\sqrt{6}} = \frac{\sqrt{6}}{12},$$

$$\sec\theta = 5, \csc\theta = \frac{5}{2\sqrt{6}} = \frac{5\sqrt{6}}{12}.$$

c. The hypotenuse is $\sqrt{16+9}=5$, and

$$\sin\theta=\frac{4}{5},\ \cos\theta=\frac{3}{5},\ \cot\theta=\frac{3}{4},\ \sec\theta=\frac{5}{3},\ \csc\theta=\frac{5}{4}.$$

d. The missing side is $\sqrt{9-4}=\sqrt{5}$, and

$$\sin\theta=\frac{2}{3},\ \cos\theta=\frac{\sqrt{5}}{3},\ \tan\theta=\frac{2}{\sqrt{5}}=\frac{2\sqrt{5}}{5},$$

$$\cot\theta=\frac{\sqrt{5}}{2},\ \sec\theta=\frac{3}{\sqrt{5}}=\frac{3\sqrt{5}}{5}.$$

20. a. $\cos\theta=-\dfrac{\sqrt{5}}{3},\ \tan\theta=-\dfrac{2\sqrt{5}}{5},\ \cot\theta=-\dfrac{\sqrt{5}}{2},\ \sec\theta=-\dfrac{3\sqrt{5}}{5},\ \csc\theta=\frac{3}{2}.$

b. $\sin\theta=-\dfrac{2\sqrt{6}}{5},\ \tan\theta=-2\sqrt{6},\ \cot\theta=-\dfrac{\sqrt{6}}{12},\ \sec\theta=5,\ \csc\theta=-\dfrac{5\sqrt{6}}{12}.$

c. $\sin\theta=-\frac{4}{5},\ \cos\theta=-\frac{3}{5},\ \cot\theta=\frac{3}{4},\ \sec\theta=-\frac{5}{3},\ \csc\theta=-\frac{5}{4}.$

d. $\sin\theta=\frac{2}{3},\ \cos\theta=-\dfrac{\sqrt{5}}{3},\ \tan\theta=-\dfrac{2\sqrt{5}}{5},\ \cot\theta=-\dfrac{\sqrt{5}}{2},\ \sec\theta=-\dfrac{3\sqrt{5}}{5}$

21. $\sin\alpha=\frac{3}{5},\ \sin\beta=\frac{4}{5},\ \cos\alpha=\frac{4}{5},\ \cos\beta=\frac{3}{5}$

a. Since $\alpha+\beta=90^\circ$, $\sin(\alpha+\beta)=1$.

b. $\cos(\alpha+\beta)=0$

c. $\sin 2\alpha=2\sin\alpha\cos\alpha=2\frac{3}{5}\frac{4}{5}=\frac{24}{25}$

d. $\cos 2\alpha=(\cos\alpha)^2-(\sin\alpha)^2=\frac{16}{25}-\frac{9}{25}=\frac{7}{25}$

e. $\sin\left(\frac{\alpha}{2}\right)=\sqrt{\frac{1-\cos\alpha}{2}}=\sqrt{\frac{1-4/5}{2}}=\sqrt{\frac{1}{10}}=\frac{\sqrt{10}}{10}$

f. $\cos\left(\frac{\alpha}{2}\right)=\sqrt{\frac{1+\cos\alpha}{2}}=\sqrt{\frac{1+4/5}{2}}=\sqrt{\frac{9}{10}}=\frac{3}{\sqrt{10}}=\frac{3\sqrt{10}}{10}$

22. $\sin\alpha=\frac{\sqrt{8}}{3},\ \sin\beta=\frac{1}{3},\ \cos\alpha=\frac{1}{3},\ \cos\beta=\frac{\sqrt{8}}{3}$

a. Since $\alpha+\beta=90^\circ$, $\sin(\alpha+\beta)=1$.

b. $\cos(\alpha+\beta)=0$

c. $\sin 2\alpha=2\sin\alpha\cos\alpha=2\frac{\sqrt{8}}{3}\frac{1}{3}=\frac{2\sqrt{8}}{9}$

d. $\cos 2\alpha=(\cos\alpha)^2-(\sin\alpha)^2=\frac{1}{9}-\frac{8}{9}=-\frac{7}{9}$

e. $\sin\left(\frac{\alpha}{2}\right)=\sqrt{\frac{1-\cos\alpha}{2}}=\sqrt{\frac{1-1/3}{2}}=\sqrt{\frac{1}{3}}=\frac{1}{\sqrt{3}}=\frac{\sqrt{3}}{3}$

f. $\cos\left(\frac{\alpha}{2}\right)=\sqrt{\frac{1+\cos\alpha}{2}}=\sqrt{\frac{1+1/3}{2}}=\sqrt{\frac{2}{3}}=\frac{\sqrt{2}}{\sqrt{3}}=\frac{\sqrt{6}}{3}$

23. If h denotes the height of the building, then $\tan 12^\circ = \frac{h}{1}$ so $h = \tan 12^\circ \approx 0.21$ miles. In feet, $h = 5280 \tan 12^\circ \approx 1122.3$ feet.

24. Let ℓ denote the length of the box. The diagonal of the front (and back)of the box is $\sqrt{3^2 + 4^2} = 5$. Since $\sin\theta = \frac{5}{13}$, the diagonal of the box is 13. So $\ell^2 + 5^2 = 13^2$, which implies that $\ell = 12$.

25. If h denotes the height of the cliff, then

$$\tan 70^\circ = \frac{h}{40} \text{ so } h = 40 \tan 70^\circ \approx 109.9 \text{ feet.}$$

26. We have

$$\cos 42^\circ = \frac{AC}{AB} = \frac{300}{AB} \text{ so } AB = \frac{300}{\cos 42^\circ} \approx 404,$$

so 404 feet of pipe are needed.

27. Let x be the distance from the base of the tower to the campfire. Then

$$\cot 10.5^\circ = \frac{x}{80} \quad \text{so} \quad x = 80 \cot 10.5^\circ \approx 431.64,$$

so the campfire is approximately 431.64 feet from the base of the tower.

Exercise Set 4.8 (Page 246)

1. $y = \arccos\left(\frac{1}{2}\right)$ implies $\cos y = \frac{1}{2}$ with $0 \le y \le \pi$, so $y = \arccos\left(\frac{1}{2}\right) = \frac{\pi}{3}$.

2. $y = \arcsin\left(-\frac{\sqrt{3}}{2}\right)$ implies $\sin y = -\frac{\sqrt{3}}{2}$ with $-\frac{\pi}{2} \le y \le \frac{\pi}{2}$, so $y = \arcsin\left(-\frac{\sqrt{3}}{2}\right) = -\frac{\pi}{3}$.

3. $y = \arcsin(1)$ implies $\sin y = 1$ with $-\frac{\pi}{2} \le y \le \frac{\pi}{2}$, so $y = \arcsin(1) = \frac{\pi}{2}$.

4. $y = \arccos(-1)$ implies $\cos y = -1$ with $0 \le y \le \pi$, so $y = \arccos(-1) = \pi$.

5. $y = \arccos\left(-\frac{\sqrt{3}}{2}\right)$ implies $\cos y = -\frac{\sqrt{3}}{2}$ with $0 \le y \le \pi$, so $y = \arccos\left(-\frac{\sqrt{3}}{2}\right) = \frac{5\pi}{6}$.

6. $y = \arcsin\left(\frac{\sqrt{2}}{2}\right)$ implies $\sin y = \frac{\sqrt{2}}{2}$ with $-\frac{\pi}{2} \le y \le \frac{\pi}{2}$, so $y = \arcsin\left(\frac{\sqrt{2}}{2}\right) = \frac{\pi}{4}$.

7. $y = \arctan\left(\sqrt{3}\right)$ implies $\tan y = \sqrt{3}$ with $-\frac{\pi}{2} < y < \frac{\pi}{2}$, so $y = \arctan\left(\sqrt{3}\right) = \frac{\pi}{3}$.

8. $y = \arctan\left(\frac{\sqrt{3}}{3}\right)$ implies $\tan y = \frac{\sqrt{3}}{3}$ with $-\frac{\pi}{2} < y < \frac{\pi}{2}$, so $y = \arctan\left(\frac{\sqrt{3}}{3}\right) = \frac{\pi}{6}$.

9. $y = \arctan(-1)$ implies $\tan y = -1$ with $-\frac{\pi}{2} < y < \frac{\pi}{2}$, so $y = \arctan(-1) = -\frac{\pi}{4}$.

10. $y = \arctan(0)$ implies $\tan y = 0$ with $-\frac{\pi}{2} < y < \frac{\pi}{2}$, so $y = \arctan(0) = 0$.

11. $y = \operatorname{arcsec}\left(\sqrt{2}\right)$ implies $\sec y = \sqrt{2}$ implies $\cos y = \frac{1}{\sqrt{2}} = \frac{\sqrt{2}}{2}$ with $0 \le y < \frac{\pi}{2}$ or $\frac{\pi}{2} < y \le \pi$, so $y = \operatorname{arcsec}\left(\sqrt{2}\right) = \frac{\pi}{4}$.

12. $y = \operatorname{arcsec}\left(-\frac{2\sqrt{3}}{3}\right)$ implies $\sec y = -\frac{2\sqrt{3}}{3}$ implies $\cos y = -\frac{3}{2\sqrt{3}} = -\frac{\sqrt{3}}{2}$ with $0 \le y < \frac{\pi}{2}$ or $\frac{\pi}{2} < y \le \pi$, so $y = \operatorname{arcsec}\left(-\frac{2\sqrt{3}}{3}\right) = \frac{5\pi}{6}$.

13. $y = \arccos(2)$ implies $\cos y = 2$, which can never occur since $-1 \le \cos y \le 1$.

14. $y = \operatorname{arcsec}(0)$ implies $\sec y = \frac{1}{\cos y} = 0$, which can never occur.

15. Since $\cos(\arccos x) = x$ for x in $[-1, 1]$, we have $\cos\left(\arccos\left(\frac{1}{2}\right)\right) = \frac{1}{2}$.

16. Since $\sin(\arcsin x) = x$ for x in $[-1, 1]$, we have $\sin\left(\arcsin\left(\frac{\sqrt{3}}{2}\right)\right) = \frac{\sqrt{3}}{2}$.

17. The $\arccos\left(\frac{1}{2}\right)$ is the angle in $[0, \pi]$ whose cosine is $\frac{1}{2}$, so $\arccos\left(\frac{1}{2}\right) = \frac{\pi}{3}$ and $\sin\left(\arccos\left(\frac{1}{2}\right)\right) = \sin\left(\frac{\pi}{3}\right) = \frac{\sqrt{3}}{2}$.

18. The $\arcsin\left(\frac{\sqrt{2}}{2}\right)$ is the angle in $\left[-\frac{\pi}{2}, \frac{\pi}{2}\right]$ whose sine is $\frac{\sqrt{2}}{2}$, so $\arcsin\left(\frac{\sqrt{2}}{2}\right) = \frac{\pi}{4}$ and $\cos\left(\arcsin\left(\frac{\sqrt{2}}{2}\right)\right) = \cos\left(\frac{\pi}{4}\right) = \frac{\sqrt{2}}{2}$.

19. The $\arcsin\left(\frac{\sqrt{3}}{2}\right)$ is the angle in $\left[-\frac{\pi}{2}, \frac{\pi}{2}\right]$ whose sine is $\frac{\sqrt{3}}{2}$, so $\arcsin\left(\frac{\sqrt{3}}{2}\right) = \frac{\pi}{3}$ and $\tan\left(\arcsin\left(\frac{\sqrt{3}}{2}\right)\right) = \tan\left(\frac{\pi}{3}\right) = \sqrt{3}$.

20. The $\arcsin\left(-\frac{1}{2}\right)$ is the angle in $\left[-\frac{\pi}{2}, \frac{\pi}{2}\right]$ whose sine is $-\frac{1}{2}$, so $\arcsin\left(-\frac{1}{2}\right) = -\frac{\pi}{6}$ and $\tan\left(\arcsin\left(-\frac{1}{2}\right)\right) = \tan\left(-\frac{\pi}{6}\right) = -\frac{\sqrt{3}}{3}$.

21. Since $\arcsin(\sin x) = x$ for x in $\left[-\frac{\pi}{2}, \frac{\pi}{2}\right]$, we have $\arcsin\left(\sin\left(-\frac{\pi}{2}\right)\right) = -\frac{\pi}{2}$.

22. Since $\arccos(\cos x) = x$ for x in $[0, \pi]$, and $\frac{2\pi}{3}$ is in this interval $\arccos\left(\cos\left(\frac{2\pi}{3}\right)\right) = \frac{2\pi}{3}$.

23. The $\arccos(\cos x) = x$ for x in $[0, \pi]$, but since $-\frac{\pi}{4}$ does not lie in this interval, the identity can not be applied. However,

$$\cos\left(-\frac{\pi}{4}\right) = \cos\left(\frac{\pi}{4}\right) \text{ implies } \arccos\left(\cos\left(-\frac{\pi}{4}\right)\right) = \arccos\left(\cos\left(\frac{\pi}{4}\right)\right) = \frac{\pi}{4}.$$

24. The $\arcsin(\sin x) = x$ for x in $\left[-\frac{\pi}{2}, \frac{\pi}{2}\right]$, but since $x = \frac{3\pi}{4}$ does not lie in the proper interval, the identity can not be applied. However,

$$\sin\frac{3\pi}{4} = \frac{\sqrt{2}}{2} = \sin\frac{\pi}{4} \text{ implies } \arcsin\left(\sin\frac{3\pi}{4}\right) = \arcsin\left(\sin\frac{\pi}{4}\right) = \frac{\pi}{4}.$$

25. Since $\arctan(\tan x) = x$ for x in $\left(-\frac{\pi}{2}, \frac{\pi}{2}\right)$, and $\frac{\pi}{3}$ is in this interval, $\arctan\left(\tan\left(\frac{\pi}{3}\right)\right) = \frac{\pi}{3}$.

26. Since $\arctan(\tan x) = x$ for x in $\left(-\frac{\pi}{2}, \frac{\pi}{2}\right)$, and $-\frac{\pi}{4}$ is in this interval, $\arctan\left(\tan\left(-\frac{\pi}{4}\right)\right) = -\frac{\pi}{4}$.

27. The $\arctan(\tan x) = x$ for x in $\left(-\frac{\pi}{2}, \frac{\pi}{2}\right)$, but since $\frac{7\pi}{6}$ does not lie in this interval, the identity can not be applied. However,

$$\tan\left(\frac{7\pi}{6}\right) = \tan\left(\frac{\pi}{6}\right) \text{ implies } \arctan\left(\tan\left(\frac{7\pi}{6}\right)\right) = \arctan\left(\tan\left(\frac{\pi}{6}\right)\right) = \frac{\pi}{6}.$$

28. The arctan $(\tan x) = x$ for x in $\left(-\frac{\pi}{2}, \frac{\pi}{2}\right)$, but since $\frac{2\pi}{3}$ does not lie in this interval, the identity can not be applied. However,

$$\tan\left(\frac{2\pi}{3}\right) = \tan\left(-\frac{\pi}{3}\right) \text{ implies } \arctan\left(\tan\left(\frac{2\pi}{3}\right)\right) = \arctan\left(\tan\left(-\frac{\pi}{3}\right)\right) = -\frac{\pi}{3}.$$

29. Let $t = \arcsin\left(\frac{3}{5}\right)$ so $\sin t = \frac{3}{5}$. The triangle in the figure has $\sin t = \frac{3}{5}$ which implies the side adjacent t is $\sqrt{25-9} = 4$ so

$$\cos t = \cos\left(\arcsin\left(\frac{3}{5}\right)\right) = \frac{4}{5}.$$

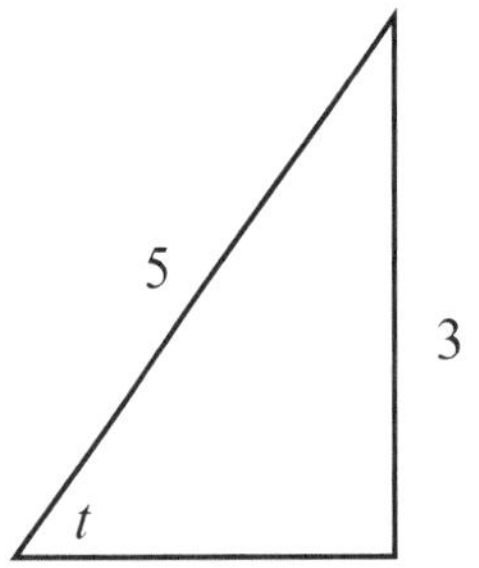

30. Let $t = \arccos\left(\frac{5}{13}\right)$ so $\cos t = \frac{5}{13}$. The triangle in the figure has $\cos t = \frac{5}{13}$ which implies the side opposite t is $\sqrt{169-25} = 12$ so

$$\sin t = \sin\left(\arccos\left(\frac{5}{13}\right)\right) = \frac{12}{13}.$$

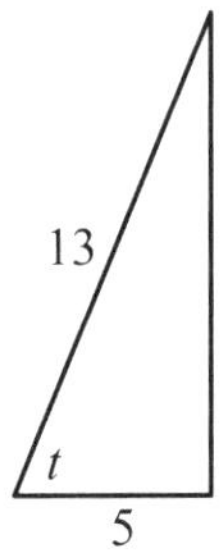

31. Let $t = \arccos\left(\frac{4}{5}\right)$ so $\cos t = \frac{4}{5}$. The triangle in the figure has $\cos t = \frac{4}{5}$ which implies the side opposite t is $\sqrt{25-16} = 3$ so

$$\tan t = \tan\left(\arccos\left(\frac{4}{5}\right)\right) = \frac{3}{4}.$$

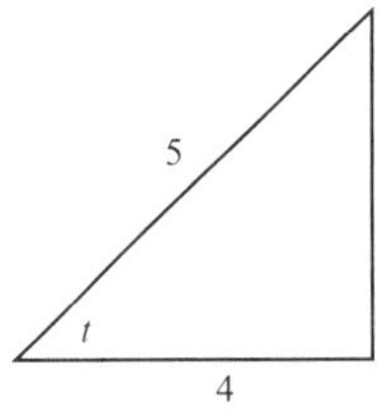

32. Let $t = \arcsin\left(\frac{12}{13}\right)$ so $\sin t = \frac{12}{13}$. The triangle in the figure has $\sin t = \frac{12}{13}$ which implies the side adjacent t is $\sqrt{169-144} = 5$ so

$$\tan t = \tan\left(\arcsin\left(\frac{12}{13}\right)\right) = \frac{12}{5}.$$

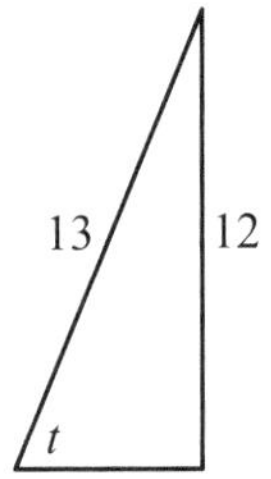

33. Let $t = \arctan 4$ so $\tan t = 4$. The triangle in the figure has $\tan t = 4$ which implies the hypotenuse is $\sqrt{16+1} = \sqrt{17}$ so

$$\cos t = \cos(\arctan 4) = \frac{1}{\sqrt{17}} = \frac{\sqrt{17}}{17}.$$

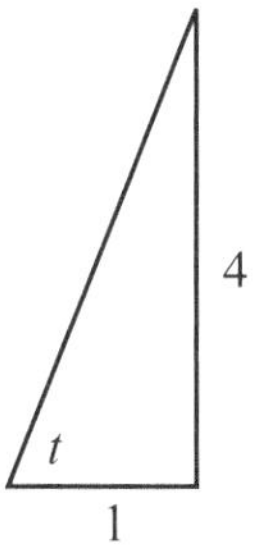

34. Let $t = \arctan 3$ so $\tan t = 3$. The triangle in the figure has $\tan t = 3$ which implies the hypotenuse is $\sqrt{9+1} = \sqrt{10}$ so

$$\sin t = \sin(\arctan 3) = \frac{3}{\sqrt{10}} = \frac{3\sqrt{10}}{10}.$$

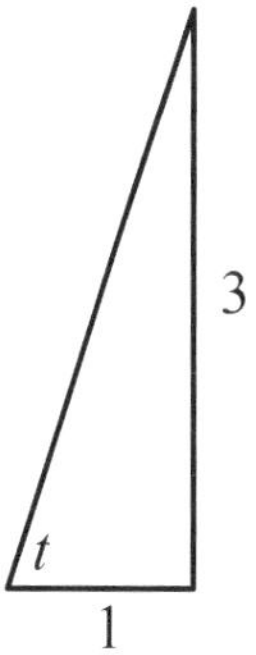

35. Apply the formula $\cos(a+b) = \cos a \cos b - \sin a \sin b$, where $a = \arcsin(3/5)$ and $b = \arccos(4/5)$. Then

$$\begin{aligned}\cos\left(\arcsin\left(\frac{3}{5}\right) + \arccos\left(\frac{4}{5}\right)\right) &= \cos\left(\arcsin\left(\frac{3}{5}\right)\right)\cos\left(\arccos\left(\frac{4}{5}\right)\right) \\ &\quad - \sin\left(\arcsin\left(\frac{3}{5}\right)\right)\sin\left(\arccos\left(\frac{4}{5}\right)\right) \\ &= \frac{4}{5}\cdot\frac{4}{5} - \frac{3}{5}\cdot\frac{3}{5} = \frac{7}{25}.\end{aligned}$$

36. Apply the formula

$$\tan(a+b) = \frac{\tan a + \tan b}{1 - \tan a \tan b},$$

where $a = \arcsin(1/3)$ and $b = \arccos(1/2)$. First

$$a = \arcsin(1/3) \text{ implies } \sin a = 1/3, \cos a = \frac{2\sqrt{2}}{3}, \tan a = \frac{1}{2\sqrt{2}} = \frac{\sqrt{2}}{4},$$

and

$$b = \arccos(1/2) \text{ implies } \cos b = 1/2, \sin b = \sqrt{3}/2, \tan b = \sqrt{3}.$$

So

$$\tan(\arcsin(1/3) + \arccos(1/2)) = \frac{\tan(\arcsin(1/3)) + \tan(\arccos(1/2))}{1 - \tan(\arcsin(1/3))\tan(\arccos(1/2))}$$

$$=\frac{\frac{\sqrt{2}}{4}+\sqrt{3}}{1-\frac{\sqrt{2}}{4}\sqrt{3}}=\frac{\frac{\sqrt{2}+4\sqrt{3}}{4}}{\frac{4-\sqrt{6}}{4}}=\frac{\sqrt{2}+4\sqrt{3}}{4-\sqrt{6}}.$$

37. Let $y=\arcsin(-x)$. Then

$$-x=\sin y \text{ implies } x=-\sin y=\sin(-y) \text{ so } -y=\arcsin x \text{ and } y=-\arcsin x.$$

38. Let $y=\arccos(-x)$. Then

$$-x=\cos y \text{ implies } x=-\cos y=\cos(\pi-y) \text{ so } \pi-y=\arccos x \text{ and } y=\pi-\arccos x.$$

39. Let $y=\arccos x$. Then $\cos y=x$. A right triangle with one angle y, adjacent side x, and hypotenuse 1 has opposite side $\sqrt{1-x^2}$ so $\cos y=x$, and $\sin y=\sqrt{1-x^2}$. So $y=\arcsin\left(\sqrt{1-x^2}\right)$.

40. Let $y=\arctan x$. Then $\tan y=x$. A right triangle with one angle y, opposite side x, and adjacent side 1 has hypotenuse $\sqrt{1+x^2}$ so $\tan y=x$, and $\sin y=\frac{x}{\sqrt{1+x^2}}$. So $y=\arcsin\left(\frac{x}{\sqrt{1+x^2}}\right)$.

41. Let $y=\arcsin x$. Then $\sin y=x$. A right triangle with one angle y, opposite side x, and hypotenuse 1 has adjacent side $\sqrt{1-x^2}$ so $\sin y=x$, and $\tan y=\frac{x}{\sqrt{1-x^2}}$. So $\tan(\arcsin x)=\frac{x}{\sqrt{1-x^2}}$.

42. By the Pythagorean Identity,

$$(\cos(\arcsin x))^2+(\sin(\arcsin x))^2=1,$$

and $\sin(\arcsin x)=x$, so

$$(\cos(\arcsin x))^2=1-x^2 \text{ so } \cos(\arcsin x)=\pm\sqrt{1-x^2}.$$

Since the range of the arcsine function is $\left[-\frac{\pi}{2},\frac{\pi}{2}\right]$, where the cosine is nonnegative, $\cos(\arcsin x))=\sqrt{1-x^2}$.

43. a. We have

$$(\tan x)^2-\tan x-2=(\tan x+1)(\tan x-2)=0,$$

which implies $\tan x=-1$ or $\tan x=2$, so $x=\arctan(-1)$, $x=\arctan(2)$.

b. We have $x=\arctan(-1)=-\frac{\pi}{4}\approx-0.785$ or $x=\arctan(2)\approx 1.107$.

44. **a.** We have

$$6(\cos x)^2 - \cos x - 5 = (6\cos x + 5)(\cos x - 1) = 0,$$

which implies $\cos x = -\frac{5}{6}$ or $\cos x = 1$, so $x = \arccos\left(-\frac{5}{6}\right)$, $x = \arccos(1)$. But on $\left[\frac{\pi}{2}, \pi\right]$, the cosine is never 1, so $x = \arccos(1)$ is not a solution.

b. We have $x = \arccos\left(-\frac{5}{6}\right) \approx 2.556$ or $x = \arccos(1) = 0$.

45. The illustration indicates that $\tan\theta = \frac{x}{4}$.

46. If γ is the angle subtended by the bottom of the picture and the horizontal line from the eye of the observer, then

$$\tan(\theta + \gamma) = \frac{a+b}{x}, \qquad \text{so } \theta + \gamma = \arctan\left(\frac{a+b}{x}\right).$$

But $\tan\gamma = \frac{b}{x}$, so

$$\gamma = \arctan\left(\frac{b}{x}\right) \quad \text{and} \quad \theta = \arctan\left(\frac{a+b}{x}\right) - \arctan\left(\frac{b}{x}\right).$$

Exercise Set 4.9 (Page 260)

1. Since $\alpha = 35°$, $b = 15$, $c = 25$, we have

$$a^2 = b^2 + c^2 - 2bc\cos\alpha = 15^2 + 25^2 - 2(15)(25)\cos 35° \text{ so } a = \sqrt{850 - 750\cos 35°} \approx 15.4,$$

and

$$b^2 = a^2 + c^2 - 2ac\cos\beta \text{ implies } \cos\beta = \frac{15^2 - (15.4)^2 - 25^2}{-2(15.4)(25)}.$$

So

$$\beta = \arccos\left(\frac{15^2 - (15.4)^2 - 25^2}{-2(15.4)(25)}\right) \approx 0.59619 = 0.59619\left(\frac{180}{\pi}\right)^{\circ} \approx 34.2°,$$

and $\gamma \approx 180 - 35 - 34.2 = 110.8°$.

2. Since $\gamma = 105°$, $a = 16$, $b = 22$ we have

$$c = \sqrt{16^2 + 22^2 - 2(16)(22)\cos 105°} \approx 30.4,$$

and

$$a^2 = b^2 + c^2 - 2bc\cos\alpha \text{ implies } \cos\alpha = \frac{16^2 - (22)^2 - (30.4)^2}{-2(22)(30.4)}.$$

So

$$\alpha = \arccos\left(\frac{16^2 - (22)^2 - (30.4)^2}{-2(22)(30.4)}\right) \approx 0.53285 = 0.53285\left(\frac{180}{\pi}\right)^\circ \approx 30.5°,$$

and $\beta \approx 180 - 105 - 30.5 = 44.5°$.

3. Since $\beta = 30°$, $a = 25$, $c = 32$ we have

$$b = \sqrt{25^2 + 32^2 - 2(25)(32)\cos 30°} \approx 16.2,$$

and

$$a^2 = b^2 + c^2 - 2bc\cos\alpha \text{ implies } \cos\alpha = \frac{25^2 - (16.2)^2 - (32)^2}{-2(16.2)(32)}.$$

So

$$\alpha = \arccos\left(\frac{25^2 - (16.2)^2 - (32)^2}{-2(16.2)(32)}\right) \approx 0.87895 = 0.87895\left(\frac{180}{\pi}\right)^\circ \approx 50.4°,$$

and $\gamma = 180 - 30 - 50.4 = 99.6°$.

4. Since $\alpha = 60°$, $b = 50$, $c = 35$ we have

$$a = \sqrt{50^2 + 35^2 - 2(50)(35)\cos 60°} \approx 44.4,$$

and

$$b^2 = a^2 + c^2 - 2ac\cos\beta \text{ implies } \cos\beta = \frac{50^2 - (44.4)^2 - 35^2}{-2(44.4)(35)}.$$

So

$$\beta = \arccos\left(\frac{50^2 - (44.4)^2 - 35^2}{-2(44.4)(35)}\right) \approx 1.3448 = 1.3448\left(\frac{180}{\pi}\right)^\circ \approx 77.1°$$

and $\gamma = 180 - 60 - 77.1 = 42.9°$.

5. Since $\alpha = 50°$, $\beta = 76°$, and $c = 100$, we have $\gamma = 180 - 50 - 76 = 54°$, and

$$\frac{\sin\alpha}{a} = \frac{\sin\gamma}{c} \text{ so } a = \frac{100\sin 50°}{\sin 54°} \approx 94.7; \quad \frac{\sin\beta}{b} = \frac{\sin\gamma}{c} \text{ so } b = \frac{100\sin 76°}{\sin 54°} \approx 119.9.$$

6. Since $\alpha = 60°$, $\beta = 56°$, and $a = 25$, we have $\gamma = 180 - 60 - 56 = 64°$, and

$$\frac{\sin\alpha}{a} = \frac{\sin\beta}{b} \text{ so } b = \frac{25\sin 56°}{\sin 60°} \approx 23.9; \quad \frac{\sin\alpha}{a} = \frac{\sin\gamma}{c} \text{ so } c = \frac{25\sin 64°}{\sin 60°} \approx 25.9.$$

7. Since $\beta = 100°$, $\gamma = 30°$, and $c = 20$, we have $\alpha = 180 - 100 - 30 = 50°$, and

$$\frac{\sin\alpha}{a} = \frac{\sin\gamma}{c} \text{ so } a = \frac{20\sin 50°}{\sin 30°} \approx 30.6; \quad \frac{\sin\beta}{b} = \frac{\sin\gamma}{c} \text{ so } b = \frac{20\sin 100°}{\sin 30°} \approx 39.4.$$

8. Since $\alpha = 65°$, $\gamma = 50°$, and $b = 10$, we have $\beta = 180 - 65 - 50 = 65°$, and

$$\frac{\sin\alpha}{a} = \frac{\sin\beta}{b} \text{ so } a = \frac{10\sin 65°}{\sin 65°} = 10; \quad \frac{\sin\beta}{b} = \frac{\sin\gamma}{c} \text{ so } c = \frac{10\sin 50°}{\sin 65°} \approx 8.5.$$

9. Since $\alpha = 120°$, $b = 3$, and $c = 5$, using the Law of Cosines we have

$$a^2 = b^2 + c^2 - 2bc\cos\alpha = 3^2 + 5^2 - 2(3)(5)\cos 120° \text{ so}$$
$$a = \sqrt{34 - 30\cos 120°} = 7.$$

Then by the Law of Sines,

$$\frac{\sin\alpha}{a} = \frac{\sin\beta}{b} \text{ so } \beta = \arcsin\left(\frac{3\sin 120°}{7}\right) \approx 21.8°,$$

and $\gamma = 180 - 120 - 21.8 = 38.2°$.

10. Since $\beta = 48.5°$, $a = 10.5$, and $c = 16$, using the Law of Cosines we have

$$b^2 = a^2 + c^2 - 2ac\cos\beta = (10.5)^2 + 16^2 - 2(10.5)(16)\cos 48.5° \text{ so}$$
$$b = \sqrt{366.25 - 336\cos 48.5°} \approx 12.$$

Then by the Law of Sines,

$$\frac{\sin\alpha}{a} = \frac{\sin\beta}{b} \text{ so } \alpha = \arcsin\left(\frac{10.5\sin 48.5°}{12}\right) \approx 41°,$$

and $\gamma = 180 - 48.5 - 41 = 90.5°$.

11. Since $a = 8$, $b = 9$, and $c = 13$, by the Law of Cosines we have

$$\cos\alpha = \frac{8^2 - (9)^2 - (13)^2}{-2(9)(13)} \approx 0.795,$$

so

$$\alpha = \arccos(0.795) \approx 0.65 = 0.65\left(\frac{180}{\pi}\right)^\circ \approx 37.4^\circ.$$

By the Law of Cosines,

$$\gamma = \arccos\left(\frac{13^2 - 9^2 - 8^2}{-2(8)(9)}\right) \approx 1.738\left(\frac{180}{\pi}\right) \approx 99.50^\circ,$$

so $\beta = 180 - 37.4 - 99.5 = 43.1^\circ$.

12. Since $a = 3$, $b = 5$, and $c = 7$, by the Law of Cosines we have

$$\cos\alpha = \frac{3^2 - 5^2 - 7^2}{-2(5)(7)} \text{ so } \alpha = \arccos\left(\frac{3^2 - 5^2 - 7^2}{-2(5)(7)}\right) \approx 0.38025$$
$$= 0.38025\left(\frac{180}{\pi}\right)^\circ \approx 21.8^\circ.$$

By the Law of Sines,

$$\frac{\sin\alpha}{a} = \frac{\sin\beta}{b} \text{ so } \beta = \arcsin\left(\frac{5\sin 21.8^\circ}{3}\right) \approx 38.2^\circ,$$

and $\gamma = 180 - 21.8 - 38.2 = 120^\circ$.

13. In order for a triangle to satisfy the conditions $a = 3$, $b = 10$, $\alpha = 25.4^\circ$, by the Law of Sines,

$$\frac{\sin\alpha}{a} = \frac{\sin\beta}{b} \text{ so } \sin\beta = \frac{b\sin\alpha}{a} = \frac{10\sin 25.4^\circ}{3} \approx 1.4.$$

Since the sine of an angle is never greater than 1, no such triangle can exist.

14. In order for a triangle to satisfy the conditions, $a = 3$, $b = 10$, $\alpha = 54^\circ$, by the Law of Sines,

$$\frac{\sin\alpha}{a} = \frac{\sin\beta}{b} \text{ so } \sin\beta = \frac{b\sin\alpha}{a} = \frac{10\sin 54^\circ}{3} \approx 2.7\,.$$

Since the sine of an angle is never greater than 1, no such triangle can exist.

15. Since $a = 125, b = 150$, and $\alpha = 55°$, by the Law of Sines,

$$\frac{\sin\alpha}{a} = \frac{\sin\beta}{b} \text{ so } \sin\beta = \frac{b\sin\alpha}{a} = \frac{150\sin 55°}{125}.$$

There are two angles between 0° and 180° that satisfy this condition. One such angle is

$$\beta = \arcsin\left(\frac{150\sin 55°}{125}\right) \approx 1.386 = 1.386\left(\frac{180}{\pi}\right)^{\circ} \approx 79.4°.$$

A second angle is $\beta_1 = 180 - 79.4 = 100.6°$.

Using β :

$$\gamma = 180 - 55 - 79.4 = 45.6° \text{ and } \frac{\sin\alpha}{a} = \frac{\sin\gamma}{c} \text{ so } c = \frac{125\sin 45.6°}{\sin 55°} \approx 109.$$

Using β_1 :

$$\gamma = 180 - 55 - 100.6 = 24.4° \text{ and } \frac{\sin\alpha}{a} = \frac{\sin\gamma}{c} \text{ so } c = \frac{125\sin 24.4°}{\sin 55°} \approx 63.$$

16. Since $a = 4, b = 5$, and $\alpha = 53°$, by the Law of Sines,

$$\frac{\sin\alpha}{a} = \frac{\sin\beta}{b} \text{ so } \sin\beta = \frac{b\sin\alpha}{a} = \frac{5\sin 53°}{4}.$$

There are two angles between 0° and 180° that satisfy this condition. One such angle is

$$\beta = \arcsin\left(\frac{5\sin 53°}{4}\right) \approx 1.5124 = 1.5124\left(\frac{180}{\pi}\right)^{\circ} \approx 86.7°.$$

A second angle is $\beta_1 = 180 - 86.7 = 93.3°$.

Using β :

$$\gamma = 180 - 53 - 86.7 = 40.3° \text{ and } \frac{\sin\alpha}{a} = \frac{\sin\gamma}{c} \text{ so } c = \frac{4\sin 40.3°}{\sin 53°} \approx 3.2.$$

Using β_1 :

$$\gamma = 180 - 53 - 93.3 = 33.7° \text{ and } \frac{\sin\alpha}{a} = \frac{\sin\gamma}{c} \text{ so } c = \frac{4\sin 33.7°}{\sin 53°} \approx 2.8.$$

17. By Heron's Formula, the area of the triangle with sides $a = 10, b = 14, c = 18$, is

$$A = \frac{1}{4}\sqrt{P(P - 2a)(P - 2b)(P - 2c)},$$

where

$$P = a + b + c = 10 + 14 + 18 = 42.$$

So

$$\begin{aligned} A =& \frac{1}{4}\sqrt{42(42-20)(42-28)(42-36)} = \frac{1}{4}\sqrt{42(22)(14)(6)} \\ =& \frac{1}{4}84\sqrt{11} = 21\sqrt{11} \approx 69.7 \text{ centimeters}^2. \end{aligned}$$

18. By Heron's Formula, the area of the triangle with sides $a = 300$, $b = 125$, $c = 275$, is

$$A = \frac{1}{4}\sqrt{P(P-2a)(P-2b)(P-2c)},$$

where

$$P = a + b + c = 300 + 125 + 275 = 700.$$

So

$$\begin{aligned} A =& \frac{1}{4}\sqrt{700(700-600)(700-250)(700-550)} = \frac{1}{4}\sqrt{700(100)(450)(150)} \\ =& 3750\sqrt{21} \approx 17184.7 \text{ cm}^2. \end{aligned}$$

19. Since point B is directly northwest of point C, the angle made by the line segment connecting C to B and the horizontal is 45°. Then angle ACB is $180 - 45 = 135°$. By the Law of Cosines

$$\overline{AB}^2 = 2^2 + 3^2 - 2(2)(3)\cos 135° \text{ so } \overline{AB} = \sqrt{13 - 12\cos 135°} \approx 4.6 \text{ miles.}$$

a. The approximate cost of construction directly between points A and B is

$$125000\sqrt{13 - 12\cos 135°} \approx \$579,400.00.$$

The approximate cost of construction from A to C, then C to B is $(3 + 2)(100000) =$ \$500, 000.00. So the engineers should select the route that avoids the swamp.

b. Let P denote the cost per mile for construction through C. If the total cost of construction from A to B to C is to equal the cost directly from A to B, then

$$5P = 579400 \text{ so } P = \frac{579400}{5} = 115880.$$

So if the cost per mile through C is approximately \$115, 800.00, then the cost of either alternative is about the same.

20. If d denotes the distance across the pond, by the Law of Cosines,

$$d^2 = 900^2 + 1200^2 - 2(900)(1200)\cos 110°,$$

and

$$d = \sqrt{900^2 + 1200^2 - 2(900)(1200)\cos 110°} \approx 1728.8\text{feet}.$$

21. The angle at point B is $180 - 105 - 42 = 33$, so by the Law of Sines

$$\frac{\sin 33°}{45} = \frac{\sin 42°}{\overline{AB}} \text{ so } \overline{AB} = \frac{45 \sin 42°}{\sin 33°} \approx 55.3 \text{ feet}.$$

22. A time t when the cutter and the boat meet is shown in the figure.
By the Law of Sines

$$\frac{\sin 48°}{25t} = \frac{\sin\theta}{15t} \text{ so } \sin\theta = \frac{15t}{25t}\sin 48° = \frac{3}{5}\sin 48°,$$

and

$$\theta = \arcsin\left(\frac{3}{5}\sin 48°\right) \approx 26.5°,$$

which is the bearing the cutter should travel. To find the time when the cutter intercepts the boat we have, $180 - 48 - 26.5 = 105.5°$, and

$$\frac{\sin 105.5°}{23} = \frac{\sin 48°}{25t} \text{ so } t = \frac{23 \sin 48°}{25 \sin 105.5°} \approx 0.71 \text{ hours}.$$

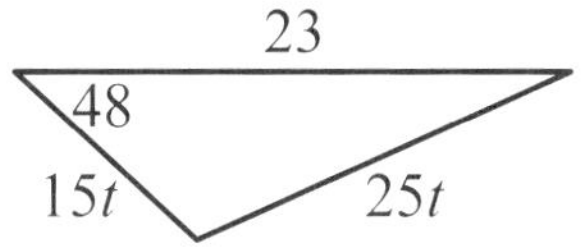

23. At 12:00 am, two hours after the ships have left port, the first ship has traveled 40 miles and the second ship 50 miles. The angle between the two ships is $62 + 75 = 137°$. If d denotes the distance between the ships, by the Law of Cosines,

$$d^2 = 40^2 + 50^2 - 2(40)(50)\cos 137°,$$

and

$$d = \sqrt{40^2 + 50^2 - 2(40)(50)\cos 137^\circ} \approx 83.8 \text{ miles.}$$

24. By the Law of Cosines, the distance d, between the two planes is

$$d^2 = 150^2 + 100^2 - 2(150)(100)\cos 50^\circ,$$

and

$$d = \sqrt{150^2 + 100^2 - 2(150)(100)\cos 50^\circ} \approx 115 \text{ miles.}$$

25. To use the Law of Sines to find the distance d from the fire to the tower B, first find the missing angle, which is $180 - 50 - 63 = 67^\circ$. Then

$$\frac{\sin 67^\circ}{10} = \frac{\sin 63^\circ}{d} \text{ so } d = \frac{10 \sin 63^\circ}{\sin 67^\circ} \approx 9.7 \text{ miles.}$$

26. The perimeter of the triangular plot is $200 + 300 + 450 = 950$, so by Heron's Formula, the area of the parcel of land is

$$A = \frac{1}{4}\sqrt{950(950 - 400)(950 - 600)(950 - 900)} = \frac{1}{4}\sqrt{950(550)(350)(50)}$$
$$= 625\sqrt{1463} \approx 23906 \text{ feet}^2.$$

Since one acre is 43,560 square feet, the value of the parcel is

$$\frac{23906}{43560} \cdot 2000 \approx \$1,098.00.$$

Review Exercises for Chapter 4 (Page 262)

1. a. $P(t) = \left(\frac{1}{2}, \frac{\sqrt{3}}{2}\right)$ **b.** $\frac{\pi}{3}$ **c.** $\cos\frac{\pi}{3} = \frac{1}{2}$, $\sin\frac{\pi}{3} = \frac{\sqrt{3}}{2}$, $\tan\frac{\pi}{3} = \sqrt{3}$, $\cot\frac{\pi}{3} = \frac{1}{\sqrt{3}} = \frac{\sqrt{3}}{3}$, $\sec\frac{\pi}{3} = 2$, $\csc\frac{\pi}{3} = \frac{2}{\sqrt{3}} = \frac{2\sqrt{3}}{3}$

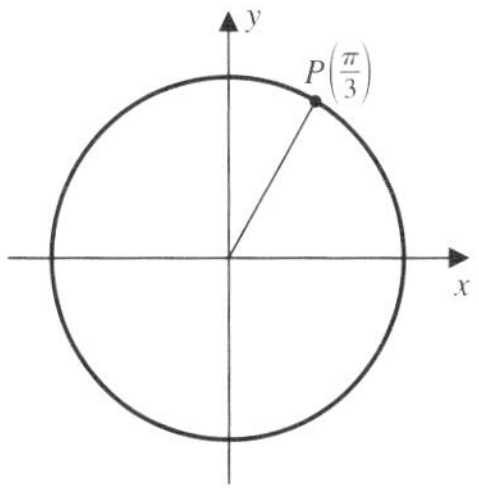

2. a. $P(t) = \left(\frac{\sqrt{3}}{2}, -\frac{1}{2}\right)$ **b.** $\frac{\pi}{6}$ **c.** $\cos\frac{11\pi}{6} = \frac{\sqrt{3}}{2}$, $\sin\frac{11\pi}{6} = -\frac{1}{2}$, $\tan\frac{11\pi}{6} = -\frac{1}{\sqrt{3}} = -\frac{\sqrt{3}}{3}$, $\cot\frac{11\pi}{6} = -\frac{3}{\sqrt{3}} = -\sqrt{3}$, $\sec\frac{11\pi}{6} = \frac{2}{\sqrt{3}} = \frac{2\sqrt{3}}{3}$, $\csc\frac{11\pi}{6} = -2$

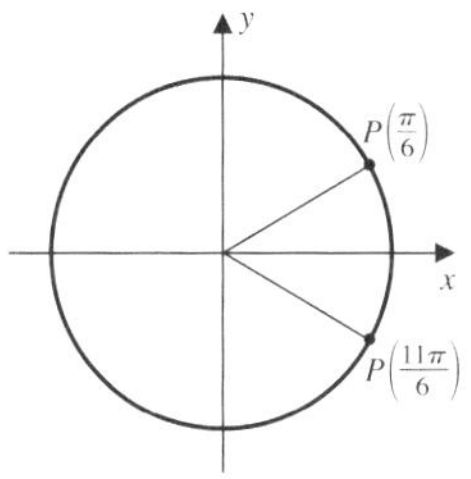

3. a. $P(t) = \left(-\frac{\sqrt{2}}{2}, -\frac{\sqrt{2}}{2}\right)$ **b.** $\frac{\pi}{4}$ **c.** $\cos\frac{5\pi}{4} = -\frac{\sqrt{2}}{2}$, $\sin\frac{5\pi}{4} = -\frac{\sqrt{2}}{2}$, $\tan\frac{5\pi}{4} = 1$, $\cot\frac{5\pi}{4} = 1$, $\sec\frac{5\pi}{4} = -\frac{2}{\sqrt{2}} = -\sqrt{2}$, $\csc\frac{5\pi}{4} = -\frac{2}{\sqrt{2}} = -\sqrt{2}$

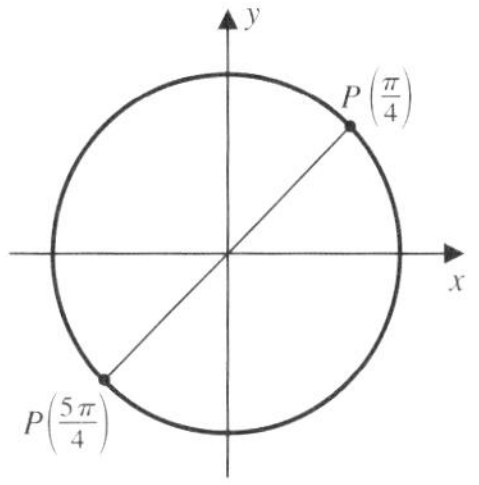

4. a. $P(t) = \left(-\frac{1}{2}, \frac{\sqrt{3}}{2}\right)$ **b.** $\frac{\pi}{3}$ **c.** $\cos\frac{8\pi}{3} = -\frac{1}{2}$, $\sin\frac{8\pi}{3} = \frac{\sqrt{3}}{2}$, $\tan\frac{8\pi}{3} = -\sqrt{3}$, $\cot\frac{8\pi}{3} = -\frac{1}{\sqrt{3}} = -\frac{\sqrt{3}}{3}$, $\sec\frac{8\pi}{3} = -2$, $\csc\frac{8\pi}{3} = \frac{2}{\sqrt{3}} = \frac{2\sqrt{3}}{3}$

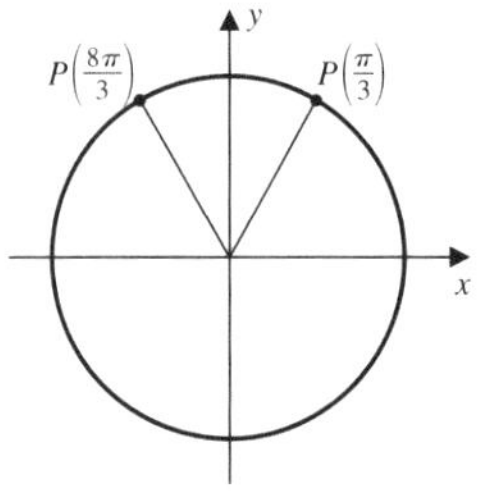

5. a. $P(t) = \left(-\frac{\sqrt{2}}{2}, \frac{\sqrt{2}}{2}\right)$ **b.** $\frac{\pi}{4}$ **c.** $\cos\left(-\frac{21\pi}{4}\right) = -\frac{\sqrt{2}}{2}$, $\sin\left(-\frac{21\pi}{4}\right) = \frac{\sqrt{2}}{2}$, $\tan\left(-\frac{21\pi}{4}\right) = -1$, $\cot\left(-\frac{21\pi}{4}\right) = -1$, $\sec\left(-\frac{21\pi}{4}\right) = -\frac{2}{\sqrt{2}} = -\sqrt{2}$, $\csc\left(-\frac{21\pi}{4}\right) = \frac{2}{\sqrt{2}} = \sqrt{2}$

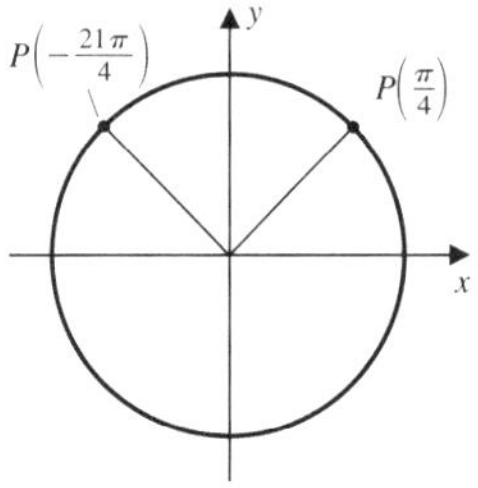

6. a. $P(t) = \left(\frac{1}{2}, \frac{\sqrt{3}}{2}\right)$ **b.** $\frac{\pi}{3}$ **c.** $\cos\left(-\frac{23\pi}{3}\right) = \frac{1}{2}$, $\sin\left(-\frac{23\pi}{3}\right) = \frac{\sqrt{3}}{2}$, $\tan\left(-\frac{23\pi}{3}\right) = \sqrt{3}$, $\cot\left(-\frac{23\pi}{3}\right) = \frac{1}{\sqrt{3}} = \frac{\sqrt{3}}{3}$, $\sec\left(-\frac{23\pi}{3}\right) = 2$, $\csc\left(-\frac{23\pi}{3}\right) = \frac{2}{\sqrt{3}} = \frac{2\sqrt{3}}{3}$

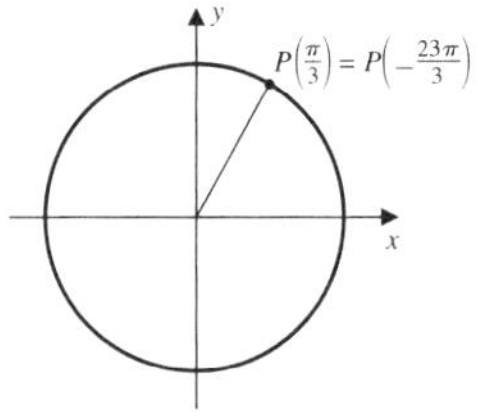

7. If $\cos t = \frac{3}{5}$ and $\frac{3\pi}{2} < t < 2\pi$, then $\sin t < 0$, and

$$(\cos t)^2 + (\sin t)^2 = 1 \text{ implies } (\sin t)^2 = 1 - \frac{9}{25} = \frac{16}{25} \text{ so } \sin t = \pm\sqrt{\frac{16}{25}} = \pm\frac{4}{5},$$

so $\sin t = -\frac{4}{5}$. Then

$$\tan t = -\frac{4}{3}, \quad \cot t = -\frac{3}{4}, \quad \sec t = \frac{5}{3}, \quad \csc t = -\frac{5}{4}.$$

8. If $\sin t = -\frac{1}{2}$ and $\cos t < 0$, then

$$(\cos t)^2 + (\sin t)^2 = 1 \text{ implies } (\cos t)^2 = 1 - \frac{1}{4} = \frac{3}{4} \text{ so } \cos t = -\frac{\sqrt{3}}{2}.$$

Then

$$\tan t = \frac{1}{\sqrt{3}} = \frac{\sqrt{3}}{3}, \quad \cot t = \sqrt{3}, \quad \sec t = -\frac{2}{\sqrt{3}} = -\frac{2\sqrt{3}}{3}, \quad \csc t = -2.$$

9. If $\tan t = \frac{1}{4}$ and $0 < t < \frac{\pi}{2}$, then $\cos t > 0$ and $\sin t > 0$, and

$$(\tan t)^2 + 1 = (\sec t)^2 \text{ implies } (\sec t)^2 = 1 + \frac{1}{16} = \frac{17}{16} \text{ so } \sec t = \frac{1}{\cos t} = \sqrt{\frac{17}{16}} = \frac{\sqrt{17}}{4}.$$

Then

$$\cos t = \frac{4}{\sqrt{17}} = \frac{4\sqrt{17}}{17}, \quad \frac{1}{4} = \tan t = \frac{\sin t}{\cos t} \text{ so } \sin t = \frac{1}{4}\frac{4\sqrt{17}}{17} = \frac{\sqrt{17}}{17},$$

and

$$\cot t = 4, \quad \csc t = \frac{17}{\sqrt{17}} = \sqrt{17}.$$

10. If $\sec t = -4$ and $\frac{\pi}{2} < t < \pi$, then $\cos t < 0$ and $\sin t > 0$ and $\cos t = -\frac{1}{4}$, so

$$(\cos t)^2 + (\sin t)^2 = 1 \text{ implies } (\sin t)^2 = 1 - \frac{1}{16} = \frac{15}{16} \text{ so } \sin t = \sqrt{\frac{15}{16}} = \frac{\sqrt{15}}{4}.$$

Then

$$\tan t = -\sqrt{15}, \quad \cot t = -\frac{1}{\sqrt{15}} = -\frac{\sqrt{15}}{15}, \quad \csc t = \frac{4}{\sqrt{15}} = \frac{4\sqrt{15}}{15}.$$

11. If $0 \le x \le \pi$, then $0 \le \frac{x}{3} \le \frac{\pi}{3}$ and $\cos\frac{x}{3} = \frac{1}{2}$ implies $\frac{x}{3} = \frac{\pi}{3}$ so $x = \pi$.

12. If $0 \le x \le \pi$, then $0 \le 4x \le 4\pi$ and $\sin 4x = -\frac{\sqrt{3}}{2}$ implies
$4x = \frac{4\pi}{3}, \frac{5\pi}{3}, \frac{10\pi}{3}, \frac{11\pi}{3}$ so $x = \frac{\pi}{3}, \frac{5\pi}{12}, \frac{5\pi}{6}, \frac{11\pi}{12}$.

13. Since

$$2(\cos x)^2 - 3\cos x + 1 = (2\cos x - 1)(\cos x - 1) = 0 \text{ implies } \cos x = \frac{1}{2}, \cos x = 1 \text{ so } x = \frac{\pi}{3}, x = 0.$$

14. Since

$$2(\cos x)^2 - \sin x - 1 = 2(1 - (\sin x)^2) - \sin x - 1 = 2(\sin x)^2 + \sin x - 1 = (2\sin x - 1)(\sin x + 1) = 0$$

$$\text{implies } \sin x = \frac{1}{2}, \sin x = -1 \text{ so } x = \frac{\pi}{6}, x = \frac{5\pi}{6}.$$

Note $\sin x = -1$ has no solutions for $0 \le x \le \pi$.

15. Since

$$(\tan x)^3 - 4\tan x = \tan x[(\tan x)^2 - 4] = \tan x(\tan x - 2)(\tan x + 2) = 0,$$

we have

$$\tan x = 0, \tan x = 2, \tan x = -2 \text{ so } x = 0, \pi, x = \arctan 2 \approx 1.1, x = \arctan(-2) = \pi - \arctan(2) \approx 2.03.$$

Note $x = \arctan(-2)$ is not in $[0, \pi]$.

16. Since

$$2\tan x - 3\cot x = 2\frac{\sin x}{\cos x} - 3\frac{\cos x}{\sin x} = \frac{2(\sin x)^2 - 3(\cos x)^2}{\cos x \sin x} = 0,$$

we have

$$2(\sin x)^2 - 3(\cos x)^2 = 2[1 - (\cos x)^2] - 3(\cos x)^2 = 2 - 5(\cos x)^2 = 0.$$

So

$$(\cos x)^2 = \frac{2}{5} \text{ implies } \cos x = \pm\sqrt{\frac{2}{5}} \text{ so } x = \arccos\sqrt{\frac{2}{5}} \approx 0.9,$$

and

$$x = \arccos\left(-\sqrt{\frac{2}{5}}\right) = \pi - \arccos\sqrt{\frac{2}{5}} \approx 2.26.$$

17. We have

$$\cot x - \csc x = \frac{\cos x}{\sin x} - \frac{1}{\sin x} = \frac{\cos x - 1}{\sin x} = 1 \text{ implies } \cos x - 1 = \sin x \text{ so } \cos x - \sin x = 1,$$

so $x = 0, \frac{3\pi}{2}$. But there are no solutions since $\cot x$ is not defined at $x = 0$ and $x = \frac{3\pi}{2}$ is not in the interval $[0, \pi]$.

18. Since $\cos 2x + \sin x = 0$ implies that

$$(\cos x)^2 - (\sin x)^2 + \sin x = 1 - (\sin x)^2 - (\sin x)^2 + \sin x = 2(\sin x)^2 - \sin x - 1 = (2\sin x + 1)(\sin x - 1) = 0$$

and we have $\sin x = -\frac{1}{2}$ and $\sin x = 1$. Hence $x = \frac{\pi}{2}$ since $\sin t > 0$ on $[0, \pi]$.

19. The function $f(x) = (\sin x)^3$ is odd since

$$f(-x) = (\sin(-x))^3 = (-\sin x)^3 = -(\sin x)^3 = -f(x).$$

20. The function $f(x) = \sin x \cos x$ is odd since

$$f(-x) = \sin(-x)\cos(-x) = -\sin x \cos x = -f(x).$$

21. The function $f(x) = x(\sin x)^3$ is even since

$$f(-x) = -x(\sin(-x))^3 = -x(-\sin x)^3 = -x[-(\sin x)^3] = x(\sin x)^3 = f(x).$$

22. The function $f(x) = x^5 + \cos x$ is neither odd nor even since

$$f(-x) = (-x)^5 + \cos(-x) = -x^5 + \cos x \neq -f(x) = -x^5 - \cos x, \text{ and } f(-x) \neq f(x).$$

23. Since (a) and (c) have period $\frac{2\pi}{2} = \pi$ and (b) and (d) have period $\frac{2\pi}{1/2} = 4\pi$, we have (a) or (c) is either (i) or (ii) and (b) or (d) is either (iii) or (iv).

a. (ii) since $\sin 2\left(0 + \frac{\pi}{2}\right) = \sin \pi = 0$.

b. (iv) since the curve has the form of a sine wave shifted to the left $\frac{\pi}{2}$ units.

c. (i). **d.** (iii).

24. **a.** (iii) **b.** (iv) **c.** (ii) **d.** (i)

25. $y = 5\sin\frac{x}{2}$

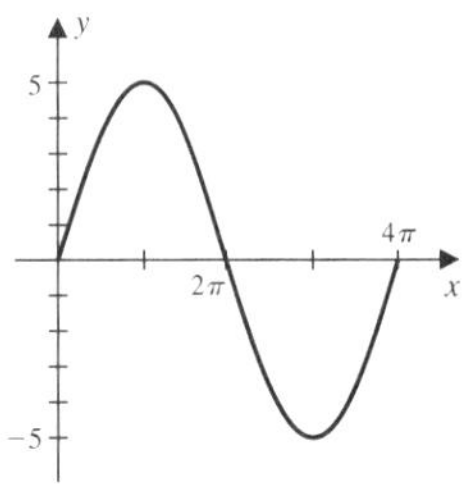

26. $y = 2\cos 2\pi x$

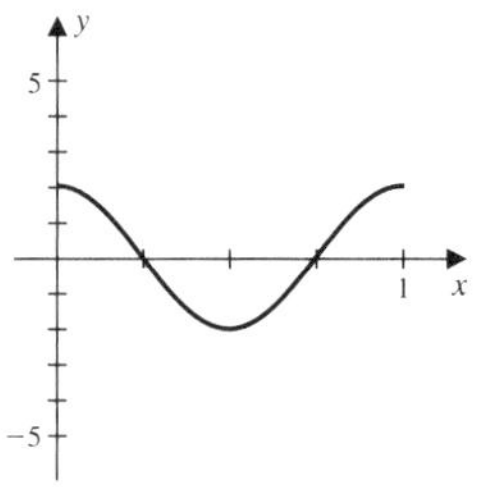

27. $y = -3\cos 2x$

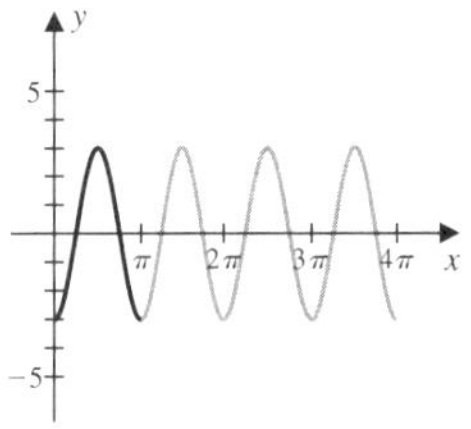

28. $y = 2 + 4\sin 4x$

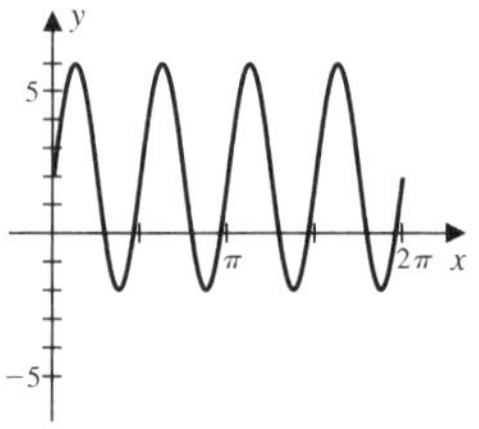

29. $y = \cos(2x - \pi) = \cos 2\left(x - \frac{\pi}{2}\right)$

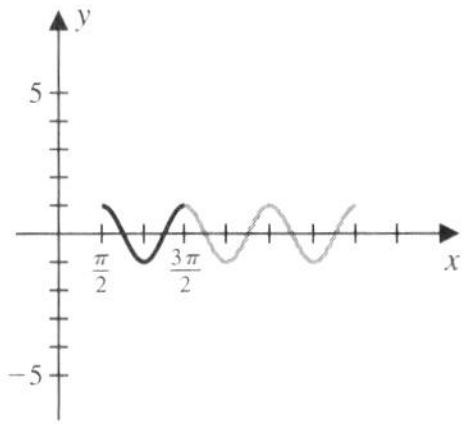

30. $y = -2\cos\left(3x - \frac{\pi}{2}\right) = -2\cos 3\left(x - \frac{\pi}{6}\right)$

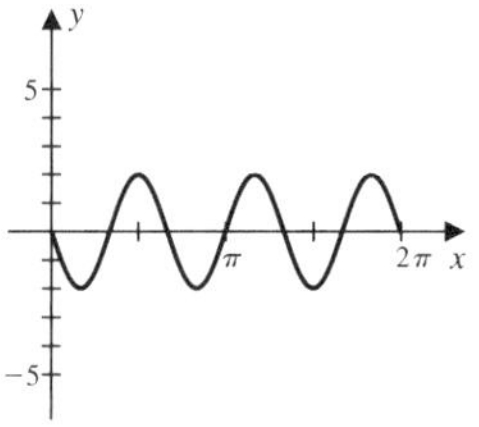

31. $y = \cot(x + \pi/6)$

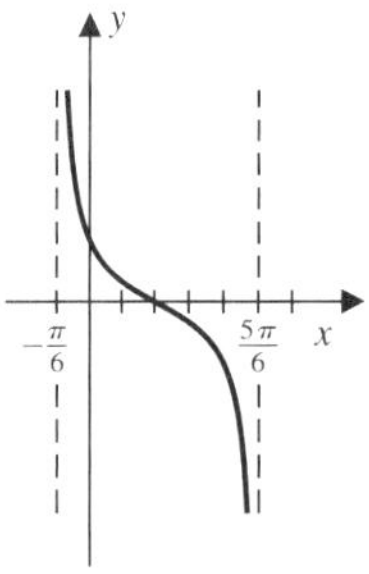

32. $y = -\tan(x - \pi/2)$

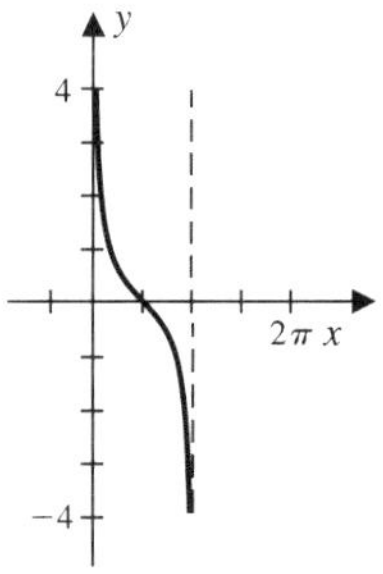

33. $y = \sec 4\pi x$

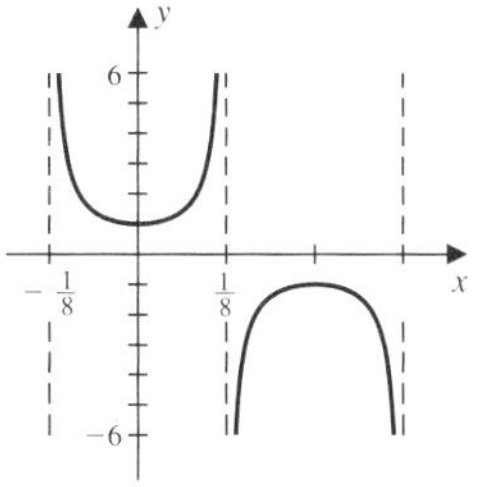

34. $y = -2\csc(x - \pi/4)$

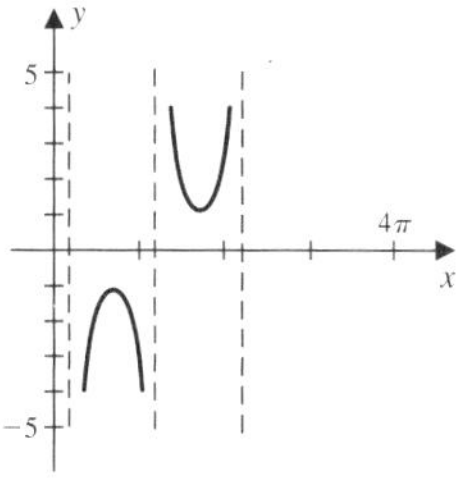

35. **a.** The curve has an equation of the form $y = A\sin(Bx)$, where the amplitude is 1 and the period is $\frac{\pi}{2}$. So

$$A = 1,\ \frac{2\pi}{B} = \frac{\pi}{2} \text{ so } B = 4 \text{ and } y = \sin 4x.$$

b. The curve has the form $y = A\cos B(x + C)$, with

$$A = 1,\ \frac{2\pi}{B} = \frac{\pi}{2} \text{ so } B = 4,\ C = -\frac{\pi}{8} \text{ so } y = \cos 4\left(x - \frac{\pi}{8}\right).$$

36. **a.** The curve has an equation of the form $y = A\sin B(x + C)$, where the amplitude is 2, the period is 2, and the phase shift is to the left $\frac{1}{2}$ unit. So

$$A = 2,\ \frac{2\pi}{B} = 2 \text{ so } B = \pi,\ C = \frac{1}{2} \text{ and } y = 2\sin\pi\left(x + \frac{1}{2}\right).$$

b. The curve has the form $y = A\cos Bx$, with

$$A = 2,\ B = \pi,\ \text{so } y = 2\cos\pi x.$$

37. **a.** The curve has an equation of the form $y = A\sin B(x + C)$, where the amplitude is 4, the period is 2, and the curve is shifted to the left $\frac{1}{2}$ unit. So

$$A = 4,\ \frac{2\pi}{B} = 2 \text{ so } B = \pi,\ C = \frac{1}{2} \text{ and } y = 4\sin\pi\left(x + \frac{1}{2}\right).$$

b. The curve has the form $y = A\cos Bx$, with

$$A = 2,\ B = \pi,\ \text{so } y = 4\cos\pi x.$$

38. **a.** The curve has an equation of the form $y = A\sin Bx + D$ where the curve is centered vertically on the line $y = 2.5$. So it has been shifted upward 2.5 units, the amplitude is 1.5, and the period is 4. Hence

$$D = 2.5,\ A = 1.5,\ \frac{2\pi}{B} = 4 \text{ so } B = \frac{\pi}{2},\ \text{and } y = 2.5 + 1.5\sin\frac{\pi}{2}x.$$

b. The curve has an equation of the form $y = A\cos B(x + C) + D$, with

$$A = 1.5,\ B = \frac{\pi}{2},\ C = -1,\ D = 2.5,\ \text{so } y = 2.5 + 1.5\cos\frac{\pi}{2}(x - 1).$$

39. For $\alpha = 60°$, $\overline{AC} = 13$ we have

a.

$$\begin{aligned}\sin 60° &= \frac{\overline{BC}}{13} \text{ so } \overline{BC} = 13\sin 60° = \frac{13\sqrt{3}}{2};\\ \cos 60° &= \frac{\overline{AB}}{13} \text{ so } \overline{AB} = 13\cos 60° = \frac{13}{2};\\ \gamma &= 90 - 60 = 30°.\end{aligned}$$

b. We have $\sin\alpha = \sqrt{3}/2$, $\cos\alpha = 1/2$, $\tan\alpha = \sqrt{3}$, $\cot\alpha = \sqrt{3}/3$, $\sec\alpha = 2$, and $\csc\alpha = 2\sqrt{3}/3$.

40. For $\alpha = 45°$, $\overline{BC} = 12$ we have

a.

$$\begin{aligned}\tan 45° &= \frac{12}{\overline{AB}} \text{ so } \overline{AB} = \frac{12}{\tan 45°} = 12;\\ \sin 45° &= \frac{12}{\overline{AC}} \text{ so } \overline{AC} = \frac{12}{\sin 45°} = \frac{12}{\sqrt{2}/2} = \frac{24}{\sqrt{2}} = 12\sqrt{2};\\ \gamma &= 90 - 45 = 45°.\end{aligned}$$

b. We have $\sin\alpha = \sqrt{2}/2$, $\cos\alpha = \sqrt{2}/2$, $\tan\alpha = 1$, $\cot\alpha = 1$, $\sec\alpha = \sqrt{2}$, and $\csc\alpha = \sqrt{2}$.

41.

$$\begin{aligned}\cos\left(\frac{5\pi}{12}\right) &= \cos\left(\frac{3\pi}{12} + \frac{2\pi}{12}\right) = \cos\left(\frac{\pi}{4} + \frac{\pi}{6}\right)\\ &= \cos\frac{\pi}{4}\cos\frac{\pi}{6} - \sin\frac{\pi}{4}\sin\frac{\pi}{6} = \frac{\sqrt{2}}{2}\frac{\sqrt{3}}{2} - \frac{\sqrt{2}}{2}\frac{1}{2}\\ &= \frac{\sqrt{2}}{4}\left(\sqrt{3} - 1\right)\end{aligned}$$

42.

$$\begin{aligned}\sin\left(\frac{7\pi}{12}\right) &= \sin\left(\frac{9\pi}{12} - \frac{2\pi}{12}\right) = \sin\left(\frac{3\pi}{4} - \frac{\pi}{6}\right)\\ &= \sin\frac{3\pi}{4}\cos\frac{\pi}{6} - \cos\frac{3\pi}{4}\sin\frac{\pi}{6} = \frac{\sqrt{2}}{2}\frac{\sqrt{3}}{2} - \left(-\frac{\sqrt{2}}{2}\right)\frac{1}{2}\\ &= \frac{\sqrt{2}}{4}\left(\sqrt{3} + 1\right)\end{aligned}$$

43.

$$\begin{aligned}\cos\left(-\frac{13\pi}{12}\right) &= \cos\left(\frac{13\pi}{12}\right) = \cos\left(\frac{9\pi}{12}+\frac{4\pi}{12}\right) = \cos\left(\frac{3\pi}{4}+\frac{\pi}{3}\right)\\ &= \cos\frac{3\pi}{4}\cos\frac{\pi}{3} - \sin\frac{3\pi}{4}\sin\frac{\pi}{3} = -\frac{\sqrt{2}}{2}\frac{1}{2} - \frac{\sqrt{2}}{2}\frac{\sqrt{3}}{2}\\ &= -\frac{\sqrt{2}}{4}\left(\sqrt{3}+1\right)\end{aligned}$$

44.

$$\begin{aligned}\tan\left(-\frac{\pi}{12}\right) &= -\tan\left(\frac{\pi}{12}\right) = -\tan\left(\frac{3\pi}{12}-\frac{2\pi}{12}\right) = -\tan\left(\frac{\pi}{4}-\frac{\pi}{6}\right)\\ &= -\frac{\tan\frac{\pi}{4}-\tan\frac{\pi}{6}}{1+\tan\frac{\pi}{4}\tan\frac{\pi}{6}} = -\frac{1-\sqrt{3}/3}{1+1(\sqrt{3}/3)}\\ &= -\frac{\frac{3-\sqrt{3}}{3}}{\frac{3+\sqrt{3}}{3}} = -\frac{3-\sqrt{3}}{3+\sqrt{3}} = \frac{\sqrt{3}-3}{\sqrt{3}+3}\end{aligned}$$

45. $\sin\frac{\pi}{8} = \sqrt{\frac{1-\cos\frac{\pi}{4}}{2}} = \sqrt{\frac{1-\sqrt{2}/2}{2}} = \sqrt{\frac{2-\sqrt{2}}{4}} = \frac{\sqrt{2-\sqrt{2}}}{2}$

46. $\cos\frac{5\pi}{12} = \sqrt{\frac{1+\cos\frac{5\pi}{6}}{2}} = \sqrt{\frac{1-\sqrt{3}/2}{2}} = \sqrt{\frac{2-\sqrt{3}}{4}} = \frac{\sqrt{2-\sqrt{3}}}{2}$

47. $\tan\frac{7\pi}{12} = \frac{\sin\frac{7\pi}{12}}{\cos\frac{7\pi}{12}} = \frac{\sqrt{\frac{1-\cos\frac{7\pi}{6}}{2}}}{-\sqrt{\frac{1+\cos\frac{7\pi}{6}}{2}}} = \frac{\sqrt{\frac{1+\sqrt{3}/2}{2}}}{-\sqrt{\frac{1-\sqrt{3}/2}{2}}} = -\frac{\sqrt{2+\sqrt{3}}}{\sqrt{2-\sqrt{3}}} = -\sqrt{7+4\sqrt{3}}$

48. $\cot\frac{3\pi}{8} = \frac{\cos\frac{3\pi}{8}}{\sin\frac{3\pi}{8}} = \frac{\sqrt{\frac{1+\cos\frac{3\pi}{4}}{2}}}{\sqrt{\frac{1-\cos\frac{3\pi}{4}}{2}}} = \frac{\sqrt{\frac{1-\sqrt{2}/2}{2}}}{\sqrt{\frac{1+\sqrt{2}/2}{2}}} = \frac{\sqrt{\frac{2-\sqrt{2}}{4}}}{\sqrt{\frac{2+\sqrt{2}}{4}}} = \frac{\sqrt{2-\sqrt{2}}}{\sqrt{2+\sqrt{2}}}$

49. $\sin(t-\pi) = \sin t\cos\pi - \cos t\sin\pi = (\sin t)(-1) - (\cos t)(0) = -\sin t$

50. $\cos(t-\pi) = \cos t\cos\pi + \sin t\sin\pi = (\cos t)(-1) + (\sin t)(0) = -\cos t$

51. $\sin\left(\frac{3\pi}{2}-t\right) = \sin\frac{3\pi}{2}\cos t - \cos\frac{3\pi}{2}\sin t = (-1)(\cos t) - (0)(\sin t) = -\cos t$

52. $\cos\left(\frac{3\pi}{2} - t\right) = \cos\frac{3\pi}{2}\cos t + \sin\frac{3\pi}{2}\sin t = (0)(\cos t) + (-1)(\sin t) = -\sin t$

53. $(\cos 3x)^2 = \frac{1+\cos 6x}{2} = \frac{1}{2} + \frac{1}{2}\cos 6x$

54. $(\sin 4x)^2 = \frac{1-\cos 8x}{2} = \frac{1}{2} - \frac{1}{2}\cos 8x$

55. $(\sin x)^4 = ((\sin x)^2)^2 = \left(\frac{1-\cos 2x}{2}\right)^2 = \frac{1}{4}\left(1 - 2\cos 2x + (\cos 2x)^2\right)$

$= \frac{1}{4}\left(1 - 2\cos 2x + \frac{1+\cos 4x}{2}\right) = \frac{3}{8} - \frac{1}{2}\cos 2x + \frac{1}{8}\cos 4x$

56.

$(\sin 2x)^2(\cos 2x)^2 = \left(\frac{1-\cos 4x}{2}\right)\left(\frac{1+\cos 4x}{2}\right) = \frac{1}{4}(1 - (\cos 4x)^2) = \frac{1}{4}\left(1 - \frac{1+\cos 8x}{2}\right) = \frac{1}{8} - \frac{1}{8}\cos 8x$

57. $\sin 4t \cos 5t = \frac{1}{2}[\sin(4t + 5t) + \sin(4t - 5t)] = \frac{1}{2}[\sin 9t - \sin t]$

58. $\cos 6t \sin 8t = \frac{1}{2}[\sin(6t + 8t) - \sin(6t - 8t)] = \frac{1}{2}[\sin 14t + \sin 2t]$

59. $\cos 2t \cos 4t = \frac{1}{2}[\cos(2t + 4t) + \cos(2t - 4t)] = \frac{1}{2}[\cos 6t + \cos 2t]$

60. $\sin 3t \sin 5t = \frac{1}{2}[\cos(3t - 5t) - \cos(3t + 5t)] = \frac{1}{2}[\cos 2t - \cos 8t]$

61. $\sin 2t + \sin 6t = 2\sin\frac{2t+6t}{2}\cos\frac{2t-6t}{2} = 2\sin 4t \cos 2t$

62. $\sin 3t + \sin 7t = 2\sin\frac{3t+7t}{2}\cos\frac{3t-7t}{2} = 2\sin 5t\cos 2t$

63. $\cos 4t + \cos 2t = 2\cos\frac{4t+2t}{2}\cos\frac{4t-2t}{2} = 2\cos 3t\cos t$

64. $\cos 5t - \cos 3t = -2\sin\frac{5t+3t}{2}\sin\frac{5t-3t}{2} = -2\sin 4t\sin t$

65. $(\cos x)^4 - (\sin x)^4 = ((\cos x)^2 - (\sin x)^2)((\cos x)^2 + (\sin x)^2) = (\cos x)^2 - (\sin x)^2 = \cos 2x$

66. $(\sin x - \cos x)^2 = (\sin x)^2 - 2\sin x\cos x + (\cos x)^2 = 1 - 2\sin x\cos x = 1 - \sin 2x$

67.

$$\begin{aligned}\frac{\sin x}{1-\cos x} &= \frac{\sin x}{1-\cos x}\cdot\frac{1+\cos x}{1+\cos x} = \frac{\sin x(1+\cos x)}{1-(\cos x)^2} \\ &= \frac{\sin x(1+\cos x)}{(\sin x)^2} = \frac{1+\cos x}{\sin x} = \cot x + \csc x\end{aligned}$$

68.

$$\begin{aligned}\frac{\cos x}{1-\tan x}+\frac{\sin x}{1-\cot x} &= \frac{\cos x}{1-\frac{\sin x}{\cos x}}+\frac{\sin x}{1-\frac{\cos x}{\sin x}} \\ &= \frac{(\cos x)^2}{\cos x-\sin x}+\frac{(\sin x)^2}{\sin x-\cos x} = \frac{(\cos x)^2-(\sin x)^2}{\cos x-\sin x} \\ &= \frac{(\cos x-\sin x)(\cos x+\sin x)}{\cos x-\sin x} = \cos x+\sin x\end{aligned}$$

69. $\sin\left(\arctan\sqrt{3}\right) = \sin\left(\frac{\pi}{3}\right) = \frac{\sqrt{3}}{2}$

70. Let $x = \arcsin\left(\frac{4}{5}\right)$ so $\sin x = \frac{4}{5}$. Then

$$(\cos t)^2 + (\sin t)^2 = 1 \text{ implies } (\cos t)^2 = 1 - \frac{16}{25} = \frac{9}{25} \text{ so } \cos t = \pm\frac{3}{5},$$

and since $-\frac{\pi}{2} \le x \le \frac{\pi}{2}$ we have $\cos x > 0$, so $\cos x = \cos\left(\arcsin\left(\frac{4}{5}\right)\right) = \frac{3}{5}$. Or draw a triangle with an angle x with opposite side 4 and hypotenuse 5.

71. Since $\sin(\arcsin x) = x$ for x in $[-1, 1]$, we have

$$\begin{aligned}\sin\left(\arcsin\left(\frac{3}{5}\right)-\arcsin\left(\frac{5}{13}\right)\right) &= \sin\left(\arcsin\left(\frac{3}{5}\right)\right)\cos\left(\arcsin\left(\frac{5}{13}\right)\right) \\ &\quad -\cos\left(\arcsin\left(\frac{3}{5}\right)\right)\sin\left(\arcsin\left(\frac{5}{13}\right)\right) \\ &= \frac{3}{5}\cos\left(\arcsin\left(\frac{5}{13}\right)\right)-\cos\left(\arcsin\left(\frac{3}{5}\right)\right)\frac{5}{13} \\ &= \frac{3}{5}\left(\frac{12}{13}\right)-\frac{4}{5}\left(\frac{5}{13}\right) = \frac{16}{65}.\end{aligned}$$

72. We have

$$\tan\left(\arccos\left(\frac{1}{4}\right)+\arcsin\left(\frac{1}{2}\right)\right)=\frac{\tan\left(\arccos\left(\frac{1}{4}\right)\right)+\tan\left(\arcsin\left(\frac{1}{2}\right)\right)}{1-\tan\left(\arccos\left(\frac{1}{4}\right)\right)\tan\left(\arcsin\left(\frac{1}{2}\right)\right)}$$
$$=\frac{\sqrt{15}+\frac{\sqrt{3}}{3}}{1-\sqrt{15}\frac{\sqrt{3}}{3}}=\frac{\frac{3\sqrt{15}+\sqrt{3}}{3}}{\frac{3-\sqrt{15}\sqrt{3}}{3}}=\frac{3\sqrt{15}+\sqrt{3}}{3-\sqrt{45}}.$$

73. Since $\alpha = 25^\circ$, $b = 12$, $c = 20$, and $a^2 = b^2 + c^2 - 2bc\cos\alpha$, we have

$$a=\sqrt{12^2+20^2-2(12)(20)\cos 25^\circ}\approx 10.4.$$

Also,

$$\frac{\sin\beta}{b}=\frac{\sin\alpha}{a}\beta=\arcsin\left(\frac{12\sin 25^\circ}{10.4}\right)\approx 0.50938=0.50938\left(\frac{180}{\pi}\right)^\circ\approx 29.2^\circ,$$

and $\gamma = 180 - 25 - 29.2 = 125.8^\circ$.

74. Since $\alpha = 30^\circ$, $\beta = 100^\circ$, and $c = 25$, we have $\gamma = 180 - 30 - 100 = 50^\circ$, and

$$\frac{\sin\alpha}{a}=\frac{\sin\gamma}{c}\text{ so }a=\frac{25\sin 30^\circ}{\sin 50^\circ}\approx 16.3;\quad \frac{\sin\beta}{b}=\frac{\sin\gamma}{c}\text{ so }b=\frac{25\sin 100^\circ}{\sin 50^\circ}\approx 32.1.$$

75. Since $a = 6$, $b = 8$, and $c = 10$, we have $a^2 + b^2 = c^2$ and $\gamma = 90^\circ$. Also,

$$\cos\alpha=\frac{6^2-8^2-10^2}{-2(8)(10)}=\frac{4}{5},$$

so

$$\alpha=\arccos\left(\frac{4}{5}\right)\approx 0.6435=0.6435\left(\frac{180}{\pi}\right)^\circ\approx 36.9^\circ,$$

and $\beta = 90^\circ - \alpha \approx 53.1^\circ$.

76. Since $a = 65$, $\beta = 30^\circ$, and $c = 35$, we have

$$b=\sqrt{65^2+35^2-2(65)(35)\cos 30^\circ}\approx 38.9,$$

$$\frac{\sin\alpha}{a}=\frac{\sin\beta}{b}\text{ so }\alpha=\arcsin\left(\frac{65\sin 30^\circ}{38.9}\right)\approx 0.989=0.989\left(\frac{180}{\pi}\right)^\circ\approx 56.7^\circ,$$

and $\gamma = 180 - \beta - \alpha = 180 - 30 - 56.7 = 93.3^\circ$.

77. Since $\beta = 76°$, $\gamma = 50°$, and $b = 10.5$, we have $\alpha = 180 - 76 - 50 = 54°$, and

$$\frac{\sin\alpha}{a} = \frac{\sin\beta}{b} \text{ so } a = \frac{10.5\sin 54°}{\sin 76°} \approx 8.8; \quad \frac{\sin\gamma}{c} = \frac{\sin\beta}{b} \text{ so } c = \frac{10.5\sin 50°}{\sin 76°} \approx 8.3.$$

78. Since $a = 8$, $c = \sqrt{3}$, and $\gamma = 45°$, we have

$$\frac{\sin\alpha}{a} = \frac{\sin\gamma}{c} \text{ so } \sin\alpha = \frac{8\sin 45°}{\sqrt{3}} = \frac{8\left(\frac{\sqrt{2}}{2}\right)}{\sqrt{3}} = \frac{4\sqrt{2}}{\sqrt{3}} > 1,$$

so there is no such triangle.

79. The points of intersection of the curves $y = \sin x$ and $y = x^2$ occur at approximately $x = 0.9$, and $x = 0$.

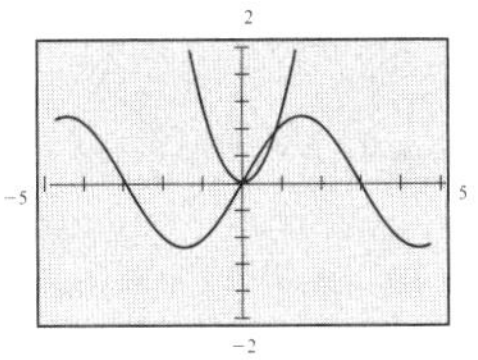

80. The point of intersection of the curves $y = \cos x$ and $y = x^3$ occurs at approximately $x = 0.9$.

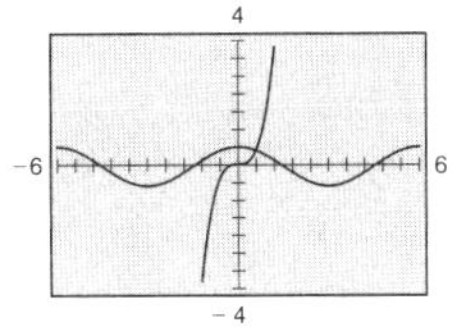

81. The period of $f(x) = 4\cos(125x)$ is $\frac{2\pi}{125} \approx 0.05$, and since the amplitude is 4, a reasonable viewing rectangle is $\left[-\frac{2\pi}{125}, \frac{2\pi}{125}\right] \times [-4, 4]$.

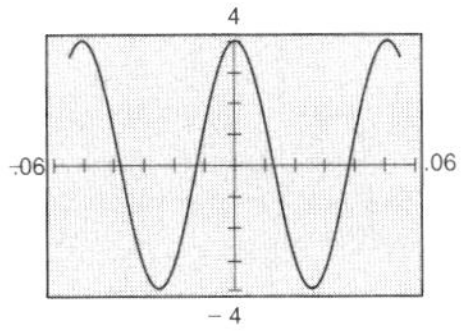

82. Since the period of $y = 10\sin(25x)$ is $\frac{2\pi}{25} \approx 0.25$ and since the amplitude is 10, a reasonable viewing rectangle for $y = 10\sin(25x)$ is $\left[-\frac{2\pi}{25}, \frac{2\pi}{25}\right] \times [-10, 10]$. A reasonable viewing rectangle for $f(x) = x - 10\sin(25x)$ is $\left[-\frac{2\pi}{25}, \frac{2\pi}{25}\right] \times [-15, 15]$.

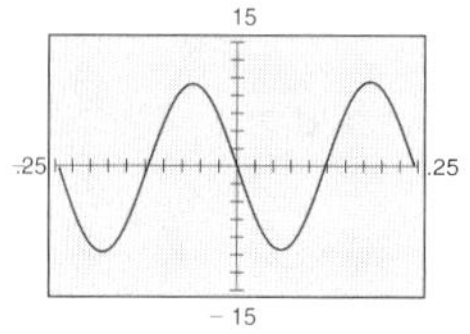

83. Since the area of a trapezoid is the product of its height and the average of its bases, we have

$$A = \frac{1}{2}[b + (b + 2b\cos\theta)]b\sin\theta = b^2(1 + \cos\theta)\sin\theta.$$

84. Since the triangle describing the position of the ships is a right triangle, the Pythagorean Theorem can be used to find the distance, d, between the two ships. Since ship A is traveling at 20 miles/hour and ship B at 35 miles/hour, two hours after leaving port the ships are 40 miles and 70 miles, respectively, from port. So

$$d = \sqrt{40^2 + 70^2} = \sqrt{6500} = 10\sqrt{65} \approx 80.6 \text{ miles}.$$

The bearing θ of ship B from ship A can be computed from

$$\sin\theta \approx \frac{70}{80.6} \text{ so } \theta \approx \arcsin\left(\frac{70}{80.6}\right) \approx 1.0521 = 1.0521\left(\frac{180}{\pi}\right)^\circ \approx 60.3^\circ.$$

85. If the plane is traveling at 380 miles/hour, then after two and one half hours the plane is 950 mi from the airport. Its bearing from the airport is 150° measured clockwise from north, or 60° measured clockwise from east. Then the plane is

$$950\cos 60^\circ = 950\left(\frac{1}{2}\right) = 475$$

miles east of the airport and

$$950\sin 60^\circ = 950\frac{\sqrt{3}}{2} \approx 822.7$$

miles south of the airport.

86. If the angle of depression from the top of the lighthouse to the ship is 28°, then the line of sight from the ship to the top of the lighthouse is also 28°. If the distance from the lighthouse to the ship is denoted by x, then

$$\tan 62^\circ = \frac{x}{200} \text{ so } x = 200\tan 62^\circ \approx 376 \text{ feet.}$$

Chapter 4 Exercises for Calculus (Page 264)

1. The graph indicates that the minimum occurs at $x = -2$ with a minimum value of approximately -1.1. The maximum occurs at $x = 3$ with a maximum value of approximately 2.9.

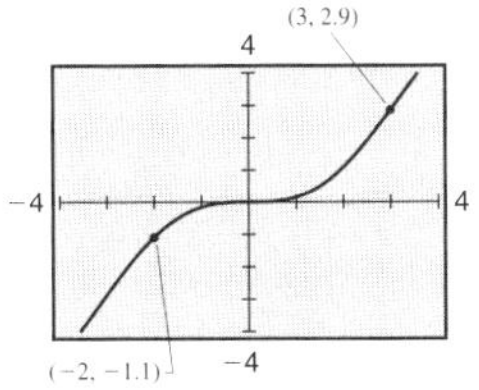

2. The graph indicates that the minimum occurs at $x \approx -2.4$ with a minimum value of approximately -1.4. The maximum occurs at $x \approx 0.8$ with a maximum value of approximately 1.4.

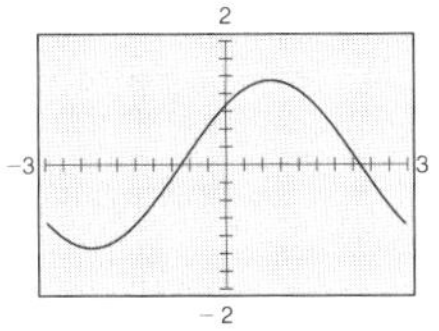

3. **a.** $\sqrt{a^2 - u^2} = \sqrt{a^2 - (a\sin t)^2} = \sqrt{a^2(1-(\sin t)^2)} = a\sqrt{(\cos t)^2} = a\cos t$

b. $\sqrt{u^2 + a^2} = \sqrt{(a\tan t)^2 + a^2} = \sqrt{a^2((\tan t)^2 + 1)} = a\sqrt{(\sec t)^2} = a\sec t$

c. $\sqrt{u^2 - a^2} = \sqrt{(a\sec t)^2 - a^2} = \sqrt{a^2((\sec t)^2 - 1)} = a\sqrt{(\tan t)^2} = a\tan t$

d. $\frac{\sqrt{u^2-a^2}}{u} = \frac{\sqrt{(a\sec t)^2 - a^2}}{a\sec t} = \frac{a\tan t}{a\sec t} = \sin t$

4. The area is the product of the length and width, so

$$A = (2r\cos\theta)(r\sin\theta) = r^2(2\cos\theta\sin\theta) = r^2\sin 2\theta.$$

5. Let x be the length of the line segment connecting the two vertices that lie approximately east-west of one another. By the Law of Cosines,

$$x^2 = 10^2 + 7^2 - 2(10)(7)\cos 95^\circ \text{ so } x = \sqrt{10^2 + 7^2 - 2(10)(7)\cos 95^\circ} \approx 13.$$

The perimeter of the upper triangle is $P_u = 10 + 7 + 13 = 30$ so by Heron's Formula, the area of the upper triangle is

$$A_u = \frac{1}{4}\sqrt{30(30-20)(30-14)(30-26)} = (\frac{1}{4})80\sqrt{3} = 20\sqrt{3} \approx 34.64.$$

For the lower triangle, we have $P_l = 9 + 12 + 13 = 34$. By Heron's Formula, the area of the lower triangle is

$$A_l = \frac{1}{4}\sqrt{34(34-18)(34-24)(34-26)} = (\frac{1}{4})16\sqrt{170} \approx 52.15.$$

The area of the quadrilateral is then $A_u + A_l \approx 34.64 + 52.15 = 86.79$ feet^2.

6. Let x denote the distance BC from B to the point C where the height meets the ground. Then $\cot\alpha = \frac{d+x}{h}$ and $\cot\beta = \frac{x}{h}$. So

$$\cot\alpha - \cot\beta = \frac{d+x}{h} - \frac{x}{h} \text{ implies } \cot\alpha - \cot\beta = \frac{d}{h} \text{ so } h = \frac{d}{\cot\alpha - \cot\beta}.$$

7. If α is the angle of the line of sight from ship A to the plane, β the angle from ship B to the plane, and d is the distance from A to B,then from Exercise 6,

$$33000 = \frac{d}{\cot\alpha - \cot\beta} = \frac{d}{\cot 32^\circ - \cot 47^\circ} \text{ so } d = 33000(\cot 32^\circ - \cot 47^\circ) \approx 22038.$$

So the two ships are approximately 22000 feet apart, rounded to the nearest 100 feet.

8. The bearing that the pilot should fly in going from city A to B, is the same as angle ABC. Denote angle ABC by α. By the Law of Cosines,

$$\alpha = \arccos\left(\frac{65^2 - 95^2 - 100^2}{-2(95)(100)}\right) = \arccos\left(\frac{65^2 - 95^2 - 100^2}{-2(95)(100)}\right)\left(\frac{180}{\pi}\right)^\circ \approx 39^\circ.$$

So the pilot should fly 129° southwest, measured counterclockwise from north.

9. Let d be the distance between A and C and let α be the angle ACB. Then $\alpha = 180 - 80 - 59 = 41^\circ$.By the Law of Sines

$$\frac{\sin 41^\circ}{300} = \frac{\sin 59^\circ}{d} \text{ so } d = \frac{300\sin 59^\circ}{\sin 41^\circ} \approx 392 \text{ feet.}$$

10. a. From the graphs of $y = \frac{\sin x}{x}$ and $y = \frac{\cos x - 1}{x}$ we can conjecture that

$\frac{\sin x}{x}$ approaches 1 as x approaches 0,

and

$\frac{\cos x - 1}{x}$ approaches 0 as x approaches 0.

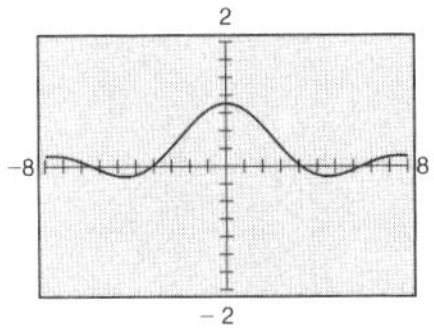

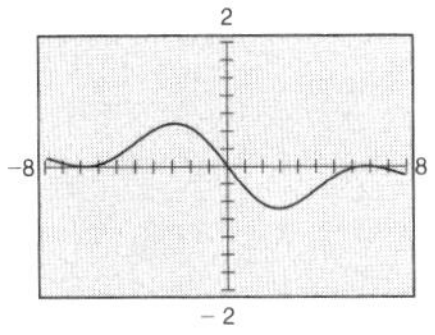

b. Let $f(x) = \sin x$ and $g(x) = \cos x$. Then

$$\begin{aligned}\frac{f(x+h) - f(x)}{h} &= \frac{\sin(x+h) - \sin x}{h} \\ &= \frac{\sin(x)\cos(h) + \cos(x)\sin(h) - \sin(x)}{h} \\ &= \frac{\sin x(\cos(h) - 1) + \cos(x)\sin(h)}{h} \\ &= \sin x\left(\frac{\cos(h) - 1}{h}\right) + \cos x\left(\frac{\sin(h)}{h}\right),\end{aligned}$$

and

$$\begin{aligned}\frac{g(x+h) - g(x)}{h} &= \frac{\cos(x+h) - \cos x}{h} \\ &= \frac{\cos(x)\cos(h) - \sin(x)\sin(h) - \cos(x)}{h} \\ &= \frac{\cos x(\cos(h) - 1) - \sin(x)\sin(h)}{h} \\ &= \cos x\left(\frac{\cos(h) - 1}{h}\right) - \sin x\left(\frac{\sin(h)}{h}\right).\end{aligned}$$

c. Since from part (a) $\frac{\cos(h)-1}{h} \to 0$ as $h \to 0$ and $\frac{\sin(h)}{h} \to 1$ as $h \to 0$,

$\frac{f(x+h) - f(x)}{h}$ approaches $\cos x$ as h approaches 0,

and

$\frac{g(x+h) - g(x)}{h}$ approaches $-\sin x$ as h approaches 0.

11. a. Let b denote the base of the inscribed triangle and h the height. Since

$$\sin\frac{\pi}{n}=\frac{b/2}{1}=\frac{b}{2}\text{ so } b=2\sin\frac{\pi}{n},$$

and since

$$\cos\frac{\pi}{n}=\frac{h}{1}\text{ so } h=\cos\frac{\pi}{n},$$

the area of each triangle is

$$A=\frac{1}{2}\left(2\sin\frac{\pi}{n}\right)\left(\cos\frac{\pi}{n}\right)=\frac{1}{2}(2\sin\frac{\pi}{n}\cos\frac{\pi}{n})=\frac{1}{2}\sin\frac{2\pi}{n}.$$

b. Since the area of a circle of radius 1 is known to be π we would expect $\frac{n}{2}\sin\frac{2\pi}{n}$ to better approximate the area as n increases. So as n approaches ∞ we expect $\frac{n}{2}\sin\frac{2\pi}{n}$ will approach π.

12. a. Since

$$\tan\frac{\pi}{n}=\frac{b/2}{1}=\frac{b}{2}\text{ implies } b=2\tan\frac{\pi}{n},\text{ so } A=\frac{1}{2}\left(2\tan\frac{\pi}{n}\right)(1)=\tan\frac{\pi}{n}.$$

b. As n approaches ∞, $n\tan\frac{\pi}{n}$ approaches the area of the circle π.

Chapter 4 Chapter Test (Page 266)

1. False. The number of degrees in $\frac{11\pi}{4}$ radians is 495°.
2. True.
3. False. The number of radians in 240° is $\frac{4\pi}{3}$.
4. True.
5. True.
6. True.
7. True.
8. False. If $\sin\theta>0$ and the $\sec\theta>0$, then θ is in quadrant I.
9. True.
10. False. If $\csc\theta>0$ and the $\cos\theta>0$, then θ is in quadrant I.
11. True.
12. False. If $\csc\theta<0$, then θ is in quadrant III or in quadrant IV.
13. False. If $\theta=\frac{5\pi}{6}$, then $\sin\theta=\frac{1}{2}$.
14. True.
15. False. If $\theta=-\frac{2\pi}{3}$, then $\cos\theta=-\frac{1}{2}$.
16. True.

17. True.

18. True.

19. True.

20. True.

21. True.

22. True.

23. True.

24. True.

25. True.

26. False. The missing side of the triangle has length $a = 8\cos 30°$.

27. True.

28. False. The hypotenuse of the triangle has length $x = 4\csc 30°$.

29. True.

30. True.

31. True.

32. True.

33. False. If $\cos\theta = \frac{4}{5}$ and θ lies in quadrant IV, then $\sin\theta = -\frac{3}{5}$.

34. True.

35. False. If $\sec\theta = \frac{3}{2}$ and $\tan\theta < 0$, then

$$\sin 2\theta = 2\sin\theta\cos\theta = 2\left(-\frac{\sqrt{5}}{3}\right)\frac{2}{3} = -\frac{4\sqrt{5}}{9}.$$

36. True.

37. False. For θ in the interval $[0, 2\pi]$, we have 2θ in $[0, 4\pi]$. So θ satisfies $\cos 2\theta = \frac{1}{2}$ when

$$\theta = \frac{\pi}{6}, \frac{5\pi}{6}, \frac{7\pi}{6}, \frac{11\pi}{6}.$$

38. False. For θ in the interval $[0, 2\pi]$, we have 3θ in $[0, 6\pi]$. So θ satisfies $\sin 3\theta = \frac{1}{2}$ when

$$\theta = \frac{\pi}{18}, \frac{5\pi}{18}, \frac{13\pi}{18}, \frac{17\pi}{18}, \frac{25\pi}{18}, \frac{29\pi}{18}.$$

39. False. The amplitude of the curve shown in the figure is 2.

40. True.

41. True.

42. True.

43. False. The period of the curve $y = 1 + 2\sin 2x$ is π.

44. True.

45. True.

46. False. The amplitude of the curve $y = -2\cos\frac{1}{2}x$ is 2. The amplitude must be positive.

47. True.

48. False. The graph of $y = \frac{1}{2}\sin 2x$ is obtained from the graph of $y = \sin x$ by a horizontal compression by a factor of 2 and a vertical compression by a factor of 2.

49. True.

50. False. The graph of $y = -\sin(x + \pi)$ is obtained by shifting the graph of $y = \sin x$ to the left π units and reflecting the result about the x-axis.

51. False. The graph of $y = \frac{1}{2}\sin\left(x - \frac{\pi}{4}\right)$ is obtained from the graph of $y = \sin x$ from a vertical compression by a factor of 2 followed by a shift to the right of $\frac{\pi}{4}$ units.

52. False. The graph of

$$y = \sin\left(2x + \frac{\pi}{2}\right) = \sin 2\left(x + \frac{\pi}{4}\right)$$

is obtained from the graph of $y = \sin x$ by a horizontal compression by a factor of 2, followed by a shift to the left $\frac{\pi}{4}$ units.

53. False. The graph of $y = 1 + \sin(2x - 3)$ is obtained from the graph of $y = \sin x$ from a horizontal compression by a factor of 2, followed by a shift to the right 3/2 units and upward 1 unit.

54. True.

55. False. The only values of θ in the interval $[0, 2\pi]$ that satisfy $2\cos^2\theta - \cos\theta - 1 = 0$ are

$$\theta = \frac{2\pi}{3}, \frac{4\pi}{3}, 0, \text{ and } 2\pi.$$

56. False. The only values of θ in the interval $[0, 2\pi]$ that satisfy $2\sin 2\theta = 4\sin\theta$ are $\theta = 0, \pi, 2\pi$.

57. True.

58. True.

59. False. For all θ, $\cos\left(\theta - \frac{\pi}{2}\right) = \sin\theta$.

60. False. For all θ, $\sin\left(\theta - \frac{\pi}{2}\right) = -\cos\theta$.

61. True.

62. False. Only when x is in the interval $[-\pi/2, \pi/2]$ is it true that $\arcsin(\sin(x)) = x$. This is because the range of the arcsine function is $[-\pi/2, \pi/2]$.

63. True.

64. False. Since the values of $\arcsin x$ are in the interval $[-\frac{\pi}{2}, \frac{\pi}{2}]$, we have $\arcsin(-\frac{1}{2}) = -\frac{\pi}{6}$.

65. True.

66. True.

67. False. The domain of the function $f(x) = \arcsin(2x)$ is $[-1/2, 1/2]$.

68. True.

69. True.

70. False. From the figure, we have $\beta = \arcsin\left(\frac{10}{4}\sin\alpha\right)$.

Exercise Set 5.2 (Page 282)

1. $f(x) = 2^x - 3$

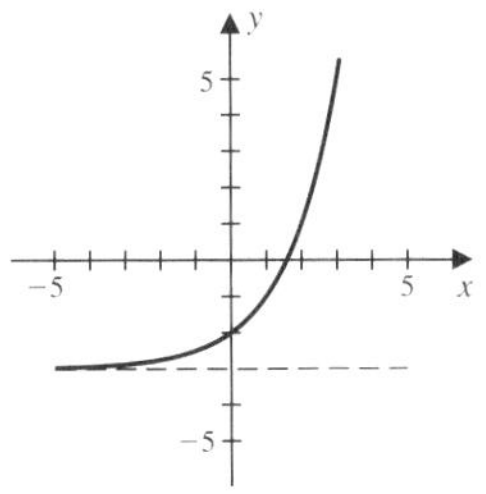

2. $f(x) = 2^{x-3} - 1$

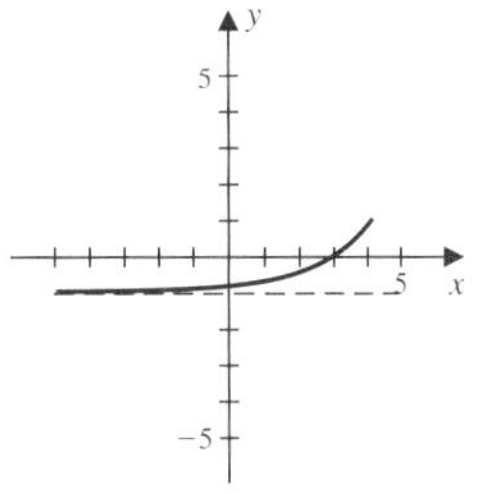

3. $f(x) = -4^x$

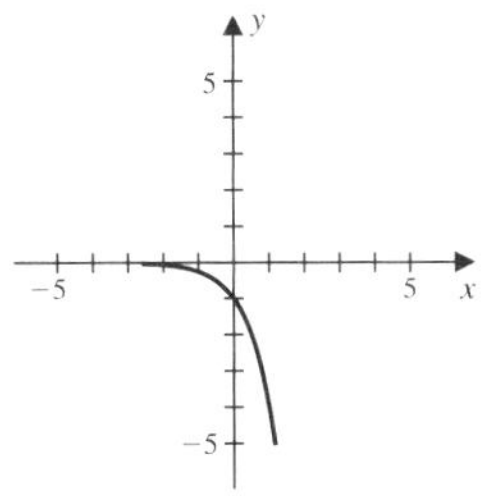

4. $f(x) = 10^{-x}$

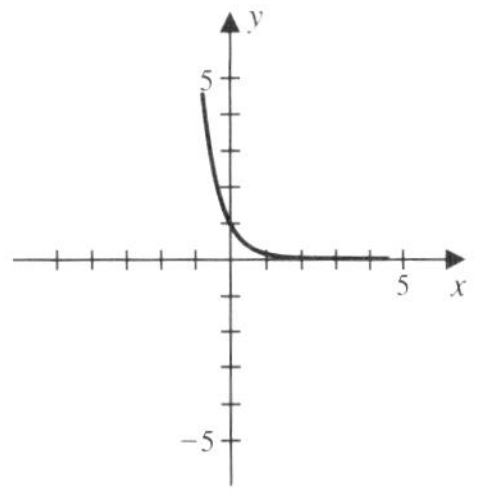

5. $f(x) = 2 \cdot \left(\frac{1}{3}\right)^{x-1} + 1$

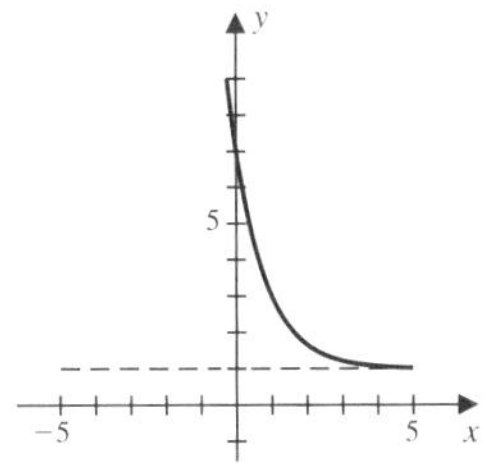

6. $f(x) = -4 \cdot \left(\frac{1}{3}\right)^{x+1} - 1$

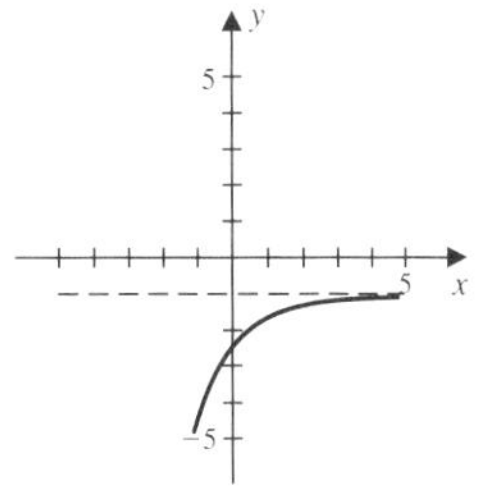

7. $f(x) = -e^{x-1}$

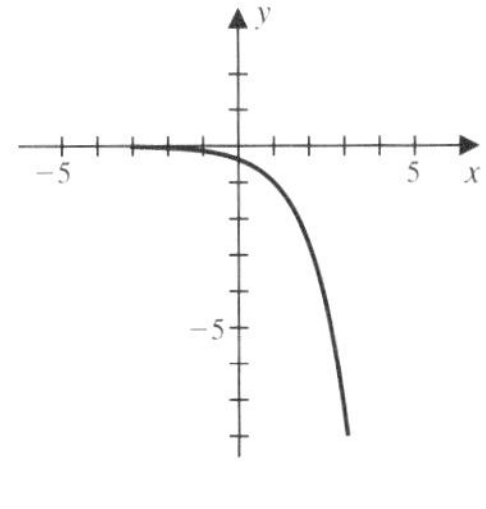

8. $f(x) = e^{x-2} + 1$

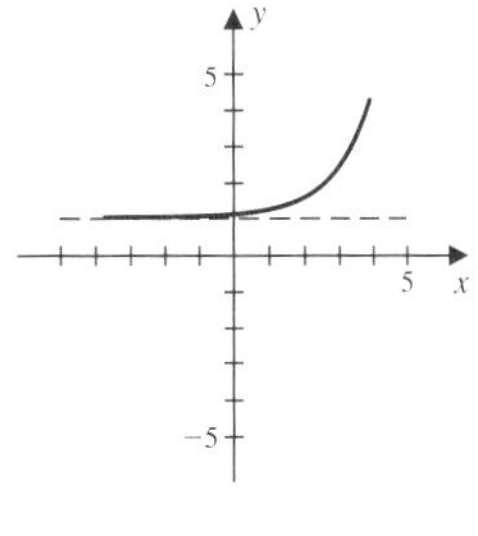

9. $f(x) = 3 - e^{-(x-1)}$

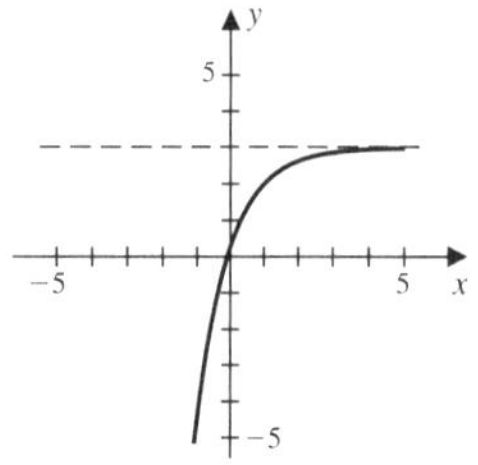

10. $f(x) = e^{-x} + 2$

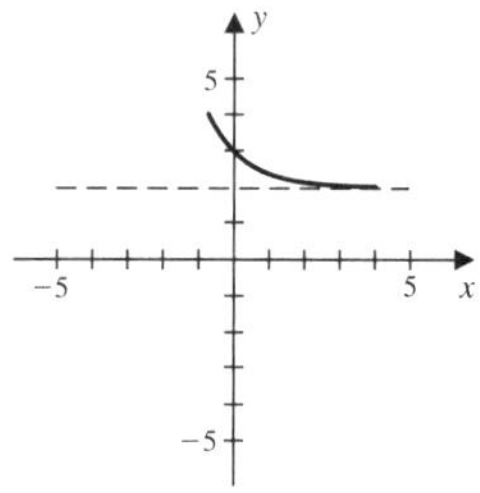

11. $f(x) = e^{2x}$

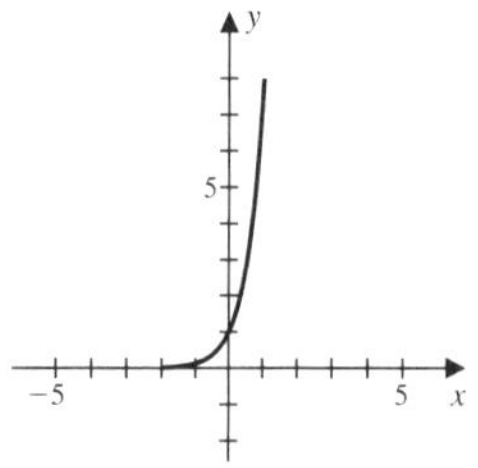

12. $f(x) = e^{2(x-1)} - 1$

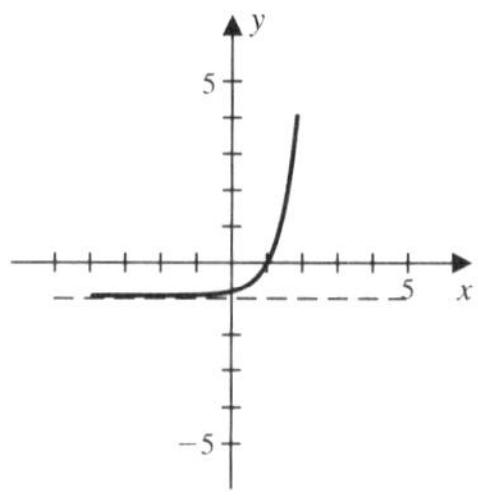

13. $f(x) = e^{|x|}$

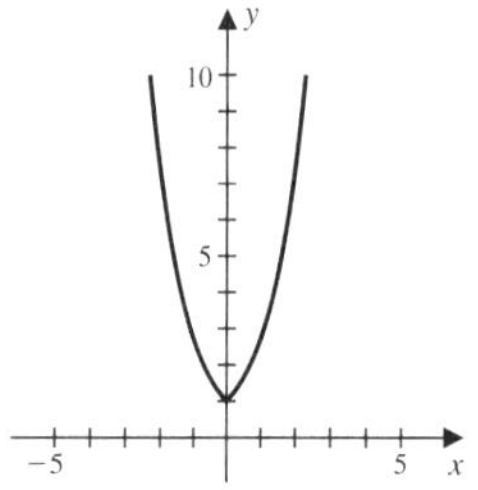

14. $f(x) = e^{-|x|}$

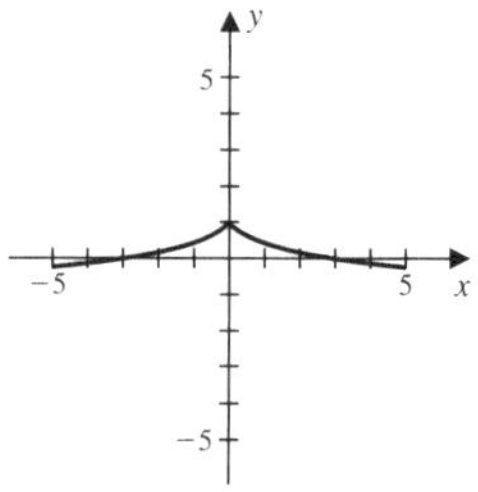

15. The graph shows that $e^x > 1$ when $x > 0$.

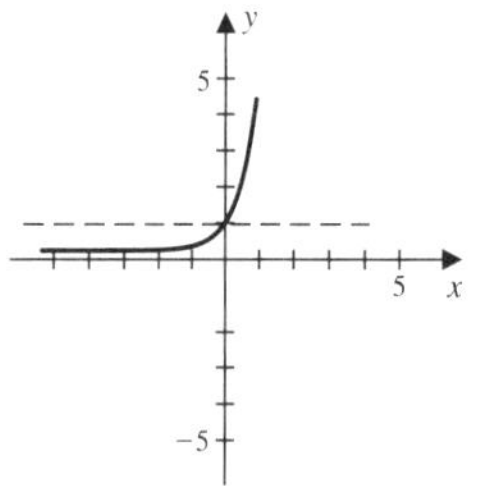

16. The graph shows that $2^x \leq 8$ when $x \leq 3$

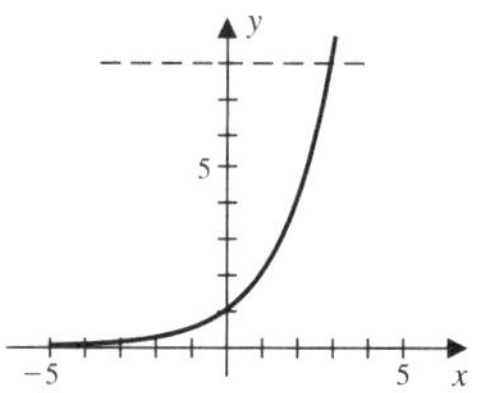

17. $\left(\frac{1}{2}\right)^x = \frac{1}{2^x} \geq 4$ implies $2^x \leq \frac{1}{4}$, so $x \leq -2$

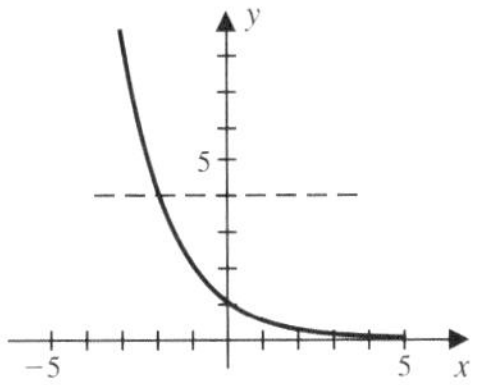

18. $\left(\frac{1}{3}\right)^x = \frac{1}{3^x} < 9$ implies $3^x > \frac{1}{9}$, so $x > -2$

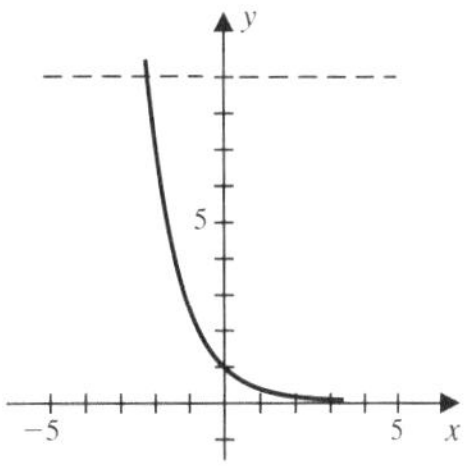

19. **a.** ii **b.** iii **c.** iv **d.** i

20. The figure shows $f(x) = ce^{x-1}$, for $c = 1, 2, -1, -2$.

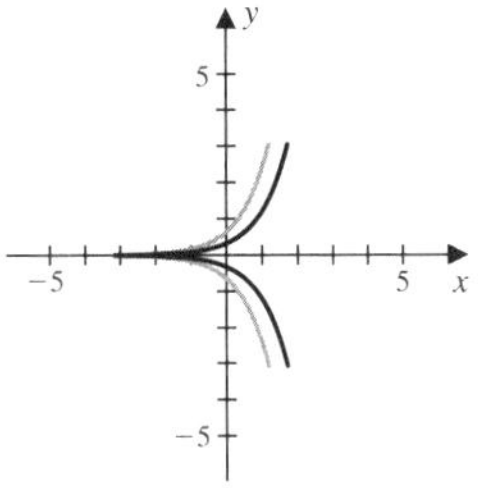

21. The points of intersection of $y = e^{x-2}$ and $y = x$ are $x \approx 0.2$, $x \approx 3.2$.

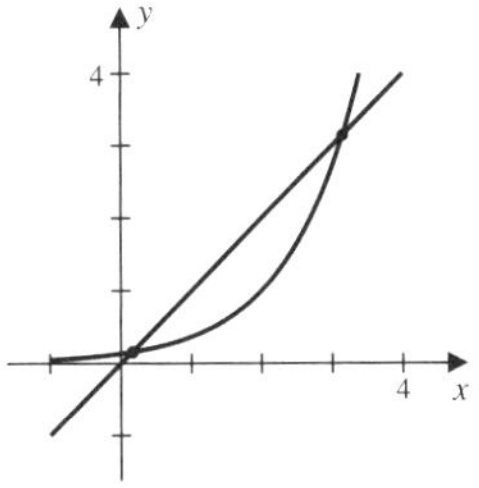

22. The point of intersection of $y = e^x$ and $y = x^2$ is $x \approx -0.7$.

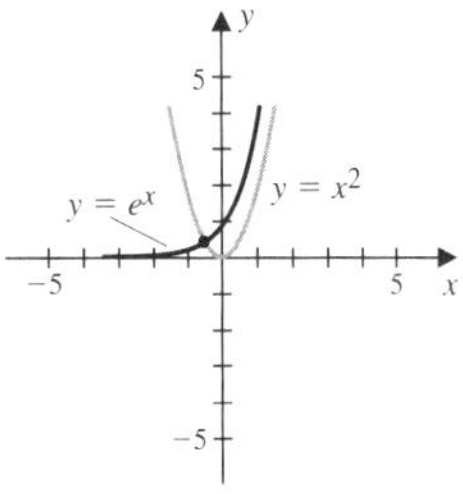

23. The points of intersection of $y = e^{-x}$ and $y = (x-2)^2$ are $x \approx 1.5$, $x \approx 2.3$, $x = -3.3$.

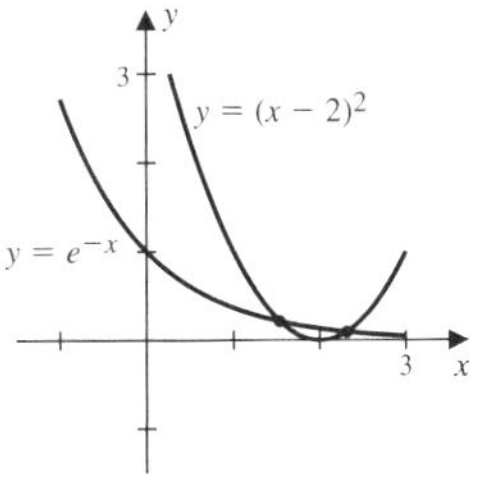

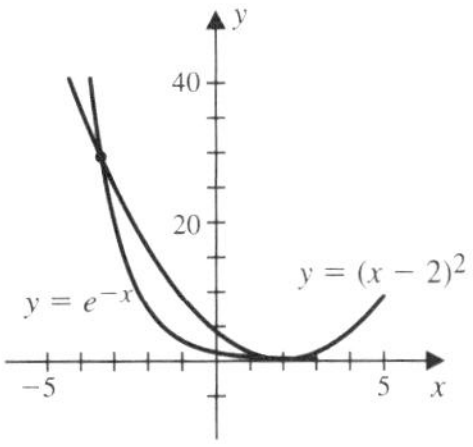

24. The points of intersection of $y = xe^x$ and $y = x^2 + 4x + 2$ are $x \approx -3.4$, $x \approx -0.7$, $x \approx 1.9$.

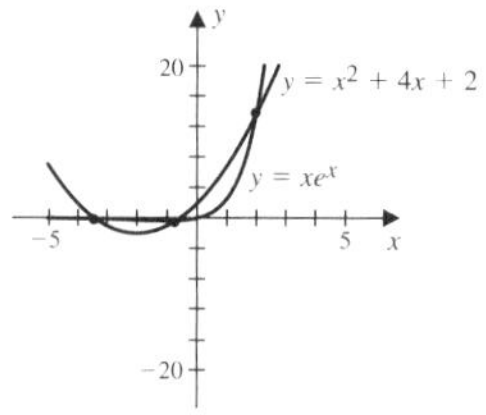

25. a. $f(x) = xe^x$
b. Increasing: $(-1, \infty)$; decreasing: $(-\infty, -1)$.

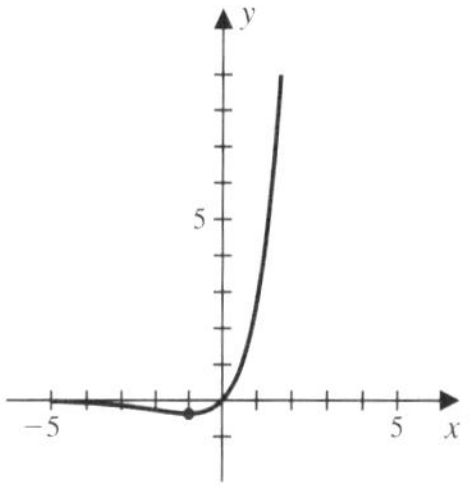

26. a. $f(x) = \frac{e^x}{x}$
b. Increasing: $(1, \infty)$; decreasing: $(-\infty, 0) \cup (0, 1)$.

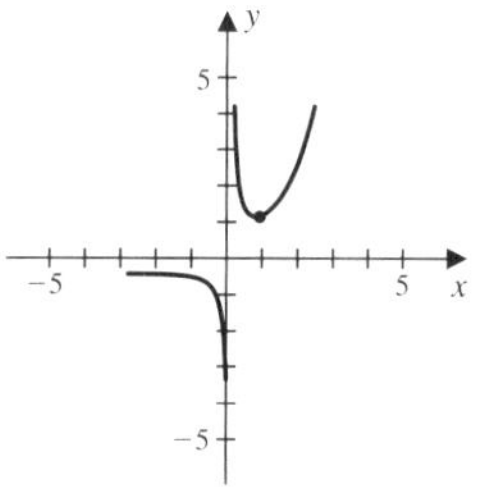

27. a. $f(x) = e^{-x^2-x}$
b. Increasing: $(-\infty, -0.5)$; decreasing: $(-0.5, \infty)$.

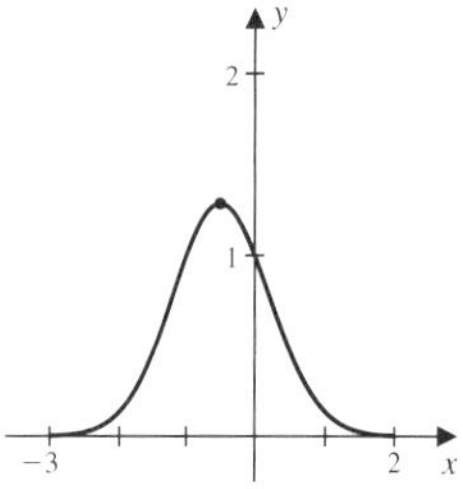

28. a. $f(x) = e^{x^3-x}$
b. Increasing: $(-\infty, -0.6) \cup (0.6, \infty)$; decreasing: $(-0.6, 0.6)$.

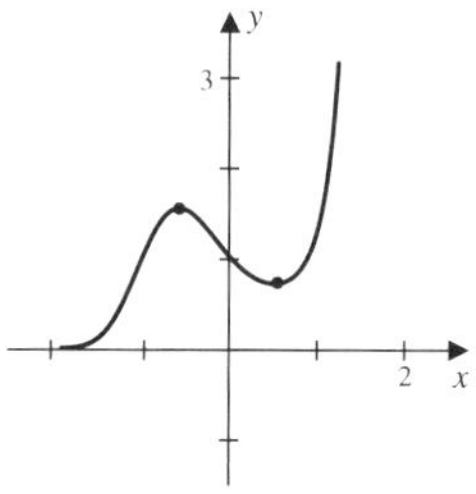

29. **a.** $f(x) = e^x - e^{-x}$
b. Increasing for all x.

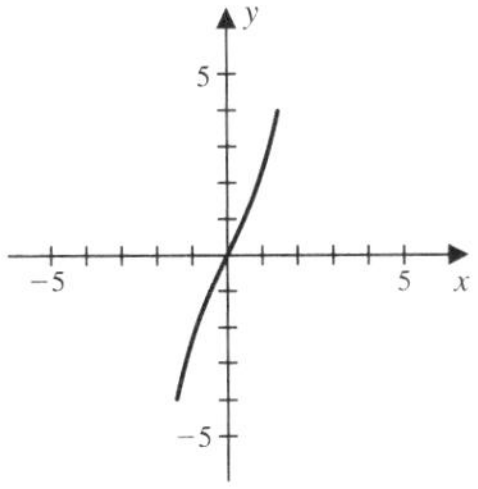

30. **a.** $f(x) = x + e^x$
b. Increasing for all x.

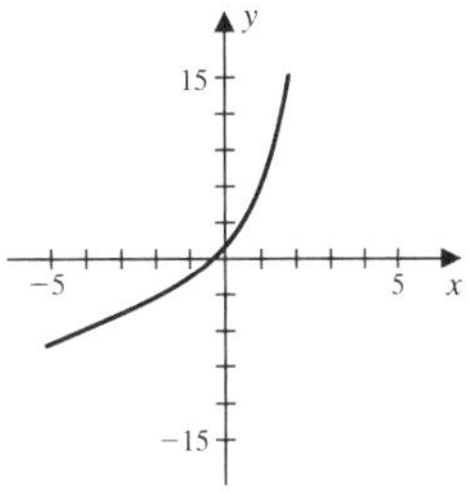

31. Let $3^x = e^{kx}$. If $x = 1$, then $3 = e^k$. The graphs of $y = e^x$ and $y = 3$ intersect when $x \approx 1.1$, so $k \approx 1.1$.

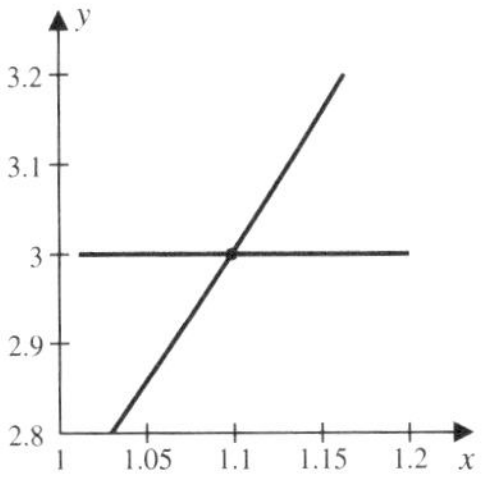

32. **a.** We have

n	1	5	10	10^2	10^3	10^4	10^5
$(1 + 1/n)^n$	2	2.4883	2.5937	2.7048	2.7169	2.7181	2.7182

b. The graph is shown.

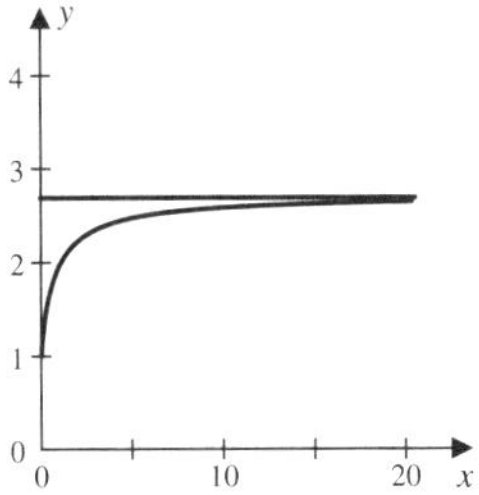

33. a. The function $f(x) = 2^x$ eventually grows much faster than $g(x) = x^5$. In Figure (iii) we see $f(x) = 2^x$ crosses $g(x) = x^5$ and remains above $y = g(x)$ from that point on.

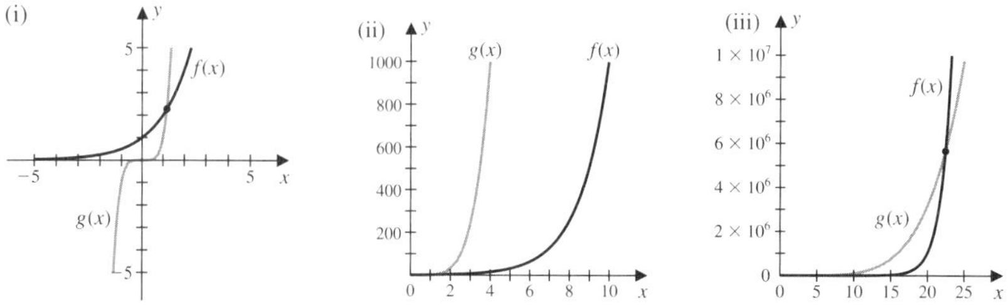

b. The points of intersection of the two graphs occur when $x \approx 1.2$ and $y \approx 2^{1.2}$, and when $x \approx 22.4$ and $y \approx 2^{22.4}$. These values of x are approximate solutions to $2^x = x^5$.

34. a. The function $f(x) = e^x$ eventually grows much faster than $g(x) = x^{10}$. In Figure (iii) we see $f(x) = e^x$ crosses $g(x) = x^{10}$ and remains above $y = g(x)$ from that point on.

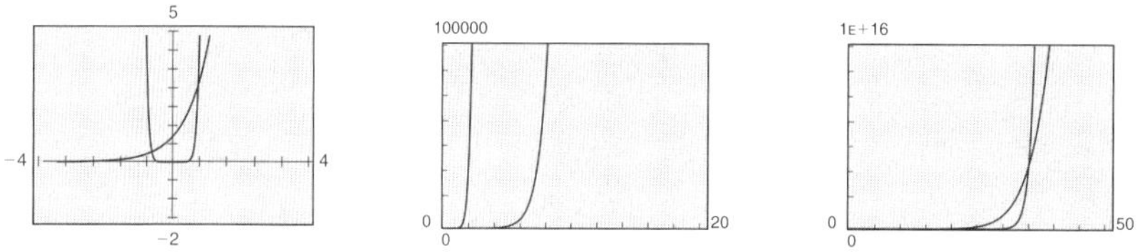

b. The points of intersection of the two graphs occur at $x \approx -0.9$ and $y \approx e^{-0.9}$, when $x \approx 1.1$, and $y \approx e^{1.1}$, and when $x \approx 35.8$, and $y \approx e^{35.8}$. These values of x are approximate solutions to $e^x = x^{10}$.

35. The value of the CD at 6.5% interest compounded n times per year and which matures in 5 years is $A_n(5) = 5000\left(1 + \frac{0.065}{n}\right)^{5n}$.

Interest Compounded	Value of CD
(a) annually: $n = 1$	\$6850.43
(b) monthly: $n = 12$	\$6914.09
(c) daily: $n = 365$	\$6919.95
(d) continuously: $A_c(5) = 5000e^{(0.065)5}$	\$6920.15

36. The value of the \$1000 at the end of 8 years which is compounded n times per year with an interest rate of 10% is $A_n(8) = 1000\left(1 + \frac{0.1}{n}\right)^{8n}$.

Interest Compounded	Value after 8 years
Annually: $n = 1$	$A_1(8) = 1000(1+0.1)^8 = \$2143.59$
Semiannually: $n = 2$	$A_2(8) = 1000\left(1+\frac{0.1}{2}\right)^{16} = \$2182.88$
Quarterly: $n = 4$	$A_4(8) = 1000\left(1+\frac{0.1}{4}\right)^{32} = \$2203.76$
Monthly: $n = 12$	$A_{12}(8) = 1000\left(1+\frac{0.1}{12}\right)^{96} = \$2218.18$
Weekly: $n = 52$	$A_{52}(8) = 1000\left(1+\frac{0.1}{52}\right)^{416} = \$2223.83$
Daily: $n = 365$	$A_{365}(8) = 1000\left(1+\frac{0.1}{365}\right)^{2920} = \$2225.30$
Hourly: $n = 8760$	$A_{8760}(8) = 1000\left(1+\frac{0.1}{8760}\right)^{70080} = \$2225.53$
Continuously	$A_c(8) = 1000e^{0.8} = \$2225.54$

37. If \$10, 000.00 is invested and the interest is compounded quarterly, after 5 years and an interest rate of

a. 8%, the value of the investment is

$$A_4(5) = 10000\left(1+\frac{0.08}{4}\right)^{20} = \$14,859.47;$$

b. 6.5%, the value of the investment is

$$A_4(5) = 10000\left(1+\frac{0.065}{4}\right)^{20} = \$13,804.20;$$

c. 6%, the value of the investment is

$$A_4(5) = 10000\left(1+\frac{0.06}{4}\right)^{20} = \$13,468.55;$$

d. 5.5%, the value of the investment is

$$A_4(5) = 10000\left(1+\frac{0.055}{4}\right)^{20} = \$13,140.66.$$

38. An initial investment of A_0 will have a return when

a. compounded annually at 8% of

$$A_0(1+0.08)^t = A_0(1.08)^t;$$

b. compounded semiannually at 7.5% of

$$A_0\left(1+\frac{0.075}{2}\right)^{2t} = A_0\left((1.0375)^2\right)^t = A_0(1.07640625)^t;$$

c. compounded continuously at 7% of

$$A_0 e^{0.07t} \approx A_0(1.072508)^t.$$

Part (a) gives the best return.

39. Solve $10000 = A_0\left(1+\frac{0.08}{2}\right)^{5(2)}$ for A_0, which gives $A_0 = \frac{10000}{(1+0.04)^{10}} \approx \$6756.00$.

40. Solve $50000 = A_0 e^{(0.085)7}$ for A_0, which gives $A_0 = \frac{50000}{e^{0.595}} \approx \$27.578.00$.

Exercise Set 5.3 (Page 292)

1. $x = \log_3 3^5$ implies $3^x = 3^5$ so $x = 5$

2. $x = \log_2 32$ implies $2^x = 32$ so $x = 5$

3. $x = \log_4 64$ implies $4^x = 64$ so $x = 3$

4. $x = \log_8 2^{12}$ implies $8^x = 2^{12}$ or $2^{3x} = 2^{12}$ so $x = 4$

5. $x = \log_4 2$ implies $4^x = 2$ so $x = \frac{1}{2}$

6. $x = \log_9 3$ implies $9^x = 3$ so $x = \frac{1}{2}$

7. $x = \log_{10} 0.001$ implies $10^x = 0.001$ or $10^x = \frac{1}{1000}$ so $x = -3$

8. $x = \log_{10} 10000$ implies $10^x = 10000$ so $x = 4$

9. $x = \log_2 \frac{1}{16}$ implies $2^x = \frac{1}{16}$ so $x = -4$

10. $x = \log_4 \frac{1}{2}$ implies $4^x = \frac{1}{2}$ so $x = -\frac{1}{2}$

11. Since $e^{\ln x} = x$ for $x > 0$ we have $e^{\ln 5} = 5$.

12. Since $a^{\log_a x} = x$ for $x > 0$ we have $5^{\ln_5 6} = 6$.

13. Since $\ln e^x = x$ for all x we have $\ln e^{\pi} = \pi$.

14. Since $\log_a a^x = x$ for all x we have $\log_2 2^{\sqrt{3}} = \sqrt{3}$.

15. Since $e^{\ln x} = x$ for $x > 0$ we have $e^{2\ln \pi} = e^{\ln \pi^2} = \pi^2$.

16. Since $e^{\ln x} = x$ for $x > 0$ we have $e^{-1/2 \ln 16} = e^{\ln(16)^{-1/2}} = e^{\ln \frac{1}{4}} = \frac{1}{4}$.

17. $\ln x(x+2) = \ln x + \ln(x+2)$

18. $\ln \frac{1}{x} = \ln 1 - \ln x = 0 - \ln x = -\ln x$

19. $\log_3 \frac{x^4}{x+1} = \log_3 x^4 - \log_3(x+1) = 4\log_3 x - \log_3(x+1)$

20. $\log_2(2x-1)^5 = 5\log_2(2x-1)$

21. $\ln \frac{3x^2}{(x+3)^4} = \ln 3x^2 - \ln(x+3)^4 = \ln 3 + \ln x^2 - 4\ln(x+3) = \ln 3 + 2\ln x - 4\ln(x+3)$

22. $\ln \frac{x\sqrt[3]{x^2}}{(x+2)^3} = \ln\left(xx^{2/3}\right) - \ln(x+2)^3 = \ln x^{5/3} - 3\ln(x+2) = \frac{5}{3}\ln x - 3\ln(x+2)$

23. We have

$$\log_3 \frac{(3x+2)^{3/2}(x-1)^3}{x\sqrt{x+1}} = \log_3\left[(3x+2)^{3/2}(x-1)^3\right] - \log_3\left(x\sqrt{x+1}\right)$$

$$
\begin{aligned}
&= \log_3(3x+2)^{3/2} + \log_3(x-1)^3 - \log_3 x - \log_3(x+1)^{1/2} \\
&= \frac{3}{2}\log_3(3x+2) + 3\log_3(x-1) - \log_3 x - \frac{1}{2}\log_3(x+1).
\end{aligned}
$$

24. We have

$$
\begin{aligned}
\ln \frac{x^2\sqrt{x+1}}{\sqrt[3]{x^2+2x+1}} &= \ln\left(x^2\sqrt{x+1}\right) - \ln(x^2+2x+1)^{1/3} \\
&= 2\ln x + \frac{1}{2}\ln(x+1) - \frac{1}{3}\ln(x^2+2x+1) \\
&= 2\ln x + \frac{1}{2}\ln(x+1) - \frac{1}{3}\ln(x+1)^2 \\
&= 2\ln x + \frac{1}{2}\ln(x+1) - \frac{2}{3}\ln(x+1) \\
&= 2\ln x - \frac{1}{6}\ln(x+1).
\end{aligned}
$$

25. We have

$$
\begin{aligned}
\ln\sqrt{x\sqrt{x+1}} &= \ln\left(x(x+1)^{1/2}\right)^{1/2} = \frac{1}{2}\ln\left(x(x+1)^{1/2}\right) \\
&= \frac{1}{2}\left[\ln x + \ln(x+1)^{1/2}\right] = \frac{1}{2}\ln x + \frac{1}{4}\ln(x+1).
\end{aligned}
$$

26. We have

$$
\begin{aligned}
\ln\sqrt{x\sqrt{\frac{x+1}{x+2}}} &= \ln\left(x\left(\frac{x+1}{x+2}\right)^{1/2}\right)^{1/2} = \frac{1}{2}\ln\left(x\left(\frac{x+1}{x+2}\right)^{1/2}\right) \\
&= \frac{1}{2}\ln x + \frac{1}{2}\ln\left(\frac{x+1}{x+2}\right)^{1/2} \\
&= \frac{1}{2}\ln x + \frac{1}{4}\ln\frac{x+1}{x+2} = \frac{1}{2}\ln x + \frac{1}{4}\ln(x+1) - \frac{1}{4}\ln(x+2).
\end{aligned}
$$

27. $\ln x + 2\ln(x+1) = \ln x + \ln(x+1)^2 = \ln\left(x(x+1)^2\right)$

28. $2\ln(x+2) + 3\ln(x-1) = \ln(x+2)^2 + \ln(x-1)^3 = \ln\left((x+2)^2(x-1)^3\right)$

29. $\frac{1}{2}\ln x - 2\ln(x-1) = \ln x^{1/2} - \ln(x-1)^2 = \ln \dfrac{\sqrt{x}}{(x-1)^2}$

30. $2\ln x - \frac{1}{3}\ln(x+1) = \ln x^2 - \ln(x+1)^{1/3} = \ln \dfrac{x^2}{\sqrt[3]{x+1}}$

31. $\ln(x-1) + \frac{1}{2}\ln x - 2\ln x = \ln(x-1)\sqrt{x} - \ln x^2 = \ln \dfrac{(x-1)\sqrt{x}}{x^2} = \ln \dfrac{x-1}{x^{3/2}}$

32.

$$\begin{aligned}\ln(x^2+x+1) - 3\ln(x+2) + \ln x &= \ln(x^2+x+1) - \ln(x+2)^3 + \ln x \\ &= \ln \frac{x^2+x+1}{(x+2)^3} + \ln x \\ &= \ln \frac{(x^2+x+1)x}{(x+2)^3} = \ln \frac{x^3+x^2+x}{(x+2)^3}\end{aligned}$$

33. $\log_3 x = 4$ implies $3^4 = x$ so $x = 81$

34. $\log_2 x = 5$ implies $2^5 = x$ so $x = 32$

35. $\log_2(3x-4) = 3$ implies $2^3 = 3x - 4$ or $3x = 12$ so $x = 4$

36. $\log_3(2-x) = 2$ implies $3^2 = 2 - x$ or $-x = 7$ so $x = -7$

37. $\log_x 4 = 2$ implies $x^2 = 4$ or $x = \pm 2$ so $x = 2$

38. $\log_x 3 = \frac{1}{3}$ implies $x^{\frac{1}{3}} = 3$ or $x = 3^3$ so $x = 27$

39. $\ln(2-x) = 4$ implies $e^{\ln(2-x)} = e^4$ or $2 - x = e^4$ so $x = 2 - e^4$

40. $1 - \ln(3x+2) = 0$ implies $\ln(3x+2) = 1$ or $e^{\ln(3x+2)} = e$
so $3x + 2 = e$ implies $3x = e - 2$ and $x = \frac{e-2}{3}$

41. $\ln 2 + \ln(x+1) = \ln(4x-7)$ implies $\ln 2(x+1) = \ln(4x-7)$
so $2x + 2 = 4x - 7$ implies $2x = 9$ and $x = \frac{9}{2}$

42. $2\ln x = \ln 4 + \ln(x+3)$ implies $\ln x^2 = \ln 4(x+3)$
so $x^2 = 4x + 12$ implies $0 = x^2 - 4x - 12 = (x-6)(x+2)$ and $x = 6, x = -2$
Hence $x = 6$, since $\ln x$ is not defined at $x = -2$, so $x = -2$ is not a solution.

43. $2\ln x = \ln(4x+6) - \ln 2$ implies $\ln x^2 = \ln \frac{4x+6}{2}$
so $x^2 = 2x + 3$ implies $x^2 - 2x - 3 = (x-3)(x+1) = 0$ and $x = 3, x = -1$
Hence $x = 3$, since $\ln x$ and $\ln(4x-6)$ are not defined at $x = -1$, so $x = -1$ is not a solution.

44. $\ln x + \ln(x-1) = \ln 2$ implies $\ln x(x-1) = \ln 2$
so $x^2 - x = 2$ implies $x^2 - x - 2 = (x-2)(x+1) = 0$ and $x = 2, x = -1$
Hence $x = 2$, since $\ln x$ and $\ln(x-1)$ are not defined at $x = -1$, is not a solution.

45. $\ln(2x-1) - \ln(x-1) = \ln 5$ implies $\ln \frac{2x-1}{x-1} = \ln 5$
so $\frac{2x-1}{x-1} = 5$ implies $2x - 1 = 5x - 5$ so $3x = 4$ and $x = \frac{4}{3}$

46. $2\ln(x+2) - \ln x = \ln 8$ implies $\ln \frac{(x+2)^2}{x} = \ln 8$
so $\frac{(x+2)^2}{x} = 8$ implies $(x+2)^2 = 8x$ so $0 = x^2 - 4x + 4 = (x-2)^2$ and $x = 2$

47. $\log_3(2x^2 + 17x) = 2$ implies $2x^2 + 17x = 9$ or $0 = 2x^2 + 17x - 9 = (2x-1)(x+9)$
so $2x - 1 = 0, x = -9$ implies $x = \frac{1}{2}, x = -9$

48. $\log_2(5x^2 - 8x) = 2$ implies $5x^2 - 8x = 4$
so $5x^2 - 8x - 4 = (5x+2)(x-2) = 0$ implies $x = -\frac{2}{5}, x = 2$.

49. $4^x = 3$ implies $\ln 4^x = \ln 3$ so $x \ln 4 = \ln 3$ and $x = \frac{\ln 3}{\ln 4} = \log_4 3$

50. $5^{2x-1} = 2$ implies $\ln 5^{2x-1} = \ln 2$ or $(2x-1)\ln 5 = \ln 2$
so $2x - 1 = \frac{\ln 2}{\ln 5}$ and $x = \frac{1}{2}\left[\frac{\ln 2}{\ln 5} + 1\right] = \frac{1}{2}[1 + \log_5 2]$.

51. $e^{2x} = 3^{x-4}$ implies $\ln e^{2x} = \ln 3^{x-4}$ or $2x \ln e = (x-4)\ln 3$
so $2x = (\ln 3)x - 4\ln 3$ implies $(\ln 3 - 2)x = 4\ln 3$ and $x = \frac{4\ln 3}{\ln 3 - 2}$

52. $2^{x-2} = e^{x/2}$ implies $(x-2)\ln 2 = x\ln 2 - 2\ln 2 = \frac{x}{2}$ or $2x\ln 2 - 4\ln 2 = x$ so $2x\ln 2 - x = 4\ln 2$ implies $x(2\ln 2 - 1) = 4\ln 2$ and $x = \frac{4\ln 2}{2\ln 2 - 1}$

53. $2\cdot 3^{-x} = 2^{3x}$ implies that $\ln\left(2\cdot 3^{-x}\right) = \ln 2^{3x}$ or $\ln 2 - x\ln 3 = 3x\ln 2$ so $\ln 2 = 3x\ln 2 + x\ln 3 = x(3\ln 2 + \ln 3)$ and $x = \frac{\ln 2}{3\ln 2 + \ln 3}$

54. $3e^{-x} = 4^{3x-1}$ implies $\ln\left(3e^{-x}\right) = \ln 4^{3x-1}$ or $\ln 3 + \ln e^{-x} = (\ln 4)(3x-1)$ so $-x + \ln 3 = (3\ln 4)x - \ln 4$ implies $(3\ln 4 + 1)x = \ln 3 + \ln 4$ and $x = \frac{\ln 3 + \ln 4}{3\ln 4 + 1}$

55. $y = \log_2(x-3)$

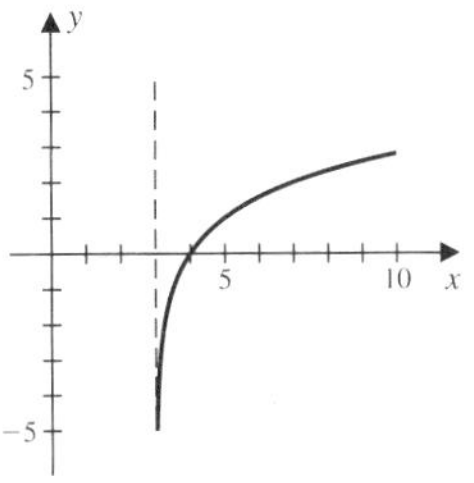

56. $y = -\log_2(x+4)$

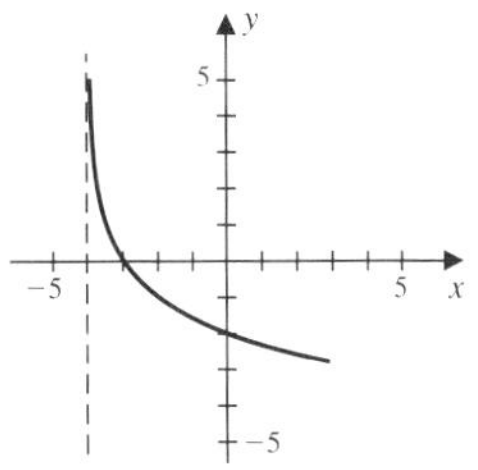

57. $y = 2 - \log_2(x-1)$

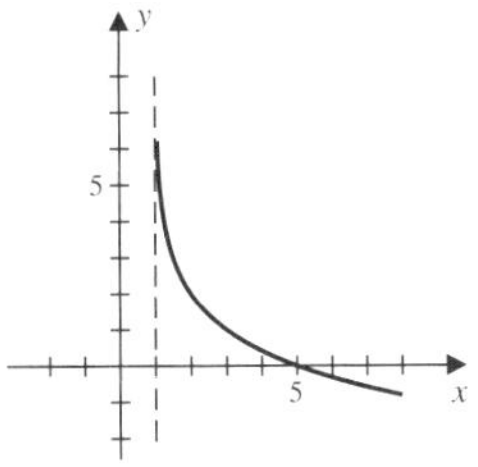

58. $y = \log_3(x-2) + 1$

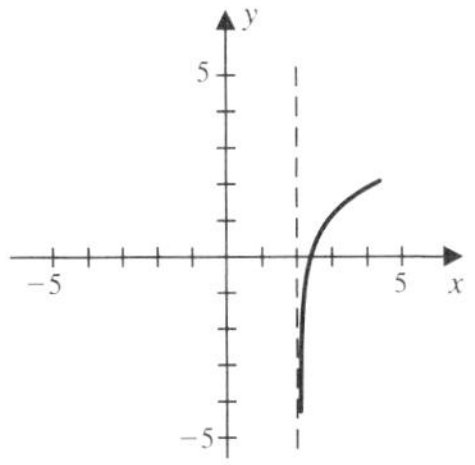

59. $y = 2\ln(x+1) - 3$

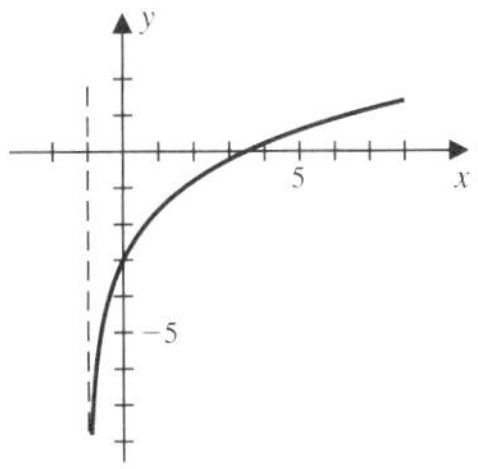

60. $y = -2\ln(x-1) + 1$

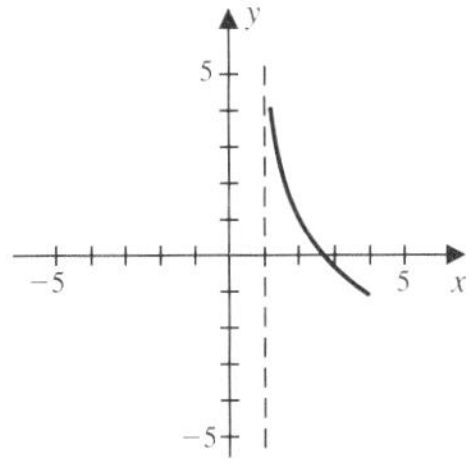

61. $y = \ln(-x)$

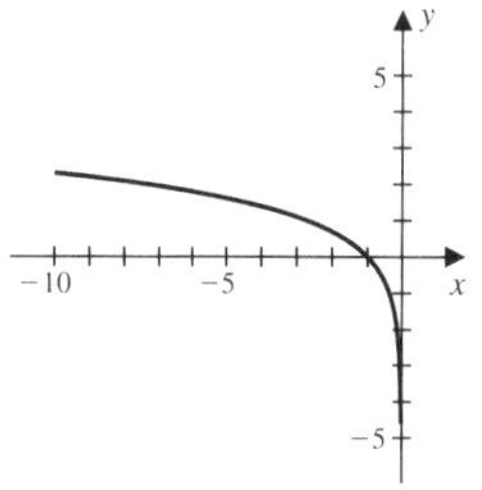

62. $y = \ln(3 - x) = \ln(-(x - 3))$

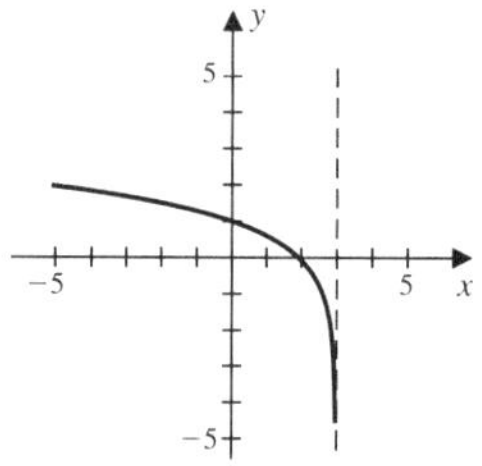

63. $y = |\ln x|$

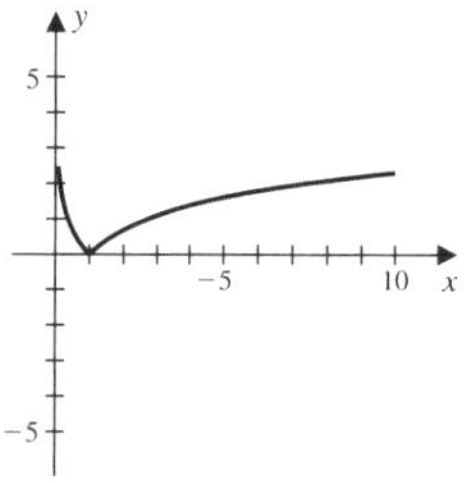

64. $y = \ln|x|$

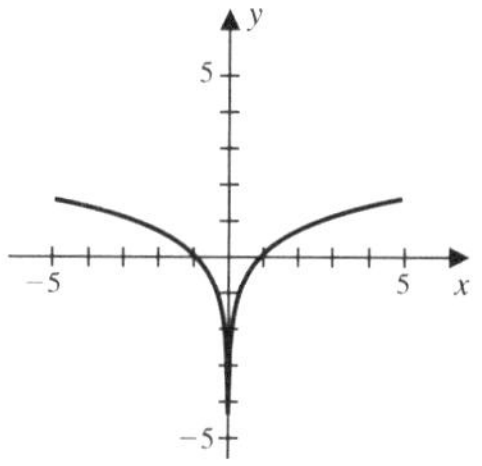

65. **a.** iii **b.** iv **c.** i **d.** ii

66. The figure shows $f(x) = c\ln(x - 1) + 1$, for $c = 1, 2, -1, -2$.

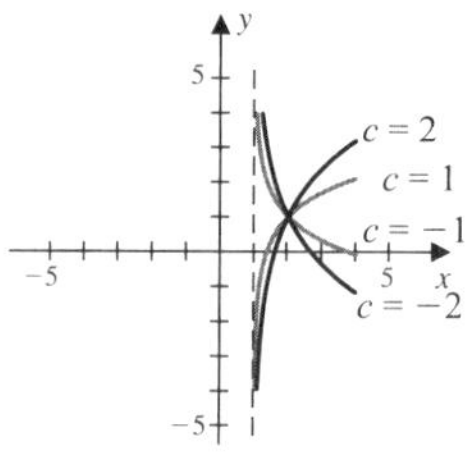

67. $y = \ln(4 - x^2)$

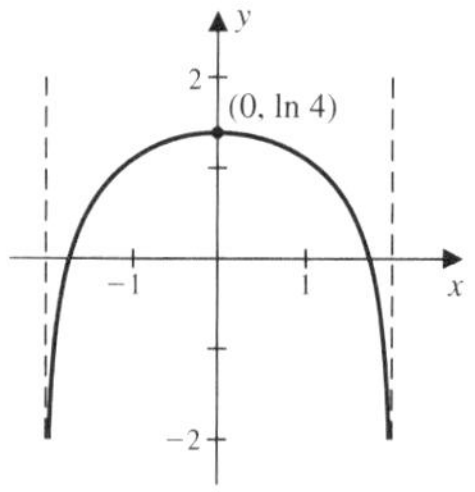

68. $y = x^2 - \ln x$

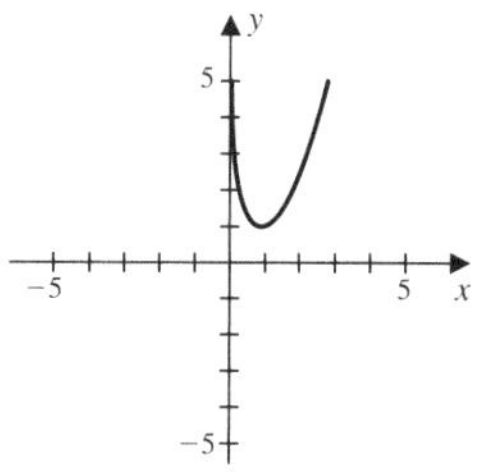

69. $y = (\ln x)/x$

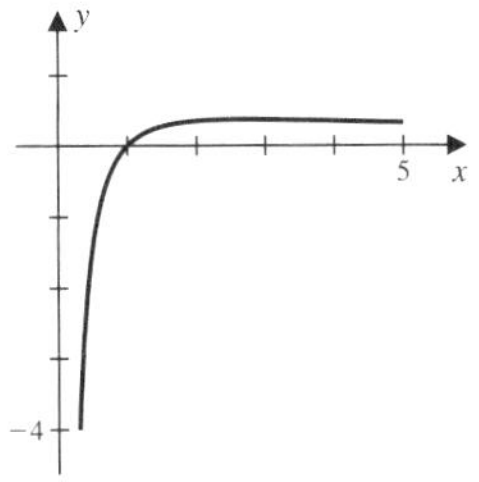

70. $y = \ln|x^2 - 1|$

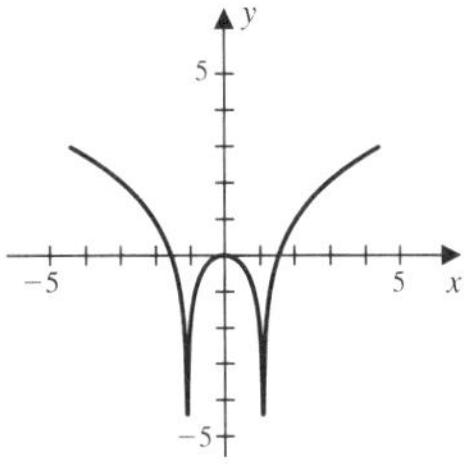

71. The function $g(x) = \sqrt[n]{x}$ grows more rapidly than $f(x) = a + \ln x$ for all $n > 0$.

72. We have $3^x = e^{kx}$ implies $\ln 3^x = \ln e^{kx}$ so $x \ln 3 = kx$ and $k = \ln 3 \approx 1.1$. In Exercise 31 of Section 4.2, the approximated value of k was 1.1.

Exercise Set 5.4 (Page 298)

1. We have $Q(t) = Q_0 e^{kt}$ with $Q_0 = 1000$.

a. Since the bacteria doubles every 4 hours,

$$2(1000) = Q(4) = 1000e^{4k} \text{ implies } e^{4k} = 2 \text{ so } 4k = \ln 2 \text{ and } k = \frac{\ln 2}{4},$$

and $Q(t) = 1000e^{\frac{\ln 2}{4}t}$.

b. $Q(7) = 1000e^{\frac{\ln 2}{4}7} \approx 3364$

c. $1000e^{\frac{\ln 2}{4}t} = 20000$ implies $e^{\frac{\ln 2}{4}t} = 20$ so $\frac{\ln 2}{4}t = \ln 20$ and $t = \frac{4 \ln 20}{\ln 2} \approx 17.3$ hours

2. We have $Q(t) = Q_0 e^{kt}$ with $Q_0 = 200$.

a. Since the bacteria doubles every 22 minutes, which is $\frac{22}{60} = \frac{11}{30}$ hours,

$$2(200) = Q\left(\frac{11}{30}\right) = 200e^{\frac{11}{30}k} \text{ implies } e^{\frac{11}{30}k} = 2 \text{ so } \frac{11}{30}k = \ln 2 \text{ and } k = \frac{30 \ln 2}{11},$$

and $Q(t) = 200e^{\frac{30 \ln 2}{11}t}$.

b. $Q(10) = 200e^{\frac{30 \ln 2}{11}10} \approx 32,429,404,090 \approx 3.24 \times 10^{10}$

c. $200e^{\frac{30 \ln 2}{11}t} = 10000$ implies $e^{\frac{30 \ln 2}{11}t} = 50$ so $\frac{30 \ln 2}{11}t = \ln 50$ and $t = \frac{11 \ln 50}{30 \ln 2} \approx 2.1$ hours

3. We have $Q(t) = Q_0 e^{kt}$ with $Q_0 = 500$.

a. Since after 5 hours there are 4000 bacteria,

$$4000 = Q(5) = 500e^{5k} \text{ implies } e^{5k} = 8 \text{ or } \ln e^{5k} = \ln 8 \text{ so } 5k = \ln 8 \text{ and } k = \frac{\ln 8}{5},$$

and $Q(t) = 500e^{\frac{\ln 8}{5}t}$.

b. $Q(6) = 500e^{\frac{\ln 8}{5}6} \approx 6063$

c. $500e^{\frac{\ln 8}{5}t} = 15000$ implies $e^{\frac{\ln 8}{5}t} = 30$ so $\frac{\ln 8}{5}t = \ln 30$ and $t = \frac{5\ln 30}{\ln 8} \approx 8.2$ hours

d. $500e^{\frac{\ln 8}{5}t} = 1000$ implies $e^{\frac{\ln 8}{5}t} = 2$ so $\frac{\ln 8}{5}t = \ln 2$ and $t = \frac{5\ln 2}{\ln 8} \approx 1.7$ hours

4. We have $Q(t) = Q_0 e^{kt}$ with $Q_0 = 8500$.

a. Since after 2 hours there was a 15% increase in the size of the population, that is, an increase of $0.15(8500) = 1275$, we have

$$9775 = Q(2) = 8500e^{2k} \text{ implies } e^{2k} = 1.15 \text{ or } \ln e^{2k} = \ln 1.15 \text{ so } 2k = \ln 1.15 \text{ and } k = \frac{\ln 1.15}{2},$$

and $Q(t) = 8500e^{\frac{\ln 1.15}{2}t}$.

b. $Q(8) = 8500e^{\frac{\ln 1.15}{2}8} \approx 14867$

c. $8500e^{\frac{\ln 1.15}{2}t} = 35000$ implies $e^{\frac{\ln 1.15}{2}t} = \frac{35000}{8500} = \frac{70}{17}$ so $\frac{\ln 1.15}{2}t = \ln\frac{70}{17}$ and $t = \frac{2\ln\frac{70}{17}}{\ln 1.15} \approx 20.3$ hours

d. $3Q_0 = Q_0 e^{\frac{\ln 1.15}{2}t}$ implies $e^{\frac{\ln 1.15}{2}t} = 3$ so $\frac{\ln 1.15}{2}t = \ln 3$ and $t = \frac{2\ln 3}{\ln 1.15} \approx 15.7$ hours

5. We have $Q(t) = Q_0 e^{kt}$ with $Q_0 = 50$.

a. If the half life is 578 hours,

$$25 = Q(578) = 50e^{578k} \text{ implies } e^{578k} = \frac{1}{2} \text{ so } 578k = \ln\frac{1}{2} = -\ln 2 \text{ and } k = -\frac{\ln 2}{578},$$

and $Q(t) = 50e^{-\frac{\ln 2}{578}t}$.

b. $Q(100) = 50e^{-\frac{\ln 2}{578}100} \approx 44.35$ milligrams

c. $50e^{-\frac{\ln 2}{578}t} = 10$ implies $e^{-\frac{\ln 2}{578}t} = \frac{1}{5}$ so $-\frac{\ln 2}{578}t = \ln\frac{1}{5} = \ln 1 - \ln 5 = -\ln 5$ and $t = \frac{578\ln 5}{\ln 2} \approx$ 1342.1 hours

6. We have $Q(t) = Q_0 e^{kt}$. If the half life of the substance is 8 years, then

$$Q_0 e^{8k} = \frac{1}{2}Q_0 \text{ implies } e^{8k} = \frac{1}{2} \text{ so } 8k = \ln\frac{1}{2} = \ln 1 - \ln 2 = -\ln 2 \text{ and } k = -\frac{\ln 2}{8},$$

and $Q(t) = Q_0 e^{-\frac{\ln 2}{8}t}$.

a. $Q(t) = 200e^{-\frac{\ln 2}{8}t}$.

b. $Q(15) = 200e^{-\frac{\ln 2}{8}15} \approx 54.5$ grams.

c. If 90% of the sample decays, there will be only 10% remaining, so solve

$$Q_0e^{-\frac{\ln 2}{8}t} = 0.1Q_0 \text{ implies } e^{-\frac{\ln 2}{8}t} = \frac{1}{10} \text{ and } -\frac{\ln 2}{8}t = \ln\frac{1}{10} = -\ln 10.$$

So $t = \frac{8\ln 10}{\ln 2} \approx 26.6$ years.

7. We have $Q(t) = Q_0e^{kt}$. If the culture doubles in size in 2 hours, then

$$2Q_0 = Q(2) = Q_0e^{2k} \text{ implies } e^{2k} = 2 \text{ so } 2k = \ln 2 \text{ and } k = \frac{\ln 2}{2},$$

and $Q(t) = Q_0e^{\frac{\ln 2}{2}t}$. The culture will triple in size when,

$$Q_0e^{\frac{\ln 2}{2}t} = 3Q_0 \text{ implies } e^{\frac{\ln 2}{2}t} = 3 \text{ so } \frac{\ln 2}{2}t = \ln 3 \text{ and } t = \frac{2\ln 3}{\ln 2} \approx 3.2 \text{ hours.}$$

8. We have $Q(t) = Q_0e^{kt}$. If the culture doubles in size in 4 hours, then

$$2Q_0 = Q(4) = Q_0e^{4k} \text{ implies } e^{4k} = 2 \text{ so } 4k = \ln 2 \text{ and } k = \frac{\ln 2}{4},$$

and $Q(t) = Q_0e^{\frac{\ln 2}{4}t}$.

a. $Q(8) = Q_0e^{\frac{\ln 2}{4}8} = Q_0e^{2\ln 2} = Q_0e^{\ln 4} = 4Q_0$

b. $Q(16) = Q_0e^{\frac{\ln 2}{4}16} = Q_0e^{4\ln 2} = Q_0e^{\ln 16} = 16Q_0$

9. We have $Q(t) = 200e^{kt}$. If the mass decays to 180 grams in 2 years, then

$$Q(2) = 200e^{2k} = 180 \text{ implies } e^{2k} = \frac{180}{200} = \frac{9}{10} \text{ and } 2k = \ln\frac{9}{10} = \ln 9 - \ln 10.$$

So $k = \frac{\ln 9 - \ln 10}{2}$ and $Q(t) = 200e^{\frac{\ln 9 - \ln 10}{2}t}$.

The half life is the time required for half the substance to decay, so

$$100 = 200e^{\frac{\ln 9-\ln 10}{2}t} \text{ implies } e^{\frac{\ln 9-\ln 10}{2}t} = \frac{1}{2} \text{ and } \frac{\ln 9 - \ln 10}{2}t = \ln\frac{1}{2} = -\ln 2.$$

So $t = -\frac{2\ln 2}{\ln 9 - \ln 10} \approx 13.16$ years.

10. We have $Q(t) = Q_0 e^{kt}$. If the substance decays 3% in 5 years, then after 5 years there remains 97% of the initial amount. So

$$0.97Q_0 = Q(5) = Q_0 e^{5k} \text{ implies } e^{5k} = 0.97 \text{ so } 5k = \ln(0.97) \text{ and } k = \frac{\ln(0.97)}{5},$$

and $Q(t) = Q_0 e^{\frac{\ln(0.97)}{5}t}$. The half life is the time required for half the substance to decay, so

$$\frac{1}{2}Q_0 = Q_0 e^{\frac{\ln(0.97)}{5}t} \text{ implies } e^{\frac{\ln(0.97)}{5}t} = \frac{1}{2} \text{ and } \frac{\ln(0.97)}{5}t = \ln\frac{1}{2} = -\ln 2.$$

So $t = -\frac{5\ln 2}{\ln(0.97)} \approx 113.8$ years.

11. a. Let $Q(t)$ represent the population t years after 1950. Since the initial population is the 1950 statistic, $Q(0) = 2555$. Since 1960 is ten years after the initial date of 1950,

$$3040 = Q(10) = 2555e^{10k} \text{ implies } e^{10k} = \frac{3040}{2555} \text{ and } 10k = \ln\frac{3040}{2555}.$$

So $k = \frac{1}{10}\ln\frac{3040}{2555}$ and $Q(t) = 2555e^{\frac{1}{10}\ln\frac{3040}{2555}t}$.

The population in 2050, which is 100 years after the initial year of 1950, is

$$Q(100) = 2555e^{\frac{1}{10}\left(\ln\frac{3040}{2555}\right)100} \approx 14528 \text{ million.}$$

b. Since

$$Q(0) = 5275 \text{ implies } 6079 = Q(10) = 5275e^{10k} \text{ we have } e^{10k} = \frac{6079}{5275} \text{ and } k = \frac{1}{10}\ln\frac{6079}{5275},$$

and

$$Q(t) = 5275e^{\frac{1}{10}\ln\frac{6079}{5275}t}.$$

The population in 2050, which is 60 years after the initial year of 1990, is

$$Q(60) = 5275e^{\frac{1}{10}\left(\ln\frac{6079}{5275}\right)60} \approx 12356 \text{ million.}$$

12. We have $Q(t) = 6.3e^{kt}$. If the population grows by 2% in one year, then the population after 1 year is, $1.02(6.3) = 6.426$. Then

$$6.426 = Q(1) = 6.3e^{k} \text{ implies } e^{k} = \frac{6.426}{6.3} \text{ and } k = \ln\frac{6.426}{6.3},$$

and $Q(t) = 6.3e^{\ln\frac{6.426}{6.3}t}$.

a. $2Q_0 = Q_0 e^{\ln\frac{6.426}{6.3}t}$ implies $\ln\frac{6.426}{6.3}t = \ln 2$ and $t = \frac{\ln 2}{\ln\frac{6.426}{6.3}} \approx 35$ years

b. $3Q_0 = Q_0 e^{\ln\frac{6.426}{6.3}t}$ implies $\ln\frac{6.426}{6.3}t = \ln 3$ and $t = \frac{\ln 3}{\ln\frac{6.426}{6.3}} \approx 55.5$ years

13. We have $A(t) = 10000e^{0.08t}$.

a. $20000 = 10000e^{0.08t}$ implies $2 = e^{0.08t}$ so $0.08t = \ln 2$ and $t = \frac{\ln 2}{0.08} = 8.7$ years

b. $30000 = 10000e^{0.08t}$ implies $3 = e^{0.08t}$ so $0.08t = \ln 3$ and $t = \frac{\ln 3}{0.08} = 13.7$ years

14. We have $A(t) = 25000e^{0.07t}$.

a. $50000 = 25000e^{0.07t}$ implies $2 = e^{0.07t}$ so $0.07t = \ln 2$ and $t = \frac{\ln 2}{0.07} \approx 9.9$ years

b. $75000 = 25000e^{0.07t}$ implies $3 = e^{0.07t}$ so $0.07t = \ln 3$ and $t = \frac{\ln 3}{0.07} \approx 15.7$ years

15. We have $A(t) = 10000e^{it}$. Then
$25000 = A(5) = 10000e^{5i}$ implies $e^{5i} = \frac{5}{2}$ so $5i = \ln \frac{5}{2}$ and $i = \frac{1}{5} \ln \frac{5}{2} \approx 0.18$ or 18%.

16. We have $A(t) = 20000e^{it}$. Then
$120000 = A(15) = 20000e^{15i}$ implies $e^{15i} = 6$ so $15i = \ln 6$ and $i = \frac{\ln 6}{15} \approx 0.12$ or 12%.

17. We have $T(t) = T_m + (T_0 - T_m)e^{kt}$, where $T_0 = -3°\text{C}$ and $T_m = 20°\text{C}$, so $T(t) = 20 - 23e^{kt}$. Since one minute later the temperature reads 5°C,

$$5 = T(1) = 20 - 23e^k \text{ implies } e^k = \frac{15}{23} \text{ so } k = \ln \frac{15}{23},$$

and $T(t) = 20 - 23e^{\ln \frac{15}{23} t}$. The thermometer will read 19.5°C, when

$$20 - 23e^{\ln \frac{15}{23} t} = 19.5 \text{ implies } e^{\ln \frac{15}{23} t} = \frac{0.5}{23} = \frac{1}{46} \text{ and } \ln \frac{15}{23} t = \ln \frac{1}{46} = -\ln 46.$$

So $t = -\frac{\ln 46}{\ln \frac{15}{23}} \approx 9$ minutes.

18. We have $T(t) = T_m + (T_0 - T_m)e^{kt}$, where $T_0 = 20°\text{C}$ and $T_m = 200°\text{C}$, so $T(t) = 200 - 180e^{kt}$. Since one hour later the temperature of the potato is 150°C,

$$150 = T(1) = 200 - 180e^k \text{ implies } e^k = \frac{5}{18} \text{ so } k = \ln \frac{5}{18},$$

and $T(t) = 200 - 180e^{\ln \frac{5}{18} t}$. The potato will reach 50°C, when

$$200 - 180e^{\ln \frac{5}{18} t} = 50 \text{ implies } e^{\ln \frac{5}{18} t} = \frac{15}{18} = \frac{5}{6} \text{ and } \ln \frac{5}{18} t = \ln \frac{5}{6}.$$

So $t = \frac{\ln \frac{5}{6}}{\ln \frac{5}{18}} \approx 0.14$ hours $= 8.5$ minutes.

19. By Newton's Law of Cooling the temperature of the body at time t is

$$T(t) = T_m + (T_0 - T_m)e^{kt},$$

where $T_m = 62°$F, the constant temperature of the lake, and $T_0 = 67°$C, the temperature when the body was found at 11:00 a.m. So $T(t) = 62 + 5e^{kt}$. At noon, or 1 hour after the body was found the temperature was 66°F, so

$$66 = T(1) = 62 + 5e^k \text{ implies } 5e^k = 4 \text{ so } e^k = \frac{4}{5} \text{ and } k = \ln\frac{4}{5},$$

so $T(t) = 62 + 5e^{\ln\frac{4}{5}t}$. The victim died at the time t when the temperature of the body was 98.6°, so

$$98.6 = 62 + 5e^{\ln\frac{4}{5}t} \text{ implies } e^{\ln\frac{4}{5}t} = \frac{36.6}{5} \text{ so } \ln\frac{4}{5}t = \ln\frac{36.6}{5},$$

and

$$t = \frac{\ln\frac{36.6}{5}}{\ln\frac{4}{5}} \approx -8.9 \text{ hours.}$$

So the death occurred about 8.9 hours before 11:00 a.m., at about 2:06 a.m.

20. The pressure at altitude h is given by $P(h) = P_0e^{kh}$. Since the pressure at sea level is given as 1.01×10^5, we have $P_0 = 1.01 \times 10^5$. At altitude $h = 2$, the pressure is 8.08×10^4, so

$$8.08 \times 10^4 = P(2) = (1.01 \times 10^5)e^{2k} \text{ implies } e^{2k} = \frac{8.08 \times 10^4}{1.01 \times 10^5},$$

and

$$2k = \ln\frac{8.08 \times 10^4}{1.01 \times 10^5} \text{ so } k = \frac{1}{2}\ln\frac{8.08 \times 10^4}{1.01 \times 10^5}.$$

Hence

$$P(h) = (1.01 \times 10^5)e^{\frac{1}{2}\ln\frac{8.08\times10^4}{1.01\times10^5}h}.$$

At 5 kilometers the atmospheric pressure is

$$P(5) = (1.01 \times 10^5)e^{\frac{5}{2}\ln\frac{8.08\times10^4}{1.01\times10^5}} \approx 5.8 \times 10^4.$$

Review Exercises for Chapter 5 (Page 300)

1. **a.** iii **b.** i **c.** iv **d.** ii

2. **a.** iii **b.** iv **c.** ii **d.** i

3. $f(x) = 2^{x-1} - 3$

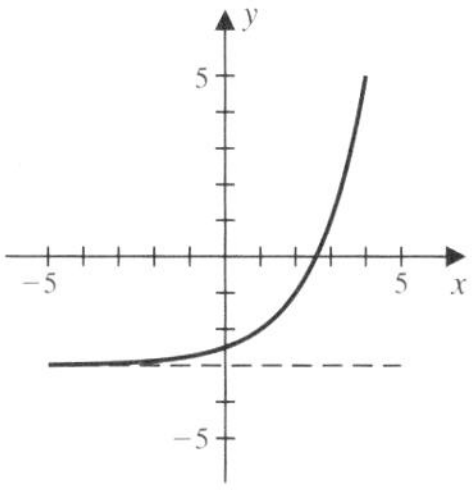

4. $f(x) = e^{x-2}$

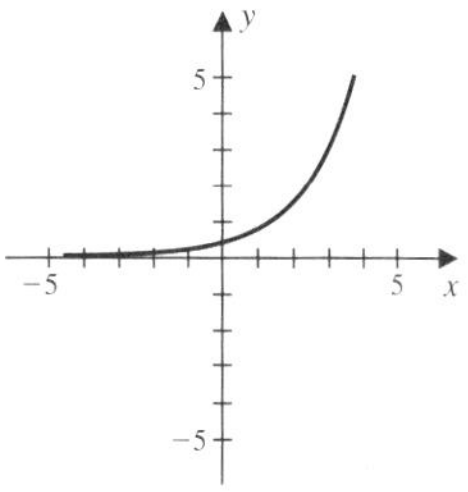

5. $f(x) = e^{-x} + 1$

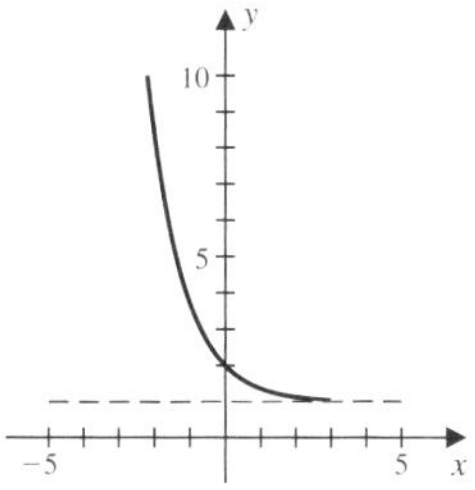

6. $f(x) = 1 - 3^{2-x} = 1 - 3^{-(x-2)}$

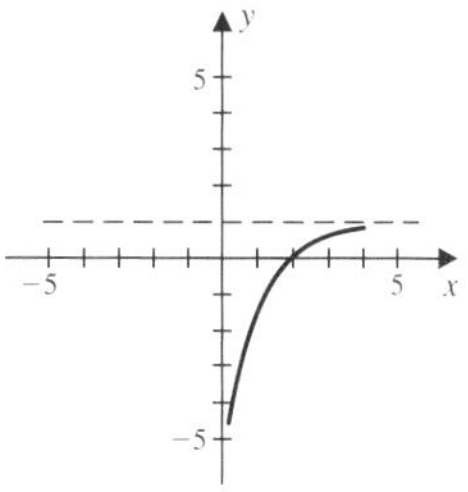

7. $y = 3e^{1-x} = 3e^{-(x-1)}$

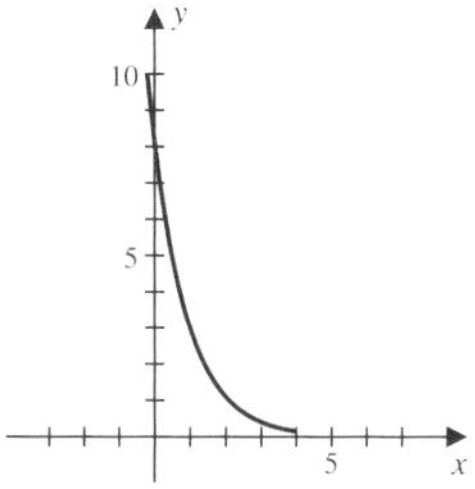

8. $f(x) = -2e^{x+1} + 1$

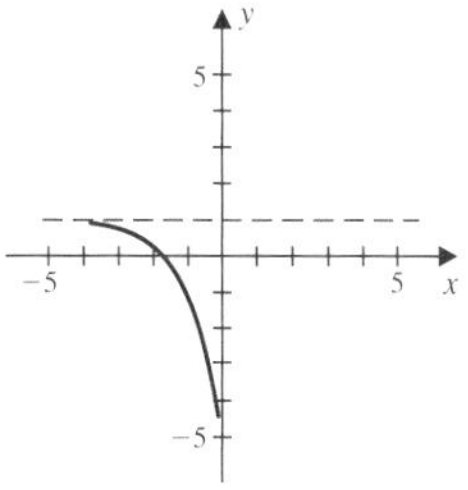

9. $f(x) = 2\ln x$

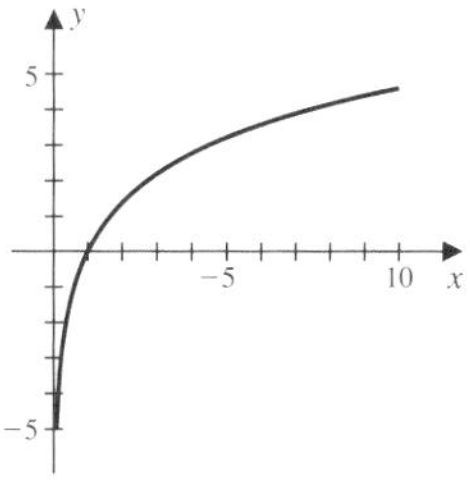

10. $f(x) = \ln(x - 2)$

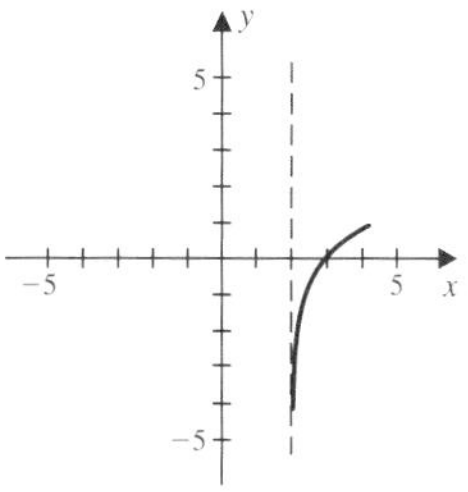

11. $f(x) = 3 - \log_2(x + 1)$

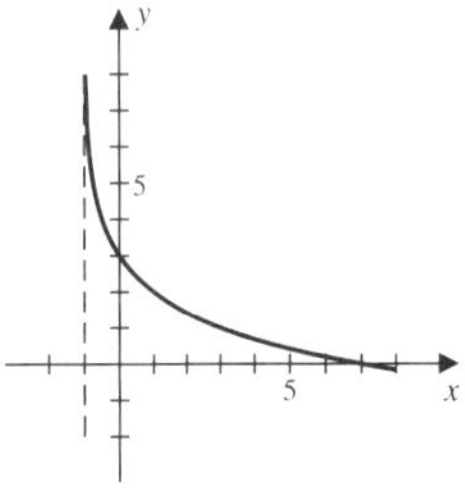

12. $f(x) = \log_{10}(3 - x) + 2 = \log_{10}(-(x - 3)) + 2$

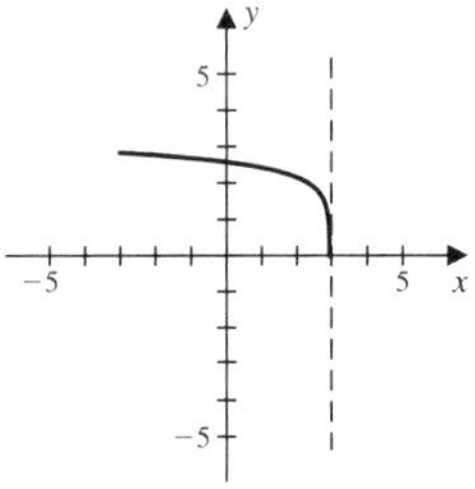

13. $f(x) = e^{-x^2+2x-1}$

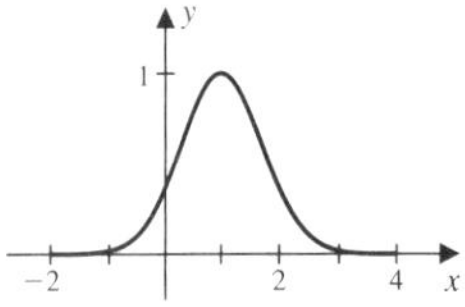

14. $f(x) = \ln x^{-2}$

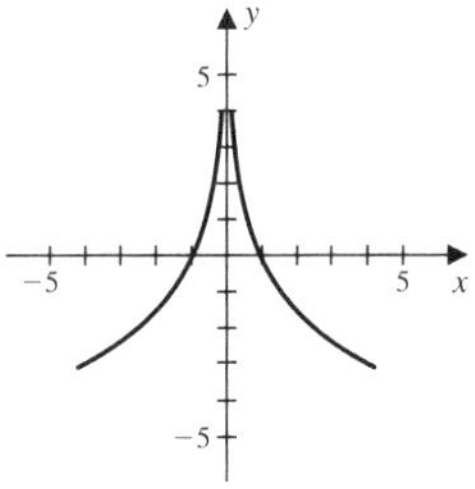

15. $x = \log_5 1$ implies $5^x = 1$ so $x = 0$

16. $x = \log_{10} 0.000001$ implies $10^x = 0.000001$ so $x = -6$

17. Since $2^{\log_2 x} = x$ for $x > 0$, we have $2^{\log_2 15} = 15$.

18. $x = \log_3 \frac{1}{81}$ implies $3^x = \frac{1}{81}$ so $3^x = \frac{1}{3^4}$ and $x = -4$

19. $x = \log_4 2$ implies $4^x = 2$ so $x = \frac{1}{2}$

20. $x = \log_2 256$ implies $2^x = 256$ so $x = 8$

21. Since $e^{\ln x} = x$ for $x > 0$, we have $e^{3 \ln 4} = e^{\ln 4^3} = 64$.

22. Since $e^{\ln x} = x$ for $x > 0$, we have

$$x = \log_5 e^{-2 \ln 5} \text{ implies } x = \log_5 e^{\ln 5^{-2}} = \log_5 \frac{1}{25} \text{ so } 5^x = \frac{1}{25} \text{ and } x = -2.$$

23. $\ln \frac{3x^2}{\sqrt{x-1}} = \ln 3x^2 - \ln(x-1)^{1/2} = \ln 3 + \ln x^2 - \frac{1}{2}\ln(x-1) = \ln 3 + 2\ln x - \frac{1}{2}\ln(x-1)$

24. $\log_2\left(\frac{x^2-1}{x^2-4}\right) = \log_2(x^2-1) - \log_2(x^2-4)$
$= \log_2(x-1)(x+1) - \log_2(x-2)(x+2)$
$= \log_2(x-1) + \log_2(x+1) - \log_2(x-2) - \log_2(x+2)$

25. $\log_{10} \frac{\sqrt{x+1}\,\sqrt[3]{x-1}}{x(x+3)^{5/2}} = \log_{10} \frac{(x+1)^{1/2}(x-1)^{1/3}}{x(x+3)^{5/2}}$
$= \log_{10}\left((x+1)^{1/2}(x-1)^{1/3}\right) - \log_{10}\left(x(x+3)^{5/2}\right)$
$= \log_{10}(x+1)^{1/2} + \log_{10}(x-1)^{1/3} - \log_{10} x - \log_{10}(x+3)^{5/2}$
$= \frac{1}{2}\log_{10}(x+1) + \frac{1}{3}\log_{10}(x-1) - \log_{10} x - \frac{5}{2}\log_{10}(x+3)$

26. $\ln\sqrt{\frac{x\sqrt{x+1}}{x+2}} = \frac{1}{2}\ln\frac{x\sqrt{x+1}}{x+2} = \frac{1}{2}\left[\ln\left(x\sqrt{x+1}\right) - \ln(x+2)\right]$
$= \frac{1}{2}\left[\ln x + \frac{1}{2}\ln(x+1) - \ln(x+2)\right] = \frac{1}{2}\ln x + \frac{1}{4}\ln(x+1) - \frac{1}{2}\ln(x+2)$

27. $\ln x + \frac{1}{3}\ln\left(x(x+1)+2\right)\ln(x-1) = \ln x + \frac{1}{3}\ln(x^2+x) + \ln(x-1)^2$
$= \ln x + \ln(x^2+x)^{1/3} + \ln(x-1)^2 = \ln\left(x(x^2+x)^{1/3}\right) + \ln(x-1)^2$
$= \ln\left(x(x-1)^2\sqrt[3]{x^2+x}\right)$

28. $\frac{1}{2}\ln(2x+1) + \ln(x-1) - \ln(x^2+1) = \ln\sqrt{2x+1} + \ln(x-1) - \ln(x^2+1)$
$= \ln\left((x-1)\sqrt{2x+1}\right) - \ln(x^2+1) = \ln\frac{(x-1)\sqrt{2x+1}}{x^2+1}$

29. $3\ln(x^3+2) + \ln 5 - \frac{1}{2}\ln(x^5-1) = \ln(x^3+2)^3 + \ln 5 - \ln\sqrt{x^5-1}$
$= \ln\left(5(x^3+2)^3\right) - \ln\sqrt{x^5-1} = \ln\frac{5(x^3+2)^3}{\sqrt{x^5-1}}$

30. $\frac{3}{2}\ln(x^2-2) - 2\ln(x+1) = \ln(x^2-2)^{3/2} - \ln(x+1)^2 = \ln\frac{(x^2-2)^{3/2}}{(x+1)^2}$

31. $\ln(2x-3) = 4$ implies $e^{\ln(2x-3)} = e^4$ so $2x-3 = e^4$ and $x = \frac{e^4+3}{2}$

32. $e^{3x-4} = 5$ implies $\ln e^{3x-4} = \ln 5$ so $3x-4 = \ln 5$ and $x = \frac{4+\ln 5}{3}$

33. We have $\ln(2x-1) + \ln(3x-2) = \ln 7$ implies $\ln\left((2x-1)(3x-2)\right) =$ $\ln 7$ so $(2x-1)(3x-2) = 7$ implies $6x^2 - 7x - 5 = (2x+1)(3x-5) = 0$ and $x = -\frac{1}{2}$ or $x = \frac{5}{3}$. Hence $x = \frac{5}{3}$, since the natural logarithms are not defined at $x = -\frac{1}{2}$.

34. $\ln(x-1)-\ln(x-3)=\ln\frac{x-1}{x-3}=1$ implies $\frac{x-1}{x-3}=e$ or $x-1=ex-3e$ so $x(e-1)=3e-1$ and $x=\frac{3e-1}{e-1}$

35. $3^x\cdot 5^{x-2}=3^{4x}$ implies $\ln\left(3^x 5^{x-2}\right)=\ln 3^{4x}$ or $\ln 3^x+\ln 5^{x-2}=4x\ln 3$ so $x\ln 3+(x-2)\ln 5=4x\ln 3$ implies $x(\ln 3+\ln 5-4\ln 3)=2\ln 5$ and $x=\frac{2\ln 5}{\ln 3+\ln 5-4\ln 3}=\frac{2\ln 5}{\ln 5-3\ln 3}$

36. We have $3\cdot 4^x=2^{2x+1}$ implies $\ln(3\cdot 4^x)=\ln 2^{2x+1}$ or $\ln 3+\ln 4^x=(2x+1)\ln 2$ so $\ln 3+x\ln 4=2x\ln 2+\ln 2$ and $x(\ln 4-2\ln 2)=\ln 2-\ln 3$. Hence there are no solutions since $\ln 4-2\ln 2=\ln 4-\ln 2^2=0$.

37. We have $2e^x x^2-e^x x=e^x$ implies $2e^x x^2-e^x x-e^x=e^x(2x^2-x-1)=2x^2-x-1=0$, since $e^x>0$ for all x. So $(2x+1)(x-1)=0$ and $x=-\frac{1}{2}$ or $x=1$.

38. We have $x\ln x-x=x(\ln x-1)=0$ implies $x=0$ or $\ln x=1$ which implies $x=0$ or $x=e$. So $x=e$, since $\ln x$ is not defined at $x=0$.

39. For $e^{x^2}=x-2$ there are no solutions since the graphs never intersect.

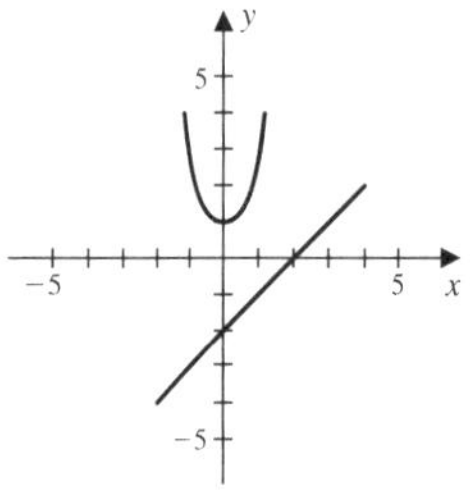

40. We have $\ln(x+1)=x^3-2$ for $x\approx 1.4$ and $x\approx -0.9$.

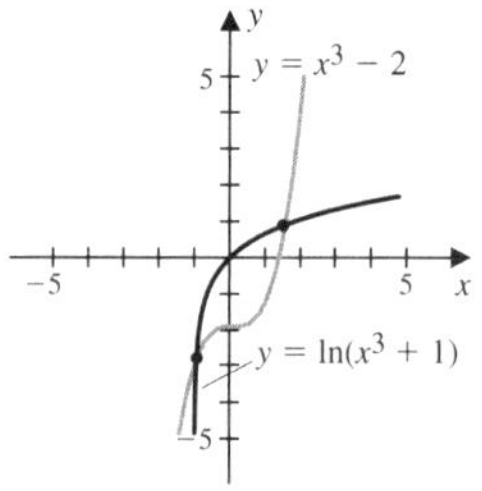

41. We have $e^x>x^4$ on $(-0.8, 1.4)\cup(8.6,\infty)$.

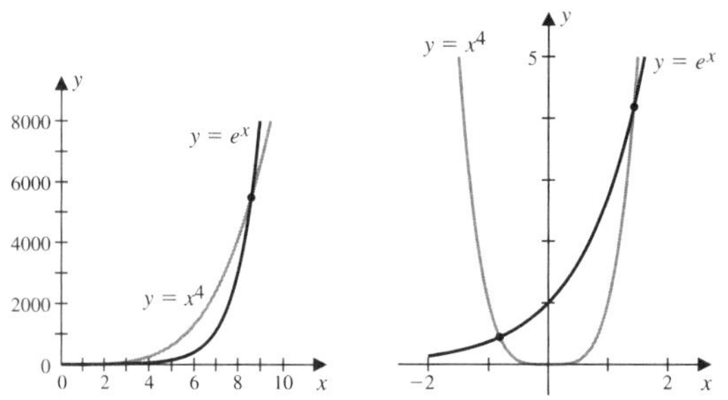

42. We have $\ln x < 2x - 3$ on $(-\infty, 0.1) \cup (1.8, \infty)$.

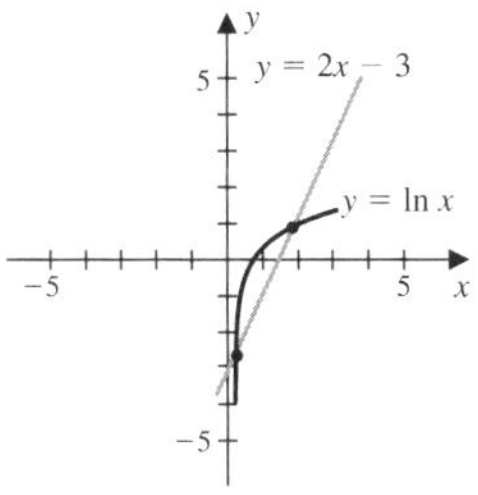

43. We have $e^{x-1} - 3 < x^5$ on $(-1.2, 14.3)$.

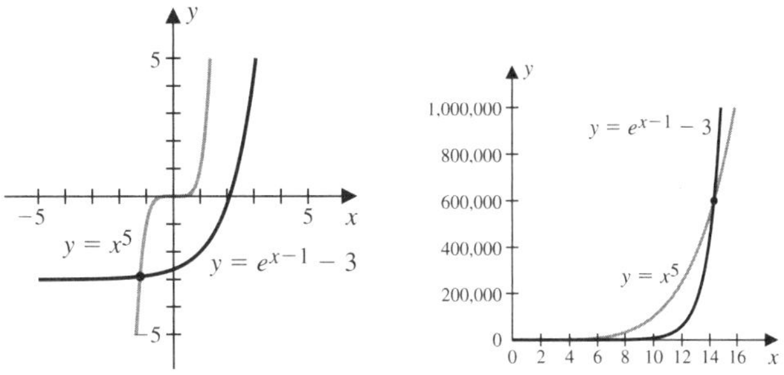

44. We have $\ln x^2 > x^3 - 2x^2 - x - 2$ on $(-\infty, -0.4) \cup (0.3, 2.8)$.

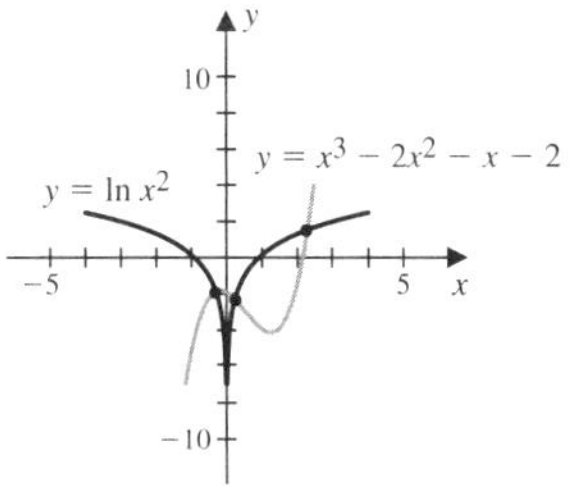

45. For $f(x) = x^2 e^{1-x^2}$, we have the following. Increasing: $(-\infty, -1) \cup (0, 1)$; decreasing: $(-1, 0) \cup (1, \infty)$; local maximums: $(-1, 1)$ and $(1, 1)$; local minimum: $(0, 0)$.

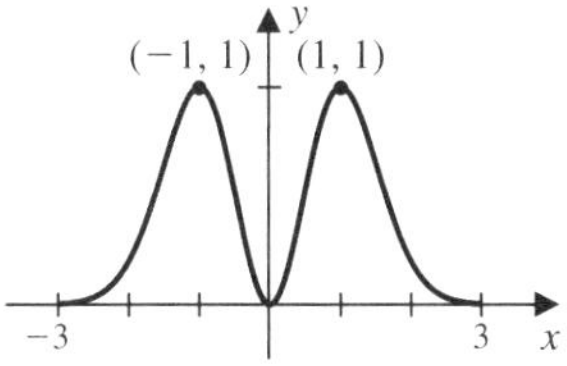

46. For $f(x) = x + \ln\left(1 - x^3\right)$, we have the following. Increasing: $(-\infty, -3) \cup (-0.7, 0.5)$; decreasing: $(-3, -0.7) \cup (0.5, 1)$; local maximums: $(0.5, 0.4)$ and $(-3, 0.3)$; local minimum: $(-0.7, -0.4)$.

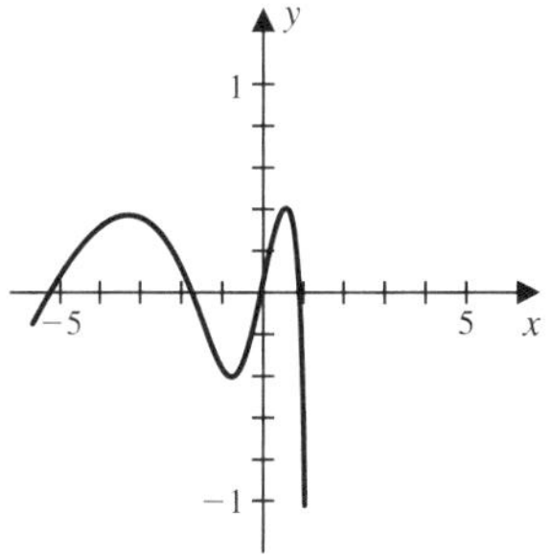

47. The value of an initial investment A_0 deposited at 6% compounded continuously is $A(t) = A_0 e^{0.06t}$. If the investment is to double,

$$2A_0 = A_0 e^{0.06t} \text{ implies } e^{0.06t} = 2 \text{ so } 0.06t = \ln 2 \text{ and } t = \frac{\ln 2}{0.06} \approx 11.6 \text{ years.}$$

48. We have $Q(t) = Q_0 e^{kt}$, and if the half life is 8.8×10^8 years

$$\frac{1}{2}Q_0 = Q(8.8 \times 10^8) = Q_0 e^{(8.8\times 10^8)k} \text{ implies } e^{(8.8\times 10^8)k} = \frac{1}{2},$$

so

$$(8.8 \times 10^8)k = \ln\frac{1}{2} = -\ln 2 \text{ implies } k = -\frac{\ln 2}{8.8 \times 10^8}, \text{ and } Q(t) = Q_0 e^{-\frac{\ln 2}{8.8\times 10^8}t}.$$

a. After 1000 years a 1 gram sample will decay to

$$Q(1000) = e^{-\frac{\ln 2}{8.8\times 10^8}1000} \approx 0.99999921 \text{ grams.}$$

So after 1000 years the amount of decay of the 1 gram sample is approximately $1 - 0.99999921 = 7.9 \times 10^{-7}$ grams.

b. If 90% of the original 1 gram mass decays, there is 10% remaining, or 0.1 grams. So, if 0.1 grams remain,

$$0.1 = Q(t) = e^{-\frac{\ln 2}{8.8\times 10^8}t} \text{ so } t = -\frac{\ln 0.1}{\ln 2}(8.8 \times 10^8) \approx 2.9 \times 10^9 \text{ years.}$$

49. We have $Q(t) = Q_0 e^{kt}$.

a. To find k use the initial information given. That is,

$$1000 = Q(1) = Q_0 e^{k(1)} \text{ implies } e^k = \frac{1000}{Q_0} \text{ so } k = \ln \frac{1000}{Q_0},$$

and

$$3000 = Q(4) = Q_0 e^{k(4)} \text{ implies } e^{4k} = \frac{3000}{Q_0} \text{ so } 4k = \ln \frac{3000}{Q_0}$$

and $4 \ln \frac{1000}{Q_0} = \ln \frac{3000}{Q_0}$. Hence

$$\left(\frac{1000}{Q_0}\right)^4 = \frac{3000}{Q_0} \text{ implies } Q_0^3 = \frac{10^{12}}{3000} = \frac{10^9}{3} \text{ so } Q_0 = \frac{10^3}{\sqrt[3]{3}}.$$

Then $k = \ln \sqrt[3]{3} = \frac{1}{3} \ln 3$, and $Q(t) = \frac{10^3}{\sqrt[3]{3}} e^{\frac{\ln 3}{3} t}$.

b. $Q(5) = \frac{10^3}{\sqrt[3]{3}} e^{\frac{\ln 3}{3} 5} \approx 4327$.

c. Since

$$20000 = \frac{10^3}{\sqrt[3]{3}} e^{\frac{\ln 3}{3} t} \text{ implies } e^{\frac{\ln 3}{3} t} = 20\sqrt[3]{3} \text{ so } \frac{\ln 3}{3} t = \ln 20\sqrt[3]{3},$$

and

$$t = \frac{3 \ln 20\sqrt[3]{3}}{\ln 3} \approx 9.2 \text{ hours.}$$

d. $3Q_0 = Q(t) = Q_0 e^{\frac{\ln 3}{3} t}$ implies $3 = e^{\frac{\ln 3}{3} t}$ so $\ln 3 = \frac{\ln 3}{3} t$ and $t = 3$ hours

50. The value of an initial investment of \$10,000.00 that matures in 8 years and pays 10 percent interest compounded n times a year is $A_n(8) = 10000\left(1 + \frac{0.1}{n}\right)^{8n}$ and if compounded continuously is $A_c(8) = 10000e^{0.8}$. Then

Interest Compounded	Value of CD
(a) annually: $A_1(8)$	\$21, 435.89
(b) monthly: $A_{12}(8)$	\$22, 181.76
(c) daily: $A_{365}(8)$	\$22, 253.00
(d) continuously: $A_c(8)$	\$22, 255.41

51. We have $A_4(t) = A_0\left(1 + \frac{0.1}{4}\right)^{4t}$, and the time it takes for the investment to double can be found from

$$2A_0 = A_0\left(1 + \frac{0.1}{4}\right)^{4t} \text{ implies } 2 = \left(1 + \frac{0.1}{4}\right)^{4t} \text{ so } \ln 2 = 4t \ln\left(1 + \frac{0.1}{4}\right)$$

and $t = \dfrac{\ln 2}{4\ln\left(1+\frac{0.1}{4}\right)} \approx 7$ years.

52. We have $A_{365}(t) = A_0\left(1+\frac{0.1}{365}\right)^{365t}$, and the time it takes for the investment to triple can be found from

$$3A_0 = A_0\left(1+\frac{0.1}{365}\right)^{365t} \text{ implies } 3 = \left(1+\frac{0.1}{365}\right)^{365t} \text{ so } \ln 3 = 365t\ln\left(1+\frac{0.1}{365}\right)$$

and $t = \dfrac{\ln 3}{365\ln\left(1+\frac{0.1}{365}\right)} \approx 11$ years.

53. We have $A_c(t) = A_0e^{0.09t}$, and the time it takes for the investment to double can be found from

$$2A_0 = A_0e^{0.09t} \text{ implies } 2 = e^{0.09t} \text{ so } \ln 2 = 0.09t \text{ and } t = \frac{\ln 2}{0.09} \approx 7.7 \text{ years.}$$

54. We have $A_c(t) = A_0e^{0.1t}$, and the time it takes for the investment to triple can be found from

$$3A_0 = A_0e^{0.1t} \text{ implies } 3 = e^{0.1t} \text{ so } \ln 3 = 0.1t \text{ and } t = \frac{\ln 3}{0.1} \approx 11 \text{ years.}$$

Chapter 5 Exercises for Calculus (Page 301)

1. a. The graphs of $f(x) = 2\ln x$ and $g(x) = e^{\frac{x}{2}}$ are reflections of one another through $y = x$, so $f = g^{-1}$.

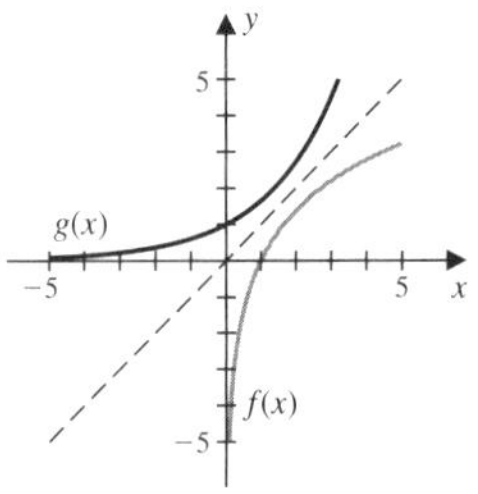

b. The graphs of $f(x) = \ln \frac{x}{2}$ and $g(x) = e^{2x}$ are not reflections of one another through $y = x$, so $f \neq g^{-1}$.

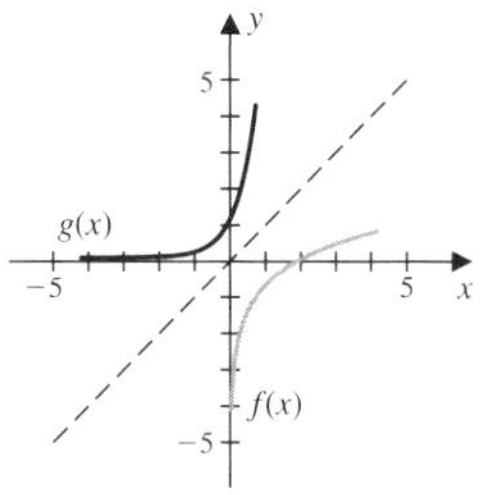

c. Neither $f(x) = \ln |x|$ nor $g(x) = e^{|x|}$ are $1 - 1$ functions, and hence can not have inverses.

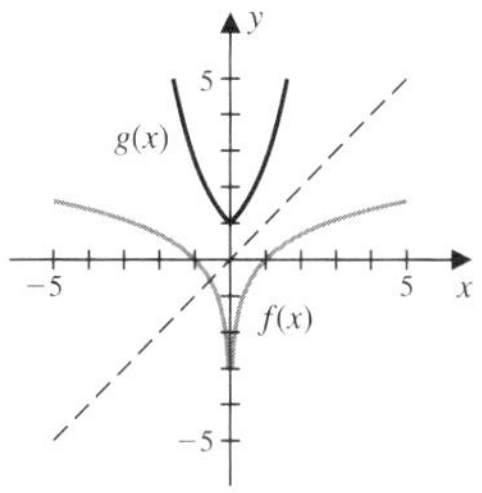

d. The graphs of $f(x) = -\ln x$ and $g(x) = e^{-x}$ are reflections of one another through $y = x$, so $f = g^{-1}$.

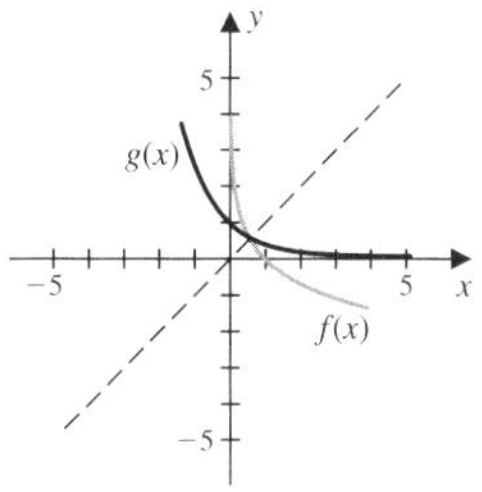

e. The graphs of $f(x) = 1 + \ln x$ and $g(x) = e^{x-1}$ are reflections of one another through $y = x$, so $f = g^{-1}$.

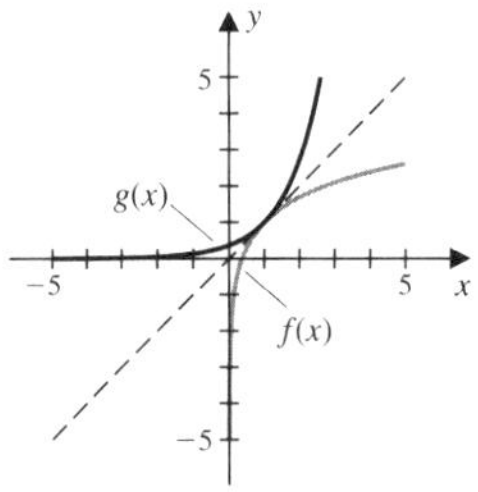

f. The graphs of $f(x) = 2\ln x$ and $g(x) = \frac{1}{2}e^x$ are not reflections of one another through $y = x$, so $f \neq g^{-1}$.

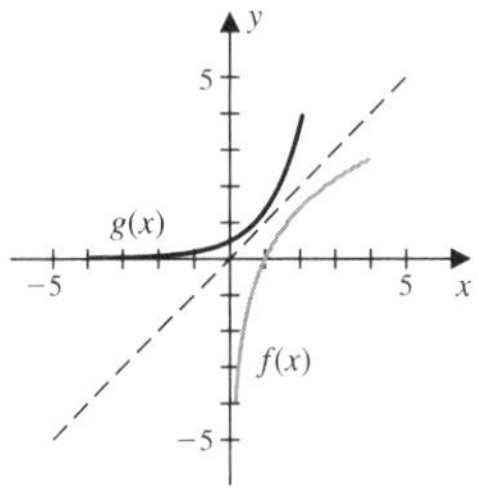

2. Let $f(x) = x^n$, $g(x) = a^x$, and $a > 0$.

<u>$a > 1$:</u> as x approaches ∞, $f(x)$ approaches ∞ and $g(x)$ approaches ∞, with $g(x)$ eventually growing faster than $f(x)$.

<u>$0 < a < 1$:</u> as x approaches ∞, $f(x)$ approaches ∞ and $g(x)$ approaches 0.

3. To order the functions according to how fast they grow as x approaches ∞, plot pairs of functions together for large x to place them in order. For example, to order $\ln x$, x^x and e^{3x}, first plot $y = \ln x$ and $y = x^x$ to see that $y = x^x$ grows faster. Then compare $y = e^{3x}$ and $y = x^x$ to see that $y = x^x$ grows faster. Finally compare $y = e^{3x}$ and $y = \ln x$ to see that $y = e^{3x}$ grows faster, so the ordering of these three functions is $\ln x$, e^{3x}, x^x. The complete ordering from smallest to largest as x approaches ∞ is

$$\frac{x^{10}}{e^x}, \quad \frac{1}{x^4}, \quad \ln x, \quad x^{1/20}, \quad x^{20}, \quad e^{3x}, \quad \frac{e^{6x}}{x^8}, \quad x^x.$$

4. We have $y = 3e^{x-2} = e^{\ln 3}e^{x-2} = e^{x-(2-\ln 3)}$, so the graph of $y = 3e^{x-2}$ is just a horizontal shift of the graph of $y = e^x$. Since $2 - \ln 3 > 0$, the shift is to the right.

5. We have $y = 3e^{x-2} = 3e^x e^{-2} = \frac{3}{e^2}e^x$, so the graph of $y = 3e^{x-2}$ is just a vertical scaling of the graph of $y = e^x$. Since $0 < \frac{3}{e^2} < 1$, it is a vertical compression.

6. We have $y = 3 + \ln 2x = \ln e^3 + \ln 2x = \ln 2e^3 x$, so the graph of $y = 3 + \ln 2x$ is a just a horizontal scaling of the graph of $y = \ln x$. Since $2e^3 > 1$, it is a horizontal compression.

7. We have $y = 3 + \ln 2x = 3 + \ln 2 + \ln x = (3 + \ln 2) + \ln x$, so the graph of $y = 3 + \ln 2x$ is a just a vertical translation of the graph of $y = \ln x$. Since $3 + \ln 2 > 0$, the shift is upward.

8. The amount of carbon 14 present in the bone t years after the animal died is $Q(t) = Q_0 e^{kt}$. Since the half life of radioactive carbon 14 is about 5730 years,

$$\frac{1}{2}Q_0 = Q(5730) = Q_0 e^{5730k} \text{ implies } e^{5730k} = \frac{1}{2} \text{ so } 5730k = -\ln 2, \text{ and } k = -\frac{\ln 2}{5730},$$

and $Q(t) = Q_0 e^{-\frac{\ln 2}{5730}t}$. The age of the bone is the time t, when there is 78% of the carbon still remaining, so

$$0.78Q_0 = Q_0 e^{-\frac{\ln 2}{5730}t} \text{ implies } 0.78 = e^{-\frac{\ln 2}{5730}t} \text{ so } -\frac{\ln 2}{5730}t = \ln(0.78).$$

And $t = -\frac{5730\ln(0.78)}{\ln 2} \approx 2054$ years.

9. If the interest on an initial investment of A_0 dollars is compounded continuously at a fixed rate of $r\%$, then the value after t years is $A(t) = A_0 e^{\frac{r}{100}t}$. Then the time at which the investment doubles is given by,

$$2A_0 = A_0 e^{\frac{r}{100}t} \text{ implies } 2 = e^{\frac{r}{100}t} \text{ so } \frac{r}{100}t = \ln 2 \text{ and } t = \frac{100\ln 2}{r} \approx \frac{70}{r}.$$

So $\frac{70}{r}$ is a reasonable estimate for the time it takes for the investment to double in value. For example, if the interest rate is 8.75%, then $\frac{70}{8.75} = 8.0$.

10. The logistic equation is given by

$$P(t) = \frac{A}{1 + Be^{-Ct}},$$

where A, B, and C are positive constants.
a. The initial population is the population at $t = 0$. When $t = 0$, the term $e^{-C(0)} = 1$, so the initial population is $\frac{A}{1+B}$.
b. As t grows indefinitely the term e^{-Ct} goes to 0, and hence the limiting population is A.
c. The figure shows the graph of the function $P(t) = \frac{5000}{1+300e^{-0.6t}}$. That is, $A = 5000$, $B = 300$, and $C = 0.6$.

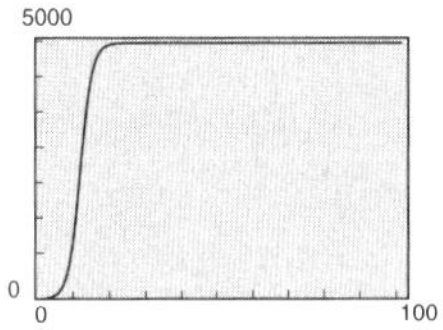

11. **a.** The concentration of the drug in the bloodstream can be modeled using exponential decay. So the concentration at time t has the form

$$C(t) = C_0 e^{kt}.$$

Since the initial concentration of the drug in the bloodstream is 20, we have $C_0 = 20$. If 3 hours later the concentration is 12,

$$12 = C(3) = 20e^{3k} \text{ implies } e^{3k} = \frac{12}{20} = \frac{3}{5} \text{ so } 3k = \ln\left(\frac{3}{5}\right) \text{ and } k = \frac{1}{3}\ln\left(\frac{3}{5}\right).$$

Then the concentration at time t is

$$C(t) = 20e^{\frac{1}{3}\ln\left(\frac{3}{5}\right)t}.$$

b. To find the half-life, set $C(t) = 10$ and solve for t. We have

$$10 = 20e^{\frac{1}{3}\ln\left(\frac{3}{5}\right)t} \text{ implies } e^{\frac{1}{3}\ln\left(\frac{3}{5}\right)t} = \frac{1}{2} \text{ so } \frac{1}{3}\ln\left(\frac{3}{5}\right)t = \ln\frac{1}{2} = -\ln 2 \text{ and } t = -\frac{3\ln 2}{\ln\left(\frac{3}{5}\right)} \approx 4.07 \text{ hours.}$$

c. Since the half-life is 5 hours, we have $C(t) = C_0 e^{kt}$ and

$$\frac{1}{2}C_0 = C(5) = C_0 e^{5k} \text{ implies } e^{5k} = \frac{1}{2} \text{ so } 5k = \ln\left(\frac{1}{2}\right) = -\ln 2 \text{ and } k = -\frac{\ln 2}{5}.$$

So

$$C(t) = C_0 e^{-\frac{\ln 2}{5}t}.$$

For a 25 kilogram dog, the amount $Q(t)$ of phenobarbital in the blood at time t is $25C(t)$. That is,

$$Q(t) = 25C_0 e^{-\frac{\ln 2}{5}t}.$$

When $t = 1$ hour, we want $Q(t) = (30)(25) = 750$. So

$$750 = Q(1) = 25C_0 e^{-\frac{\ln 2}{5}} \text{ so } C_0 = 30e^{\frac{\ln 2}{5}} \approx 34.46 \text{ mg/kg}.$$

Since the dog's weight is equivalent to 25 kilograms, the initial dose should be $(34.46)(25) = 861.5$ milligrams.

12. **a.** Plotted in the figure is the hyperbolic cosine function $\cosh x = \frac{e^x+e^{-x}}{2}$ along with $y = \frac{e^x}{2}$ and $y = \frac{e^{-x}}{2}$.

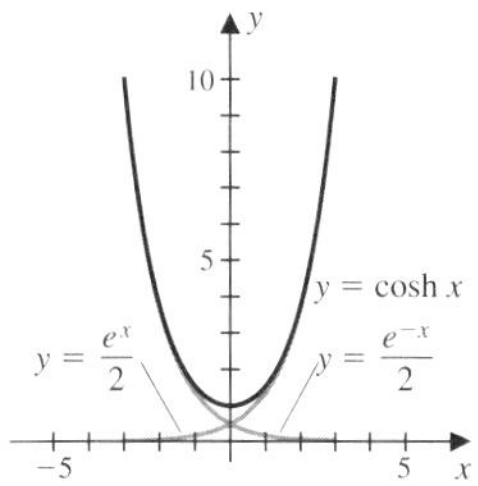

b. Plotted in the figure is the hyperbolic sine function $\sinh x = \frac{e^x-e^{-x}}{2}$ along with $y = \frac{e^x}{2}$ and $y = -\frac{e^{-x}}{2}$.

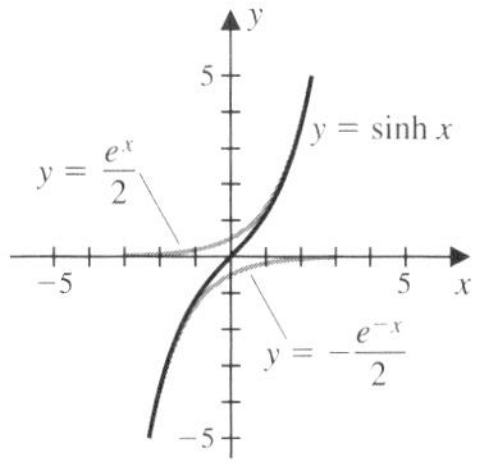

c.

$$(\cosh x)^2 - (\sinh x)^2 = \left(\frac{e^x + e^{-x}}{2}\right)^2 - \left(\frac{e^x - e^{-x}}{2}\right)^2$$
$$= \frac{1}{4}\left(e^{2x} + 2 + e^{-2x}\right) - \frac{1}{4}\left(e^{2x} - 2 + e^{-2x}\right) = 1$$

$$\sinh a \cosh b + \cosh a \sinh b = \left(\frac{e^a - e^{-a}}{2}\right)\left(\frac{e^b + e^{-b}}{2}\right) + \left(\frac{e^a + e^{-a}}{2}\right)\left(\frac{e^b - e^{-b}}{2}\right)$$
$$= \frac{1}{4}\left(e^{a+b} + e^{a-b} - e^{b-a} - e^{-a-b}\right)$$

$$+\frac{1}{4}\left(e^{a+b} - e^{a-b} + e^{b-a} - e^{-a-b}\right)$$

$$=\frac{1}{2}e^{a+b} - \frac{1}{2}e^{-a-b} = \frac{e^{a+b} - e^{-(a+b)}}{2} = \sinh(a+b)$$

d. The figure shows the graph of $y = a\cosh\left(\frac{x}{a}\right)$, for $a = 1, 2, 3, 4$. The figure indicates that as a increase the catenary becomes more flattened.

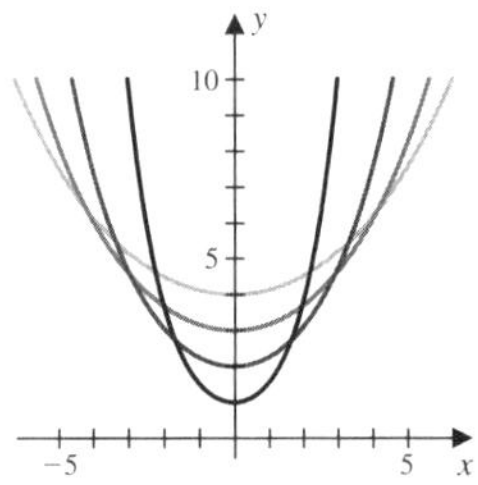

Chapter 5 Chapter Test (Page 302)

1. True.

2. True.

3. True.

4. False. The solution to the equation $\log_3 x = 2$ is 9.

5. True.

6. False. We have $e^{\ln x} = x$ only when $x > 0$.

7. False. The domain of the function $f(x) = \ln(x-2)$ is $(2, \infty)$.

8. False. The range of the function $f(x) = 1 + e^{x-2}$ is $(1, \infty)$.

9. True.

10. False. The domain of $f(x) = 3 + e^{2-x}$ is $(-\infty, \infty)$.

11. True.

12. True.

13. True.

14. False. The only solution to the equation $4^x = 5^{2x-1}$ is $x = \frac{\ln 5}{2(\ln 5 - \ln 2)}$.

15. False. The expression $\ln xy^2$ is equivalent to $\ln x + 2\ln y$.

16. True.

17. False. The only solution to the equation $\ln x + \ln(x-1) = \ln(4x+6)$ is $x = 6$, since $x = -1$ is not in the domain.

18. True.

19. True.

20. True.

21. False. The graph of $y = -1 + e^{x-2}$ is obtained by shifting the graph of $y = e^x$ to the right 2 units and downward 1 unit.

22. True.

23. True.

24. False. The graph of $y = -3 + \ln(x - 1)$ has a vertical asymptote $x = 1$.

25. False. The graph of $y = 2 + \ln(x - 1)$ is obtained by shifting the graph of $y = \ln x$ to the right 1 unit and upward 2 units.

26. True.

27. False. The blue graph is given by $y = e^{x-1}$.

28. False. The red graph is given by $y = e^{x+1}$.

29. True.

30. True.

31. True.

32. True.

33. False. The green graph is given by $y = -1 - \ln(x + 1)$.

34. True.

35. False. The inverse of $f(x) = 2^{x+1}$ is $f^{-1}(x) = -1 + \log_2 x$.

36. True.

37. False. If a bacteria culture starts with 1000 bacteria and after 3 hours there are 2500 bacteria, the number of bacteria at any time t is given by $Q(t) = 1000e^{(t \ln 2.5)/3}$.

38. True.

39. True.

40. True.

Exercise Set 6.2 (Page 313)

1. $y = 2x^2 = \frac{1}{4c}x^2$ implies $\frac{1}{4c} = 2$ so $c = \frac{1}{8}$
$V(0, 0)$; $F(0, 1/8)$; directrix, D: $y = -1/8$

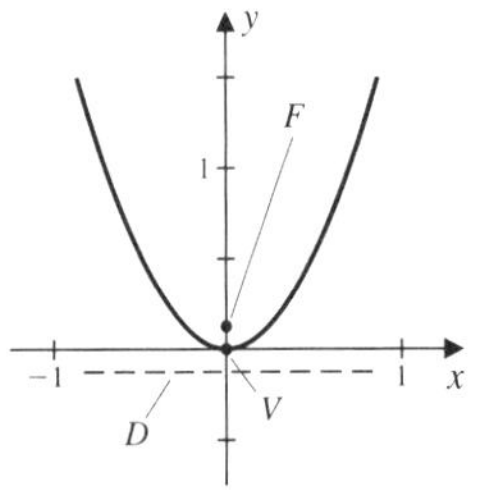

2. $16y = 9x^2$ implies $y = \frac{9}{16}x^2$
so $\frac{1}{4c} = \frac{9}{16}$ and $c = \frac{4}{9}$
$V(0, 0)$; $F(0, 4/9)$; directrix, D: $y = -4/9$

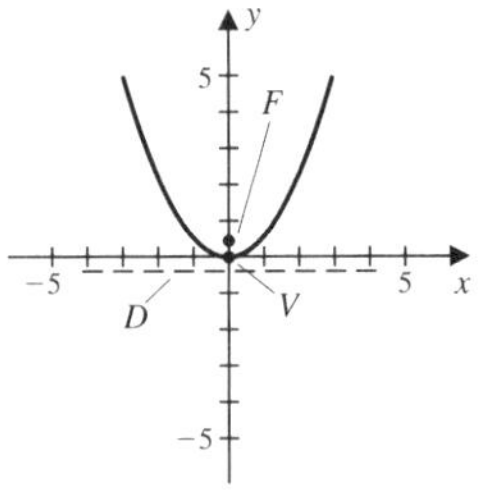

3. $9y = -16x^2$ implies $y = -\frac{16}{9}x^2$
so $\frac{1}{4c} = -\frac{16}{9}$ and $c = -\frac{9}{64}$
$V(0, 0)$; $F(0, -9/64)$; directrix, D:
$y = 9/64$

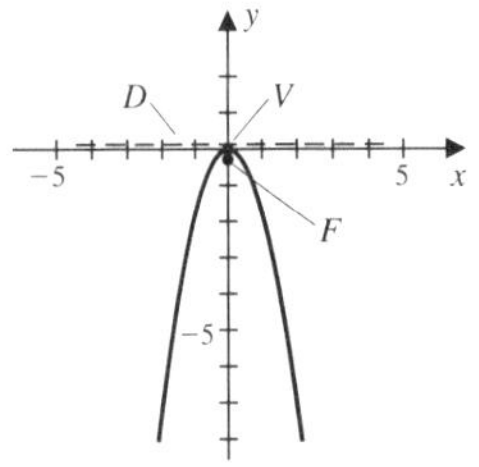

4. $y = -2x^2$ implies $\frac{1}{4c} = -2$ so $c = -\frac{1}{8}$
$V(0, 0)$; $F(0, -1/8)$; directrix, D: $y = 1/8$

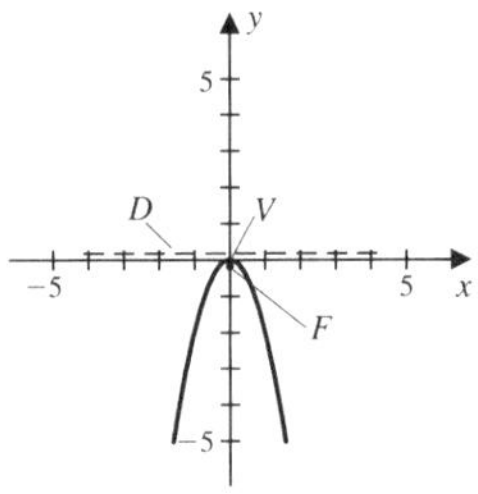

5. $y^2 = 2x$ implies $x = \frac{1}{2}y^2$ so $\frac{1}{4c} = \frac{1}{2}$
and $c = \frac{1}{2}$
$V(0, 0)$; $F(1/2, 0)$; directrix, D: $x = -1/2$

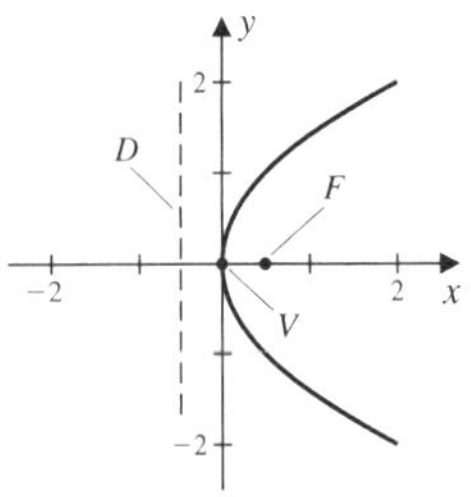

6. $9y^2 = 16x$ implies $x = \frac{9}{16}y^2$ so $\frac{1}{4c} = \frac{9}{16}$
and $c = \frac{4}{9}$
$V(0, 0)$; $F(4/9, 0)$; directrix, D: $x = -4/9$

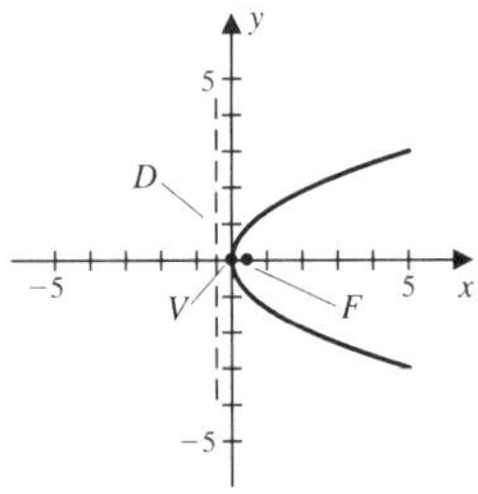

7. $9y^2 = -16x$ implies $x = -\frac{9}{16}y^2$
so $\frac{1}{4c} = -\frac{9}{16}$ and $c = -\frac{4}{9}$
$V(0, 0)$; $F(-4/9, 0)$; directrix, D: $x = 4/9$

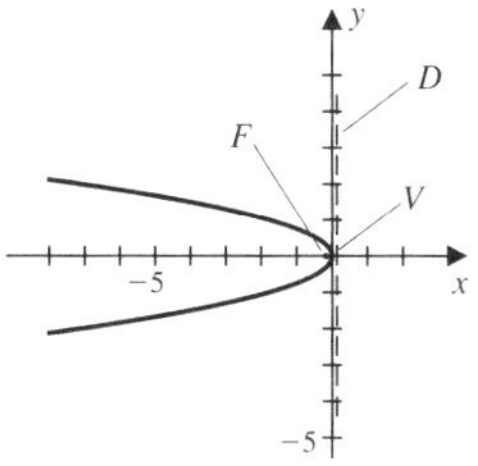

8. $y^2 = -2x$ implies $x = -\frac{1}{2}y^2$ so $\frac{1}{4c} = -\frac{1}{2}$
and $c = -\frac{1}{2}$
$V(0, 0)$; $F(-1/2, 0)$; directrix, D: $x = 1/2$

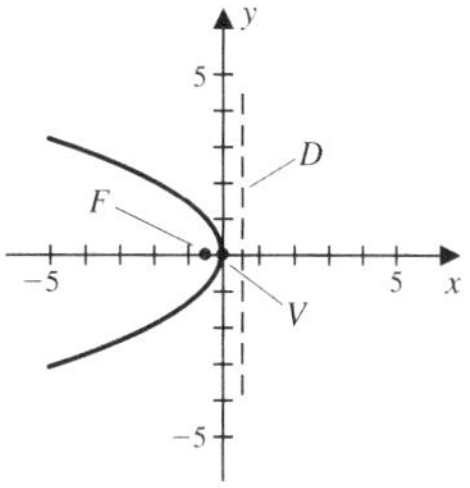

9. $x^2 - 6x + 9 = 2y$ implies $(x - 3)^2 = 2y$
or $y = \frac{1}{2}(x - 3)^2$ so $\frac{1}{4c} = \frac{1}{2}$ and $c = \frac{1}{2}$
$V(3, 0)$; $F(3, 1/2)$; directrix, D: $y = -1/2$

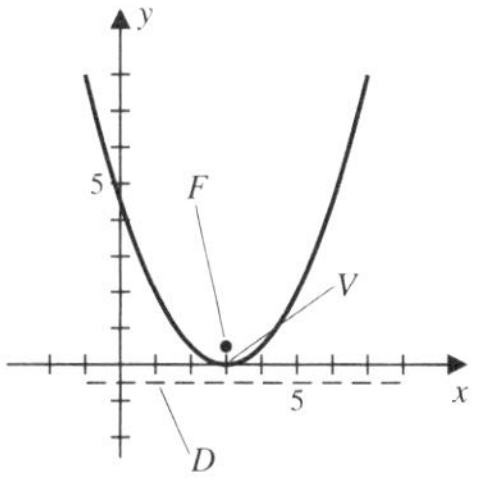

10. $x^2 + 4x + 4 = 2y$ implies $(x + 2)^2 = 2y$
or $y = \frac{1}{2}(x + 2)^2$ so $\frac{1}{4c} = \frac{1}{2}$ and $c = \frac{1}{2}$
$V(-2, 0)$; $F(-2, 1/2)$; directrix, D:
$y = -1/2$

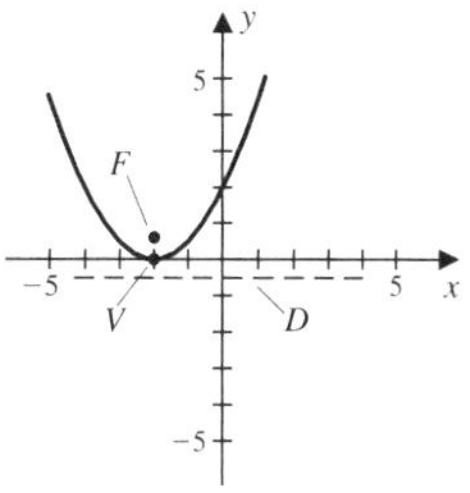

11. $x^2 - 4x - 2y + 2 = x^2 - 4x + 4 - 4 - 2y + 2 = 0$
implies $(x - 2)^2 = 2y + 2$
or $y + 1 = \frac{1}{2}(x - 2)^2$ so $\frac{1}{4c} = \frac{1}{2}$ and $c = \frac{1}{2}$
$V(2, -1)$; $F(2, -1/2)$; directrix, D:
$y = -3/2$

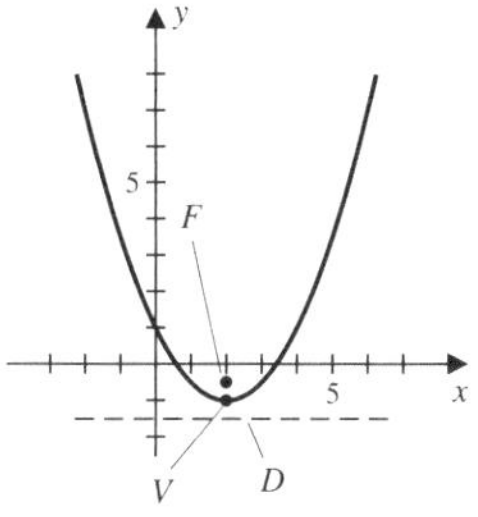

12. $x^2 - 4x + 6 + 2y = x^2 - 4x + 4 - 4 + 6 + 2y = 0$
implies $(x - 2)^2 = -2y - 2$
or $y + 1 = -\frac{1}{2}(x - 2)^2$ so $\frac{1}{4c} = -\frac{1}{2}$
and $c = -\frac{1}{2}$
$V(2, -1)$; $F(2, -3/2)$; directrix, D:
$y = -1/2$

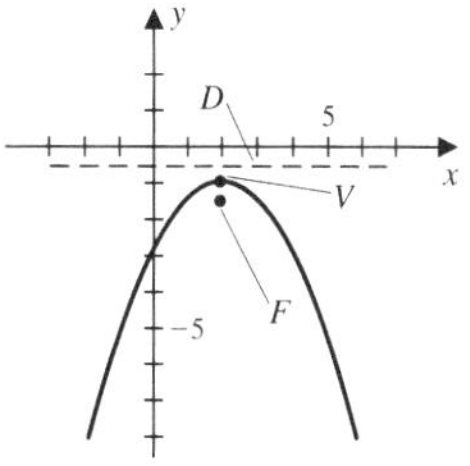

13. $y^2-8y+12=2x$ implies $y^2-8y+16-16=2x-12$ or $(y-4)^2=2x+4$ implies $x+2=\frac{1}{2}(y-4)^2$ so $\frac{1}{4c}=\frac{1}{2}$ and $c=\frac{1}{2}$
$V(-2,4)$; $F(-3/2,4)$; directrix, D: $x=-5/2$

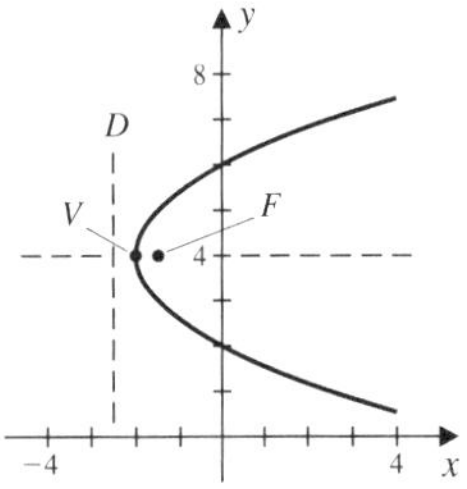

14. $y^2+6y+6-3x=0$ implies $y^2+6y+9-9=3x-6$ or $(y+3)^2=3x+3$ implies $x+1=\frac{1}{3}(y+3)^2$ so $\frac{1}{4c}=\frac{1}{3}$ and $c=\frac{3}{4}$
$V(-1,-3)$; $F(-1/4,-3)$; directrix, D: $x=-7/4$

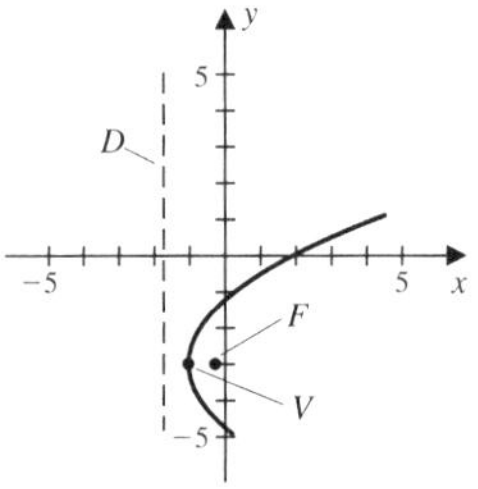

15. $2x^2+4x-9y+20=0$ implies $2(x^2+2x+1-1)=9y-20$ or $2(x+1)^2=9y-18$ implies $y-2=\frac{2}{9}(x+1)^2$ so $\frac{1}{4c}=\frac{2}{9}$ and $c=\frac{9}{8}$
$V(-1,2)$; $F(-1,25/8)$; directrix, D: $y=7/8$

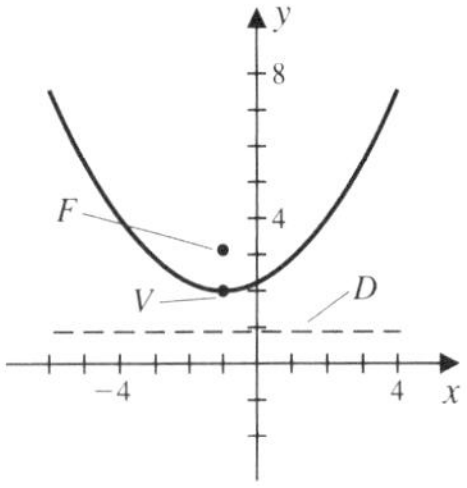

16. $2x^2-4x+3y-4=0$ implies $2(x^2-2x+1-1)=-3y+4$ or $2(x-1)^2=-3y+6$ implies $y-2=-\frac{2}{3}(x-1)^2$ so $\frac{1}{4c}=-\frac{2}{3}$ and $c=-\frac{3}{8}$
$V(1,2)$; $F(1,13/8)$; directrix, D: $y=19/8$

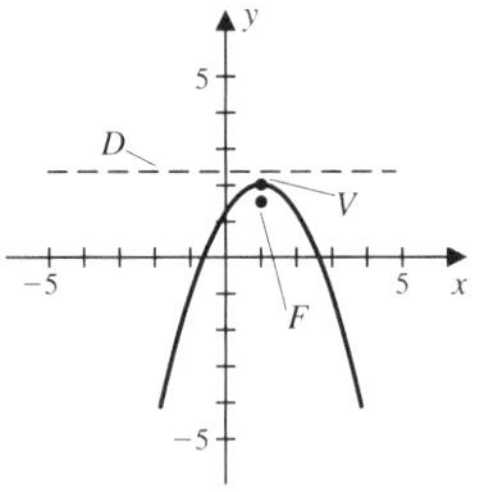

17. $3x^2-12x+4y+8=0$ implies $3(x^2-4x+4-4)=-4y-8$ or $3(x-2)^2=-4y+4$ implies $y-1=-\frac{3}{4}(x-2)^2$ so $\frac{1}{4c}=-\frac{3}{4}$ and $c=-\frac{1}{3}$
$V(2,1)$; $F(2,2/3)$; directrix, D: $y=4/3$

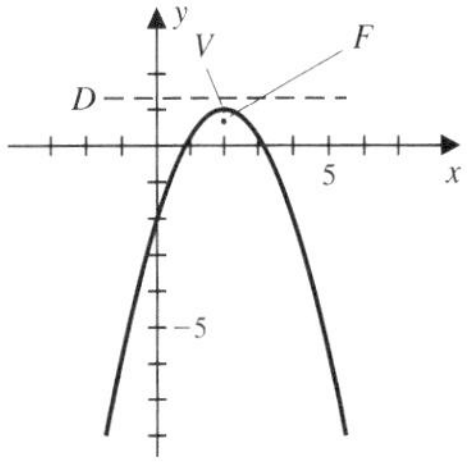

18. $3x^2-6x-2y-1=0$ implies $3(x^2-2x+1-1)=2y+1$ or $3(x-1)^2=2y+4$ implies $y+2=\frac{3}{2}(x-1)^2$ so $\frac{1}{4c}=\frac{3}{2}$ and $c=\frac{1}{6}$
$V(1,-2)$; $F(1,-11/6)$; directrix, D: $y=-13/6$

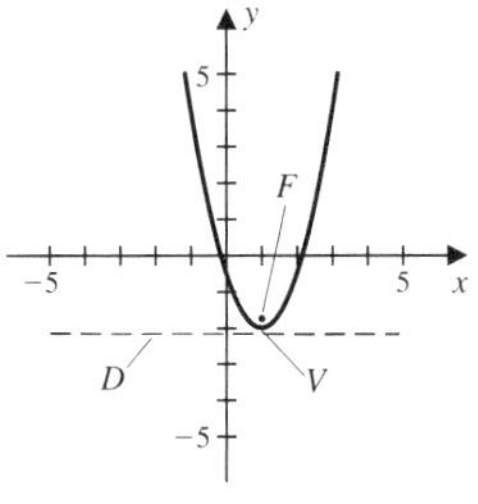

19. The parabola is in standard position with axis along the y-axis, so has equation in the form

$y = \frac{1}{4c}x^2$ and focus $(0, c)$, $c > 0$. Since the point $(2, c)$ lies on the curve, $c = \frac{1}{4c}(2)^2$ implies $c^2 = 1$ so $c = 1$. The equation is $y = \frac{1}{4}x^2$.

20. The parabola is in standard position with axis along the y-axis, so has equation in the form $y = \frac{1}{4c}x^2$ and focus $\left(0, -\frac{1}{2}\right)$. The equation is $y = \frac{1}{4(-1/2)}x^2$ or $y = -\frac{1}{2}x^2$.

21. The parabola is in standard position with axis along the x-axis, so has equation in the form $x = \frac{1}{4c}y^2$, and since the directrix is $x = -1$, the focus is $(1, 0)$. The equation is $x = \frac{1}{4(1)}y^2$ or $x = \frac{1}{4}y^2$.

22. The parabola is in standard position with axis along the x-axis, so has equation in the form $x = \frac{1}{4c}y^2$ and focus $(c, 0)$, $c < 0$. Since the point $\left(c, \frac{1}{2}\right)$ lies on the curve, $c = \frac{1}{4c}\left(\frac{1}{2}\right)^2$ implies $c^2 = \frac{1}{16}$ so $c = \pm\sqrt{\frac{1}{16}}$ and $c = -\frac{1}{4}$. The equation is $x = \frac{1}{4(-1/4)}y^2$ or $x = -y^2$.

23. Focus: $(-2, 2)$; Directrix: $y = -2$

Since the vertex lies midway between the focus and the directrix, and the distance from the focus to the directrix is 4, the vertex is $(-2, 0)$. The axis of the parabola is vertical with $c = 2$, so $y = \frac{1}{8}(x + 2)^2$.

24. Focus: $(-2, 2)$; Directrix: $x = 2$

Since the vertex lies midway between the focus and the directrix, and the distance from the focus to the directrix is 4, the vertex is $(0, 2)$. The axis of the parabola is horizontal with $c = -2$, so $x = -\frac{1}{8}(y - 2)^2$.

25. Vertex: $(-2, 2)$; Directrix: $x = 2$

Since the vertex lies midway between the focus and the directrix, and the distance from the directrix to the vertex is 4, the focus is $(-6, 2)$. The axis of the parabola is horizontal with $c = -4$, so $x + 2 = \frac{1}{4(-4)}(y - 2)^2$ or $x + 2 = -\frac{1}{16}(y - 2)^2$.

26. Vertex: $(-2, 2)$; Directrix: $y = -2$

Since the vertex lies midway between the focus and the directrix, and the distance from the directrix to the vertex is 4, the focus is $(-2, 6)$. The axis of the parabola is vertical with $c = 4$, so $y - 2 = \frac{1}{4(4)}(x + 2)^2$ or $y - 2 = \frac{1}{16}(x + 2)^2$.

27. Vertex: $(-2, 2)$; Focus: $(-2, 0)$

The distance between the vertex and the focus is 2, so the directrix is $y = 4$. The axis of the parabola is vertical and $c = -2$. The equation is $y - 2 = -\frac{1}{8}(x + 2)^2$.

28. Vertex: $(-2, 2)$; Focus: $(-2, 4)$

The distance between the vertex and the focus is 2, so the directrix is $y = 0$. The axis of the parabola is vertical and $c = 2$. The equation is $y - 2 = \frac{1}{8}(x + 2)^2$.

29. Vertex: $(-2, 2)$; Focus: $(-4, 2)$

The distance between the vertex and the focus is 2, so the directrix is $x = 0$. The axis of the parabola is horizontal and $c = -2$. The equation is $x + 2 = -\frac{1}{8}(y - 2)^2$.

30. Vertex: $(-2, 2)$; Focus: $(2, 2)$

The distance between the vertex and the focus is 4, so the directrix is $x = -6$. The axis of the parabola is horizontal and $c = 4$. The equation is $x + 2 = \frac{1}{16}(y - 2)^2$.

31. a. Vertex: $(0, 0)$; Point on parabola: $(4, 6)$

The axis is parallel to the y-axis, so since the vertex is at the origin, the axis is the y-axis, and the parabola has the form $y = \frac{1}{4c}x^2$. Since the parabola passes through the point $(4, 6)$,
$6 = \frac{1}{4c}(4)^2$ implies $6 = \frac{4}{c}$ so $c = \frac{2}{3}$, and the equation of the parabola is $y = \frac{1}{4(2/3)}x^2 = \frac{3}{8}x^2$.

b. Vertex: $(1, 0)$; Point on parabola: $(5, 6)$

The axis is parallel to the y-axis, so since the vertex is at the point $(1, 0)$, the axis is the line $x = 1$, and the parabola has the form $y = \frac{1}{4c}(x - 1)^2$. Since the parabola passes through the point $(5, 6)$,
$6 = \frac{1}{4c}(5 - 1)^2$ implies $6 = \frac{4}{c}$ so $c = \frac{2}{3}$, and the equation of the parabola is
$y = \frac{1}{4(2/3)}(x - 1)^2 = \frac{3}{8}(x - 1)^2$.

c. Vertex: $(1, 2)$; Point on parabola: $(5, 8)$

The axis is parallel to the y-axis, so since the vertex is at the point $(1, 2)$, the axis is the line $x = 1$, and the parabola has the form $y - 2 = \frac{1}{4c}(x - 1)^2$. Since the parabola passes through the point $(5, 8)$,
$8 - 2 = \frac{1}{4c}(5 - 1)^2$ implies $6 = \frac{4}{c}$ so $c = \frac{2}{3}$, and the equation of the parabola is $y - 2 = \frac{3}{8}(x - 1)^2$.

d. Vertex: $(0, 2)$; Point on parabola: $(4, 8)$

The axis is parallel to the y-axis, so since the vertex is at the point $(0, 2)$, the axis is the y-axis, and the parabola has the form $y - 2 = \frac{1}{4c}x^2$. Since the parabola passes through the point $(4, 8)$,
$8 - 2 = \frac{1}{4c}(4)^2$ implies $6 = \frac{4}{c}$ so $c = \frac{2}{3}$, and the equation of the parabola is $y - 2 = \frac{3}{8}x^2$.

32. a. Vertex: $(0, 0)$; Point on parabola: $(4, 6)$

The axis is parallel to the x-axis, so since the vertex is at the origin, the axis is the x-axis, and the parabola has the form $x = \frac{1}{4c}y^2$. Since the parabola passes through the point $(4, 6)$,
$4 = \frac{1}{4c}(6)^2$ implies $4 = \frac{9}{c}$ so $c = \frac{9}{4}$, and the equation of the parabola is $x = \frac{1}{4(9/4)}y^2 = \frac{1}{9}y^2$.

b. Vertex: $(1, 0)$; Point on parabola: $(5, 6)$

The axis is parallel to the x-axis, so since the vertex is at the point $(1, 0)$, the axis is the x-axis, and the parabola has the form $x - 1 = \frac{1}{4c}y^2$. Since the parabola passes through the point $(5, 6)$, $5 - 1 = \frac{1}{4c}(6)^2$ implies $4 = \frac{9}{c}$ so $c = \frac{9}{4}$, and the equation of the parabola is $x - 1 = \frac{1}{4(9/4)}y^2$ or $x - 1 = \frac{1}{9}y^2$.

c. Vertex: $(1, 2)$; Point on parabola: $(5, 8)$

The axis is parallel to the x-axis, so since the vertex is at the point $(1, 2)$, the axis is the line $y = 2$, and the parabola has the form $x - 1 = \frac{1}{4c}(y - 2)^2$. Since the parabola passes through the point $(5, 8)$, $5 - 1 = \frac{1}{4c}(8 - 2)^2$ implies $4 = \frac{9}{c}$ so $c = \frac{9}{4}$, and the equation of the parabola is $x - 1 = \frac{1}{4(9/4)}(y - 2)^2$ or $x - 1 = \frac{1}{9}(y - 2)^2$.

d. Vertex: $(0, 2)$; Point on parabola: $(4, 8)$

The axis is parallel to the x-axis, so since the vertex is at the point $(0, 2)$, the axis is the line $y = 2$, and the parabola has the form $x = \frac{1}{4c}(y - 2)^2$. Since the parabola passes through the point $(4, 8)$, $4 = \frac{1}{4c}(8 - 2)^2$ implies $4 = \frac{9}{c}$ so $c = \frac{9}{4}$, and the equation of the parabola is $x = \frac{1}{4(9/4)}(y - 2)^2$ or $x = \frac{1}{9}(y - 2)^2$.

33. Since the axis of the parabola is the y-axis, the vertex lies on the y-axis and hence is of the form $(0, b)$. The equation of the parabola is $y - b = \frac{1}{4c}x^2$. Since the parabola passes through $(1, 2)$, $2 - b = \frac{1}{4c}(1)^2$ implies $c = \frac{1}{4(2-b)}$, and the equation is $y - b = \frac{1}{4(1/(4(2-b)))}x^2$ or $y - b = (2 - b)x^2, b \neq 2$.

34. Since the axis of the parabola is the x-axis, the vertex lies on the x-axis and hence is of the form $(b, 0)$. The equation of the parabola is $x - b = \frac{1}{4c}y^2$. Since the parabola passes through $(1, 2)$, $1 - b = \frac{1}{4c}(2)^2$ implies $c = \frac{1}{(1-b)}$, and the equation is $x - b = \frac{1}{4(1/((1-b))}y^2$ or $x - b = \frac{1-b}{4}y^2, b \neq 1$.

35. To describe the light, use a parabola in standard position and axis along the x-axis, so the equation has the form $x = \frac{1}{4c}y^2, c > 0$. The information implies the point $(2, 2)$ lies on the parabola, so $2 = \frac{1}{4c}(2)^2$ implies $c = \frac{1}{2}$. So the focal point of the parabola is $\left(\frac{1}{2}, 0\right)$, and to produce a parallel beam of light the light source should be placed $\frac{1}{2}$ inch from the vertex of the light.

36. To describe the dish, use a parabola in standard position and axis along the y-axis, so the equation has the form $y = \frac{1}{4c}x^2, c > 0$. The information implies the point $(10, 4)$ lies on the parabola, so $4 = \frac{1}{4c}(10)^2$ implies $c = \frac{25}{4}$. So the focal point of the parabola is $\left(0, \frac{25}{4}\right)$, and the receiver should be placed $\frac{25}{4}$ inches from the vertex of the dish.

37. a. The parabolic path of the ball is a parabola with axis along the y-axis and vertex at $(0, 64)$. The equation of the path has the form $y - 64 = \frac{1}{4c}x^2$. The information implies the parabola passes through the point $(100, 64 - 16) = (100, 48)$. So

$$48 - 64 = \frac{1}{4c}(100)^2 \text{ implies } 4c = -\frac{10000}{16} = -625,$$

and

$$y - 64 = -\frac{1}{625}x^2.$$

To find where the ball hits the ground, find x when $y = 0$. That is,

$$-64 = -\frac{1}{625}x^2 \text{ implies } x^2 = 64(625) = 40000 \text{ so } x = \sqrt{40000} = 200.$$

The ball hits the ground 200 feet from the building.

b. Using the same analysis as in part (a), the equation of the parabolic path is

$$y - 1450 = -\frac{1}{625}x^2,$$

and if the distance above the ground is to be 0,

$$-1450 = -\frac{1}{625}x^2 \text{ so } x = \sqrt{(1450)(625)} \approx 952.$$

The ball hits the ground about 952 feet from the building.

38. Since the parabola describing the path of the projectile is symmetric about its axis, the vertex where the maximum height occurs will be midway between 0 and 1000. That is, the vertex is the point (500,200), and the equation of the parabolic path has the form $y - 200 = \frac{1}{4c}(x - 500)^2$. Since the projectile strikes the ground 1000 feet from the origin, the point (1000,0) is on the parabola. So

$$0 - 200 = \frac{1}{4c}(1000 - 500)^2 \text{ implies } 4c = -\frac{500^2}{200} = -1250 \text{ so } y - 200 = -\frac{1}{1250}(x - 500)^2.$$

To find when the projectile is 150 feet above the ground solve

$$150 - 200 = -\frac{1}{1250}(x - 500)^2 \text{ implies } 50(1250) = (x - 500)^2$$

$$\text{so } x - 500 = \pm\sqrt{62500} = \pm 250 \text{ and } x = 250, x = 750.$$

Exercise Set 6.3 (Page 321)

1. $\frac{x^2}{4}+y^2=1$ so $a=2, b=1, c=\sqrt{4-1}=\sqrt{3}$
Focal points: $\left(\sqrt{3},0\right), \left(-\sqrt{3},0\right)$

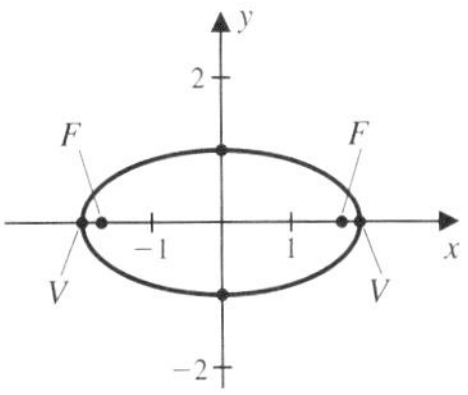

2. $\frac{x^2}{16}+y^2=1$ so $a=4, b=1, c=\sqrt{16-1}=\sqrt{15}$
Focal points: $\left(\sqrt{15},0\right), \left(-\sqrt{15},0\right)$

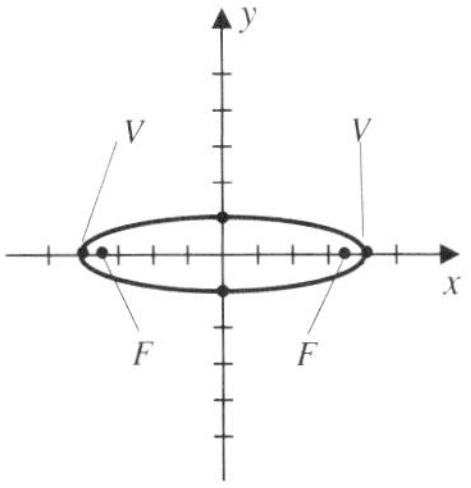

3. $x^2+\frac{y^2}{9}=1$ so $a=3, b=1, c=\sqrt{9-1}=\sqrt{8}=2\sqrt{2}$
Focal points: $\left(0,2\sqrt{2}\right), \left(0,-2\sqrt{2}\right)$

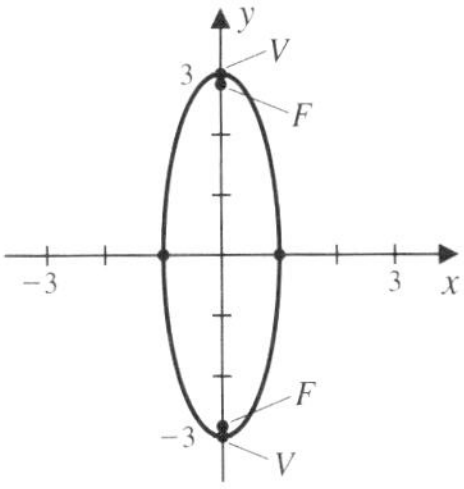

4. $\frac{x^2}{4}+\frac{y^2}{16}=1$ so $a=4, b=2, c=\sqrt{16-4}=\sqrt{12}=2\sqrt{3}$
Focal points: $\left(0,2\sqrt{3}\right), \left(0,-2\sqrt{3}\right)$

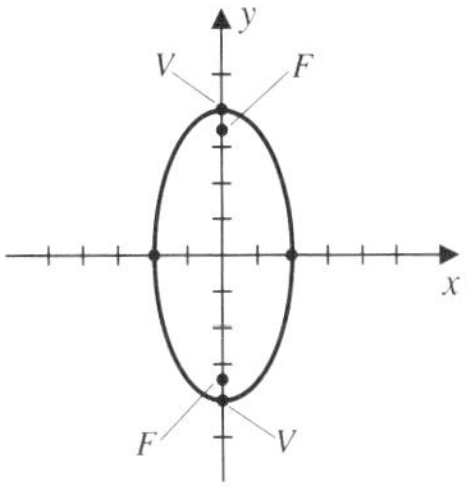

5. $16x^2+25y^2=400$ implies $\frac{x^2}{25}+\frac{y^2}{16}=1$ so $a=5, b=4, c=\sqrt{25-16}=\sqrt{9}=3$
Focal points: $(3,0), (-3,0)$

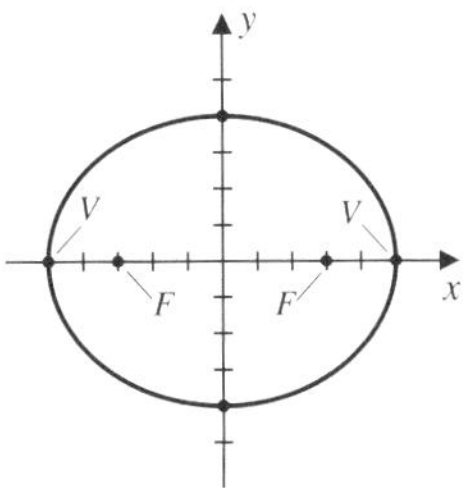

6. $25x^2+16y^2=400$ implies $\frac{x^2}{16}+\frac{y^2}{25}=1$ so $a=5, b=4, c=3$
Focal points: $(0,3), (0,-3)$

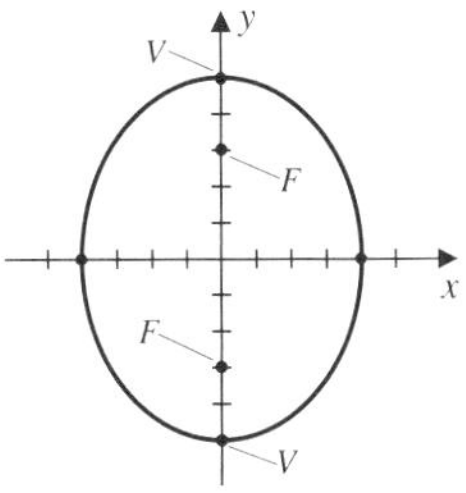

7. $3x^2 + 2y^2 = 6$ implies $\frac{x^2}{2} + \frac{y^2}{3} = 1$ so $a = \sqrt{3}, b = \sqrt{2}, c = 1$
Focal points: $(0, 1), (0, -1)$;
Vertices: $(0, \sqrt{3}), (0, -\sqrt{3})$

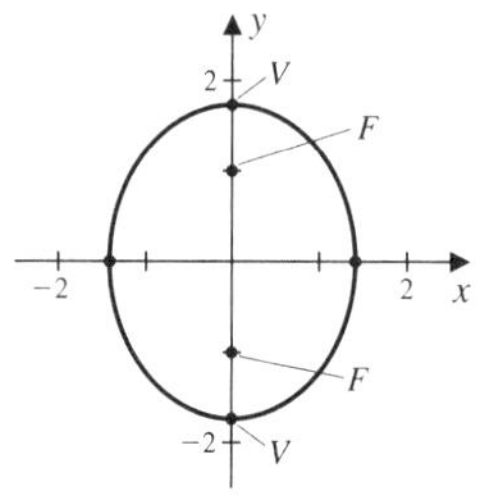

8. $3x^2 + 4y^2 = 12$ implies $\frac{x^2}{4} + \frac{y^2}{3} = 1$ so $a = 2, b = \sqrt{3}, c = 1$
Focal points: $(1, 0), (-1, 0)$;
Vertices: $(2, 0), (-2, 0)$

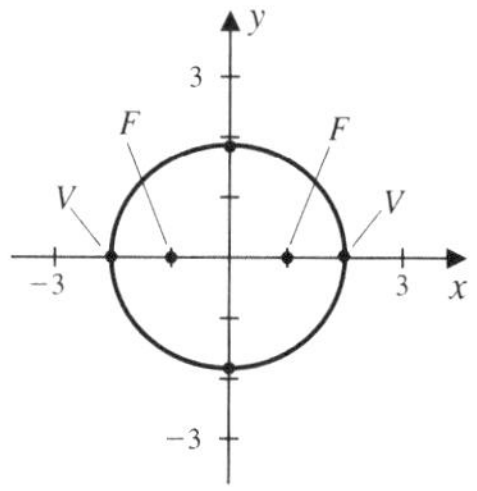

9. $4x^2 + y^2 + 16x + 12 = 0$ implies $4(x^2 + 4x + 4 - 4) + y^2 = -12$ or $4(x+2)^2 + y^2 = 4$ so $(x+2)^2 + \frac{y^2}{4} = 1$ and $a = 2, b = 1, c = \sqrt{4-1} = 3$
Focal points: $\left(-2, \sqrt{3}\right), \left(-2, -\sqrt{3}\right)$;
Center: $(-2, 0)$

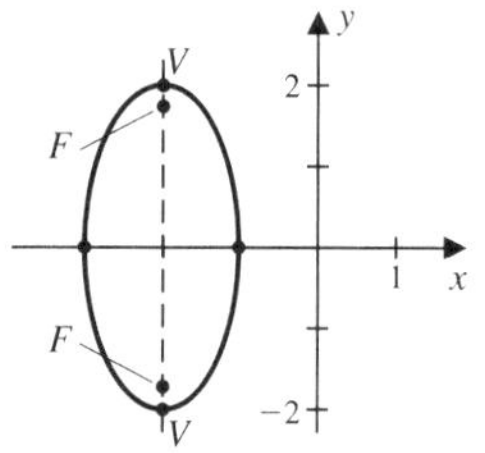

10. $2x^2 + 4y^2 - 4x - 14 = 0$ implies $2(x^2 - 2x + 1 - 1) + 4y^2 = 14$ or $2(x-1)^2 + 4y^2 = 16$ so $\frac{(x-1)^2}{8} + \frac{y^2}{4} = 1$ and $a = \sqrt{8} = 2\sqrt{2}, b = 2, c = \sqrt{8-4} = 2$
Focal points: $(3, 0), (-1, 0)$;
Vertices: $\left(1 + 2\sqrt{2}, 0\right), \left(1 - 2\sqrt{2}, 0\right)$
Center: $(1, 0)$

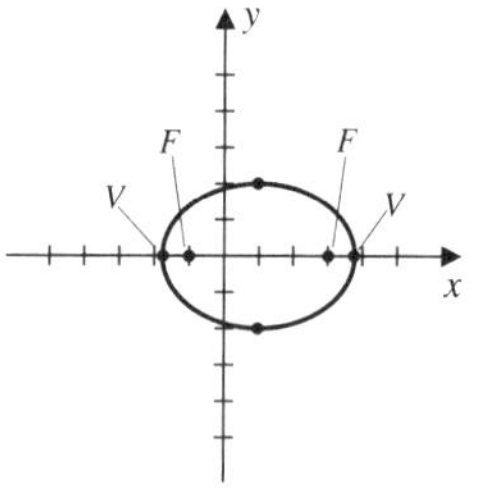

11. $x^2 + 4y^2 - 2x - 16y + 13 = 0$ implies
$x^2 - 2x + 1 - 1 + 4(y^2 - 4y + 4 - 4) = -13$
or $(x - 1)^2 + 4(y - 2)^2 = 4$
so $\frac{(x-1)^2}{4} + (y - 2)^2 = 1$
and $a = 2, b = 1, c = \sqrt{4 - 1} = \sqrt{3}$
Focal points: $(1 - \sqrt{3}, 2), (1 + \sqrt{3}, 2)$;
Vertices: $(-1, 2), (3, 2)$; Center: $(1, 2)$

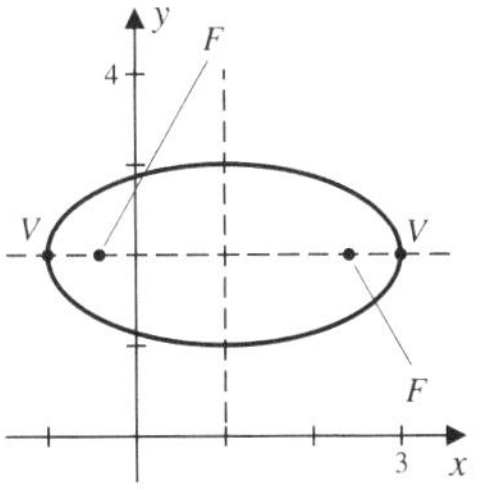

12. $4x^2 + 9y^2 - 16x + 90y + 97 = 0$
implies $4(x^2 - 4x + 4 - 4) + 9(y^2 + 10y + 25 - 25) = -97$
or $4(x - 2)^2 + 9(y + 5)^2 = 144$
so $\frac{(x-2)^2}{36} + \frac{(y+5)^2}{16} = 1$
and $a = 6, b = 4, c = \sqrt{36 - 16} = 2\sqrt{5}$
Focal points: $(2 - 2\sqrt{5}, -5), (2 + 2\sqrt{5}, -5)$;
Vertices: $(8, -5), (-4, -5)$; Center:
$(2, -5)$

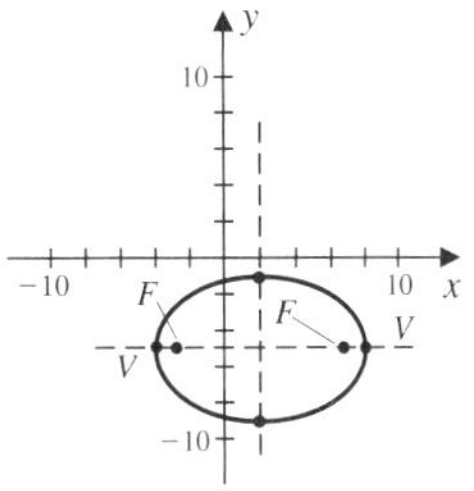

13. $2x^2 + 4y^2 + 4x - 16y + 2 = 0$
implies $2(x^2 + 2x + 1 - 1) + 4(y^2 - 4y + 4 - 4) = -2$ or $2(x + 1)^2 + 4(y - 2)^2 = 16$
so $\frac{(x+1)^2}{8} + \frac{(y-2)^2}{4} = 1$
and $a = \sqrt{8} = 2\sqrt{2}, b = 2, c = \sqrt{8 - 4} = 2$
Focal points: $(1, 2), (-3, 2)$; Center: $(-1, 2)$
Vertices: $\left(-1 + 2\sqrt{2}, 2\right), \left(-1 - 2\sqrt{2}, 2\right)$

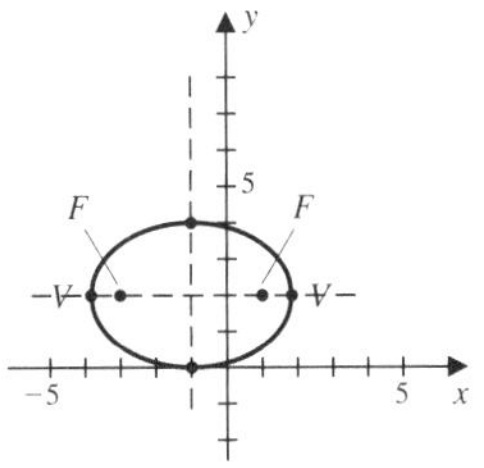

14. $3x^2 + 4y^2 + 12x + 8y + 4 = 0$
implies $3(x^2 + 4x + 4 - 4) + 4(y^2 + 2y + 1 - 1) = -4$ or $3(x + 2)^2 + 4(y + 1)^2 = 12$
so $\frac{(x+2)^2}{4} + \frac{(y+1)^2}{3} = 1$ and $a = 2, b = \sqrt{3}, c = 1$
Focal points: $(-3, -1), (-1, -1)$; Center: $(-2, -1)$
Vertices: $(-4, -1), (0, -1)$

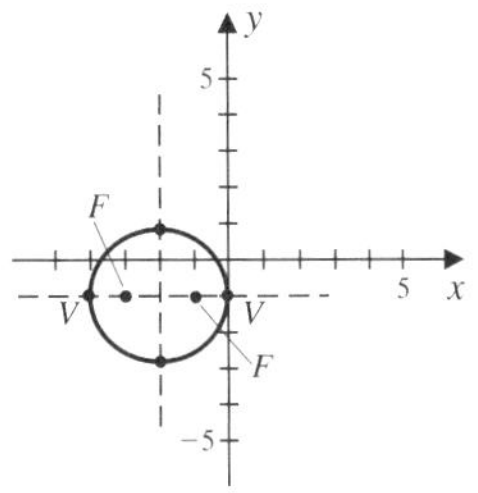

15. x-intercepts: $(\pm 4, 0)$; y-intercepts: $(0, \pm 3)$

The intercepts give the vertices of the ellipse, so the ellipse is in standard position, center at the origin, major axis along the x-axis, and $a = 4, b = 3$. The equation is $\frac{x^2}{16} + \frac{y^2}{9} = 1$.

16. x-intercepts: $(\pm 2, 0)$; y-intercepts: $(0, \pm 5)$

The intercepts give the vertices of the ellipse, so the ellipse is in standard position, center at the origin, major axis along the y-axis, and $a = 5, b = 2$. The equation is $\frac{y^2}{25} + \frac{x^2}{4} = 1$.

17. Foci: $(\pm 2, 0)$; Vertices: $(\pm 3, 0)$

Since the foci are centered about the origin and on the x-axis, the ellipse is in standard position with center at the origin and major axis on the x-axis. So

$$a = 3, c = 2, \text{ and } c^2 = a^2 - b^2 \text{ implies } 4 = 9 - b^2 \text{ so } b^2 = 5.$$

The equation is $\frac{x^2}{9} + \frac{y^2}{5} = 1$.

18. Foci: $(\pm 2, 0)$; y-intercepts: $(0, \pm 2)$

Since the foci are centered about the origin and on the x-axis, the ellipse is in standard position with center at the origin and major axis on the x-axis. So

$$b = 2, c = 2, \text{ and } c^2 = a^2 - b^2 \text{ implies } 4 = a^2 - 4 \text{ so } a^2 = 8.$$

The equation is $\frac{x^2}{8} + \frac{y^2}{4} = 1$.

19. Foci: $(0, \pm 1)$; x-intercepts: $(\pm 2, 0)$

Since the foci are centered about the origin and on the y-axis, the ellipse is in standard position with center at the origin and major axis on the y-axis. So

$$b = 2, c = 1, \text{ and } c^2 = a^2 - b^2 \text{ implies } 1 = a^2 - 4 \text{ so } a^2 = 5.$$

The equation is $\frac{y^2}{5} + \frac{x^2}{4} = 1$.

20. Foci: $(0, \pm 1)$; y-intercepts: $(0, \pm 2)$

Since the foci are centered about the origin and on the y-axis, the ellipse is in standard position with center at the origin and major axis on the y-axis. So

$$a = 2, c = 1, \text{ and } c^2 = a^2 - b^2 \text{ implies } 1 = 4 - b^2 \text{ so } b^2 = 3.$$

The equation is $\frac{y^2}{4} + \frac{x^2}{3} = 1$.

21. Length of major axis 5; Length of minor axis 3; Foci on the x-axis;

Since the ellipse is in standard position, $a = \frac{5}{2}, b = \frac{3}{2}$, and the equation is
$\frac{x^2}{25/4} + \frac{y^2}{9/4} = 1$ or $\frac{4x^2}{25} + \frac{4y^2}{9} = 1$.

22. Foci: $(0, \pm 2)$; Length of major axis: 8

Since the foci are centered about the origin and on the y-axis, the ellipse is in standard position with major axis on the y-axis, and the equation has the form $\frac{y^2}{a^2} + \frac{x^2}{b^2} = 1$. Since the foci are $(0, \pm 2)$, $c = 2$, and since the major axis has length 8, we have $2a = 8$ so $a = 4$. This gives

$$c^2 = a^2 - b^2 \text{ implies } 4 = 16 - b^2 \text{ so } b^2 = 12.$$

The equation is $\frac{y^2}{16} + \frac{x^2}{12} = 1$.

23. Foci: $(3, 0)$, $(1, 0)$; Vertex: $(0, 0)$

Since the foci are on the x-axis, the major axis of the ellipse is on the x-axis. The foci are centered about $(2, 0)$, so $c = 1$. Since a vertex is $(0, 0)$, the other vertex is $(4, 0)$, and the length of the major axis 4. This gives $a = 2$. Then

$$c^2 = a^2 - b^2 \text{ implies } 1 = 4 - b^2 \text{ so } b^2 = 3.$$

The equation is $\frac{(x-2)^2}{4} + \frac{y^2}{3} = 1$.

24. Foci: $(0, 4)$, $(0, 8)$; Vertex: $(0, 2)$

Since the foci are on the y-axis, the major axis of the ellipse is on the y-axis. The foci are centered about $(0, 6)$, so $c = 2$. Since a vertex is $(0, 2)$, the other vertex is $(0, 10)$, and the length of the major axis 8. This gives $a = 4$. Then

$$c^2 = a^2 - b^2 \text{ implies } 4 = 16 - b^2 \text{ so } b^2 = 12.$$

The equation is $\frac{x^2}{12} + \frac{(y-6)^2}{16} = 1$.

25. Focus: $(-4, 0)$; Vertices: $(-4, -2)$, $(-4, 8)$

Since the vertices are on the vertical line $x = -4$, the major axis of the ellipse is on the line $x = -4$. The vertices are 10 units apart, which is the length of the major axis, so $a = 5$. The center is $(-4, 3)$, so $c = 3$. Then

$$c^2 = a^2 - b^2 \text{ implies } 9 = 25 - b^2 \text{ so } b^2 = 16.$$

The equation is $\frac{(x+4)^2}{16} + \frac{(y-3)^2}{25} = 1$.

26. Vertices: $(2, 2)$, $(6, 2)$; Focal point: $(5, 2)$

The vertices and the focal point are on the line $y = 2$, so the major axis of the ellipse is $y = 2$. Since the vertices are centered about $(4, 2)$,

$$a = 2, c = 1 \text{ and } c^2 = a^2 - b^2 \text{ implies } 1 = 4 - b^2 \text{ so } b^2 = 3.$$

The equation is $\frac{(x-4)^2}{4} + \frac{(y-2)^2}{3} = 1$.

27. Vertices: $(3, 3)$, $(3, -1)$; Passing through: $(2, 1)$

The center is midway between the two vertices, so is $(3, 1)$, and $a = 4$. The ellipse has the form

$$\frac{(x-3)^2}{b^2} + \frac{(y-1)^2}{16} = 1.$$

If the ellipse passes through (2, 1), then

$$\frac{(2-3)^2}{b^2}+\frac{(1-1)^2}{16}=1 \text{ so } b^2=1.$$

The equation is $(x-3)^2+\frac{(y-1)^2}{16}=1$.

28. Vertices: $(-4,-2)$, $(2,-2)$; Passing through: $(-1,1)$

The center is midway between the two vertices, so is $(-1,-2)$, and $a=3$. The ellipse has the form

$$\frac{(x+1)^2}{9}+\frac{(y+2)^2}{b^2}=1.$$

If the ellipse passes through $(-1,1)$, then

$$\frac{(-1+1)^2}{9}+\frac{(1+2)^2}{b^2}=1 \text{ so } b^2=9.$$

The equation is $\frac{(x+1)^2}{9}+\frac{(y+2)^2}{9}=1$ or $(x+1)^2+(y+2)^2=9$, which is the equation of a circle.

29. Let P be the point of intersection of the latus rectum and the upper half of the ellipse. The length of the latus rectum is then $2y$, where the coordinates of $P=(c,y)$. Since the equation of the ellipse is

$$\frac{x^2}{a^2}+\frac{y^2}{b^2}=1 \text{ implies } \frac{y^2}{b^2}=1-\frac{x^2}{a^2} \text{ so } y=\pm\sqrt{b^2\left(1-\frac{x^2}{a^2}\right)}.$$

Letting $x=c$, multiplying by 2, and noting $b^2=a^2-c^2$, the length of the latus rectum is then

$$2\sqrt{b^2\left(1-\frac{c^2}{a^2}\right)}=2\sqrt{b^2\left(\frac{a^2-c^2}{a^2}\right)}=2\sqrt{\frac{b^4}{a^2}}=\frac{2b^2}{a}.$$

30. Completing the square on the x and y terms of $Ax^2+Cy^2+Dx+Ey+F=0$ gives,

$$A\left(x^2+\frac{D}{A}x\right)+C\left(y^2+\frac{E}{C}y\right)=-F$$

$$A\left(x^2+\frac{D}{A}x+\frac{D^2}{4A^2}\right)+C\left(y^2+\frac{E}{C}y+\frac{E^2}{4C^2}\right)=-F+\frac{AD^2}{4A^2}+\frac{CE^2}{4C^2}$$

$$A\left(x+\frac{D}{2A}\right)^2+C\left(y+\frac{E}{2C}\right)^2=\frac{CD^2+AE^2-4ACF}{4AC}.$$

For $A>0$ and $C>0$, if CD^2+AE^2-4ACF is

a. positive and $A \neq C$, the equation describes an ellipse,

b. positive and $A = C$, the equation describes a circle,

c. 0, the equation describes a single point,

d. less than 0, the equation is satisfied by no points.

31. Since the major axis has length 480,

$$2a = 480 \quad \text{implies} \quad a = 240 \quad \text{so} \quad a^2 = 57600,$$

and since the minor axis has length 280,

$$2b = 280 \quad \text{implies} \quad b = 140 \quad \text{so} \quad b^2 = 19600.$$

The equation is $\frac{x^2}{57600} + \frac{y^2}{19600} = 1$.

32. Since the major axis has length 5.39×10^9,

$$2a = 5.39 \times 10^9 \quad \text{so} \quad a = 2.695 \times 10^9,$$

and since the minor axis has length 1.36×10^9,

$$2b = 1.36 \times 10^9 \quad \text{so} \quad b = 0.68 \times 10^9.$$

Then $c = \sqrt{a^2 - b^2} \approx 2.608 \times 10^9$. Since the sun is at one of the foci, the closest the comet comes to the sun is

$$2.695 \times 10^9 - 2.608 \times 10^9 = 0.087 \times 10^9 = 8.7 \times 10^7 \text{ kilometers.}$$

33. The length of the major axis of the satellites orbit is the diameter of the earth, 12760, plus 160 plus 16000, which equals 28920. Then $a = \frac{28920}{2} = 14460$, $c = 14460 - 6540 = 7920$ so $e = \frac{c}{a} = \frac{7920}{14460} \approx 0.6$. Then $b^2 = a^2 - c^2 \approx 146365200$. Since $a^2 = 14460^2 = 209091600$, the equation of the orbit is $\frac{x^2}{209091600} + \frac{y^2}{146365200} = 1$.

34. The equation of the elliptical shaped window is

$$\frac{x^2}{20^2} + \frac{y^2}{12^2} = 1 \text{ implies } \frac{x^2}{400} + \frac{y^2}{144} = 1 \text{ so } y = \sqrt{144\left(1 - \frac{x^2}{400}\right)}.$$

The height of the window 16 inches in from the center is then

$$h = \sqrt{144\left(1 - \frac{16^2}{400}\right)} \approx 7.2 \text{ inches.}$$

Exercise Set 6.4 (Page 329)

1. $\frac{x^2}{4} - \frac{y^2}{9} = 1$ implies $a = 2, b = 3$
so $c^2 = a^2 + b^2 = 13$ and $c = \sqrt{13}$;
Vertices: $(2, 0), (-2, 0)$;
Foci: $\left(\sqrt{13}, 0\right), \left(-\sqrt{13}, 0\right)$;
Asymptotes: $y = \pm\frac{3}{2}x$

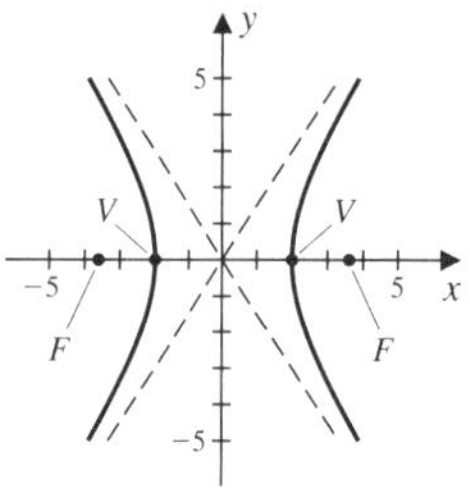

2. $\frac{x^2}{9} - \frac{y^2}{4} = 1$ implies $a = 3, b = 2$
so $c^2 = 13$ and $c = \sqrt{13}$;
Vertices: $(3, 0), (-3, 0)$;
Foci: $\left(\sqrt{13}, 0\right), \left(-\sqrt{13}, 0\right)$;
Asymptotes: $y = \pm\frac{2}{3}x$

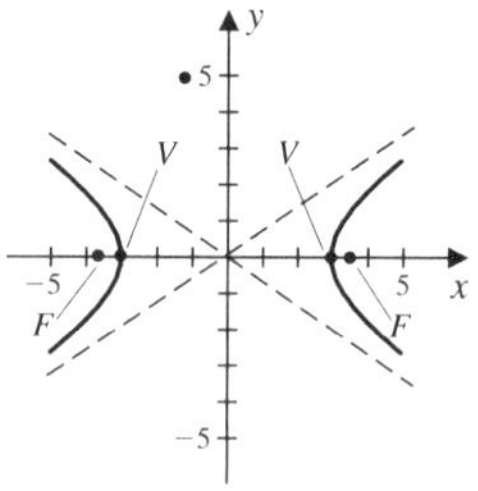

3. $\frac{y^2}{4} - \frac{x^2}{9} = 1$ implies $a = 2, b = 3$
so $c^2 = 13$ and $c = \sqrt{13}$;
Vertices: $(0, 2), (0, -2)$;
Foci: $\left(0, \sqrt{13}\right), \left(0, -\sqrt{13}\right)$;
Asymptotes: $y = \pm\frac{2}{3}x$

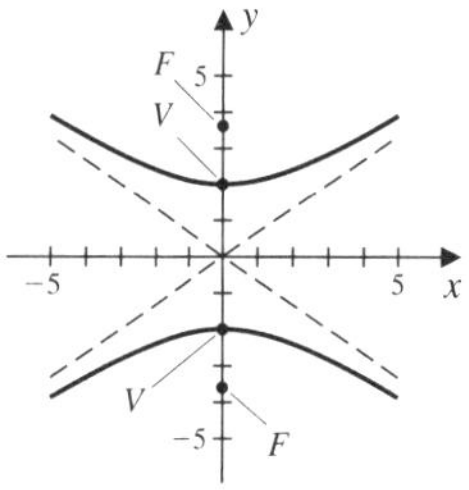

4. $\frac{y^2}{9} - \frac{x^2}{4} = 1$ implies $a = 3, b = 2$
so $c^2 = 13$ and $c = \sqrt{13}$;
Vertices: $(0, 3), (0, -3)$;
Foci: $\left(0, \sqrt{13}\right), \left(0, -\sqrt{13}\right)$;
Asymptotes: $y = \pm\frac{3}{2}x$

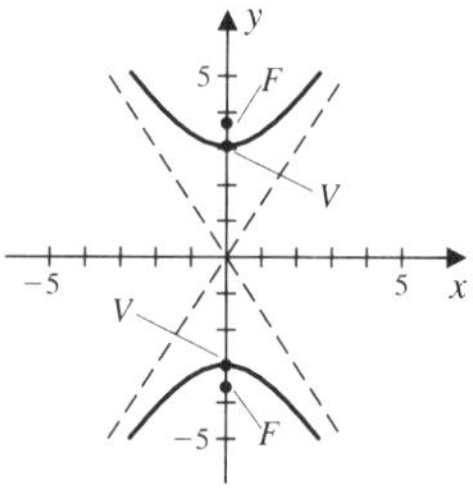

5. $x^2 - y^2 = 1$ implies $a = 1, b = 1$ so $c^2 = 2$ and $c = \sqrt{2}$;
Vertices: $(1, 0), (-1, 0)$;
Foci: $\left(\sqrt{2}, 0\right), \left(-\sqrt{2}, 0\right)$;
Asymptotes: $y = \pm x$

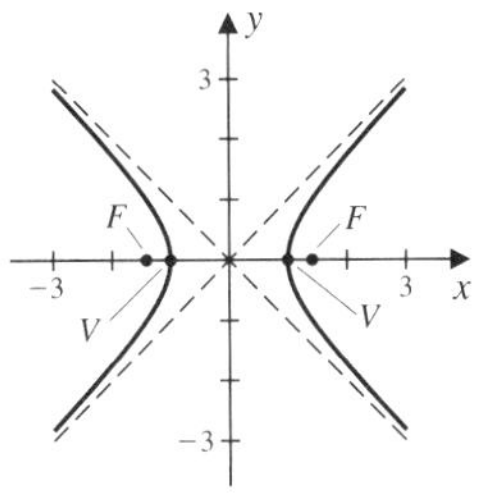

6. $y^2 - 4x^2 = 1$ implies $y^2 - \frac{x^2}{1/4} = 1$; $a = 1, b = 1/2$ so $c^2 = 1 + 1/4 = \frac{5}{4}$ and $c = \frac{\sqrt{5}}{2}$;
Vertices: $(0, 1), (0, -1)$;
Foci: $\left(0, \sqrt{5}/2\right), \left(0, -\sqrt{5}/2\right)$;
Asymptotes: $y = \pm 2x$

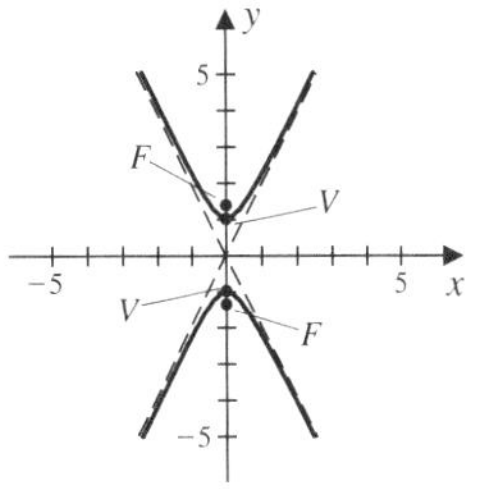

7. $9y^2 - 18y - 4x^2 = 9(y^2 - 2y + 1 - 1) - 4x^2 = 27$
implies $9(y-1)^2 - 4x^2 = 36$ or $\frac{(y-1)^2}{4} - \frac{x^2}{9} = 1$.
So $a = 2, b = 3$ implies $c^2 = 13$ so $c = \sqrt{13}$. Vertices: $(0, 3), (0, -1)$;
Foci: $\left(0, 1 + \sqrt{13}\right), \left(0, 1 - \sqrt{13}\right)$;
Asymptotes: $y - 1 = \pm\frac{2}{3}x$ or $y = \pm\frac{2}{3}x + 1$.

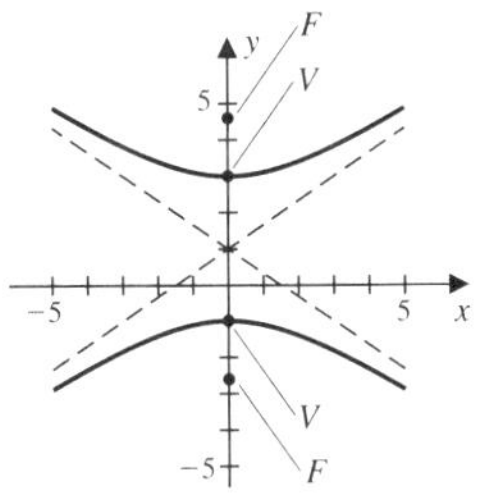

8. $x^2 + 2x - 4y^2 = x^2 + 2x + 1 - 1 - 4y^2 = 3$ implies $(x+1)^2 - 4y^2 = 4$ or $\frac{(x+1)^2}{4} - y^2 = 1$.
So $a = 2, b = 1$ implies $c^2 = 5$ so $c = \sqrt{5}$.
Vertices: $(-3, 0), (1, 0)$;
Foci: $\left(-1 + \sqrt{5}, 0\right), \left(-1 - \sqrt{5}, 0\right)$;
Asymptotes: $y = \pm\frac{1}{2}(x + 1)$

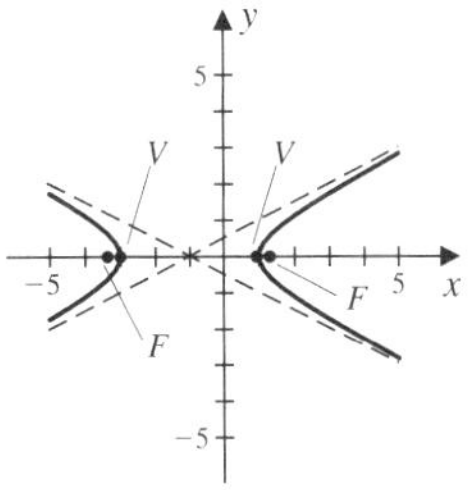

9. $3x^2 - y^2 = 6x$ implies $3(x^2 - 2x + 1 - 1) - y^2 = 0$ so $3(x-1)^2 - y^2 = 3$ or $(x-1)^2 - \frac{y^2}{3} = 1$.
So $a = 1, b = \sqrt{3}$ implies $c^2 = 4$ so $c = 2$.
Vertices: $(0,0), (2,0)$; Foci: $(3,0), (-1,0)$;
Asymptotes: $y = \pm\sqrt{3}(x-1)$

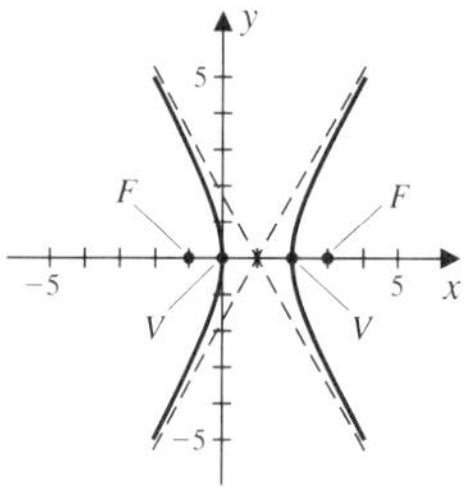

10. $2y^2 + 8y = 9x^2$ implies $2(y^2 + 4y + 4 - 4) - 9x^2 = 0$ so $2(y+2)^2 - 9x^2 = 8$ or $\frac{(y+2)^2}{4} - \frac{x^2}{8/9} = 1$.
So $a = 2, b = \frac{2\sqrt{2}}{3}$ implies $c^2 = 4 + \frac{8}{9} = \frac{44}{9}$ so $c = \frac{\sqrt{44}}{3}$.
Vertices: $(0,0), (0,-4)$;
Foci: $\left(0, -2 - \sqrt{44}/3\right), \left(0, -2 + \sqrt{44}/3\right)$;
Asymptotes: $y + 2 = \pm\frac{3\sqrt{2}}{2}x$

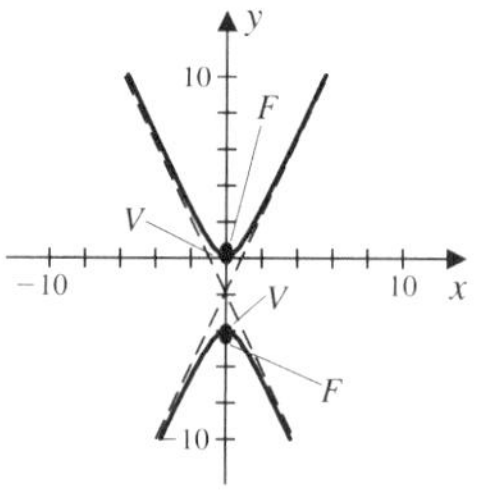

11. $9x^2 - 4y^2 - 18x - 8y = 9(x^2 - 2x + 1 - 1) - 4(y^2 + 2y + 1 - 1) = 31$ implies $9(x-1)^2 - 4(y+1)^2 = 36$ or $\frac{(x-1)^2}{4} - \frac{(y+1)^2}{9} = 1$.
So $a = 2, b = 3$ implies $c^2 = 13$ so $c = \sqrt{13}$.
Vertices: $(3,-1), (-1,-1)$;
Foci: $\left(1 - \sqrt{13}, -1\right), \left(1 + \sqrt{13}, -1\right)$;
Asymptotes: $y + 1 = \pm\frac{3}{2}(x-1)$

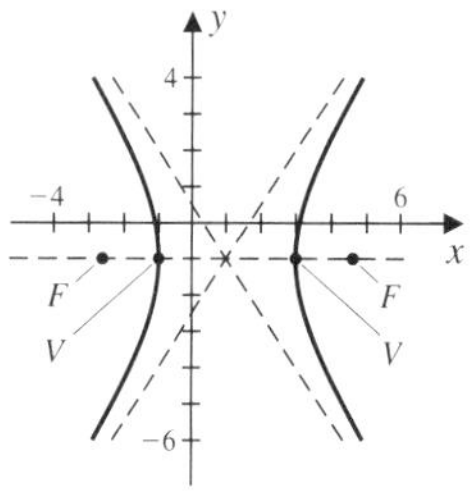

12. $y^2 - 4x^2 - 2y - 16x = y^2 - 2y + 1 - 1 - 4(x^2 + 4x + 4 - 4) = 19$ implies $(y-1)^2 - 4(x+2)^2 = 4$ or $\frac{(y-1)^2}{4} - (x+2)^2 = 1$.
So $a = 2, b = 1$ implies $c^2 = 5$ so $c = \sqrt{5}$.
Vertices: $(-2,3), (-2,-1)$;
Foci: $\left(-2, 1 + \sqrt{5}\right), \left(-2, 1 - \sqrt{5}\right)$;
Asymptotes: $y - 1 = \pm 2(x+2)$

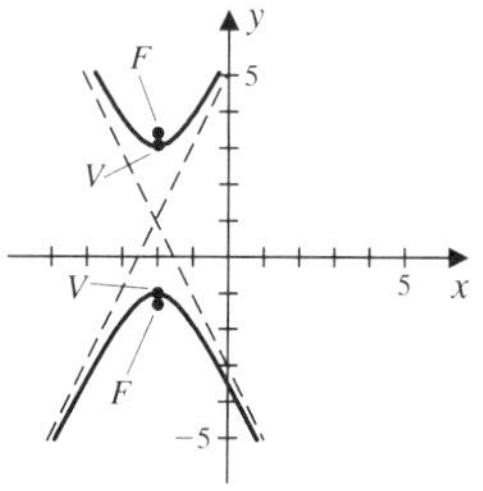

13. $9y^2 - 4x^2 - 36y + 16x - 16 = 0$ implies $9(y^2 - 4y + 4 - 4) - 4(x^2 - 4x + 4 - 4) = 16$
so $9(y-2)^2 - 4(x-2)^2 = 36$
or $\frac{(y-2)^2}{4} - \frac{(x-2)^2}{9} = 1$.
So $a = 2, b = 3,$ and $c = \sqrt{13}$.
Vertices: $(2, 4), (2, 0)$;
Foci: $\left(2, 2+\sqrt{13}\right), \left(2, 2-\sqrt{13}\right)$
Asymptotes: $y - 2 = \pm\frac{2}{3}(x-2)$

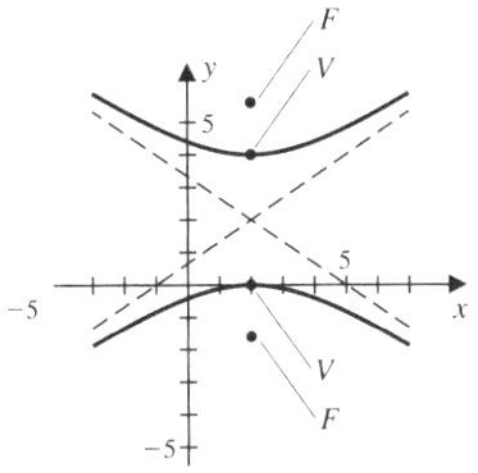

14. $4x^2 - 6y^2 - 8x + 24y - 44 = 0$ implies $4(x^2 - 2x + 1 - 1) - 6(y^2 - 4y + 4 - 4) = 44$
so $4(x-1)^2 - 6(y-2)^2 = 24$
or $\frac{(x-1)^2}{6} - \frac{(y-2)^2}{4} = 1$.
So $a = \sqrt{6}, b = 2,$ and $c = \sqrt{10}$.
Vertices: $\left(1+\sqrt{6}, 2\right), \left(1-\sqrt{6}, 2\right)$;
Foci: $\left(1+\sqrt{10}, 2\right), \left(1-\sqrt{10}, 2\right)$
Asymptotes: $y - 2 = \pm\frac{\sqrt{6}}{3}(x-1)$

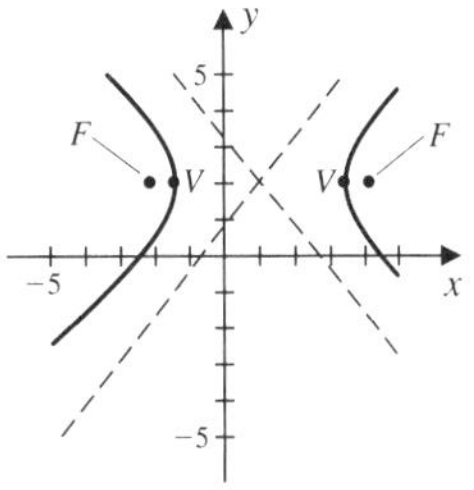

15. Foci: $(\pm 5, 0)$; Vertices: $(\pm 3, 0)$

Since the foci are centered about the origin and on the x-axis the axis is on the x-axis, $a = 3, c = 5,$ and

$$c^2 = a^2 + b^2 \text{ implies } 25 = 9 + b^2 \text{ so } b^2 = 16.$$

The equation is $\frac{x^2}{9} - \frac{y^2}{16} = 1$.

16. Foci: $(0, \pm 13)$; Vertices: $(0, \pm 12)$

Since the foci are centered about the origin and on the y-axis, the axis is on the y-axis, $a = 12, c = 13,$and

$$c^2 = a^2 + b^2 \text{ implies } 169 = 144 + b^2 \text{ so } b^2 = 25.$$

The equation is $\frac{y^2}{144} - \frac{x^2}{25} = 1$.

17. Foci: $(0, \pm 5)$; Vertices: $(0, \pm 4)$

Since the foci are centered about the origin and on the y-axis, the axis is on the y-axis, $a = 4, c = 5,$and

$$c^2 = a^2 + b^2 \text{ implies } 25 = 16 + b^2 \text{ so } b^2 = 9.$$

The equation is $\frac{y^2}{16} - \frac{x^2}{9} = 1$.

18. Foci: $(\pm 13, 0)$; Vertices: $(\pm 5, 0)$

Since the foci are centered about the origin and on the x-axis, the axis is on the x-axis, $a = 5$, $c = 13$, and

$$c^2 = a^2 + b^2 \text{ implies } 169 = 25 + b^2 \text{ so } b^2 = 144.$$

The equation is $\frac{x^2}{25} - \frac{y^2}{144} = 1$.

19. Focus: $(2, 2)$; Vertices: $(2, 1)$, $(2, -3)$

Since the focus and the vertices are on the line $x = 2$, the axis is parallel to the y-axis. Since the vertices are centered about $(2, -1)$, $a = 2$, and since the focus $(2, 2)$ is 3 units above the center $(2, -1)$, $c = 3$. Then

$$c^2 = a^2 + b^2 \text{ implies } 9 = 4 + b^2 \text{ so } b^2 = 5.$$

The equation is $\frac{(y+1)^2}{4} - \frac{(x-2)^2}{5} = 1$.

20. Focus: $(-3, 3)$; Vertices: $(-3, 0)$, $(-3, -6)$

Since the focus and the vertices are on the line $x = -3$, the axis is parallel to the y-axis. Since the vertices are centered about $(-3, -3)$, $a = 3$, and since the focus $(-3, 3)$ is 6 units above the center $(-3, -3)$, $c = 6$. Then

$$c^2 = a^2 + b^2 \text{ implies } 36 = 9 + b^2 \text{ so } b^2 = 27.$$

The equation is $\frac{(y+3)^2}{9} - \frac{(x+3)^2}{27} = 1$.

21. Foci: $(-1, 4)$, $(5, 4)$; Vertex: $(0, 4)$

Since the foci and the one vertex are on the line $y = 4$, the axis is parallel to the x-axis. Since the foci are centered about the point $(2, 4)$, $c = 3$, and since the vertex $(0, 4)$ is 2 units to the left of the center $(2, 4)$, $a = 2$. Then

$$c^2 = a^2 + b^2 \text{ implies } 9 = 4 + b^2 \text{ so } b^2 = 5.$$

The equation is $\frac{(x-2)^2}{4} - \frac{(y-4)^2}{5} = 1$.

22. Foci: $(-1, -2)$, $(-7, -2)$; Vertex: $(-2, -2)$

Since the foci and the one vertex are on the line $y = -2$, the axis is parallel to the x-axis. Since the foci are centered about the point $(-4, -2)$, $c = 3$, and since the vertex $(-2, -2)$ is 2 units to the right of the center $(-4, -2)$, $a = 2$. Then

$$c^2 = a^2 + b^2 \text{ implies } 9 = 4 + b^2 \text{ so } b^2 = 5.$$

The equation is $\frac{(x+4)^2}{4} - \frac{(y+2)^2}{5} = 1$.

23. Vertices: $(0, \pm 2)$; Passing through: $(3, 4)$

Since the vertices are centered about the origin and on the y-axis, the hyperbola is in standard position with axis on the y-axis and has the form $\frac{y^2}{a^2} - \frac{x^2}{b^2} = 1$. Then $a = 2$, and since the hyperbola passes through $(3, 4)$, we have

$$\frac{(4)^2}{4} - \frac{(3)^2}{b^2} = 1 \text{ implies } \frac{9}{b^2} = 3 \text{ so } b = \sqrt{3}.$$

The equation is $\frac{y^2}{4} - \frac{x^2}{3} = 1$.

24. Vertices: $(\pm 2, 2)$; Passing through: $(8, 8)$

The vertices are centered about the point $(0, 2)$ and lie on the line $y = 2$, so the axis is parallel to the x-axis. Since the distance from the center to the vertices is 2, $a = 2$, and the equation has the form $\frac{x^2}{4} - \frac{(y-2)^2}{b^2} = 1$. Since the hyperbola passes through $(8, 8)$, we have

$$\frac{8^2}{4} - \frac{(8-2)^2}{b^2} = 1 \text{ so } b^2 = \frac{36}{15}.$$

The equation is $\frac{x^2}{4} - \frac{15(y-2)^2}{36} = 1$.

25. Vertices: $(\pm 3, 0)$; Asymptotes: $y = \pm\frac{4}{3}x$.

Since the vertices are centered about the origin on the x-axis, the hyperbola is in standard position with axis on the x-axis. The vertices imply $a = 3$, and the asymptotes imply

$$\frac{b}{a} = \frac{4}{3} \text{ implies } \frac{b}{3} = \frac{4}{3} \text{ so } b = 4.$$

The equation is $\frac{x^2}{9} - \frac{y^2}{16} = 1$.

26. Vertices: $(\pm 3, 0)$; Asymptotes: $y = \pm\frac{3}{4}x$.

Since the vertices are centered about the origin on the x-axis, the hyperbola is in standard position with axis on the x-axis. The vertices imply $a = 3$, and the asymptotes imply

$$\frac{b}{a} = \frac{3}{4} \text{ implies } \frac{b}{3} = \frac{3}{4} \text{ so } b = \frac{9}{4}.$$

The equation is $\frac{x^2}{9} - \frac{16y^2}{81} = 1$.

27. Let $P(x, y)$ be the point of intersection of the latus rectum and the hyperbola as shown in the figure. The length is then $2y$, where the coordinates of $P = (c, y)$. Since the equation of the hyperbola is

$$\frac{x^2}{a^2} - \frac{y^2}{b^2} = 1 \text{ implies } \frac{y^2}{b^2} = \frac{x^2}{a^2} - 1 \text{ so } y = \pm\sqrt{b^2\left(\frac{x^2}{a^2} - 1\right)}.$$

Letting $x = c$, multiplying by 2, and noting $b^2 = c^2 - a^2$, the length of the latus rectum is

$$2\sqrt{b^2\left(\frac{c^2}{a^2} - 1\right)} = 2\sqrt{b^2\left(\frac{c^2 - a^2}{a^2}\right)} = 2\sqrt{\frac{b^4}{a^2}} = \frac{2b^2}{a}.$$

28. Completing the square on the x and y terms of $Ax^2 - Cy^2 + Dx + Ey + F = 0$ gives

$$A\left(x^2 + \frac{D}{A}x\right) - C\left(y^2 - \frac{E}{C}y\right) = -F$$

$$A\left(x^2 + \frac{D}{A}x + \frac{D^2}{4A^2}\right) - C\left(y^2 - \frac{E}{C}y + \frac{E^2}{4C^2}\right) = -F + \frac{AD^2}{4A^2} - \frac{CE^2}{4C^2}$$

$$A\left(x + \frac{D}{2A}\right)^2 - C\left(y - \frac{E}{2C}\right)^2 = \frac{CD^2 - AE^2 - 4ACF}{4AC}.$$

If $A > 0$ and $C > 0$, and $CD^2 - AE^2 - 4ACF$

a. is greater than 0, the equation describes a hyperbola with axis parallel to the x-axis,

b. is less than 0, the equation describes a hyperbola with axis parallel to the y-axis,

c. is equal to 0, it implies

$$y = \pm\sqrt{\frac{A}{C}}\left(x + \frac{D}{2A}\right) + \frac{E}{2C},$$

so the equation describes two intersecting lines.

29. The axes of the hyperbolas are perpendicular to each other.

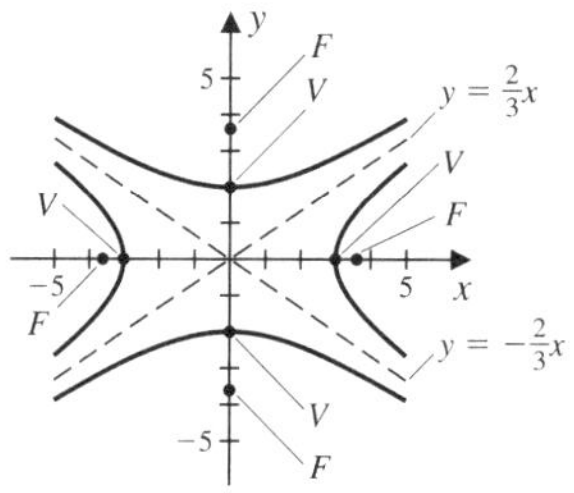

30. Let A be the eastmost station, C the westmost station, B the middle station, and S the source of the sound. If s is the speed the sound travels, and if it takes x seconds for the sound to reach station A, it takes $x + 2$ seconds to reach station C and $x + 1$ seconds to reach station B. The source is the point of intersection of the two hyperbolas described by the equations

$$d(S, C) - d(S, A) = (x + 2)s - xs = 2s \quad \text{and}$$

$$d(S, B) - d(S, A) = (x + 1)s - xs = s.$$

There are two possible locations for the point S, one to the north of the detection points and one to the south.

31. If the total cost of production and delivery from plants A and B to a destination C are denoted T_A and T_B, then

$$T_A - T_B = 130 + d(A, C) - d(B, C).$$

If $d(B, C) = d(A, C) + 130$, then $T_A - T_B = 0$, and it makes no difference from which plant the car is shipped. If $d(B, C) < d(A, C) + 130$, then

$$T_A - T_B > 0 \text{ so } T_A > T_B,$$

and the car should be shipped from plant B. If $d(B, C) > d(A, C) + 130$, then

$$T_A - T_B < 0 \text{ so } T_A < T_B,$$

and the car should be shipped from plant A.

Exercise Set 6.5 (Page 341)

1. a.

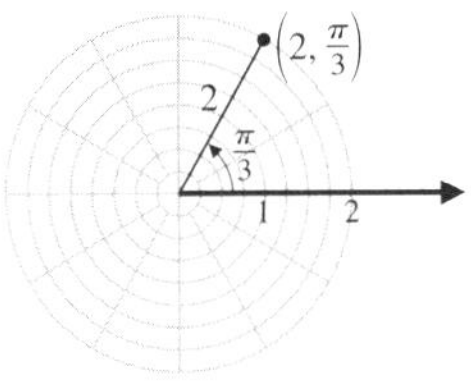

b. $x = r\cos\theta = 2\cos\frac{\pi}{3} = 1$;
$y = r\sin\theta = 2\sin\frac{\pi}{3} = \sqrt{3}$

c. $\left(2, \frac{7\pi}{3}\right), \left(-2, \frac{4\pi}{3}\right)$

2. a.

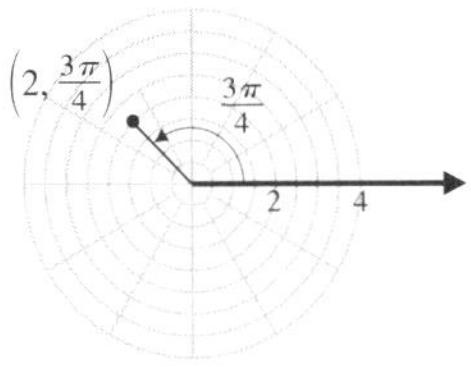

b. $x = r\cos\theta = 2\cos\frac{3\pi}{4} = -\sqrt{2}$;
$y = r\sin\theta = 2\sin\frac{3\pi}{4} = \sqrt{2}$

c. $\left(2, \frac{11\pi}{4}\right), \left(-2, \frac{7\pi}{4}\right)$

3. a.

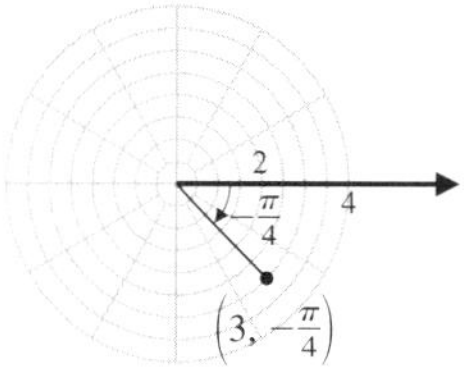

b. $x = r\cos\theta = 3\cos\left(-\frac{\pi}{4}\right) = \frac{3\sqrt{2}}{2}$;
$y = r\sin\theta = 3\sin\left(-\frac{\pi}{4}\right) = -\frac{3\sqrt{2}}{2}$

c. $\left(3, \frac{7\pi}{4}\right), \left(-3, \frac{3\pi}{4}\right)$

4. a.

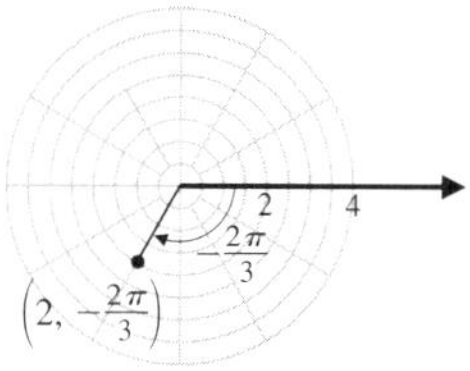

b. $x = r\cos\theta = 2\cos\left(-\frac{2\pi}{3}\right) = -1$;
$y = r\sin\theta = 2\sin\left(-\frac{2\pi}{3}\right) = -\sqrt{3}$

c. $\left(2, \frac{4\pi}{3}\right), \left(-2, \frac{\pi}{3}\right)$

5. a.

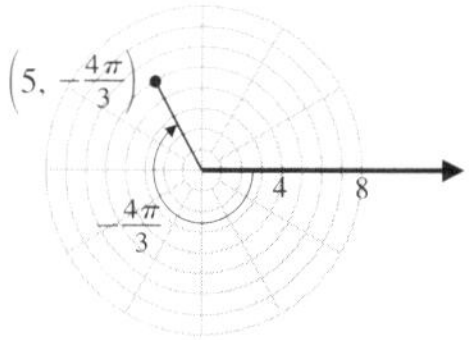

b. $x = r\cos\theta = 5\cos\left(-\frac{4\pi}{3}\right) = -\frac{5}{2}$;
$y = r\sin\theta = 5\sin\left(-\frac{4\pi}{3}\right) = \frac{5\sqrt{3}}{2}$

c. $\left(5, \frac{2\pi}{3}\right), \left(-5, \frac{5\pi}{3}\right)$

6. a.

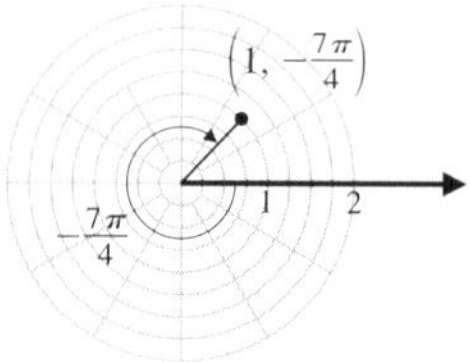

b. $x = r\cos\theta = \cos\left(-\frac{7\pi}{4}\right) = \frac{\sqrt{2}}{2}$;
$y = r\sin\theta = \sin\left(-\frac{7\pi}{4}\right) = \frac{\sqrt{2}}{2}$

c. $\left(1, \frac{\pi}{4}\right), \left(-1, \frac{5\pi}{4}\right)$

7. a.

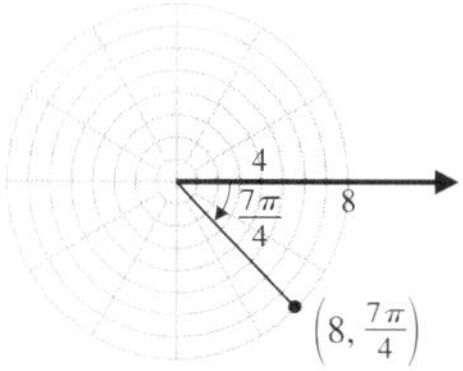

b. $x = r\cos\theta = 8\cos\left(\frac{7\pi}{4}\right) = 4\sqrt{2}$;
$y = r\sin\theta = 8\sin\left(\frac{7\pi}{4}\right) = -4\sqrt{2}$

c. $\left(8, -\frac{\pi}{4}\right), \left(-8, \frac{3\pi}{4}\right)$

8. a.

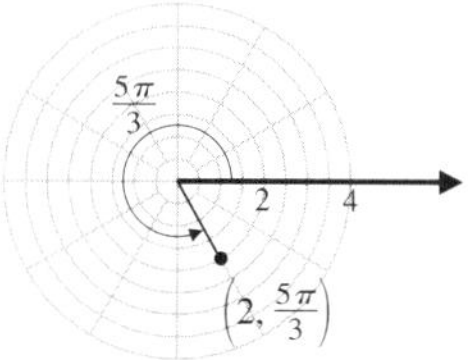

b. $x = r\cos\theta = 2\cos\left(\frac{5\pi}{3}\right) = 1$;
$y = r\sin\theta = 2\sin\left(\frac{5\pi}{3}\right) = -\sqrt{3}$

c. $\left(2, -\frac{\pi}{3}\right), \left(-2, \frac{2\pi}{3}\right)$

9. a.

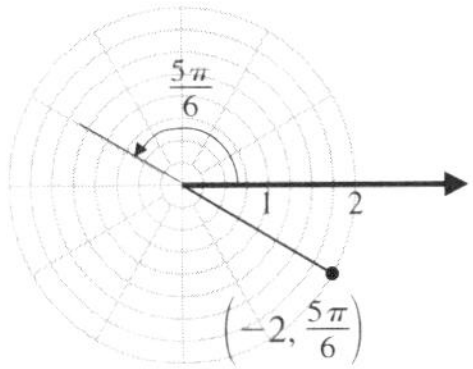

b. $x = r\cos\theta = -2\cos\left(\frac{5\pi}{6}\right) = \sqrt{3}$;

$y = r\sin\theta = -2\sin\left(\frac{5\pi}{6}\right) = -1$

c. $\left(2, \frac{11\pi}{6}\right), \left(-2, -\frac{7\pi}{6}\right)$

10. a.

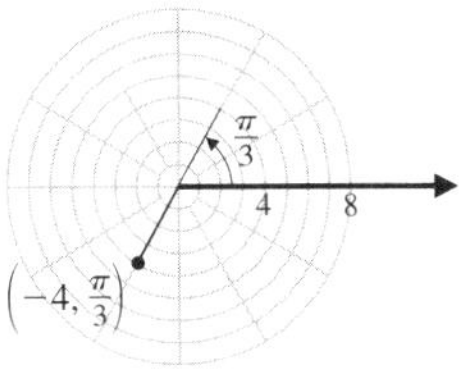

b. $x = r\cos\theta = -4\cos\left(\frac{\pi}{3}\right) = -2$;

$y = r\sin\theta = -4\sin\left(\frac{\pi}{3}\right) = -2\sqrt{3}$

c. $\left(4, \frac{4\pi}{3}\right), \left(-4, \frac{7\pi}{3}\right)$

11. a.

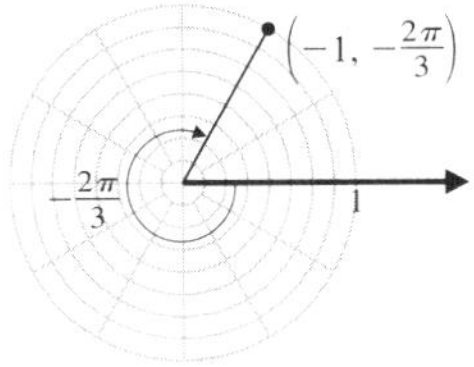

b. $x = r\cos\theta = -\cos\left(-\frac{2\pi}{3}\right) = \frac{1}{2}$;

$y = r\sin\theta = -\sin\left(-\frac{2\pi}{3}\right) = \frac{\sqrt{3}}{2}$

c. $\left(1, \frac{\pi}{3}\right), \left(-1, \frac{4\pi}{3}\right)$

12. a.

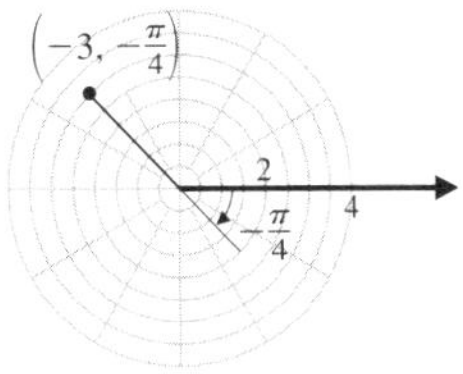

b. $x = r\cos\theta = -3\cos\left(-\frac{\pi}{4}\right) = -\frac{3\sqrt{2}}{2}$;

$y = r\sin\theta = -3\sin\left(-\frac{\pi}{4}\right) = \frac{3\sqrt{2}}{2}$

c. $\left(3, \frac{3\pi}{4}\right), \left(-3, \frac{7\pi}{4}\right)$

13. The point $(2, 0)$ lies on the positive x-axis which coincides with the ray $\theta = 0$, so the point is also $(2, 0)$ in polar coordinates.

14. The point $(-2, 0)$ lies on the negative x-axis which coincides with the ray $\theta = \pi$, so the point is $(2, \pi)$ in polar coordinates.

15. The point $(0, -4)$ lies on the negative y-axis which coincides with the ray $\theta = \frac{3\pi}{2}$, so the point is $\left(4, \frac{3\pi}{2}\right)$ in polar coordinates.

16. The point $(0, 4)$ lies on the positive y-axis which coincides with the ray $\theta = \frac{\pi}{2}$, so the point is $\left(4, \frac{\pi}{2}\right)$ in polar coordinates.

17. Since

$$x = 1, y = -\sqrt{3} \text{ so } r = \sqrt{x^2 + y^2} = \sqrt{4} = 2,$$

and since the point lies in quadrant IV,

$$\tan\theta = \frac{y}{x} = -\sqrt{3} \text{ so } \theta = \frac{5\pi}{3}.$$

So the point can be represented as $\left(2, \frac{5\pi}{3}\right)$ in polar coordinates.

18. Since

$$x = \sqrt{3},\ y = -1 \text{ so } r = \sqrt{x^2 + y^2} = \sqrt{4} = 2,$$

and since the point lies in quadrant IV,

$$\tan\theta = \frac{y}{x} = -\frac{\sqrt{3}}{3} \text{ so } \theta = \frac{11\pi}{6}.$$

So the point can be represented as $\left(2, \frac{11\pi}{6}\right)$ in polar coordinates.

19. Since

$$x = -4,\ y = 4 \text{ so } r = \sqrt{x^2 + y^2} = \sqrt{32} = 4\sqrt{2},$$

and since the point lies in quadrant II,

$$\tan\theta = \frac{y}{x} = -1 \text{ so } \theta = \frac{3\pi}{4}.$$

So the point can be represented as $\left(4\sqrt{2}, \frac{3\pi}{4}\right)$ in polar coordinates.

20. Since

$$x = -3,\ y = -3\sqrt{3} \text{ so } r = \sqrt{x^2 + y^2} = \sqrt{36} = 6,$$

and since the point lies in quadrant III,

$$\tan\theta = \frac{y}{x} = \sqrt{3} \text{ so } \theta = \frac{4\pi}{3}.$$

So the point can be represented as $\left(6, \frac{4\pi}{3}\right)$ in polar coordinates.

21. $r = 4$ implies $r^2 = 16$ so $x^2 + y^2 = 16$

22. $r = 2$ implies $r^2 = 4$ so $x^2 + y^2 = 4$

23. $\theta = \frac{3\pi}{4}$ implies $\frac{y}{x} = \tan\frac{3\pi}{4} = -1$ so $y = -x$

24. $\theta = \frac{\pi}{6}$ implies $\frac{y}{x} = \tan\frac{\pi}{6} = \frac{\sqrt{3}}{3}$ so $y = \frac{\sqrt{3}}{3}x$

25. Since $r^2 = x^2 + y^2$ and $x = r\cos\theta$, we have

$r = 2\cos\theta$ so $r^2 = 2r\cos\theta$, which implies $x^2+y^2 = 2x$ so $x^2-2x+y^2 = 0$ and $x^2-2x+1-1+y^2 = 0$,

and $(x-1)^2 + y^2 = 1$.

26. Since $r^2 = x^2 + y^2$ and $y = r\sin\theta$, we have

$r = 4\sin\theta$ so $r^2 = 4r\sin\theta$, which implies $x^2+y^2 = 4y$ so $x^2+y^2-4y = 0$ and $x^2+y^2-4y+4-4 = 0$,

and $x^2 + (y-2)^2 = 4$.

27. Since $x = r\cos\theta$ and $y = r\sin\theta$, we have

$$y = x \text{ so } r\cos\theta = r\sin\theta, \text{ which implies } \tan\theta = 1 \text{ so } \theta = \frac{\pi}{4}.$$

28. Since $x = r\cos\theta$ and $y = r\sin\theta$, we have

$$y = \sqrt{3}x \text{ so } \frac{y}{x} = \sqrt{3}, \text{ which implies } \tan\theta = \sqrt{3} \text{ so } \theta = \frac{\pi}{3}.$$

29. Since $r^2 = x^2 + y^2$, we have

$$x^2 + y^2 = 9, \text{ which implies } r^2 = 9 \text{ so } r = 3.$$

30. Since $r^2 = x^2 + y^2$, we have

$$x^2 + y^2 = 5, \text{ which implies } r^2 = 5 \text{ so } r = \sqrt{5}.$$

31. Since $r^2 = x^2 + y^2$ and $y = r\sin\theta$, we have

$$x^2 + y^2 = 2y, \text{ which implies } r^2 = 2r\sin\theta \text{ so } r = 2\sin\theta.$$

32. Since $r^2 = x^2 + y^2$ and $x = r\cos\theta$, we have

$$x^2 + y^2 = 3x, \text{ which implies } r^2 = 3r\cos\theta \text{ so } r = 3\cos\theta.$$

33. $r = 3$

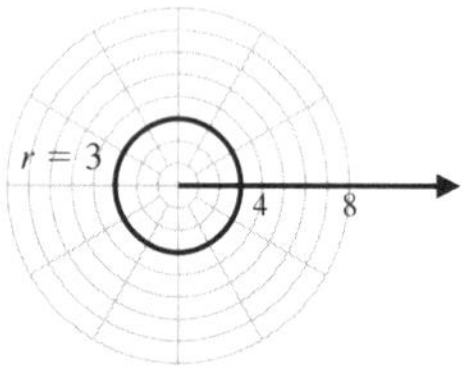

34. $r = 5\pi/3$

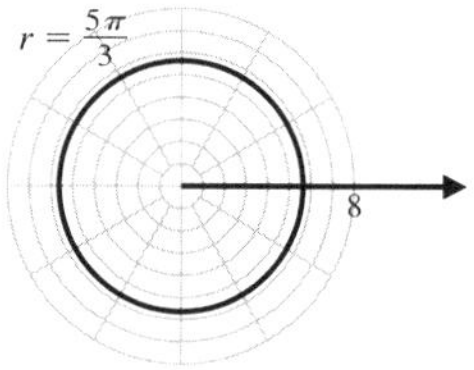

35. $\theta = 5\pi/3$

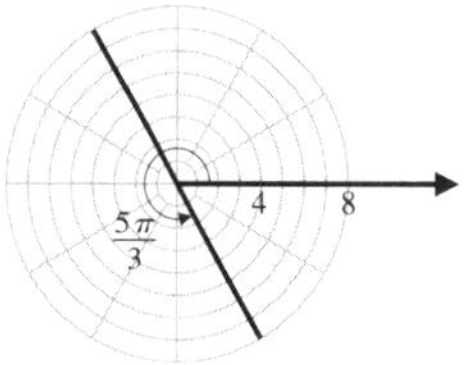

36. $\theta = 3$

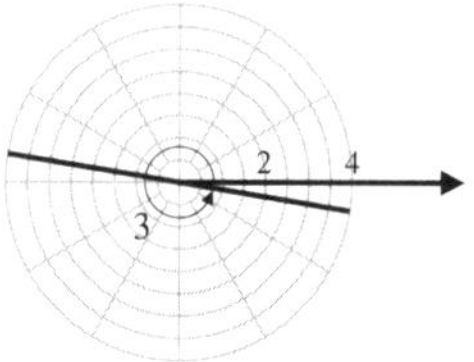

37. $r = 3\cos\theta$

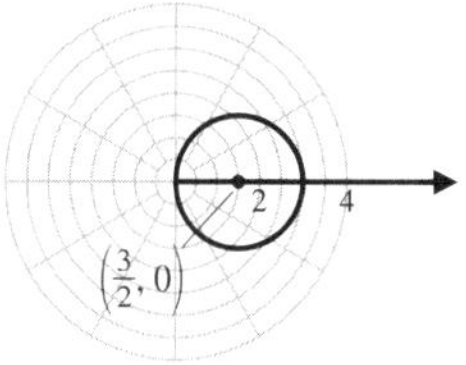

38. $r = 4\sin\theta$

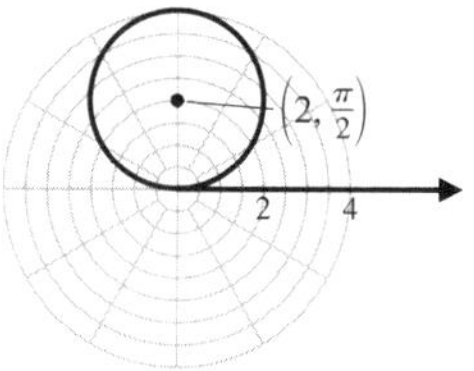

39. $r = -4\sin\theta$

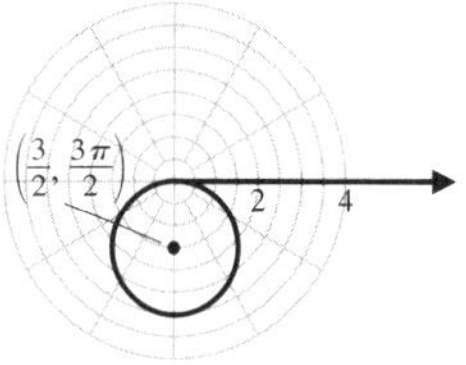

40. $r = -3\cos\theta$

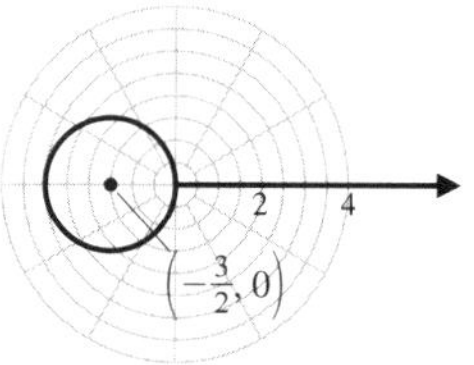

41. $r = 2 + 2\cos\theta$

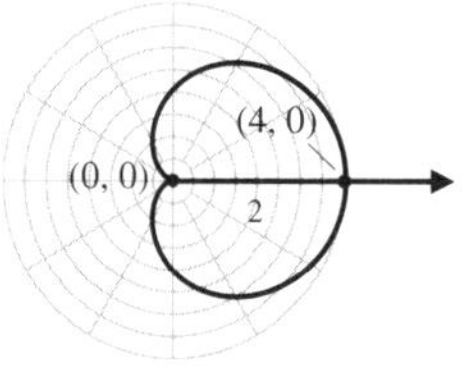

42. $r = 1 + \sin\theta$

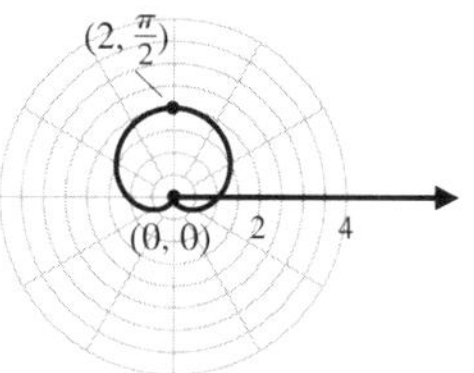

43. $r = 2 + \sin\theta$

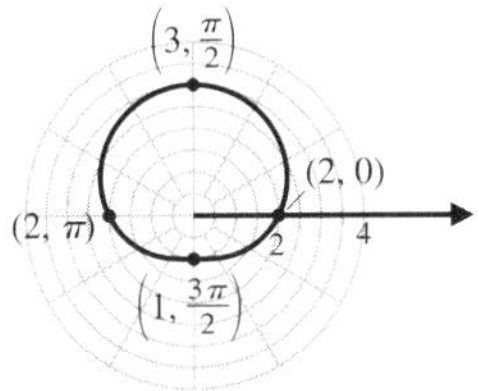

44. $r = 2 + \cos\theta$

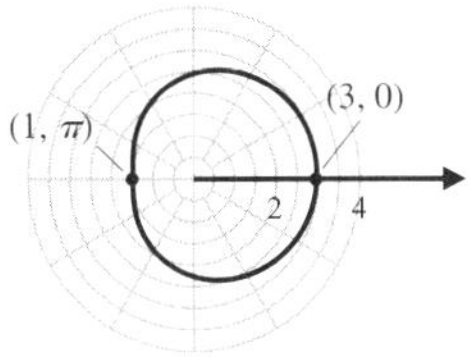

45. $r = 1 - 2\cos\theta$

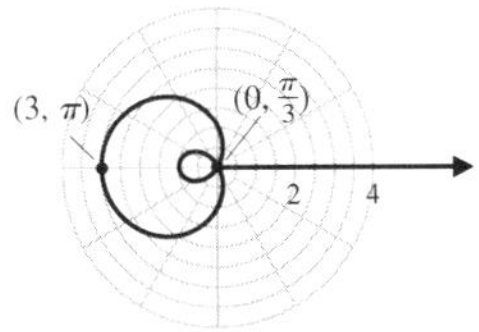

46. $r = 2 - 4\sin\theta$

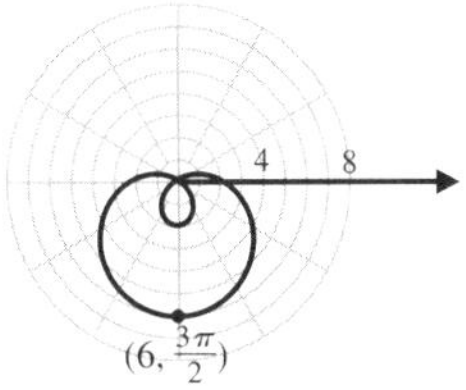

47. $r = 3\sin 3\theta$

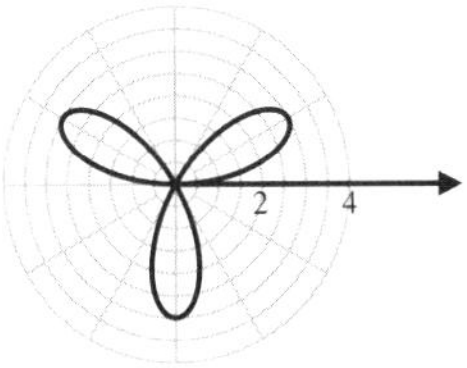

48. $r = 4\cos 3\theta$

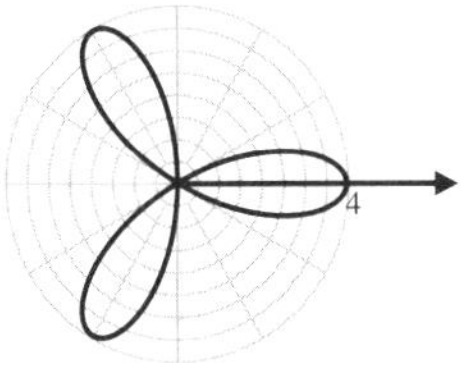

49. $r = 3\cos 2\theta$

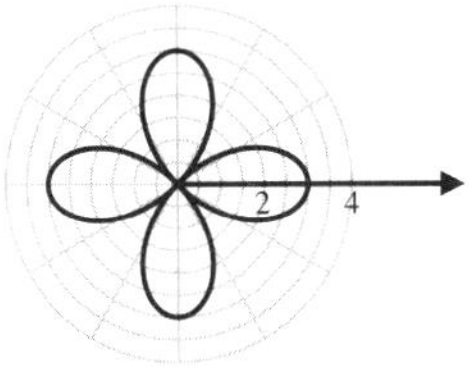

50. $r = 4\sin 2\theta$

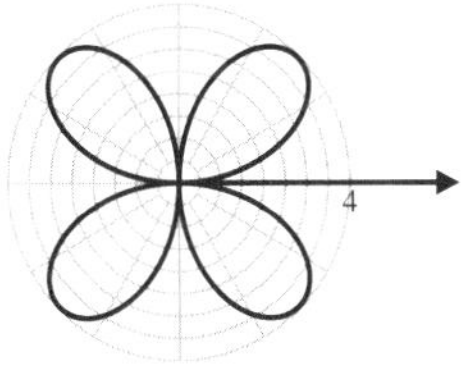

51. $r^2 = 16\cos 2\theta$

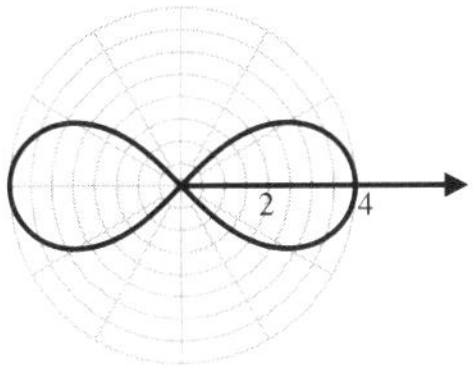

52. $r^2 = 4\sin 2\theta$

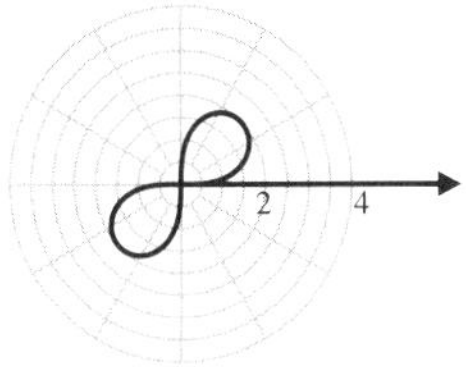

53. $r = \theta$

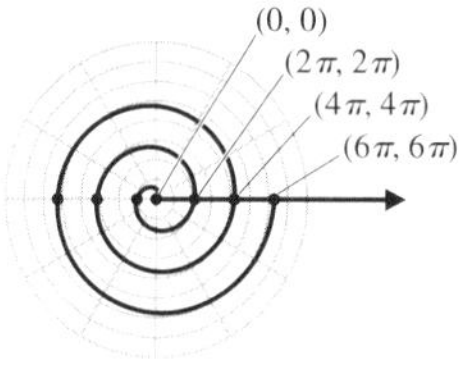

54. $r = 2^{-\theta}$

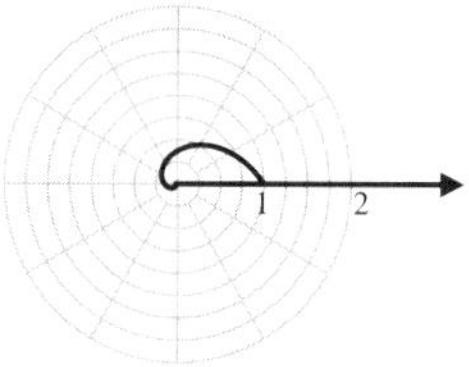

55. $r = e^{\theta}$

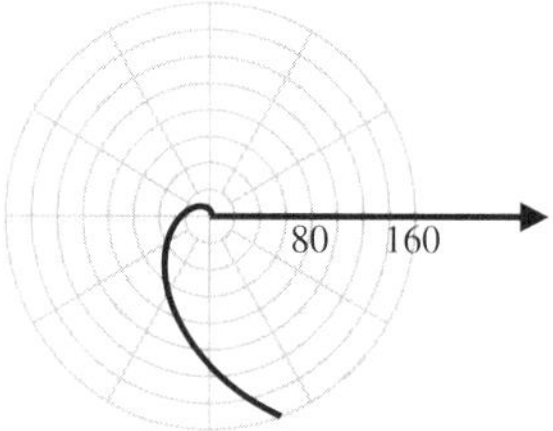

56. $r = \ln\theta$

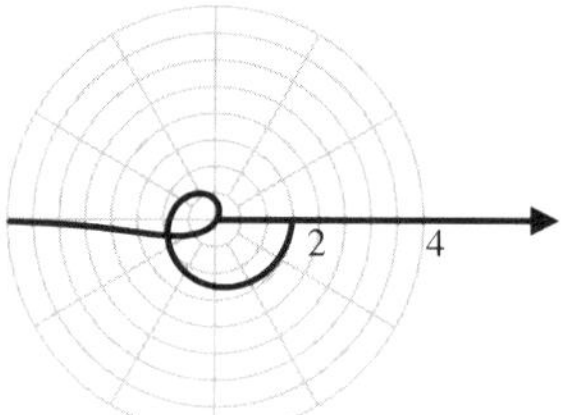

57. The graphs are all circles. The graph in (a) and (c) are symmetric about the y-axis and are reflections of each other about the x-axis. The graph in (b) and (d) are symmetric about the x-axis and are reflections of each other about the y-axis.

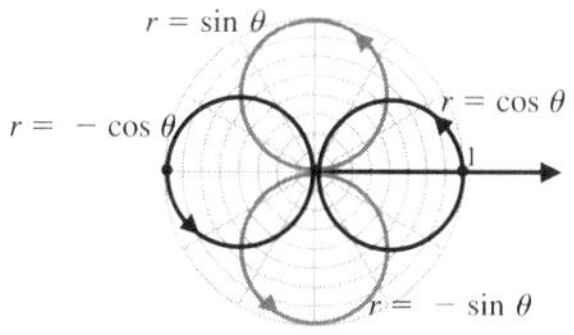

58. Graph (a) is a circle with center $\left(\frac{1}{2}, 0\right)$ and radius $\frac{1}{2}$, and graphs (b)-(d) are cardioids with axis along the x-axis which approach a circle.

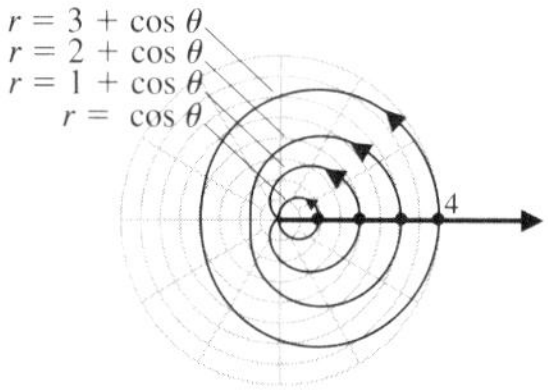

59. Graph (a) is a circle with center $\left(0, \frac{1}{2}\right)$ and radius $\frac{1}{2}$, and graphs (b)-(d) are cardioids with axis along the y-axis which approach a circle.

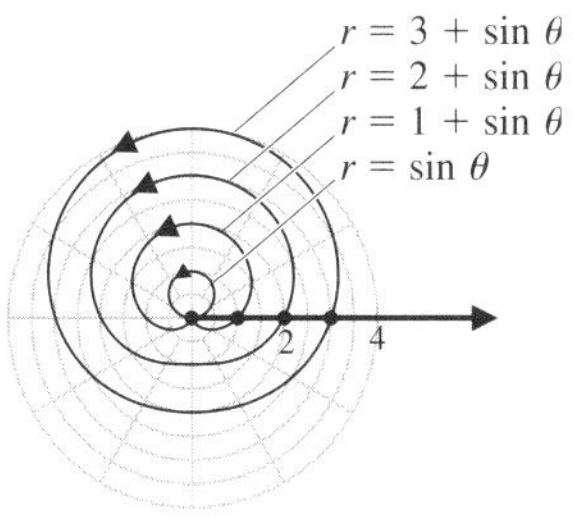

60. The graphs are all cardioids with axis along the x-axis. The graphs in (a) and (c) coincide, as do the graphs in (b) and (d). The graph in (b) is the reflection of graph (a) about the y-axis.

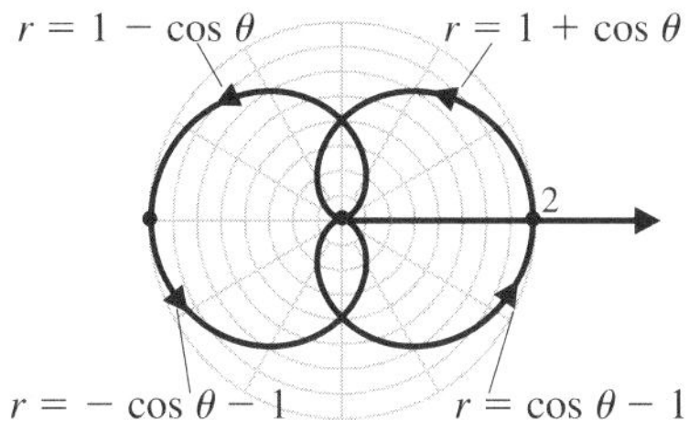

61. The graphs are all cardioids with axis along the y-axis. The graphs in (a) and (c) coincide, as do the graphs in (b) and (d). The graph in (b) is the reflection of graph (a) about the x-axis.

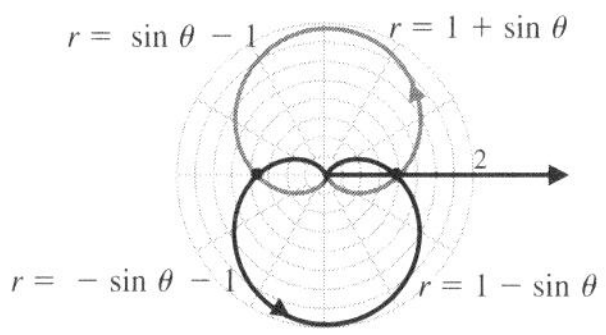

62. Graph (a) is a circle with center $\left(0, \frac{1}{2}\right)$ and radius $\frac{1}{2}$, and graphs (b)-(d) are leafed roses.

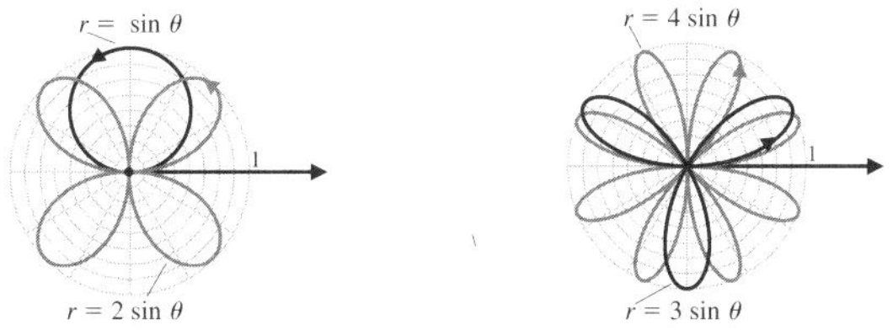

63. **a.** $(0, 0)$, $\left(90, \frac{\pi}{4}\right)$, $\left(90\sqrt{2}, \frac{\pi}{2}\right)$, $\left(90, \frac{3\pi}{4}\right)$

b. $(0, 0)$, $(90, 0)$, $\left(90\sqrt{2}, \frac{\pi}{4}\right)$, $\left(90, \frac{\pi}{2}\right)$

64. If the length of the core is a, the form of the equation of the apple is a cardioid with axis along the ray $\theta = \frac{\pi}{2}$, which coincides with the y-axis, and has equation $r = \frac{a}{2} + \frac{a}{2}\sin\theta$.

65. Setting the equations for r equal, we have $1 + 2\cos\theta = 1$ implies $2\cos\theta = 0$ so $\theta = \frac{\pi}{2}$ or $\theta = \frac{3\pi}{2}$. The figure indicates the curves also intersect at the point $(1, 0)$. The polar coordinates of the points of intersection are $\left(1, \frac{\pi}{2}\right)$, $\left(1, \frac{3\pi}{2}\right)$, $(1, 0)$. The rectangular coordinates of the points of intersection are $(0, 1)$, $(0, -1)$, $(1, 0)$.

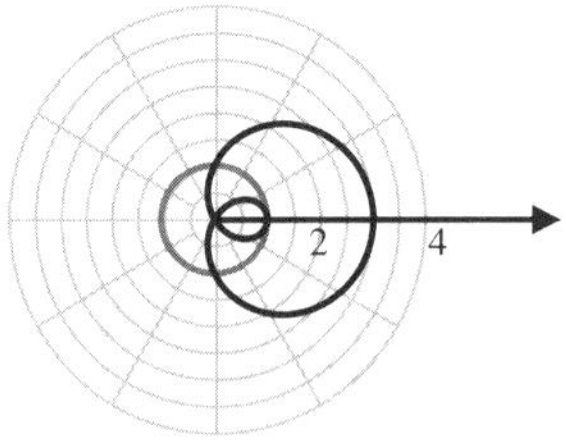

66. Setting the equations for r equal, we have $\sin 2\theta = 1$ implies $2\theta = \frac{\pi}{2}$ or $2\theta = \frac{5\pi}{2}$ so $\theta = \frac{\pi}{4}$ or $\theta = \frac{5\pi}{4}$. Note that $0 \le \theta \le 2\pi$ implies $0 \le 2\theta \le 4\pi$. The figure indicates the curves also intersect when $2\theta = \frac{3\pi}{2}$ or $2\theta = \frac{7\pi}{2}$ so $\theta = \frac{3\pi}{4}$ or $\theta = \frac{7\pi}{4}$. The polar coordinates of the points of intersection are $\left(1, \frac{\pi}{4}\right)$, $\left(1, \frac{3\pi}{4}\right)$, $\left(1, \frac{5\pi}{4}\right)$, $\left(1, \frac{7\pi}{4}\right)$. The rectangular coordinates of the points of intersection are $\left(\frac{\sqrt{2}}{2}, \frac{\sqrt{2}}{2}\right)$, $\left(-\frac{\sqrt{2}}{2}, \frac{\sqrt{2}}{2}\right)$, $\left(-\frac{\sqrt{2}}{2}, -\frac{\sqrt{2}}{2}\right)$, $\left(\frac{\sqrt{2}}{2}, -\frac{\sqrt{2}}{2}\right)$.

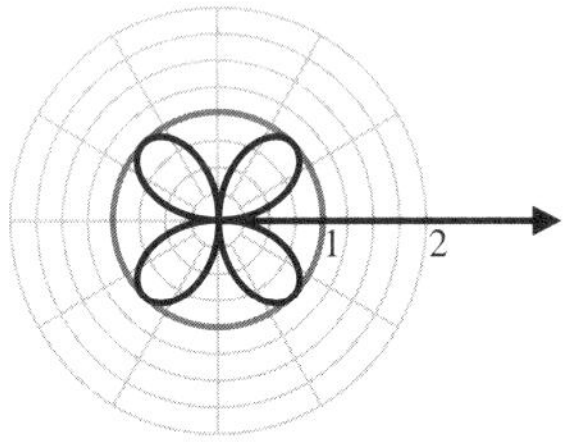

67. Setting the equations for r equal, we have
$2 - 2\sin\theta = 2\sin\theta$ implies $4\sin\theta = 2$ so $\sin\theta = \frac{1}{2}$ and $\theta = \frac{\pi}{6}$ or $\theta = \frac{5\pi}{6}$. The figure indicates the curves also intersect at the pole. The polar coordinates of the points of intersection are $\left(1, \frac{\pi}{6}\right), \left(1, \frac{5\pi}{6}\right)$, $(0, 0)$. The rectangular coordinates of the points of intersection are $\left(\frac{\sqrt{3}}{2}, \frac{1}{2}\right), \left(-\frac{\sqrt{3}}{2}, \frac{1}{2}\right)$, $(0, 0)$.

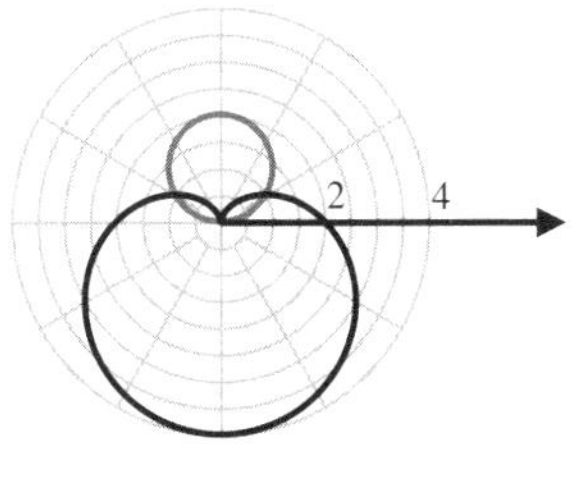

68. Setting the equations for r equal, we have
$2 + 2\cos\theta = -2\cos\theta$ implies $4\cos\theta = -2$ so $\cos\theta = -\frac{1}{2}$ and $\theta = \frac{2\pi}{3}$ or $\theta = \frac{4\pi}{3}$. The figure indicates the curves also intersect at the pole. The polar coordinates of the points of intersection are $\left(1, \frac{2\pi}{3}\right), \left(1, \frac{4\pi}{3}\right)$, $(0, 0)$. The rectangular coordinates of the points of intersection are $\left(-\frac{1}{2}, \frac{\sqrt{3}}{2}\right), \left(-\frac{1}{2}, -\frac{\sqrt{3}}{2}\right)$, $(0, 0)$.

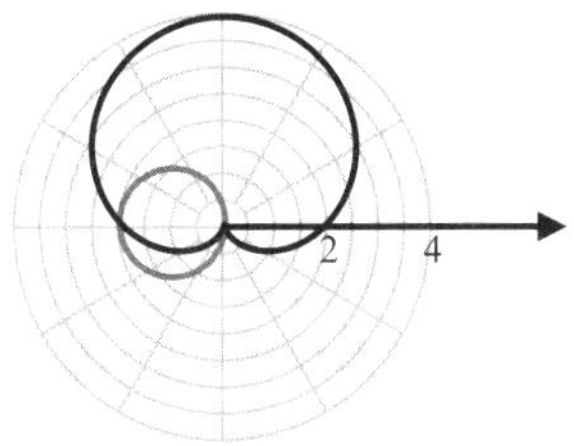

69. The graphs of $r = 1 + \sin(n\theta) + (\cos(2n\theta))^2$ for $n = 1, 2, 3, 4$ are shown below.

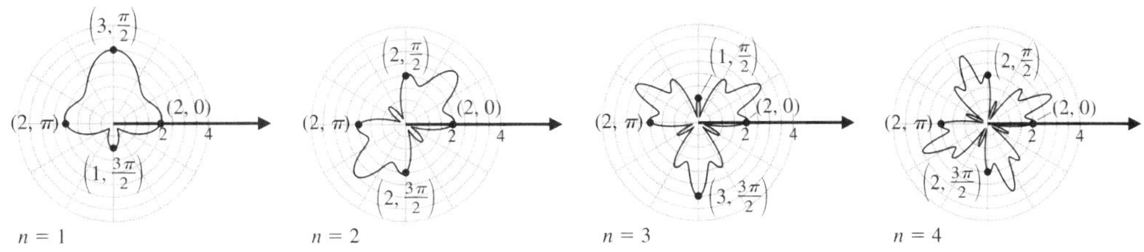

70. The graphs of $r = (\sin m\theta)(\cos n\theta)$ are shown below.

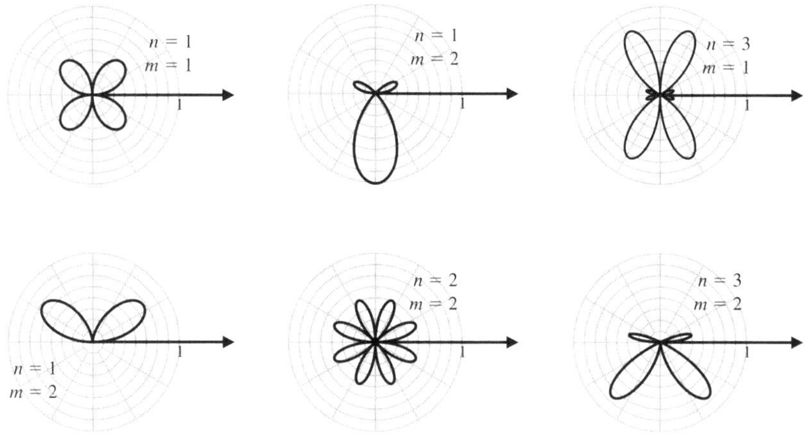

71. The graphs of $r = \sin m\theta$ and $r = |\sin m\theta|$ are shown below.

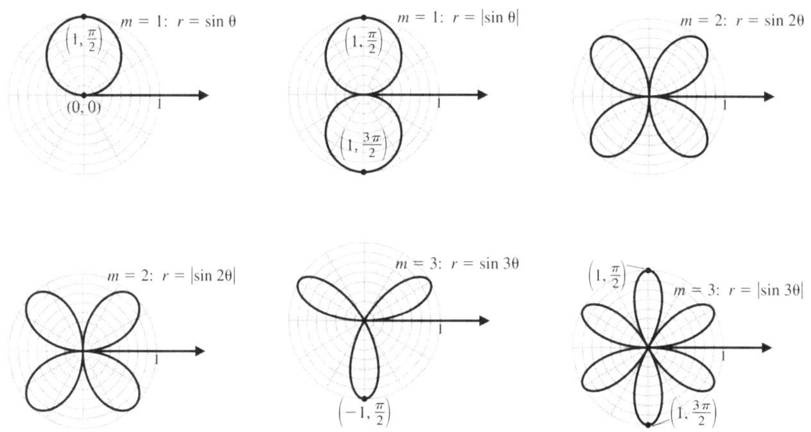

Exercise Set 6.6 (Page 348)

1. a. The equation $r = \frac{2}{1+\cos\theta}$ is in the standard form $r = \frac{ed}{1+e\cos\theta}$, where $e = 1$ implies $d = 2$, so the equation describes a parabola opening to the left with vertex $(1, 0)$, focus $(0, 0)$, and directrix along the line $x = 2$.

b. Since $r^2 = x^2 + y^2$ and $x = r\cos\theta$, we have

$$r = \frac{2}{1+\cos\theta} \text{ implies } r + r\cos\theta = 2 \text{ so } r = 2 - r\cos\theta = 2 - x \text{ and } r^2 = 4 - 4x + x^2.$$

So

$$x^2 + y^2 = 4 - 4x + x^2 \text{ implies } x - 1 = -\frac{1}{4}y^2.$$

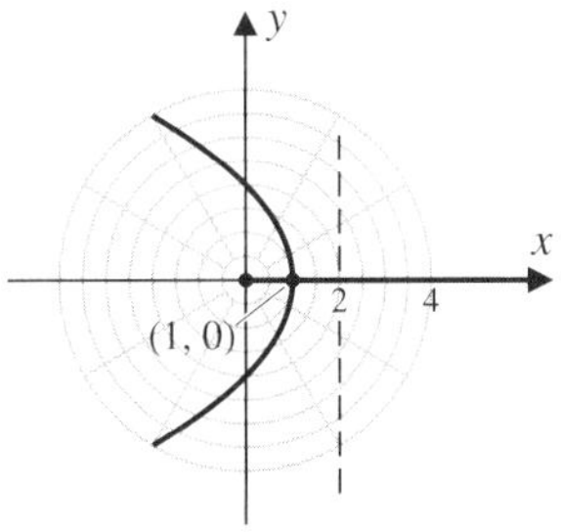

2. a. The equation

$$r = \frac{3}{4 + 4\sin\theta} = \frac{3}{4(1+\sin\theta)} = \frac{3/4}{1+\sin\theta}$$

is in the standard form $r = \frac{ed}{1+e\sin\theta}$, where $e = 1$ implies $d = \frac{3}{4}$, so the equation describes a parabola opening downward with vertex $\left(0, \frac{3}{8}\right)$, focus $(0, 0)$, and directrix along the line $y = \frac{3}{4}$.

b. Since $r^2 = x^2 + y^2$ and $y = r\sin\theta$, we have

$$r = \frac{3}{4 + 4\sin\theta} \text{ implies } 4r + 4r\sin\theta = 3 \text{ so } 4r = 3 - 4r\sin\theta = 3 - 4y \text{ and } 16r^2 = 9 - 24y + 16y^2.$$

So

$$16x^2 + 16y^2 = 9 - 24y + 16y^2 \text{ implies } 16x^2 = 24\left(\frac{9}{24} - y\right) \text{ so } y - \frac{3}{8} = -\frac{2}{3}x^2.$$

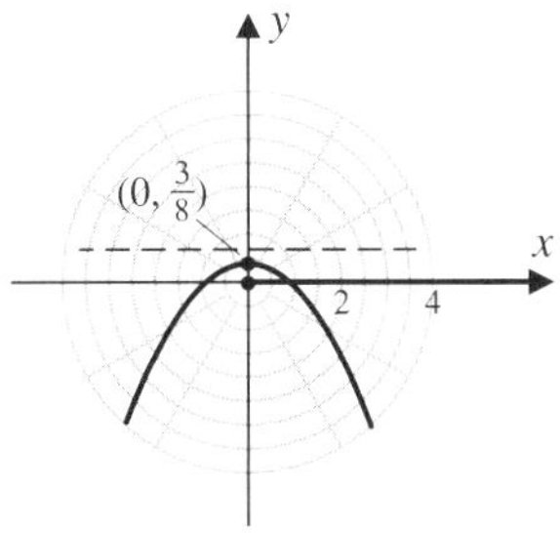

3. a. The equation

$$r = \frac{2}{2-\sin\theta} = \frac{2}{2\left(1-\frac{1}{2}\sin\theta\right)} = \frac{1}{1-\frac{1}{2}\sin\theta}$$

is in the standard form $r = \frac{ed}{1-e\sin\theta}$, where $e = \frac{1}{2}$ implies $d = 2$. Since $e < 1$, the equation describes an ellipse with major axis along the y-axis and with horizontal directrix $y = -2$.

b. Since $r^2 = x^2 + y^2$ and $y = r\sin\theta$, we have

$$r = \frac{2}{2-\sin\theta} \text{ implies } 2r - r\sin\theta = 2 \text{ so } 2r = 2 + r\sin\theta = 2 + y \text{ and } 4r^2 = 4 + 4y + y^2.$$

So

$$4x^2 + 4y^2 = 4 + 4y + y^2 \text{ implies } 4x^2 + 3y^2 - 4y = 4.$$

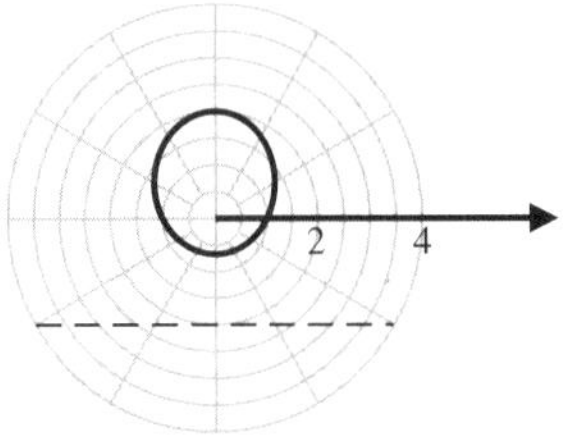

4. a. The equation

$$r = \frac{2}{4+\cos\theta} = \frac{2}{4(1+1/4\cos\theta)} = \frac{1/2}{(1+1/4\cos\theta)}$$

is in the standard form $r = \frac{ed}{1+e\cos\theta}$, where $e = \frac{1}{4}$ and $ed = \frac{1}{2}$ implies $d = 2$. Since $e < 1$, the equation describes an ellipse with major axis along the x-axis and with vertical directrix $x = 2$.

b. Since $r^2 = x^2 + y^2$ and $x = r\cos\theta$, we have

$$r = \frac{2}{4+\cos\theta} \text{ implies } 4r + r\cos\theta = 2 \text{ so } 4r = 2 - x \text{ and } 16r^2 = 4 - 4x + x^2.$$

So

$$16x^2 + 16y^2 = 4 - 4x + x^2 \text{ implies } 15x^2 + 4x + 16y^2 = 4.$$

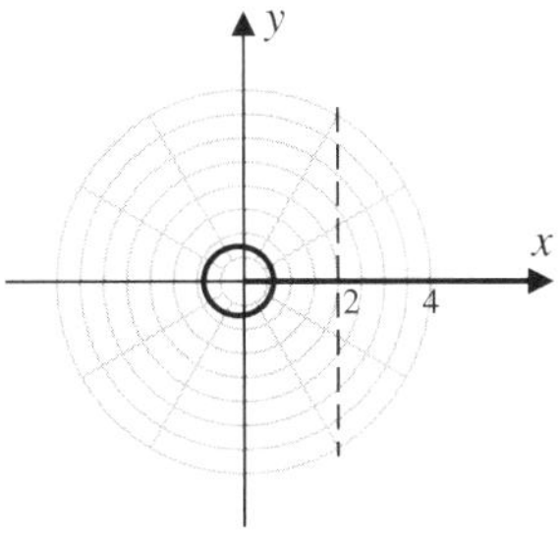

5. a. The equation $r = \frac{1}{1+2\cos\theta}$ is in the standard form $r = \frac{ed}{1+e\cos\theta}$, where $e = 2$ and $ed = 1$ implies $d = \frac{1}{2}$. Since $e > 1$, the equation describes a hyperbola with axis along the x-axis and vertical directrix $x = \frac{1}{2}$.

b. Since $r^2 = x^2 + y^2$ and $x = r\cos\theta$, we have

$$r = \frac{1}{1+2\cos\theta} \text{ implies } r + 2r\cos\theta = 1 \text{ so } r = 1 - 2x \text{ and } r^2 = 1 - 4x + 4x^2.$$

So

$$x^2 + y^2 = 1 - 4x + 4x^2 \text{ implies } 3x^2 - 4x - y^2 = -1.$$

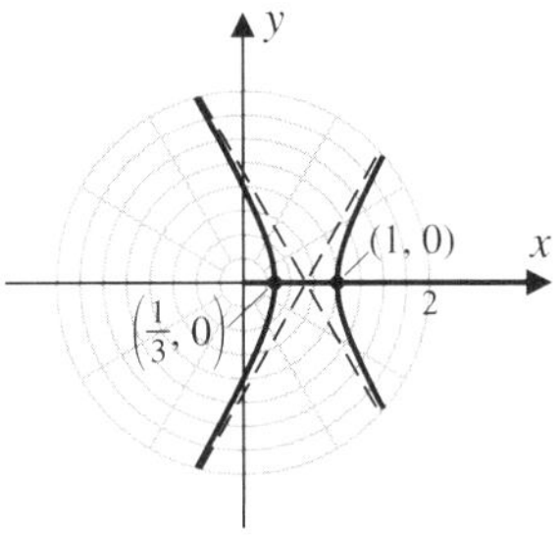

6. a. The equation $r = \frac{2}{1-2\sin\theta}$ is in the standard form $r = \frac{ed}{1-e\sin\theta}$, where $e = 2$ and $ed = 2$ implies $d = 1$. Since $e > 1$, the equation describes a hyperbola with axis along the y-axis and horizontal directrix $y = -1$.

b. Since $r^2 = x^2 + y^2$ and $y = r\sin\theta$, we have

$$r = \frac{2}{1 - 2\sin\theta} \text{ implies } r - 2r\sin\theta = 2 \text{ so } r = 2 + 2y \text{ and } r^2 = 4 + 8y + 4y^2.$$

So

$$x^2 + y^2 = 4 + 8y + 4y^2 \text{ implies } 3y^2 + 8y - x^2 = -4.$$

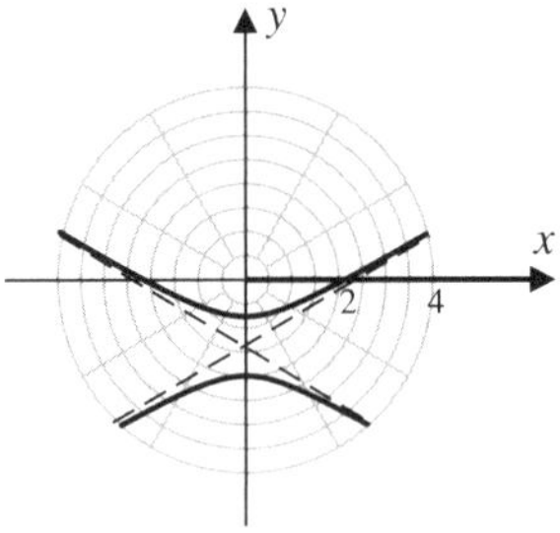

7. a. The equation $r = \frac{3}{1-2\sin\theta}$ is in the standard form $r = \frac{ed}{1-e\sin\theta}$, where $e = 2$ and $ed = 3$ implies $d = \frac{3}{2}$. Since $e > 1$, the equation describes a hyperbola with axis along the y-axis and horizontal directrix $y = -\frac{3}{2}$.

b. Since $r^2 = x^2 + y^2$ and $y = r\sin\theta$, we have

$$r = \frac{3}{1 - 2\sin\theta} \text{ implies } r - 2r\sin\theta = 3 \text{ so } r = 3 + 2y \text{ and } r^2 = 9 + 12y + 4y^2.$$

So

$$x^2 + y^2 = 9 + 12y + 4y^2 \text{ implies } 3y^2 + 12y - x^2 = -9.$$

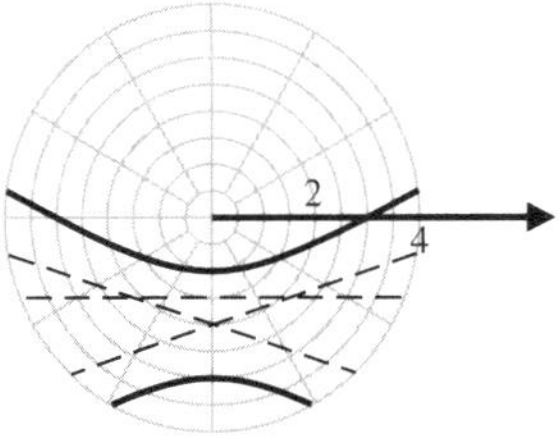

8. a. The equation

$$r = \frac{4}{2 + 3\cos\theta} = \frac{4}{2\left(1 + \frac{3}{2}\cos\theta\right)} = \frac{2}{1 + \frac{3}{2}\cos\theta}$$

is in the standard form $r = \frac{ed}{1+e\cos\theta}$, where $e = \frac{3}{2}$ and $ed = 2$ implies $d = \frac{4}{3}$. Since $e > 1$, the equation describes a hyperbola with axis along the x-axis and vertical directrix $x = \frac{4}{3}$.

b. Since $r^2 = x^2 + y^2$ and $x = r\cos\theta$, we have

$$r = \frac{4}{2 + 3\cos\theta} \text{ implies } 2r + 3r\cos\theta = 4 \text{ so } 2r = 4 - 3x \text{ and } 4r^2 = 16 - 24x + 9x^2.$$

So

$$4x^2 + 4y^2 = 16 - 24x + 9x^2 \text{ implies } 5x^2 - 24x - 4y^2 = -16.$$

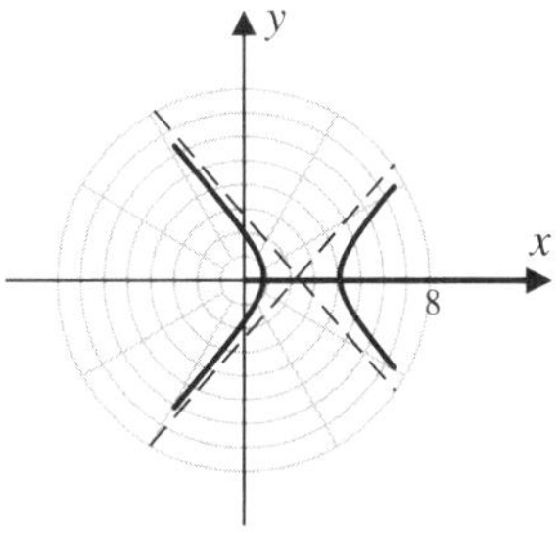

9. Since the directrix $x = -4$ is vertical and to the left of the pole, the conic has the form $r = \frac{ed}{1 - e\cos\theta}$. Since $e = 2$ and $d = 4$, we have

$$r = \frac{8}{1 - 2\cos\theta}.$$

10. Since the directrix $y = -2$ is horizontal and below the pole, the conic has the form $r = \frac{ed}{1 - e\sin\theta}$. Since $e = \frac{1}{2}$ and $d = 2$, we have

$$r = \frac{1}{1 - 1/2\sin\theta}.$$

11. Since the directrix $y = -\frac{1}{4}$ is horizontal and below the pole, the conic has the form $r = \frac{ed}{1 - e\sin\theta}$. Since $e = 1$ and $d = \frac{1}{4}$, we have

$$r = \frac{1/4}{1 - \sin\theta} = \frac{1}{4 - 4\sin\theta}.$$

12. Since the directrix $x = 1$ is vertical and to the right of the pole, the conic has the form $r = \frac{ed}{1 + e\cos\theta}$. Since $e = \frac{1}{3}$ and $d = 1$, we have

$$r = \frac{1/3}{1 + 1/3\cos\theta} = \frac{1}{3 + \cos\theta}.$$

13. Since the directrix $x = 2$ is vertical and to the right of the pole, the conic has the form $r = \frac{ed}{1+e\cos\theta}$. Since $e = 3$ and $d = 2$, we have

$$r = \frac{6}{1 + 3\cos\theta}.$$

14. Since the directrix $y = 2$ is horizontal and above the pole, the conic has the form $r = \frac{ed}{1+e\sin\theta}$. Since $e = \frac{2}{3}$ and $d = 2$, we have

$$r = \frac{4/3}{1 + \frac{2}{3}\sin\theta} = \frac{4}{3 + 2\sin\theta}.$$

15. Since the directrix $y = 1$ is horizontal and above the pole, the conic has the form $r = \frac{ed}{1+e\sin\theta}$. Since $e = \frac{1}{4}$ and $d = 1$, we have

$$r = \frac{1/4}{1 + \frac{1}{4}\sin\theta} = \frac{1}{4 + \sin\theta}.$$

16. Since the directrix $x = -3$ is vertical and to the left of the pole, the conic has the form $r = \frac{ed}{1-e\cos\theta}$. Since $e = 1$ and $d = 3$, we have

$$r = \frac{3}{1 - \cos\theta}.$$

17. The conic has equation in the form $r = \frac{ed}{1+e\cos\theta}$, and since $(1, 0)$ and $(3, \pi)$ are on the curve,

$$1 = \frac{ed}{1 + e\cos 0} = \frac{ed}{1 + e} \text{ implies } 1 + e = ed,$$

and

$$3 = \frac{ed}{1 + e\cos\pi} = \frac{ed}{1 - e} \text{ so } 3 - 3e = ed \text{ implies } 1 + e = 3 - 3e \text{ so } e = \frac{1}{2},$$

and

$$\frac{1}{2}d = 1 + \frac{1}{2} \text{ so } d = 3.$$

So

$$r = \frac{ed}{1 + e\cos\theta} = \frac{3/2}{1 + 1/2\cos\theta} = \frac{3}{2 + \cos\theta}.$$

18. Since the focus is at the origin and the vertex $(-6, 0)$ is midway between the focus and the directrix, the directrix is to the left of the pole and is the vertical line $x = -12$. So $d = 12$, and

$$r = \frac{ed}{1 - e\cos\theta} = \frac{12e}{1 - e\cos\theta}.$$

Since the curve is a parabola, $e = 1$, and the equation of the conic is

$$r = \frac{12}{1 - \cos\theta}.$$

19. Since the orbit is elliptical if we assume the axis is horizontal, the form of the equation is

$$r = \frac{ed}{1 - e\cos\theta},$$

with $e < 1$. The satellite will be at its maximum height when $\theta = 0$, so

$$\frac{ed}{1 - e} = 3960 + 560 = 4520,$$

and at its minimum height when $\theta = \pi$, so

$$\frac{ed}{1 + e} = 3960 + 145 = 4105.$$

Then

$$4520(1 - e) = ed = 4105(1 + e) \text{ implies } 8625e = 415,$$

and

$$e = \frac{415}{8625} \approx 0.048 \text{ so } ed = 4105\left(1 + \frac{415}{8625}\right) \approx 4303.$$

The equation of the orbit is

$$r = \frac{4303}{1 - 0.048\cos\theta}.$$

20. The equation of the earths orbit is of the form $r = \frac{ed}{1+e\cos\theta}$, with $e < 1$, where the earth is furthest from the sun when $\theta = \pi$ and closest to the sun when $\theta = 0$. The length of the major axis is then

$$\frac{ed}{1 + e\cos\pi} - \frac{ed}{1 + e\cos 0} = \frac{ed}{1 - e} - \frac{ed}{1 + e} = \frac{2e^2 d}{1 - e^2}.$$

If the length of the major axis is about 2.99×10^8 kilometers and $e = 0.0167$, then

$$2.99 \times 10^8 = \frac{2(0.0167)^2 d}{1-(0.0167)^2} \text{ so } d = \frac{(2.99 \times 10^8)\left(1-(0.0167)^2\right)}{2(0.0167)^2} \approx 5.4 \times 10^{11}.$$

The equation of the earth's orbit about the sun is

$$r = \frac{0.0167(5.4 \times 10^{11})}{1+0.0167\cos\theta} = \frac{9 \times 10^9}{1+0.0167\cos\theta}.$$

Exercise Set 6.7 (Page 353)

1. a. $x = 3t$, $y = \frac{t}{2}$ implies $t = \frac{x}{3}$ and $y = \frac{1}{2}\frac{x}{3} = \frac{x}{6}$.

b.

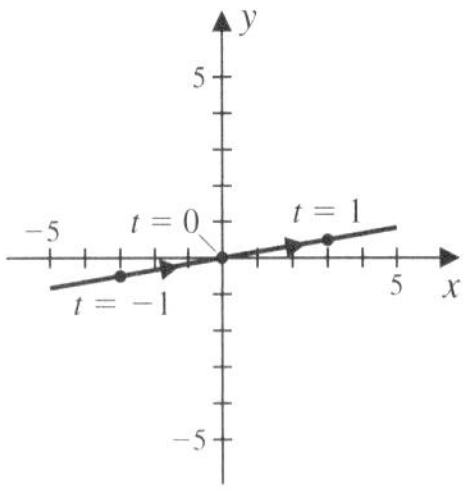

2. a. $x = 2t+1$, $y = 3t-2$ implies $t = \frac{x-1}{2}$ and $y = 3\left(\frac{x-1}{2}\right) - 2 = \frac{3}{2}x - \frac{7}{2}$.

b.

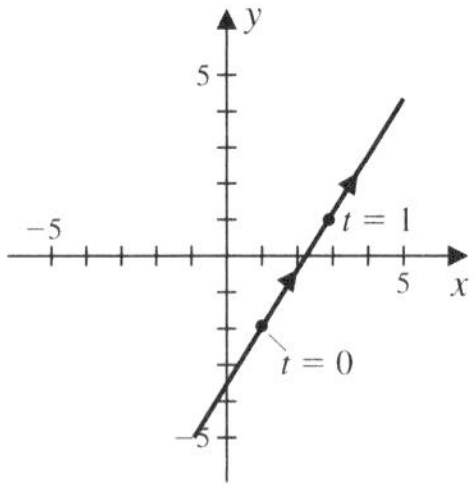

3. a. $x = \sqrt{t}$, $y = t+1$ implies $t = x^2$ and $y = x^2+1$, only for $x \geq 0$.

b.

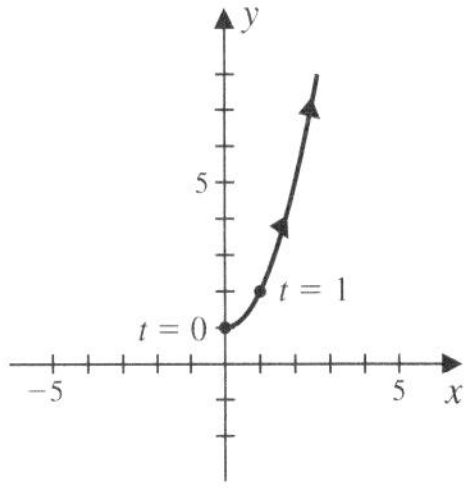

4. a. $x = t+2$, $y = -3\sqrt{t}$ implies $t = x-2$ and $y = -3\sqrt{x-2}$, for $x \geq 2$.

b.

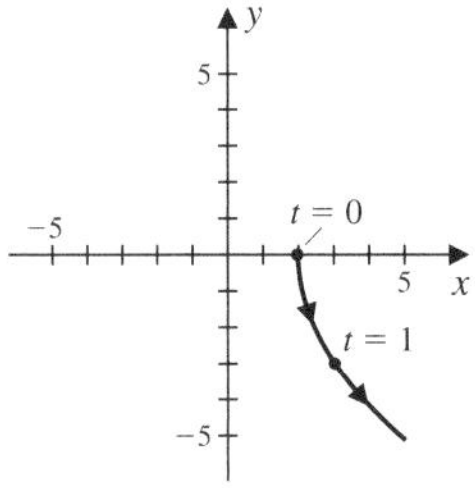

5. a. $x = \sin t$, $y = (\cos t)^2$ implies $y = 1 - (\sin t)^2 = 1 - x^2$ for $y \geq 0$.

b.

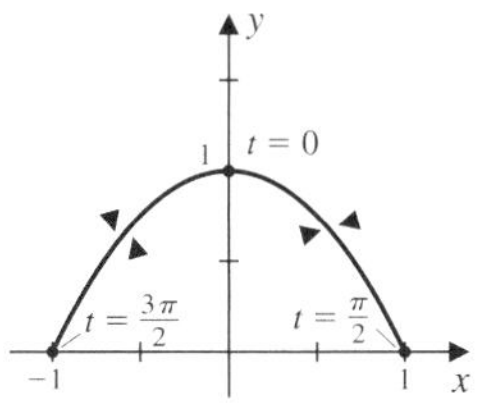

6. a. $x = (\cos t)^2$, $y = \sin t$ implies $x = 1 - (\sin x)^2 = 1 - y^2$ for $x \geq 0$.

b.

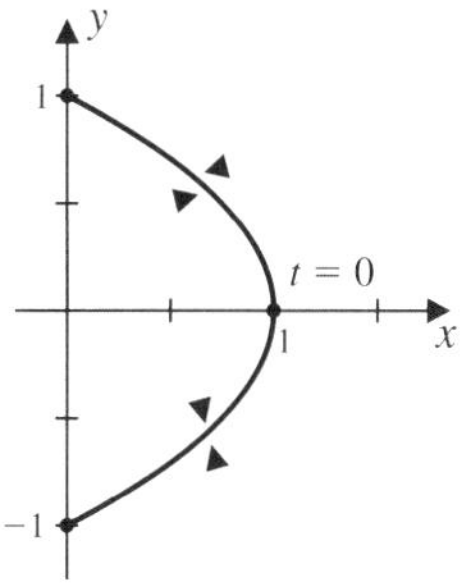

7. a. $x = \sec t$, $y = \tan t$ and $(\tan t)^2 + 1 = (\sec t)^2$. So $y^2 + 1 = x^2$ implies $x^2 - y^2 = 1$. Since $x = \sec t$, $|x| \geq 1$.

b.

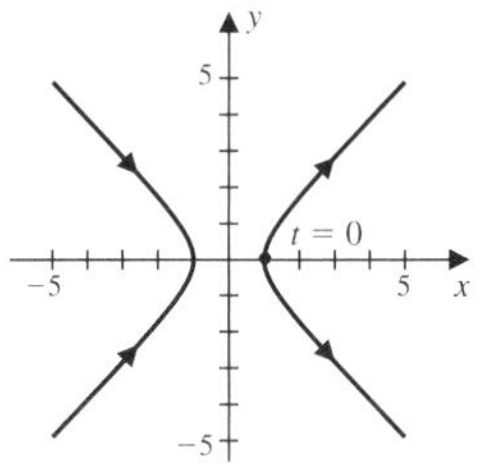

8. a. $x = 3\sin t$, $y = 4\cos t$ implies $\frac{x^2}{9} + \frac{y^2}{16} = \frac{9(\sin t)^2}{9} + \frac{16(\cos t)^2}{16}$ or $\frac{x^2}{9} + \frac{y^2}{16} = (\sin t)^2 + (\cos t)^2 = 1$ so $\frac{x^2}{9} + \frac{y^2}{16} = 1$.

b.

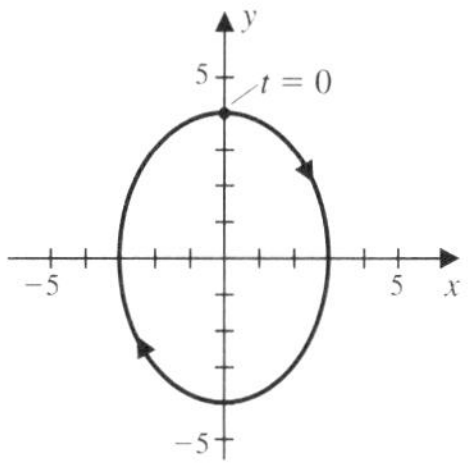

9. **a.** $x = 3\cos t$, $y = 2\sin t$ implies
$\frac{x^2}{9} + \frac{y^2}{4} = \frac{9(\cos t)^2}{9} + \frac{4(\sin t)^2}{4}$
or $\frac{x^2}{9} + \frac{y^2}{4} = (\cos t)^2 + (\sin t)^2 = 1$
so $\frac{x^2}{9} + \frac{y^2}{4} = 1$.
b.

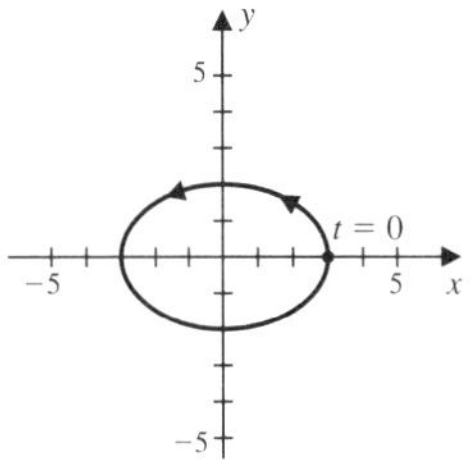

10. **a.** $x = \ln t$, $y = \ln \sqrt{t}$
implies $y = \ln t^{\frac{1}{2}} = \frac{1}{2}\ln t = \frac{1}{2}x$.
b.

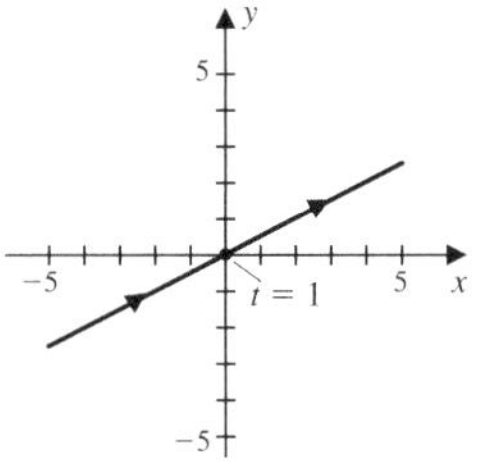

11. **a.** $x = e^t$, $y = e^{-t}$ implies $y = x^{-1} = \frac{1}{x}$, for $x > 0$.
b.

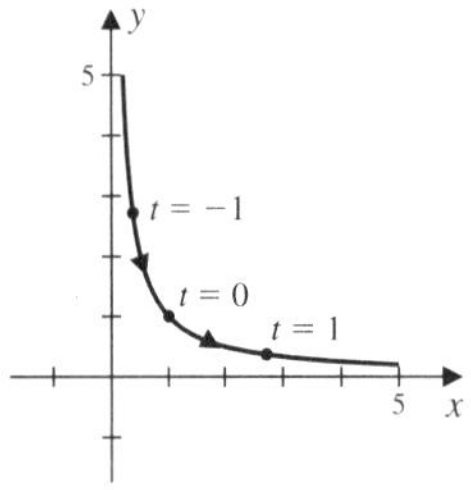

12. **a.** $x = e^t - 2$, $y = e^{2t} + 3$ implies $e^t = x + 2$
so $y = \left(e^t\right)^2 + 3 = (x+2)^2 + 3$, for $x \geq -2$.
b.

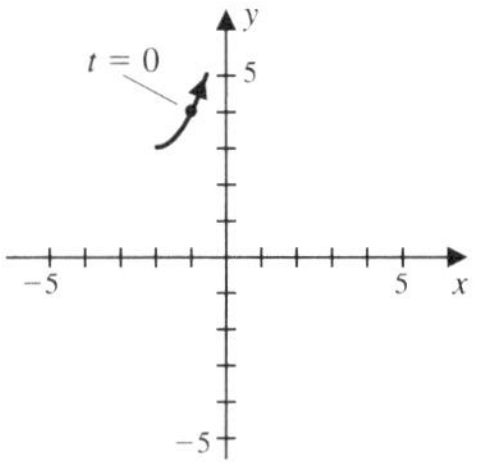

13. The graphs of the parametric equations are shown below.

a. $x = t$, $y = t^2$ implies $y = x^2$

b. $x = t^2$, $y = t$ implies $x = y^2$

c. $x = t^2$, $y = t^4$ implies $y = x^2$, $x \geq 0$

d. $x = t^4$, $y = t^2$ implies $x = y^2$, $y \geq 0$

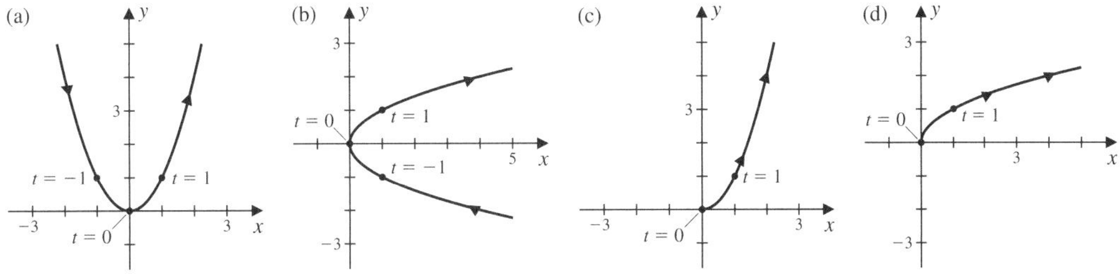

14. The graphs of the parametric equations are shown below.

a. $x = 2t + 1$, $y = 3t$ implies $t = \frac{x-1}{2}$, $y = 3\left(\frac{x-1}{2}\right) = \frac{3}{2}x - \frac{3}{2}$

b. $x = 3t + 1$, $y = 2t$ implies $t = \frac{x-1}{3}$, $y = 2\left(\frac{x-1}{3}\right) = \frac{2}{3}x - \frac{2}{3}$

c. $x = 2t$, $y = 3t + 1$ implies $t = \frac{x}{2}$, $y = \frac{3}{2}x + 1$

d. $x = 3t$, $y = 2t + 1$ implies $t = \frac{x}{3}$, $y = \frac{2}{3}x + 1$

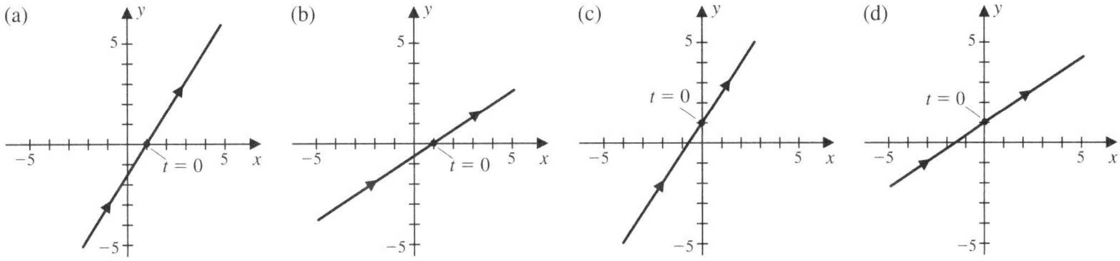

15. The graphs of the parametric equations are shown below.

a. $x = \sin t$, $y = \cos t$ implies $x^2 + y^2 = (\sin t)^2 + (\cos t)^2 = 1$

b. $x = \sin t$, $y = \cos t + 1$ implies $x^2 + (y - 1)^2 = (\sin t)^2 + (\cos t)^2 = 1$

c. $x = \cos t + 1$, $y = \sin t$ implies $(x - 1)^2 + y^2 = (\cos t)^2 + (\sin t)^2 = 1$

d. $x = \cos t$, $y = \sin t + 1$ implies $x^2 + (y-1)^2 = (\cos t)^2 + (\sin t)^2 = 1$

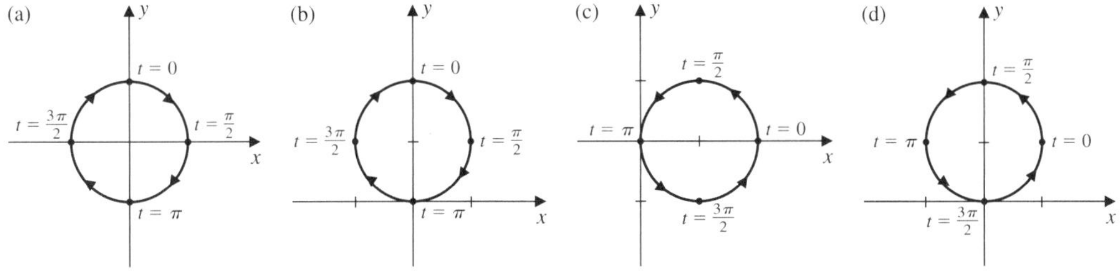

16. The graphs of the parametric equations are shown below.

a. $x = 1 - \sin t$, $y = 1 - \cos t$ implies $(x-1)^2 + (y-1)^2 = 1$

b. $x = \sin t - 1$, $y = \cos t - 1$ implies $(x+1)^2 + (y+1)^2 = 1$

c. $x = 1 - \cos t$, $y = 1 - \sin t$ implies $(x-1)^2 + (y-1)^2 = 1$

d. $x = \cos t - 1$, $y = \sin t - 1$ implies $(x+1)^2 + (y+1)^2 = 1$

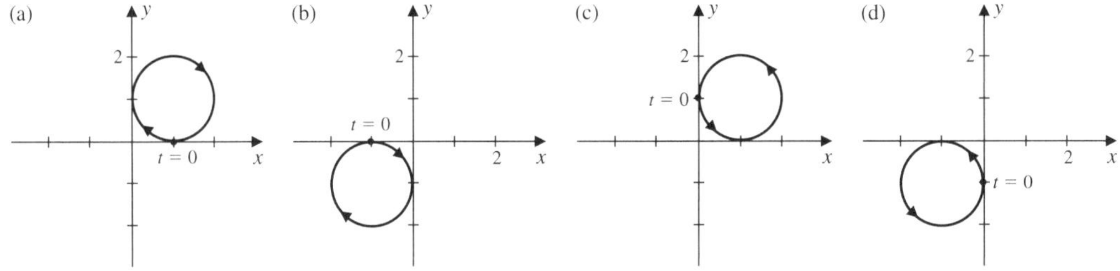

17. The graphs of the parametric equations are shown below.

a. $x = \cos t$, $y = \sin t$ implies $x^2 + y^2 = 1$

b. $x = \sin t$, $y = \cos t$ implies $x^2 + y^2 = 1$

c. $x = t$, $y = \sqrt{1 - t^2}$ implies $y = \sqrt{1 - x^2}$

d. $x = -t$, $y = \sqrt{1 - t^2}$ implies $y = \sqrt{1 - x^2}$

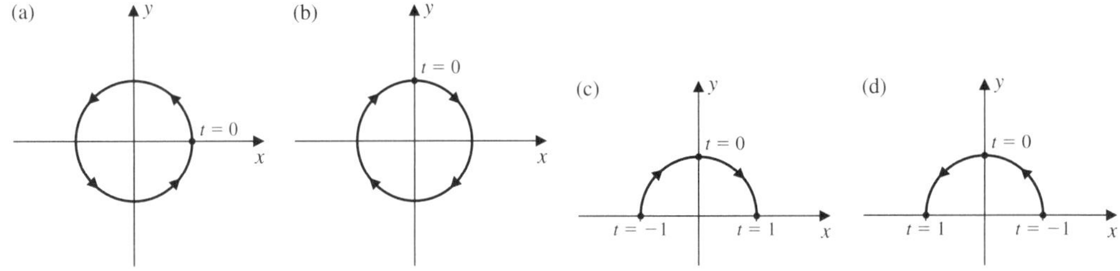

18. The graphs of the parametric equations are shown below.

a. $x = t + 1,\ y = 2t + 3$ implies $t = x - 1,\ y = 2(x - 1) + 3 = 2x + 1$

b. $x = t^2 - \frac{1}{2},\ y = 2t^2$ implies $t^2 = x + \frac{1}{2},\ y = 2\left(x + \frac{1}{2}\right) = 2x + 1,\ x \geq -\frac{1}{2}$

c. $x = \ln t,\ y = 1 + \ln t^2$ implies $y = 1 + 2\ln t = 1 + 2x$

d. $x = \sin t,\ y = 1 + 2\sin t$ implies $y = 1 + 2x,\ -1 \leq x \leq 1$

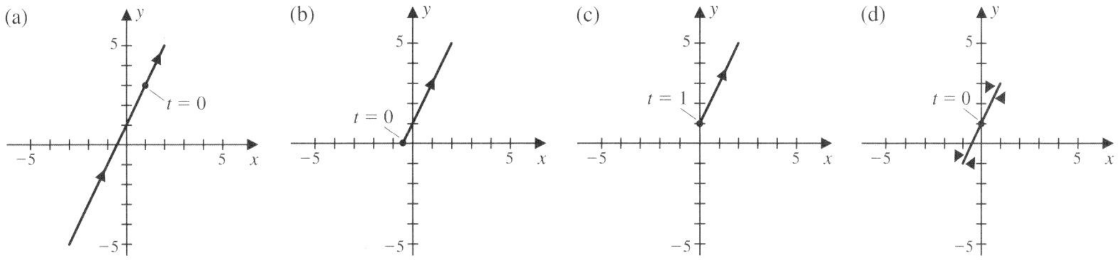

19. The graphs of the parametric equations are shown below.

a. $x = t,\ y = \ln t$ implies $y = \ln x,\ t > 0$

b. $x = e^t,\ y = t$ implies $\ln x = t,\ y = \ln x$

c. $x = t^2,\ y = 2\ln t$ implies $y = \ln t^2 = \ln x$

d. $x = \frac{1}{t},\ y = -\ln t$ implies $y = \ln t^{-1} = \ln x$

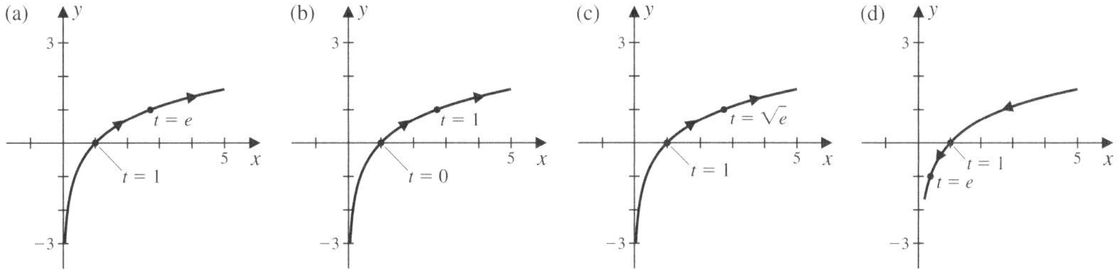

20. The graphs of the parametric equations are shown below.

a. $x = t,\ y = \frac{1}{t}$ implies $y = \frac{1}{x}$

b. $x = e^t,\ y = e^{-t}$ implies $y = \frac{1}{e^t} = \frac{1}{x},\ x > 0$

c. $x = \sin t,\ y = \csc t$ implies $y = \frac{1}{\sin t} = \frac{1}{x}$ and $x = \sin t$ implies $|x| \leq 1$ so $|y| = \left|\frac{1}{x}\right| \geq 1$

d. $x = \tan t$, $y = \cot t$ implies $y = \frac{1}{x}$

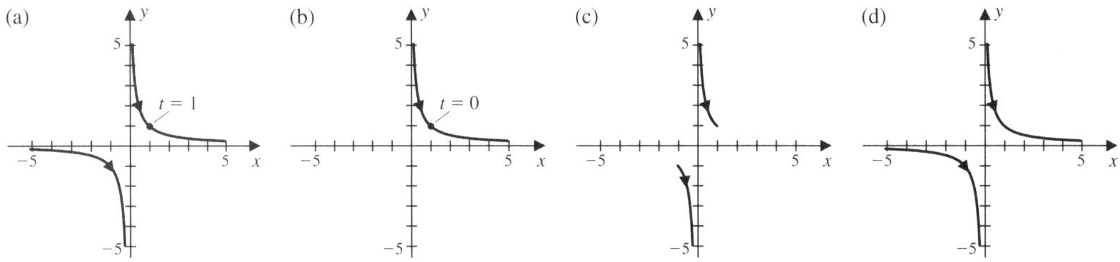

21. A rectangular equation of the line with slope $\frac{1}{3}$ and passing through $(2, -1)$ is

$$y + 1 = \frac{1}{3}(x - 2).$$

One set of parametric equations is obtained by setting

$$t = x - 2 \text{ so } x = t + 2,\ y = \frac{1}{3}t - 1.$$

22. The line passing through $(5, 3)$ and $(-2, 7)$ has slope $m = \frac{7-3}{-2-5} = -\frac{4}{7}$ and has rectangular equation

$$y - 3 = -\frac{4}{7}(x - 5).$$

One set of parametric equations is obtained by setting

$$t = x - 5 \text{ so } x = t + 5,\ y = -\frac{4}{7}t + 3.$$

23. A parabola with vertex $(1, -2)$ and passing through the points $(0, 0)$ and $(2, 0)$ has the form $y = a(x - 1)^2 - 2$. Since $(0, 0)$ is on the curve,
$0 = a(-1)^2 - 2$ implies $a = 2$ so $y = 2(x - 1)^2 - 2$. One set of parametric equations is obtained by setting

$$t = x - 1 \text{ so } x = t + 1,\ y = 2t^2 - 2.$$

24. A parabola with vertex $(2, 1)$ and passing through the points $(0, -2)$ and $(0, 4)$ has the form $x = a(y - 1)^2 + 2$. Since $(0, -2)$ is on the curve,

$0 = a(-2-1)^2 + 2$ implies $9a + 2 = 0$ so $a = -\frac{2}{9}$ and $x = -\frac{2}{9}(y-1)^2 + 2$. One set of parametric equations is obtained by setting

$$t = y - 1 \text{ so } y = t + 1,\ x = -\frac{2}{9}t^2 + 2.$$

25. **a.** $x = r\cos t,\ y = r\sin t$ **b.** $x = r\cos 2t,\ y = r\sin 2t$ **c.** $x = -r\cos 3t,\ y = -r\sin 3t$ **d.** $x = r\sin 2t,\ y = r\cos 2t$; **e.** $x = r\sin 3t,\ y = r\cos 3t$

26. **a.** Since

$$x = \frac{t^2}{1+t^3},\ y = \frac{t}{1+t^3} \text{ implies } 1 + t^3 = \frac{t}{y} \text{ so } x = \frac{t^2}{t/y} = yt \text{ and } t = \frac{x}{y},$$

we have

$$y = \frac{x/y}{1+(x/y)^3} = \frac{x/y}{\frac{y^3+x^3}{y^3}} = \frac{xy^2}{y^3+x^3} \text{ implies } y^4 + yx^3 = xy^2 \text{ so } y^3 + x^3 = xy.$$

b. The folium of Descartes is shown.

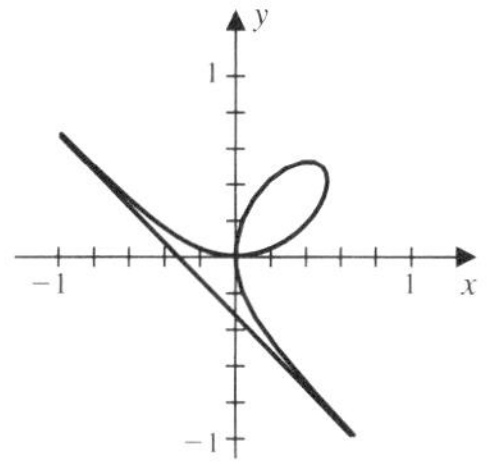

27. The curves when $a = 1, b = 1$; $a = 1, b = 2$; and $a = 2, b = 1$ are shown.

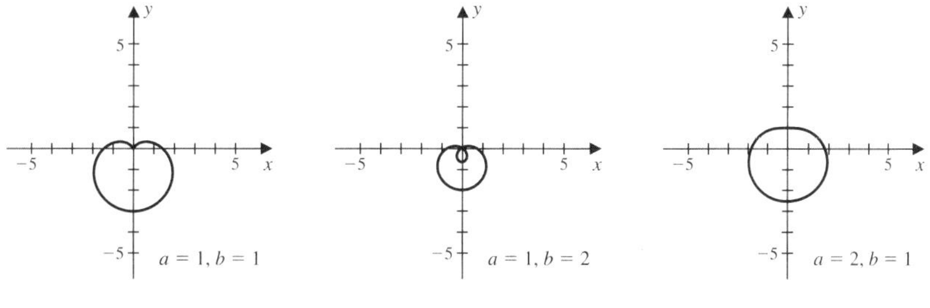

28. The parameter a determines where the curve touches the x- and y-axes.

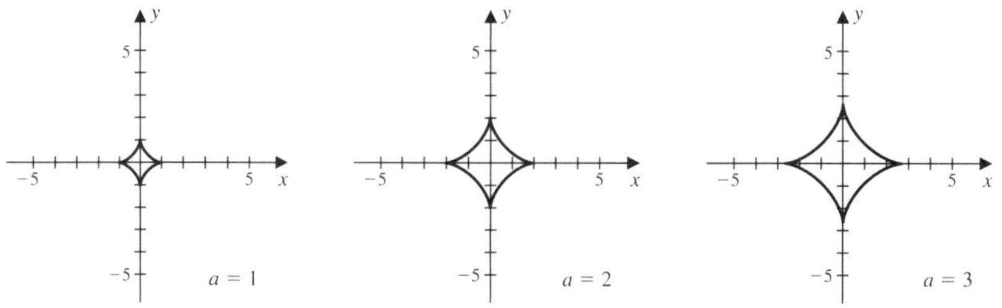

29. The parameter a determines the maximum height of the curve, $2a$, and the period of the curve, $2\pi a$, which equals the circumference of the rolling circle. The curve touches the x-axis for those t for which the y-coordinate is 0. That is,

$$a(1 - \cos t) = 0 \text{ implies } \cos t = 1 \text{ so } t = \pm 2k\pi, \text{ for } k = 0, 1, 2, ...,$$

and

$$x = a(t - \sin t) = at = \pm 2ak\pi \text{ for } k = 0, 1, 2...$$

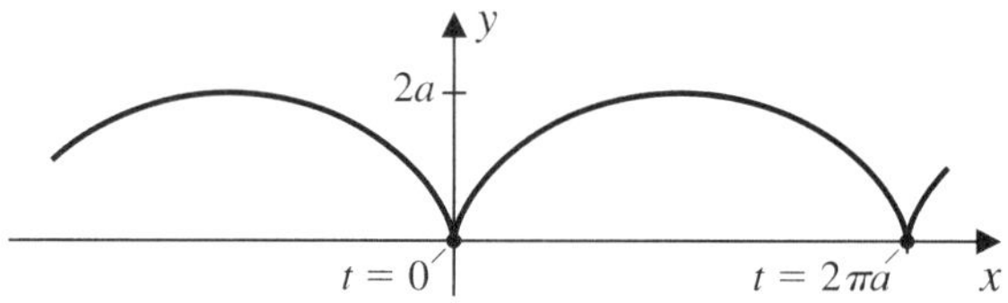

Review Exercises for Chapter 6 (Page 354)

1. $4y - x^2 = 0$ implies $y = \frac{1}{4}x^2$ so $c = 1$
Vertex: $(0, 0)$; Focus: $(0, 1)$;
Directrix, D: $y = -1$

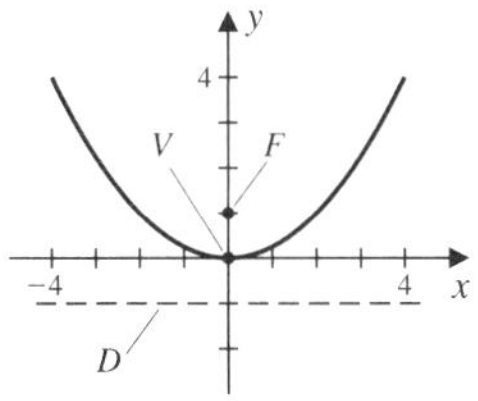

2. $y^2 + 12x = 0$ implies $x = -\frac{1}{12}y^2$ so $c = -3$
Vertex: $(0, 0)$; Focus: $(-3, 0)$;
Directrix, D: $x = 3$

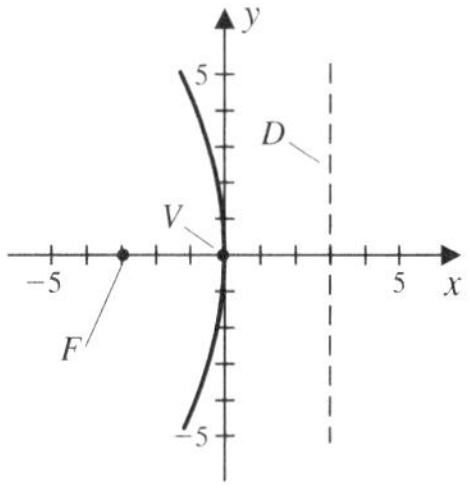

3. $4x - y^2 + 6y - 17 = 0$ implies $-(y^2 - 6y + 9 - 9) = 17 - 4x$
so $-(y-3)^2 = 8 - 4x$ and $x - 2 = \frac{1}{4}(y-3)^2$ implies $c = 1$;
Vertex: $(2, 3)$; Focus: $(3, 3)$;
Directrix, D: $x = 1$

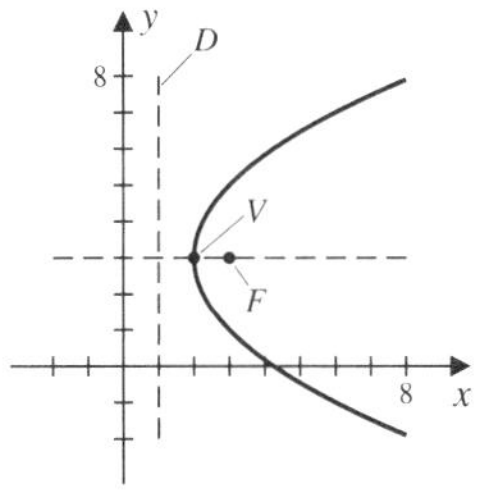

4. $x^2 + 4x + 8y - 4 = 0$ implies $x^2 + 4x + 4 - 4 = 4 - 8y$
so $(x+2)^2 = 8 - 8y$ and $y - 1 = -\frac{1}{8}(x+2)^2$ implies $c = -2$
Vertex: $(-2, 1)$; Focus: $(-2, -1)$;
Directrix, D: $y = 3$

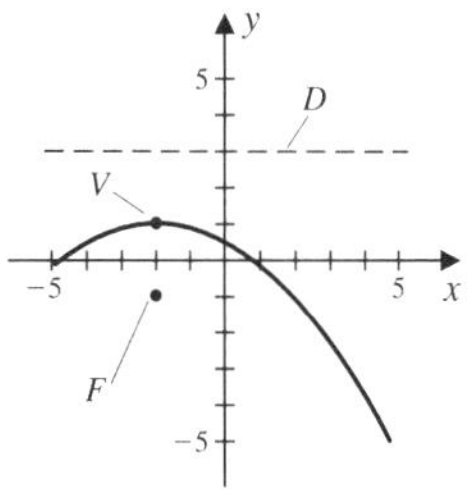

5. $x^2 + 4y^2 = 4$ implies $\frac{x^2}{4} + y^2 = 1$ so $a = 2$, $b = 1$ and $c^2 = 3$ implies $c = \sqrt{3}$
Foci: $\left(\sqrt{3}, 0\right), \left(-\sqrt{3}, 0\right)$; Vertices: $(2, 0), (-2, 0)$

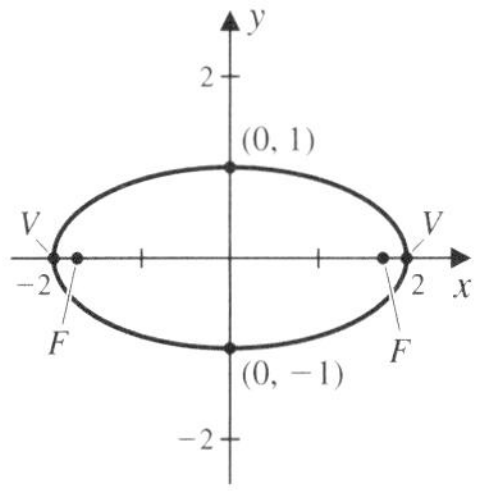

6. $4x^2 + y^2 = 16$ implies $\frac{x^2}{4} + \frac{y^2}{16} = 1$ so $a = 4$, $b = 2$ and $c^2 = 12$ implies $c = 2\sqrt{3}$
Foci: $\left(0, 2\sqrt{3}\right), \left(0, -2\sqrt{3}\right)$; Vertices: $(0, 4), (0, -4)$

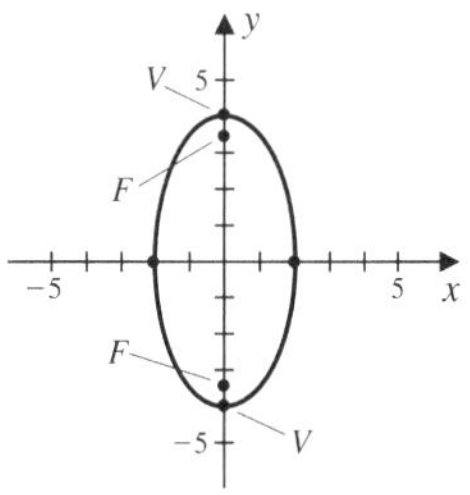

7. $4(x-1)^2+9(y+2)^2=36$ implies $\frac{(x-1)^2}{9}+\frac{(y+2)^2}{4}=1$ so $a=3, b=2$ and $c^2=5$ implies $c=\sqrt{5}$
Foci: $\left(1-\sqrt{5},-2\right),\left(1+\sqrt{5},-2\right)$;
Vertices: $(4,-2),(-2,-2)$

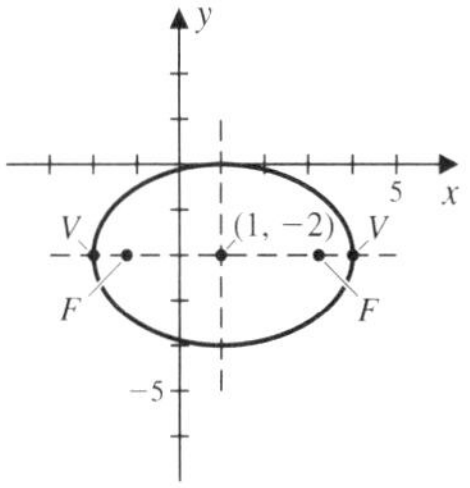

8. $2(x+1)^2+(y-1)^2=2$ implies $(x+1)^2+\frac{(y-1)^2}{2}=1$ so $a=\sqrt{2}, b=1$ and $c^2=1$ implies $c=1$
Foci: $(-1,0),(-1,2)$;
Vertices: $\left(-1,1-\sqrt{2}\right),\left(-1,1+\sqrt{2}\right)$

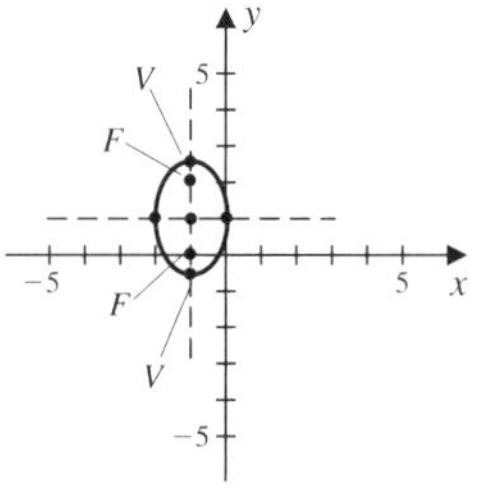

9. $x^2-2y^2=4$ implies $\frac{x^2}{4}-\frac{y^2}{2}=1$ so $a=2, b=\sqrt{2}$ and $c^2=6$ implies $c=\sqrt{6}$
Foci: $\left(\sqrt{6},0\right),\left(-\sqrt{6},0\right)$;
Vertices: $(2,0),(-2,0)$;
Asymptotes: $y=\pm\frac{\sqrt{2}}{2}x$

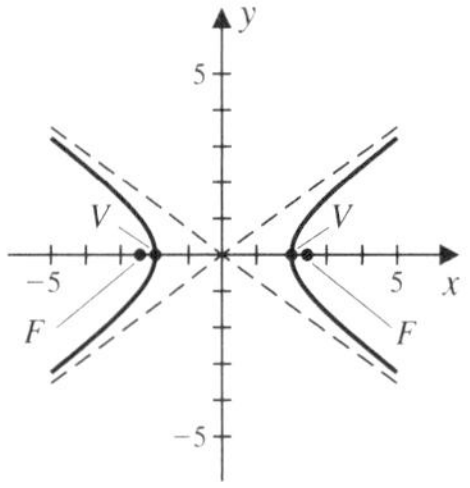

10. $4y^2-x^2=16$ implies $\frac{y^2}{4}-\frac{x^2}{16}=1$ so $a=2, b=4$ and $c^2=20$ implies $c=2\sqrt{5}$
Foci: $\left(0,2\sqrt{5}\right),\left(0,-2\sqrt{5}\right)$;
Vertices: $(0,2),(0,-2)$;
Asymptotes: $y=\pm\frac{1}{2}x$

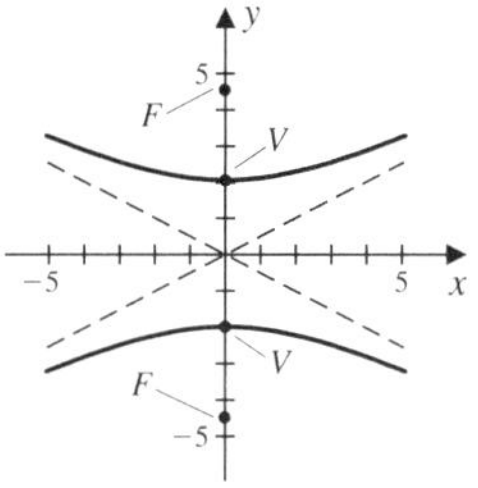

11. $2x^2 - 4x - 4y^2 + 1 = 0$ implies
$2(x^2 - 2x + 1 - 1) - 4y^2 = -1$
so $2(x-1)^2 - 4y^2 = 1$ and $a = \frac{\sqrt{2}}{2}$, $b = \frac{1}{2}$ implies $c^2 = \frac{1}{2} + \frac{1}{4} = \frac{3}{4}$ so $c = \frac{\sqrt{3}}{2}$
Foci: $\left(1 - \sqrt{3}/2, 0\right), \left(1 + \sqrt{3}/2, 0\right)$;
Vertices: $\left(1 + \sqrt{2}/2, 0\right), \left(1 - \sqrt{2}/2, 0\right)$;
Asymptotes: $y = \pm\frac{\sqrt{2}}{2}(x - 1)$

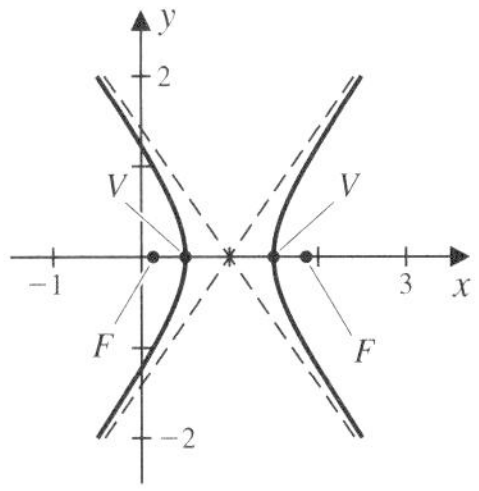

12. $9x^2 - 4y^2 - 18x + 16y - 43 = 0$ implies $9(x^2 - 2x + 1 - 1) - 4(y^2 - 4y + 4 - 4) = 43$
so $9(x-1)^2 - 4(y-2)^2 = 36$ and $\frac{(x-1)^2}{4} - \frac{(y-2)^2}{9} = 1$ implies $a = 2, b = 3$
so $c^2 = 13$ and $c = \sqrt{13}$
Foci: $\left(1 + \sqrt{13}, 2\right), \left(1 - \sqrt{13}, 2\right)$;
Vertices: $(3, 2), (-1, 2)$;
Asymptotes: $y - 2 = \pm\frac{3}{2}(x - 1)$

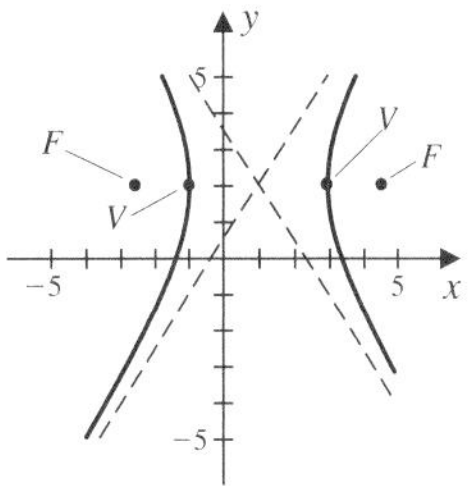

13. Parabola with equation
$x^2 = -2(y - 5)$ implies $y - 5 = -\frac{1}{2}x^2$.

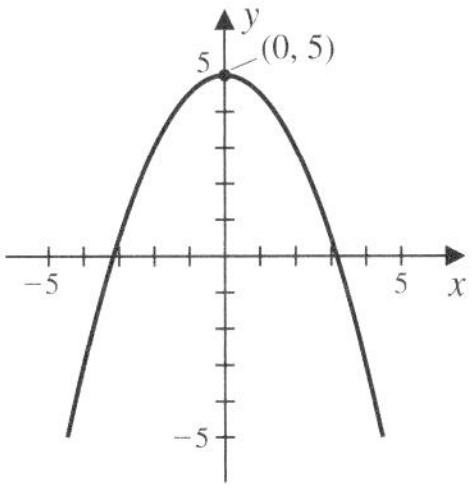

14. Ellipse with equation
$9(x+2)^2 - 4(y-5)^2 = 36$
implies $\frac{(x+2)^2}{4} - \frac{(y-5)^2}{9} = 1$.

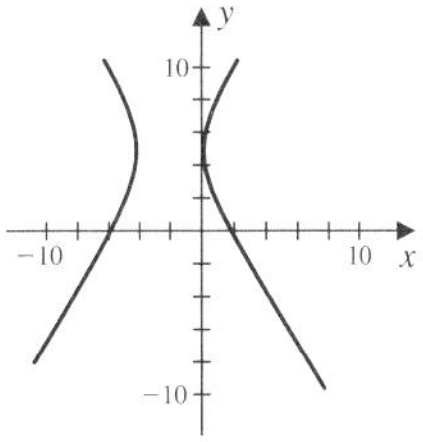

15. Ellipse with equation
$16x^2 + 25y^2 = 400$ implies $\frac{x^2}{25} + \frac{y^2}{16} = 1$.

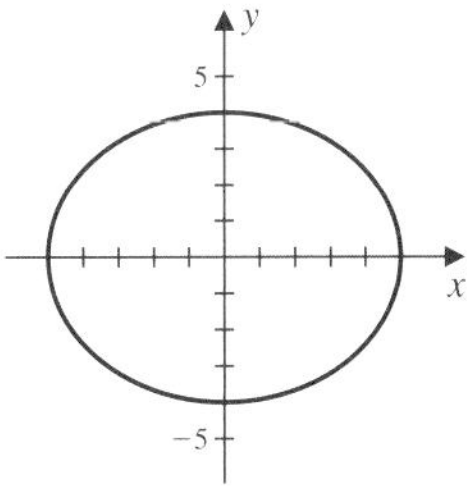

16. Parabola with equation
$y^2 = -16(x + 1)$ implies $x + 1 = -\frac{1}{16}y^2$.

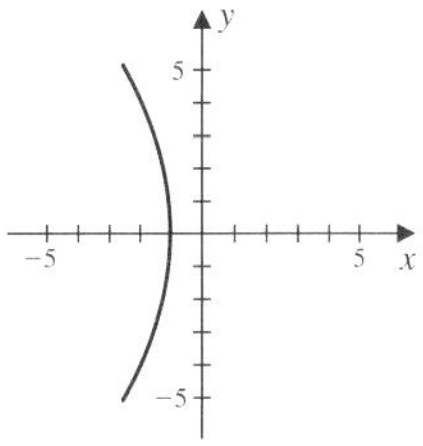

17. Parabola with equation $x^2 - 2x - 4y - 11 = 0$ implies $x^2 - 2x + 1 - 1 = 4y + 11$ so $(x - 1)^2 = 4(y + 3)$
and $y + 3 = \frac{1}{4}(x - 1)^2$.

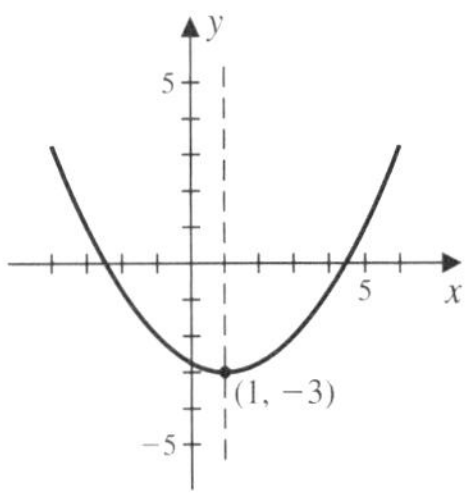

18. Parabola with equation $y^2 - 2x + 2y + 7 = 0$ implies $y^2 + 2y + 1 - 1 = 2x - 7$ so $(y + 1)^2 = 2(x - 3)$
and $x - 3 = \frac{1}{2}(y + 1)^2$.

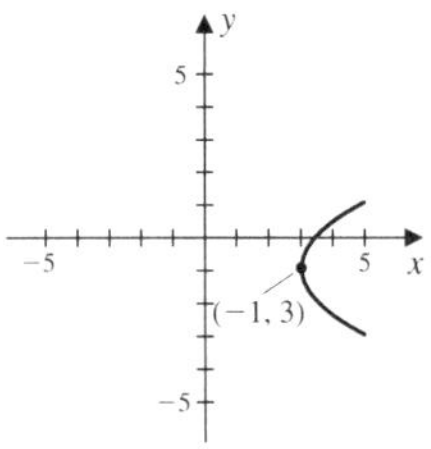

19. Hyperbola with equation
$9x^2 - 16y^2 = 144$ implies $\frac{x^2}{16} - \frac{y^2}{9} = 1$.

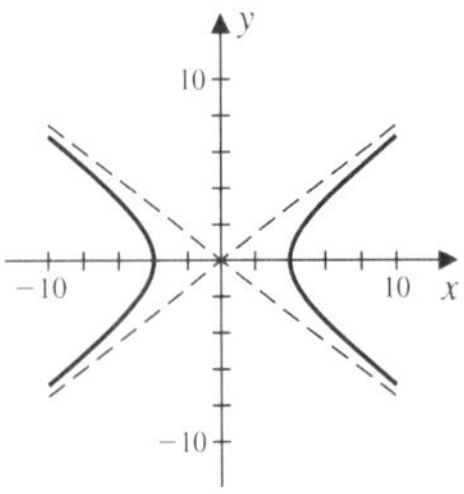

20. Ellipse with equation $9x^2 + 4y^2 - 90x - 16y + 205 = 0$
implies $9(x^2 - 10x + 25 - 25) + 4(y^2 - 4y + 4 - 4) = -205$
so $9(x - 5)^2 + 4(y - 2)^2 = 36$ and $\frac{(x-5)^2}{4} + \frac{(y-2)^2}{3} = 1$.

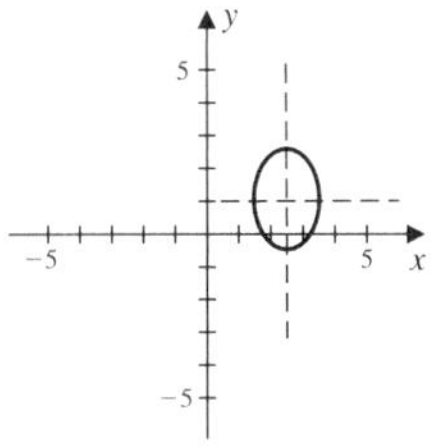

21. Hyperbola with equation $16x^2 - 64x - 25y^2 + 150y = 561$
implies $16(x^2 - 4x + 4 - 4) - 25(y^2 - 6y + 9 - 9) = 561$
so $16(x - 2)^2 - 25(y - 3)^2 = 400$ and $\frac{(x-2)^2}{25} - \frac{(y-3)^2}{16} = 1$.

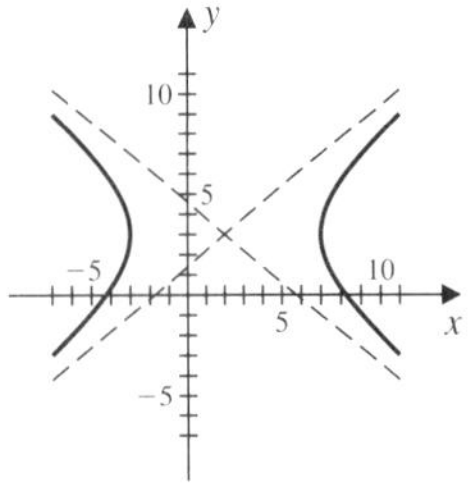

22. Hyperbola with equation $4x^2 - 9y^2 - 16x - 90y - 173 = 0$
implies $4(x^2 - 4x + 4 - 4) - 9(y^2 + 10y + 25 - 25) = 173$
so $4(x - 2)^2 - 9(y + 5)^2 = -36$ and $\frac{(y+5)^2}{4} - \frac{(x-2)^2}{9} = 1$.

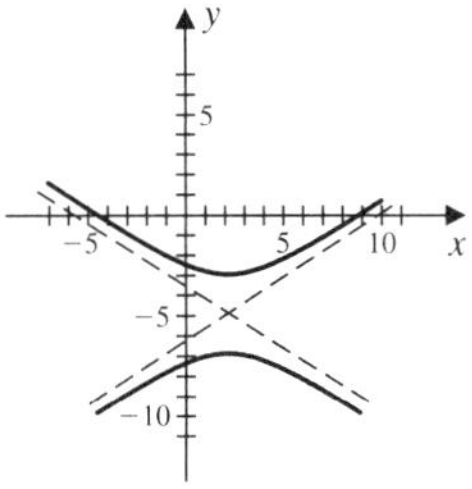

23. The vertex is midway between the focus (0, 0) and the directrix $y = 2$ so is (0, 1), and the equation has the form $y - 1 = \frac{1}{4c}x^2$. Since the distance from the focus to the vertex is 1, and the directrix is above the focus, $c = -1$, $y - 1 = -\frac{1}{4}x^2$.

24. Since the foci $(0, \pm 1)$ are on the y-axis, the major axis is the y-axis, and $c = 1$. Since the vertices are $(0, \pm 3)$, $a = 3$, and

$$c^2 = a^2 - b^2 \text{ implies } 1 = 9 - b^2 \text{ so } b^2 = 8.$$

The equation is

$$\frac{x^2}{8} + \frac{y^2}{9} = 1.$$

25. Since the foci $(\pm 3, 0)$ are centered about the origin, the hyperbola is in standard position with axis on the x-axis, and $c = 3$. Since a vertex is $(1, 0)$, $a = 1$, and

$$c^2 = a^2 + b^2 \text{ implies } 9 = 1 + b^2 \text{ so } b^2 = 8.$$

The equation is

$$x^2 - \frac{y^2}{8} = 1.$$

26. Since the parabola has a focus $(0, 1)$ the axis of the parabola is the y-axis, and since the vertex is $(0, -1)$, the parabola opens upward. So $c = 2$, and

$$y + 1 = \frac{1}{8}x^2.$$

27. Since the foci of the ellipse are $(0, \pm 5)$, the equation has the form $\frac{y^2}{a^2} + \frac{x^2}{b^2} = 1$, and $c = 5$. Since $(4, 0)$ is on the curve,

$$\frac{16}{b^2} = 1 \text{ implies } b^2 = 16.$$

Then

$$c^2 = a^2 - b^2 \text{ implies } a^2 = 25 + 16 = 41,$$

and the equation is

$$\frac{y^2}{41} + \frac{x^2}{16} = 1.$$

28. Since the directrix $x = 2$ is vertical and to the right of the pole, the conic has the form $r = \frac{ed}{1+e\cos\theta}$. Since $e = \frac{3}{4} < 1$, the conic is an ellipse, and

$$r = \frac{\left(\frac{3}{4}\right)2}{1 + \frac{3}{4}\cos\theta} = \frac{6}{4 + 3\cos\theta}.$$

29. Since the directrix $y = -2$ is horizontal and below the pole, the conic has the form $r = \frac{ed}{1-e\sin\theta}$. Since $e = 3 > 1$, the conic is a hyperbola, and

$$r = \frac{(3)\,2}{1 - 3\sin\theta} = \frac{6}{1 - 3\sin\theta}.$$

30. Since the directrix $x = -3$ is vertical and to the left of the pole, the conic has the form $r = \frac{ed}{1-e\cos\theta}$. Since $e = 1$, the conic is a parabola, and

$$r = \frac{(1)\,3}{1 - \cos\theta} = \frac{3}{1 - \cos\theta}.$$

31. $r = 4 + 4\cos\theta$

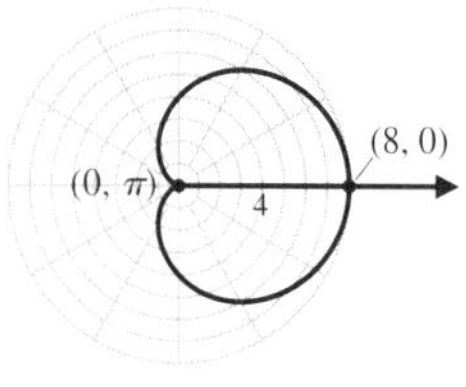

32. $r = 3 + 2\sin\theta$

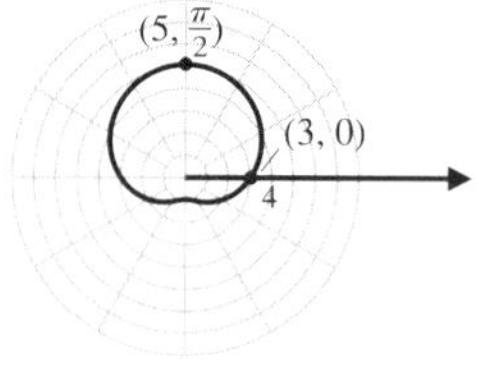

33. $r = 1 + 3\sin\theta$

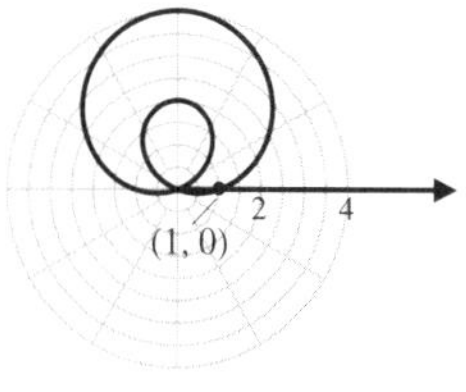

34. $r = 3 + 3\cos\theta$

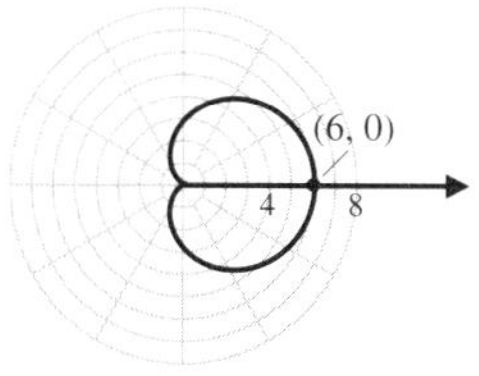

35. $r = 2\sin\theta$

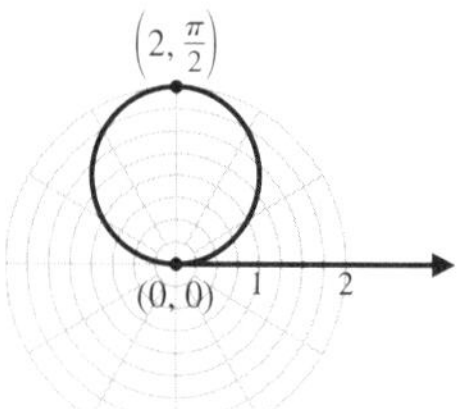

36. $r = 2$

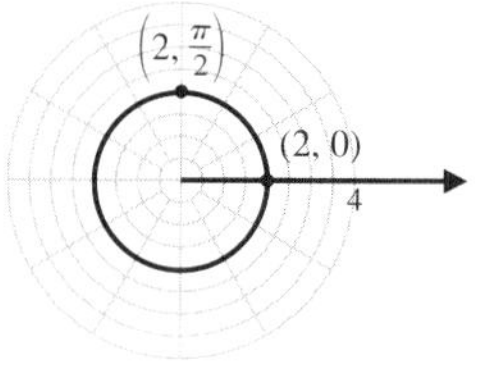

37. $r = 2\cos 2\theta$

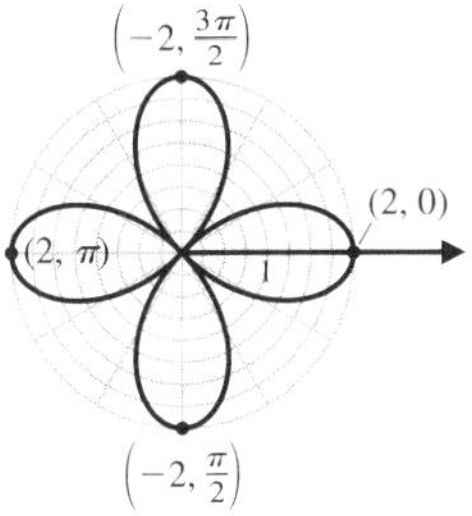

38. $r = 2\sin 3\theta$

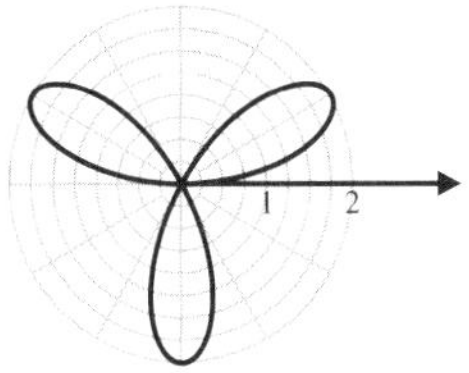

39. $\theta = 1/2$

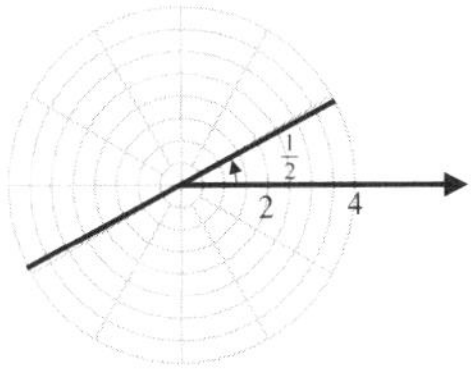

40. $\theta = -\frac{\pi}{4}$

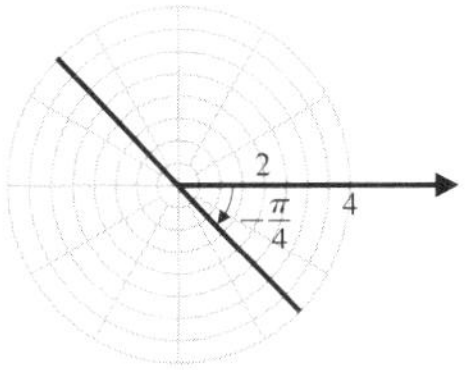

41. Since $x^2 + y^2 = r^2$ and $x = r\cos\theta$,

$$r = \frac{3}{1+\cos\theta} \text{ implies } r + r\cos\theta = 3 \text{ so } r = 3 - x \text{ and}$$
$$r^2 = 9 - 6x + x^2 \text{ implies } x^2 + y^2 = 9 - 6x + x^2 \text{ so}$$
$$y^2 = 9 - 6x = -6\left(x - \frac{3}{2}\right) \text{ or } x - \frac{3}{2} = -\frac{1}{6}y^2.$$

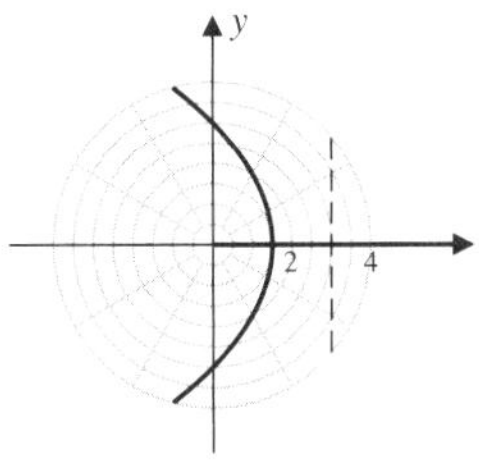

42. Since $x^2 + y^2 = r^2$ and $x = r\cos\theta$,

$$r = \frac{3}{2 + 4\cos\theta} \text{ implies } 2r + 4r\cos\theta = 3 \text{ so } 2r = 3 - 4x \text{ and}$$
$$4r^2 = 9 - 24x + 16x^2 \text{ implies } 4x^2 + 4y^2 = 9 - 24x + 16x^2 \text{ or}$$
$$12x^2 - 24x - 4y^2 = -9 = 12(x^2 - 2x + 1 - 1) - 4y^2 = -9 \text{ so } 12(x-1)^2 - 4y^2 = 3.$$

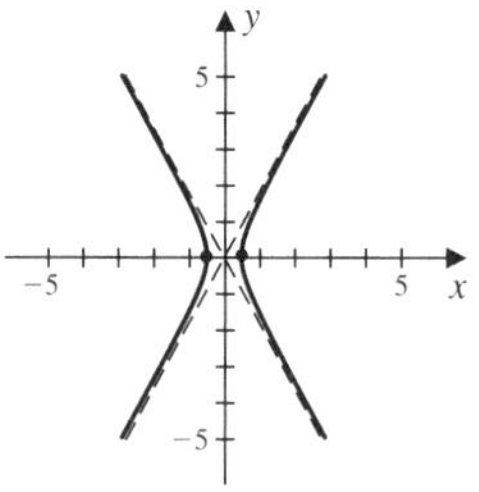

43. Since $x^2 + y^2 = r^2$ and $x = r\cos\theta$,

$$r = \frac{4}{2 - \cos\theta} \text{ implies } 2r - r\cos\theta = 4 \text{ so } 2r = 4 + x \text{ and } 4r^2 = 16 + 8x + x^2 \text{ or}$$

$$4x^2 + 4y^2 = 16 + 8x + x^2 \text{ implies } 3x^2 - 8x + 4y^2 = 16,$$

and

$$3\left(x^2 - \frac{8}{3}x + \left(\frac{4}{3}\right)^2 - \left(\frac{4}{3}\right)^2\right) + 4y^2 = 16 \text{ implies } 3\left(x - \frac{4}{3}\right)^2 + 4y^2 = \frac{64}{3}$$

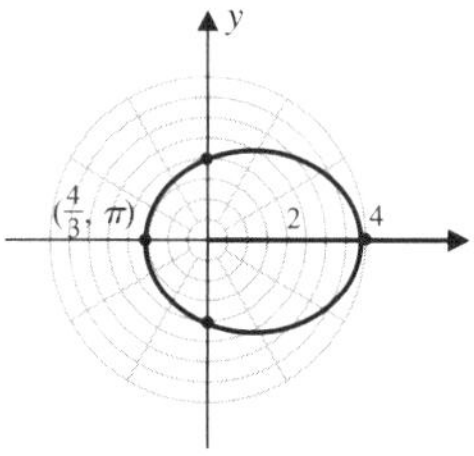

44. Since $x^2 + y^2 = r^2$ and $y = r\sin\theta$,

$$r = \frac{2}{1-\sin\theta} \text{ implies } r - r\sin\theta = 2 \text{ or } r = 2 + y \text{ so}$$
$$r^2 = 4 + 4y + y^2 \text{ implies } x^2 + y^2 = 4 + 4y + y^2 \text{ so } x^2 = 4(1+y) \text{ or}$$
$$y + 1 = \frac{1}{4}x^2.$$

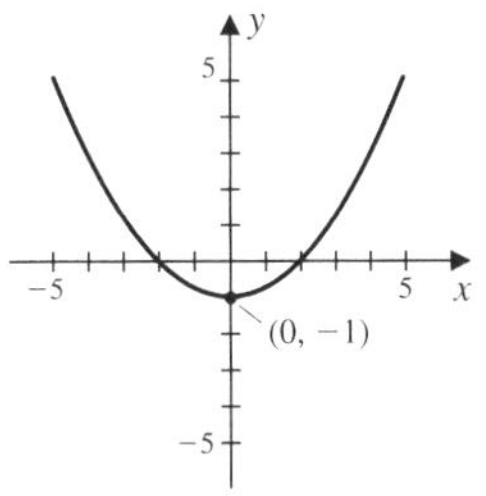

45. $x = t^2 - 1$, $y = t + 1$ implies $t = y - 1$, $x = (y-1)^2 - 1$ so $x + 1 = (y-1)^2$.

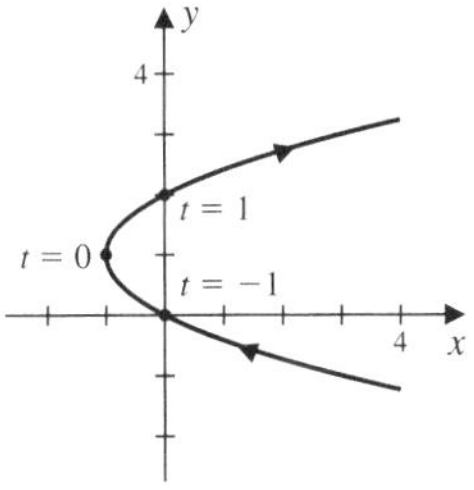

46. $x = 2\cos t$, $y = 3\sin t$ implies $\frac{x^2}{4} + \frac{y^2}{9} = \frac{4(\cos t)^2}{4} + \frac{9(\sin t)^2}{9} = (\cos t)^2 + (\sin t)^2 = 1$ so $\frac{x^2}{4} + \frac{y^2}{9} = 1$.

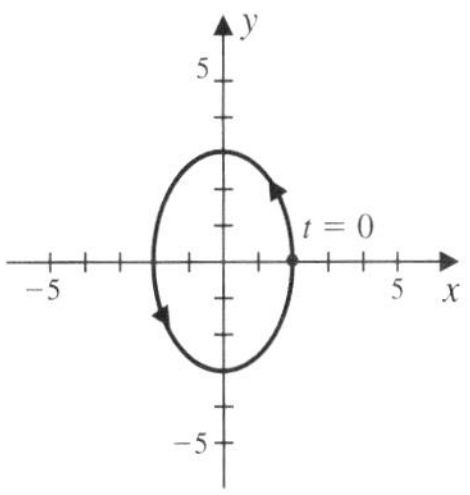

47. $x = e^t$, $y = 1 + e^{-t}$ implies $y = 1 + x^{-1}$ so $y = 1 + \frac{1}{x}$ for $x > 0$, since $x = e^t > 0$.

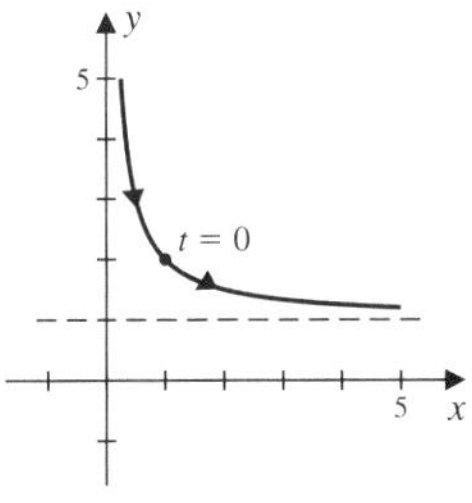

48.

$x = t + 2$, $y = (t-1)^2 + 1$ implies $t = x - 2$ and $y = (x - 2 - 1)^2 + 1 = (x-3)^2 + 1$.

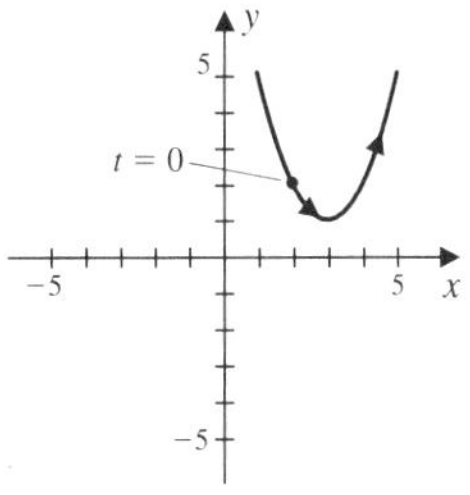

49. $x = (\sin t)^2 + 1$, $y = (\cos t)^2$ implies $x + y = (\sin t)^2 + (\cos t)^2 + 1 = 2$ so $y = -x + 2$ for $1 \le x \le 2$, $0 \le y \le 1$.

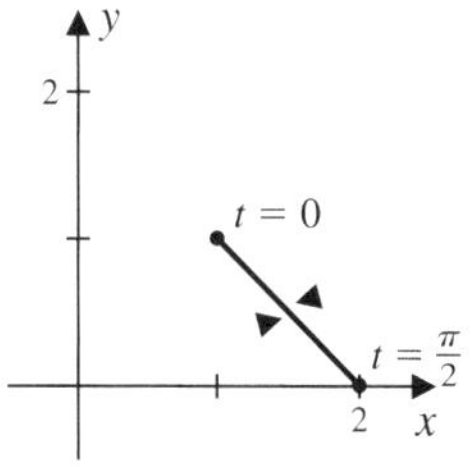

50. $x = \ln t^2$, $y = \ln t^3 + 1$ implies $x = 2 \ln t$ so $\ln t = \frac{x}{2}$ and $y = \ln t^3 + 1 = 3 \ln t + 1$ so $y = 3\left(\frac{x}{2}\right) + 1 = \frac{3}{2}x + 1$.

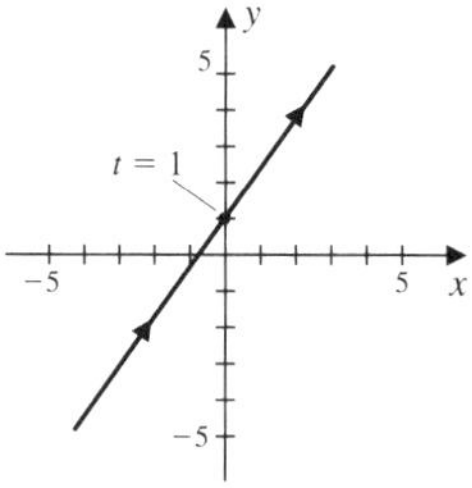

51. The number of leafs in the curves $r = 1 + \sin 2n\theta + (\cos n\theta)^2$ is $2n$.

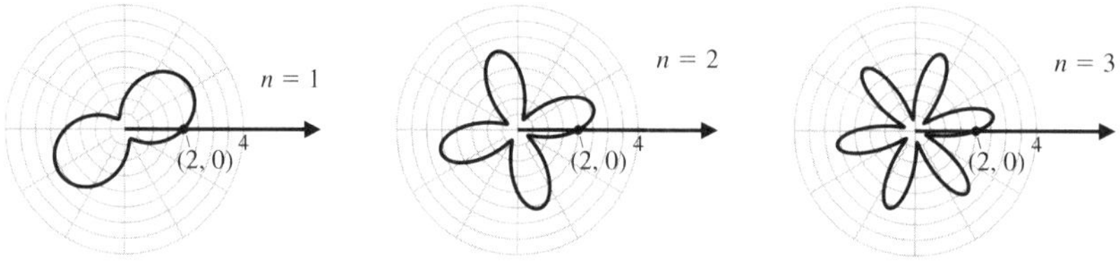

52. The graphs of $r = \sin m\theta + (\cos n\theta)^2$ are shown below.

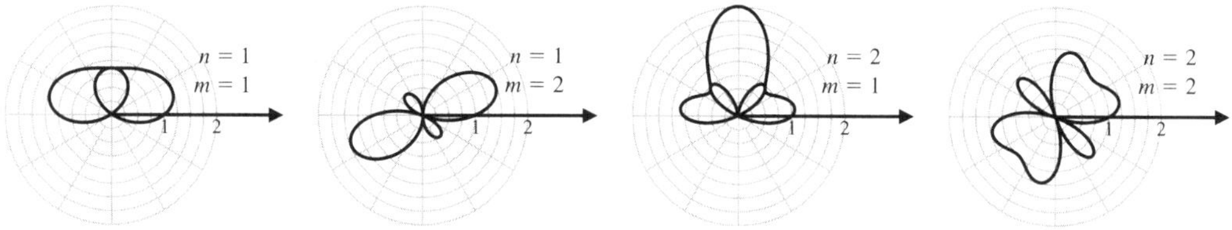

53. The curves in the figures are given by the parametric equations

$$x = (1 - a \sin t) \cos t,\ y = (1 - a \sin t) \sin t,\ \text{for } 0 \le t \le 2\pi.$$

If $a = 1$, the curve is a cardioid. If $a < 1$, the curve is a limacon without a loop. If $a > 1$, the curve is a limacon with a loop.

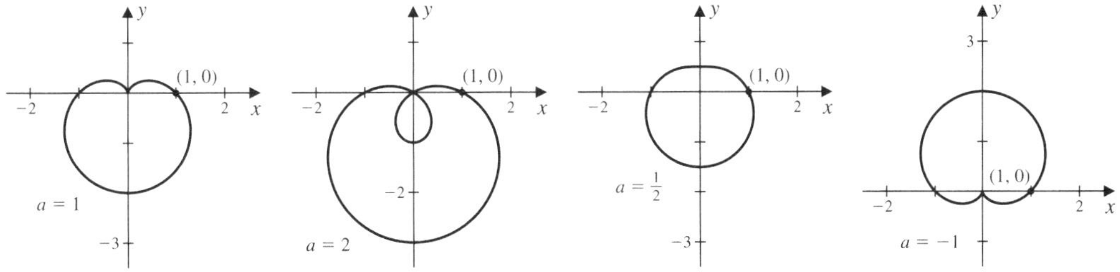

54. The hypocycloids defined by the parametric equations

$$x = (a - b) \cos t + b \cos \frac{a-b}{b} t,\ y = (a - b) \sin t - b \sin \frac{a-b}{b} t,\ \text{for } 0 \le t \le 2\pi.$$

are shown below.

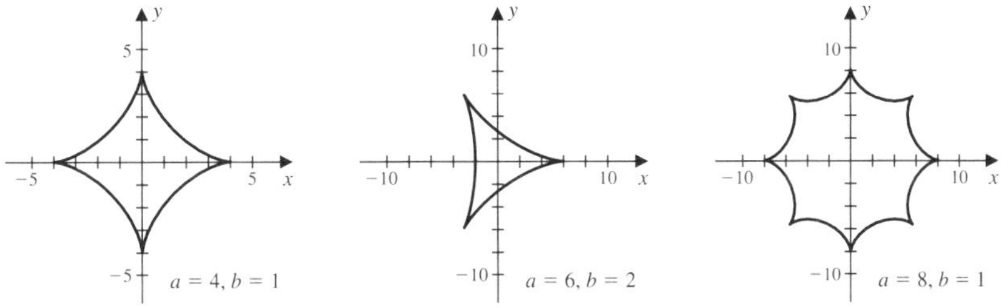

55. The epicycloids defined by the parametric equations

$$x = (a+b)\cos t - b\cos\frac{a+b}{b}t,\ y = (a+b)\sin t - b\sin\frac{a+b}{b}t,\ \text{for } 0 \le t \le 2\pi.$$

are shown below.

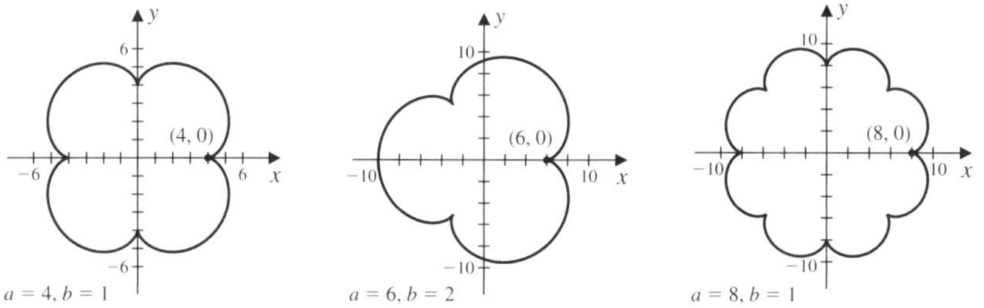

Chapter 6 Exercises for Calculus (Page 355)

1. a. The equation has the general form $y = ax^2$. Since (x_1, y_1) is on the curve, $y_1 = ax_1^2$ implies $a = \frac{y_1}{x_1^2}$, and

$$y = \frac{y_1}{x_1^2}x^2.$$

b. The equation has the general form $y - k = a(x-h)^2$. Since (x_1, y_1) is on the curve, $y_1 - k = a(x_1 - h)^2$ implies $a = \frac{y_1 - k}{(x_1-h)^2}$, and

$$y - k = \frac{y_1 - k}{(x_1 - h)^2}(x-h)^2.$$

2. a. The equation has the general form $x = ay^2$. Since (x_1, y_1) is on the curve, $x_1 = ay_1^2$ implies $a = \frac{x_1}{y_1^2}$, and

$$x = \frac{x_1}{y_1^2}y^2.$$

b. The equation has the general form $x - h = a(y-k)^2$. Since (x_1, y_1) is on the curve, $x_1 - h = a(y_1 - k)^2$ implies $a = \frac{x_1 - h}{(y_1-k)^2}$, and

$$x - h = \frac{x_1 - h}{(y_1 - k)^2}(y-k)^2.$$

3. a. The equation has the general form $y - k = a(x - h)^2$. Since $(h+1, k+1)$ is on the curve, $k + 1 - k = a(h + 1 - h)^2$ implies $a = 1$, and $y - k = (x - h)^2$.

b. The equation has the general form $x - h = a(y - k)^2$. Since $(h + 1, k + 1)$ is on the curve, $h + 1 - h = a(k + 1 - k)^2$ implies $a = 1$, and

$$x - h = (y - k)^2.$$

4. a. The equation has the form $y = ax^2 + bx + c$. If the points $(0, 1)$, $(1, 0)$ and $(2, 2)$ lie on the parabola, then

$$\begin{aligned} c &= 1 \\ 0 &= a + b + c \\ 2 &= 4a + 2b + c \end{aligned}$$

implies $c = 1, a = -1 - b, 2 = 4(-1 - b) + 2b + 1$ so $b = -\frac{5}{2}, a = \frac{3}{2}, c = 1$.

The parabola has equation

$$y = \frac{3}{2}x^2 - \frac{5}{2}x + 1.$$

b. The equation has the form $x = ay^2 + by + c$. If the points $(1, 0)$, $(0, 1)$ and $(2, 2)$ lie on the parabola, then

$$c = 1,\ \ 0 = a + b + c, \quad \text{and} \quad 2 = 4a + 2b + c$$

implies $c = 1, a = -1 - b, 2 = 4(-1 - b) + 2b + 1$ so $b = -\frac{5}{2}, a = \frac{3}{2}, c = 1$.

The parabola has equation

$$x = \frac{3}{2}y^2 - \frac{5}{2}y + 1.$$

5. The equation of the ellipse is $\frac{x^2}{4} + \frac{y^2}{9} = 1$, so $a^2 = 4, b^2 = 9$. The equation of the tangent at the point $\left(1, \frac{3\sqrt{3}}{2}\right)$ is

$$\frac{x(1)}{4} + \frac{y\left(3\sqrt{3}/2\right)}{9} = 1 \text{ implies } 9x + 6\sqrt{3}y = 36 \text{ so } y = -\frac{\sqrt{3}}{2}x + 2\sqrt{3}.$$

6. The equation of the ellipse is $x^2 + \frac{y^2}{4} = 1$, so $a^2 = 1, b^2 = 4$. Since $(3, 0)$ is on the tangent line,

$$\frac{xx_0}{a^2} + \frac{yy_0}{b^2} = 1 \text{ implies } \frac{(3)x_0}{1} + \frac{(0)y_0}{4} = 1 \text{ so } x_0 = \frac{1}{3}.$$

Since (x_0, y_0) is also on the ellipse,

$$4x_0^2 + y_0^2 = 4 \text{ implies } y_0^2 = 4 - \frac{4}{9} = \frac{32}{9} \text{ so } y_0 = \pm\frac{4\sqrt{2}}{3}.$$

So the equation of the tangent lines are

$$\frac{y_0}{4}y = -\frac{1}{3}x + 1 \text{ implies } y = -\frac{4}{3y_0}(x-3) = \pm\frac{\sqrt{2}}{2}(x-3).$$

7. The equation of the hyperbola is $\frac{x^2}{4} - \frac{y^2}{3} = 1$, so $a^2 = 4$, $b^2 = 3$. The equation of the tangent at the point $\left(2\sqrt{2}, \sqrt{3}\right)$ is

$$\frac{x(2\sqrt{2})}{4} - \frac{y\left(\sqrt{3}\right)}{3} = 1 \text{ implies } 6\sqrt{2}x - 4\sqrt{3}y = 12 \text{ so } y = \frac{\sqrt{6}}{2}x - \sqrt{3}.$$

8. The equation of the hyperbola is $\frac{x^2}{4} - \frac{y^2}{9} = 1$, so $a^2 = 4$, $b^2 = 9$. Since $(1, 0)$ is on the tangent line,

$$\frac{xx_0}{a^2} - \frac{yy_0}{b^2} = 1 \text{ implies } \frac{(1)x_0}{4} - \frac{(0)y_0}{9} = 1 \text{ so } x_0 = 4.$$

Since (x_0, y_0) is also on the hyperbola,

$$9x_0^2 - 4y_0^2 = 36 \text{ implies } -4y_0^2 = 36 - 144 = -108 \text{ so } y_0 = \pm\sqrt{27} = \pm 3\sqrt{3}.$$

The equation of the tangent lines are

$$x - \frac{yy_0}{9} = 1 \text{ implies } y = \frac{9}{y_0}(x-1) = \pm\frac{9}{3\sqrt{3}}(x-1) = \pm\sqrt{3}(x-1).$$

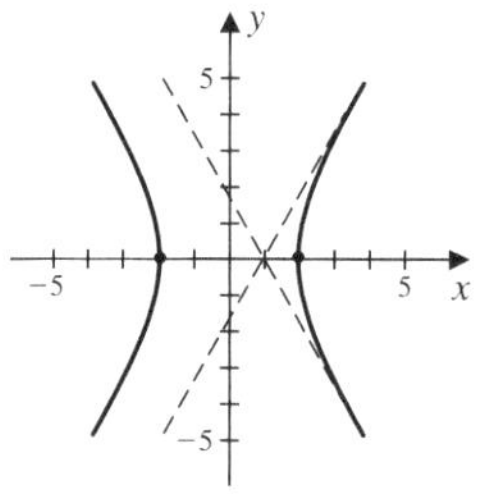

9. Since the two triangles in the figure are similar, $\frac{\sqrt{36-y^2}}{6} = \frac{x}{2}$, so the equation for the point on the ladder is

$$\sqrt{36 - y^2} = 3x \text{ implies } 9x^2 + y^2 = 36.$$

So the point moves along the portion of the ellipse $y = 3\sqrt{4 - x^2}$ from the point $(0, 6)$ to the point $(2, 0)$.

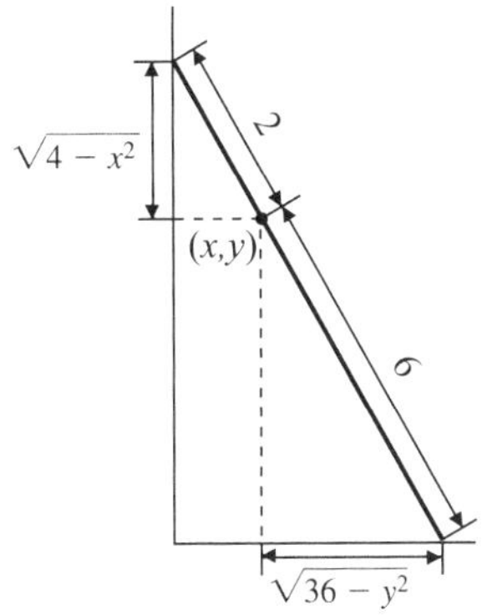

10. The cross section of the dish is a parabola with focal point $(0, c)$, and since the amplifier is in line with the edge of the dish, the point $(9, c)$ lies on the parabola. The equation of the parabola is

$$y = \frac{1}{4c}x^2 \text{ implies } c = \frac{1}{4c}(9)^2 \text{ so } c^2 = \frac{81}{4} \text{ and } c = \frac{9}{2}.$$

The depth of the dish is then $\frac{9}{2}$ inches.

11. The light source should be placed at the focal point of the parabola. Since $y = \frac{1}{4c}x^2 = \frac{1}{4}x^2$, the axis of the parabola is the y-axis, and $c = 1$, so the focal point is $(0, 1)$. Place the light source one unit along the axis from the vertex.

12. With the origin of the xy-coordinate system at the lowest point of the cable, the parabola describing the cable is in standard position with axis along the y-axis, so has equation of the form $y = \frac{1}{4c}x^2$. The point $\left(\frac{L}{2}, h\right)$ lies on the parabola, so

$$h = \frac{1}{4c}\left(\frac{L}{2}\right)^2 \text{ implies } 4ch = \frac{L^2}{4} \text{ so } 4c = \frac{L^2}{4h},$$

and the equation of the cable is $y = \frac{4h}{L^2}x^2$. The equation for the cable of the George Washington bridge, with a span of $L = 3500$ feet and a sag of $h = 316$ feet, is

$$y = \frac{4(316)}{(3500)^2}x^2 = \frac{79}{765625}x^2.$$

13. If the parabolic cross section of the mirror is a parabola of the form $x = \frac{1}{4c}y^2$, the information implies the point $(3.75, 100) = \left(\frac{15}{4}, 100\right)$ lies on the parabola, so

$$\frac{15}{4} = \frac{1}{4c}(100)^2 \text{ implies } 4c = \frac{4(100)^2}{15} = \frac{8000}{3} \text{ so } c = \frac{2000}{3}.$$

So the focal point of the mirror is $\left(\frac{2000}{3}, 0\right)$, and the observers viewing area is $\frac{2000}{3}$ inches from the center of the mirror.

14. The eccentricity of the path is 1, since the path is parabolic, and the path of the satellite has an equation of the form $r = \frac{d}{1+\cos\theta}$. If m denotes the diameter of the moon, then

$$m + 5783 = \frac{d}{1+\cos 60} = \frac{2d}{3} \text{ implies } d = \frac{3m + 3(5783)}{2},$$

and

$$m + 143 = \frac{d}{1+\cos 0} = \frac{d}{2} \text{ implies } d = 2m + 286.$$

Then

$$2m + 286 = \frac{3m}{2} + \frac{3}{2}5783 \text{ implies } \frac{1}{2}m = \frac{3}{2}5783 - 286 \text{ so } m = 16777.$$

15. Since

$$r = a\cos\theta + b\sin\theta \text{ implies } r^2 = ar\cos\theta + br\sin\theta \text{ so } x^2 + y^2 = ax + by,$$

we have

$$x^2 - ax + y^2 - by = 0 \text{ implies } x^2 - ax + \frac{a^2}{4} + y^2 - by + \frac{b^2}{4} = \frac{a^2+b^2}{4},$$

and

$$\left(x - \frac{a}{2}\right)^2 + \left(y - \frac{b}{2}\right)^2 = \frac{a^2+b^2}{4}.$$

16. The following graph of $r = 2e^{-0.2\theta}$ is a reasonable approximation for the chambered nautilus.

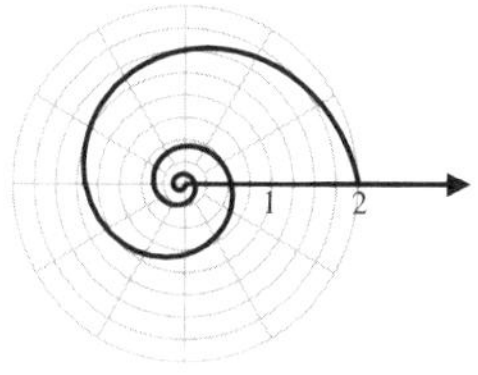

17. $r = \frac{ed}{1-e\cos\theta}$

a. The aphelion, a, occurs when $\theta = 0$ and the perihelion, p, when $\theta = \pi$. So

$$a = \frac{ed}{1 - e\cos 0} = \frac{ed}{1-e}, \qquad p = \frac{ed}{1 - e\cos\pi} = \frac{ed}{1+e},$$

and

$$\frac{a(1-e)}{e} = d = \frac{p(1+e)}{e} \text{ implies } a(1-e) = p(1+e) \text{ so } a - ae = p + pe,$$

and

$$a - p = ae + pe \text{ implies } a - p = e(a+p) \text{ so } e = \frac{a-p}{a+p}.$$

b. The length of the major axis, $2R$, satisfies

$$2R = a + p = \frac{ed}{1-e} + \frac{ed}{1+e}.$$

From part (a),

$$d = \frac{a(1-e)}{e} \text{ implies } 2R = \frac{e\left(\frac{a(1-e)}{e}\right)}{1-e} + \frac{e\left(\frac{a(1-e)}{e}\right)}{1+e} \text{ so } 2R = a + \frac{a(1-e)}{1+e},$$

and

$$2R = \frac{a(1+e) + a(1-e)}{1+e} = \frac{2a}{1+e} \text{ implies } a = R(a+e).$$

Also from part (a),

$$d = \frac{p(1+e)}{e} \text{ implies } 2R = \frac{e\left(\frac{p(1+e)}{e}\right)}{1-e} + \frac{e\left(\frac{p(1+e)}{e}\right)}{1+e} \text{ so } 2R = \frac{p(1+e)}{1-e} + p,$$

and

$$2R = \frac{p(1+e) + p(1-e)}{1-e} \text{ implies } 2R = \frac{2p}{1-e} \text{ so } p = R(1-e).$$

18. The form of the equation of the orbit is

$$r = \frac{ed}{1 - e\cos\theta}, \text{ where } e < 1.$$

The eccentricity of the orbit is $e = 0.99993$, so

$$r = \frac{0.99993d}{1 - 0.99993\cos\theta}.$$

The perihelion of 1.95×10^7 occurs when $\theta = \pi$, so

$$1.95 \times 10^7 = \frac{0.99993d}{1 - 0.99993 \cos \pi} \text{ implies } d = 1.95 \times 10^7 \left(\frac{1 + 0.99993}{0.99993} \right) \approx 3.9001 \times 10^7,$$

and $ed \approx 3.899 \times 10^7$. So

$$r = \frac{3.899 \times 10^7}{1 - 0.99993 \cos \theta}.$$

The maximum distance of Kohoutek from the sun occurs when $\theta = 0$, so the maximum distance is about

$$\frac{3.899 \times 10^7}{1 - 0.99993} \approx 5.57 \times 10^{11} \text{ miles.}$$

19. We have

$$\begin{aligned}
\sqrt{x^2 + (y-1)^2} + 1 &= \sqrt{x^2 + (y+1)^2} \\
x^2 + y^2 - 2y + 1 + 2\sqrt{x^2 + (y-1)^2} + 1 &= x^2 + y^2 + 2y + 1 \\
2\sqrt{x^2 + (y-1)^2} &= 4y - 1 \\
4(x^2 + y^2 - 2y + 1) &= 16y^2 - 8y + 1 \\
12y^2 - 4x^2 &= 3,
\end{aligned}$$

which is the equation of a hyperbola.

Chapter 6 Chapter Test (Page 357)

1. True.

2. True.

3. True.

4. False. The equation of the red parabola is $x = -\frac{1}{8}y^2$.

5. False. The vertex of the blue parabola is $(2, 1)$.

6. True.

7. False. The focus of the blue parabola is $(2, 2)$.

8. True.

9. False. The equation of the blue parabola is $y - 1 = \frac{1}{4}(x-2)^2$.

10. True.

11. False. The focal points of the blue ellipse are $\left(2\sqrt{3}, 0\right)$ and $\left(-2\sqrt{3}, 0\right)$.

12. False. The focal points of the red ellipse are $\left(2 + 2\sqrt{3}, 3\right)$ and $\left(2 - 2\sqrt{3}, 3\right)$.

13. True.

14. False. The equation of the red ellipse is $\frac{(x-2)^2}{16} + \frac{(y-3)^2}{4} = 1$. It is the same as the ellipse in the previous exercise except that it has been shifted right 2 units and upward 3 units.

15. False. The equation of the blue hyperbola is $\frac{y^2}{9} - \frac{x^2}{4} = 1$.

16. True.

17. True.

18. False. The asymptotes of the red hyperbola are $y - 2 = \pm\frac{3}{2}(x-3)$.

19. True.

20. False. The polar coordinates $\left(2, -\frac{3\pi}{4}\right)$ and $\left(2, \frac{5\pi}{4}\right)$ represent the same point, but $\left(-2, -\frac{\pi}{4}\right)$ does not. To be correct, it could be replaced by $\left(-2, \frac{\pi}{4}\right)$.

21. True.

22. False. The point $(3, -3)$ in rectangular coordinates is equivalent to the point $\left(-3\sqrt{2}, \frac{3\pi}{4}\right)$ or to $\left(3\sqrt{2}, -\frac{\pi}{4}\right)$ in polar coordinates.

23. True.

24. True.

25. False. The rectangular equation $x^2 + y^2 = 4$ has polar equation $r = 2$.

26. False. The polar equation $r = 3$ has rectangular equation $x^2 + y^2 = 9$.

27. True.

28. True.

29. False. The blue curve could have polar equation $r = 2\cos\theta$.

30. False. The red curve could have polar equation $r = 2\sin\theta$.

31. True.

32. True.

33. False. The yellow curve could have polar equation $r = 1 + 2\cos\theta$.

34. False. The yellow curve could have polar equation $r = 2 + \cos\theta$.

35. True.

36. False. The polar equation $r = \frac{5}{5 - \sin\theta}$ describes an ellipse with horizontal directrix.

37. False. The polar curves $r = 1 + 2\cos\theta$ and $r = 1$ intersect at three points. These points have rectangular coordinates $(0, 0)$, $(0, 1)$, and $(0, -1)$.

38. False. The polar curves $r = \cos\theta$ and $r = \sin\theta$ intersect at the two points $\left(\frac{\sqrt{2}}{2}, \frac{\pi}{4}\right)$ and $(0, 0)$.

39. True.

40. False. The parametric equations $x = 2t - 1$, $y = 3t + 2$ describe a line with slope $\frac{3}{2}$ and y-intercept $\frac{7}{2}$.

41. True.

42. False. The parametric equations $x = e^t - 1$, $y = e^{2t} + 1$ describe a parabola with vertex $(-1, 1)$ and opening upward.

43. True.

44. False. The parametric equations $x = 1 + 2\cos 2t, \quad y = 2 + 2\sin 2t$ describe the circle with center $(1, 2)$ and radius 2, tracing the circle counterclockwise.

45. False. The parametric equations $x = \cos t, \quad y = 2\sin t$ describe an ellipse with rectangular equation $4x^2 + y^2 = 4$.

46. True.

47. True.

48. True.

49. False. The parametric equations $x = e^t, \quad y = e^{-t}$ and the rectangular equation $y = \frac{1}{x}$ describe the same curve when $x > 0$, but the graph of $y = \frac{1}{x}$ is also defined when $x < 0$.

50. False. The parametric equations $x = \sqrt{t}, \quad y = t + 1$ and the rectangular equation $y = x^2 + 1$ describe the same curve when $x \geq 0$, but the graph of $y = x^2 + 1$ is also defined when $x < 0$.